Waste-to-Resources 2021

Matthias Kuehle-Weidemeier

WASTE-TO-RESOURCES 2021

9TH INTERNATIONAL SYMPOSIUM CIRCULAR ECONOMY, MBT, MRF AND RECYCLING

Resources and Energy from Waste

Proceedings
original language edition

18th -20th of May 2021

Organisers

www.icp-ing.de

www.wasteconsult.de

Supported by

Cuvillier Verlag

Bibliografische Information der Deutschen Nationalbibliothek
Die Deutsche Nationalbibliothek verzeichnet diese Publikation in der
Deutschen Nationalbibliografie; detaillierte bibliographische Daten sind im Internet
über http://dnb.d-nb.de abrufbar.
1. Aufl. - Göttingen: Cuvillier, 2021

Nonnenstieg 8, 37075 Göttingen
Telefon: 0551-54724-0
Telefax: 0551-54724-21
www.cuvillier.de

1. Auflage, 2021
Gedruckt auf umweltfreundlichem, säurefreiem Papier aus nachhaltiger
Forstwirtschaft.

ISBN 978-3-7369-7500-2
eISBN 978-3-7369-6500-3

ICP - Ingenieurgesellschaft Prof. Czurda and Partner mbH

Ingenieurgesellschaft Prof. Czurda and Partner mbH is an independent and internationally recognized Engineering Consultancy based in Karlsruhe, Germany, in 1990. The company's core business is the provision of advisory services concerning solid waste management (collection, transport, recycling, treatment and/ or disposal). These services may cover the following project stages:

1. Analysis - Analysis of current situation, assessment of current constraints, detailed technical due diligence report, review of available studies, environmental and social assessment,

2. Planning - Detailed design of the infrastructure, feasibility assessment, determination of construction methods, operating principles, technical specifications and bills of quantities, as well as experience in the preparation and evaluation of technical solutions, solid waste plans, strategies, concepts, feasibility studies and economic analysis.

3. Procurement – Preparation and evaluation of tender documents, offers, bids, support during procurement, contracts (including FIDIC-based ones) for the construction works and the supply of services/ goods, as well as support and training to the Client during procurement procedures,

4. Supervision - Coordination of works and delivery of services/ goods, permanent supervision of works, trials and tests, quality control, acceptance of all performances and approval of invoices.

5. Training - Support during facility commissioning, training on installations and equipment operation and training concerning previous measures.

Moreover, ICP has experience in brownfield remediation, construction quality assurance, geodesy, geo-technics, geology and hydro-geology.

ICP has been contracted by private organisations as well as public authorities, international development banks, or development aid institutions (KfW, EBRD, EIB, World Bank, EC etc.). Its consultants understand the requirements and needs of the private clients as well as the committee- based budget systems of the public authorities and their corresponding decision making processes. ICP is familiar with the working methods of development agencies and investment institutions.

ICP's membership in the "*Verband Beratender Ingenieure*" (the German Association of Consulting Engineers) as well as numerous specialist organisations and working groups in Germany ensure the continuous professional development of its staff. Due to the involvement of own consultants/ staff in several expert committees and working groups, ICP staff is permanently updated with the latest developments in their specialist fields.

The application of the most modern software also contributes to maintaining the highest quality standards with state of the art technology for ICP clients. An example: all design and modelling in ground works or for landfills is untaken with three dimensional modelling of ground profiles. This guarantees the highest degree of accuracy for the calculation of earth volume quantities.

ICP has implemented ISO 9001 standards to guarantee quality assurance of its services, from the analysis to the training stage. The ISO 9001 certification aids the company in meeting the ever changing needs of you, our customer, with consistent quality.

Fields of activity and services provided

Solid Waste Management

- Preparation of solid waste plans, strategies, concepts, feasibility studies and economic analysis;
- Landfill technology (design and construction supervision of new landfills or landfill phased extensions, including sludge landfills);

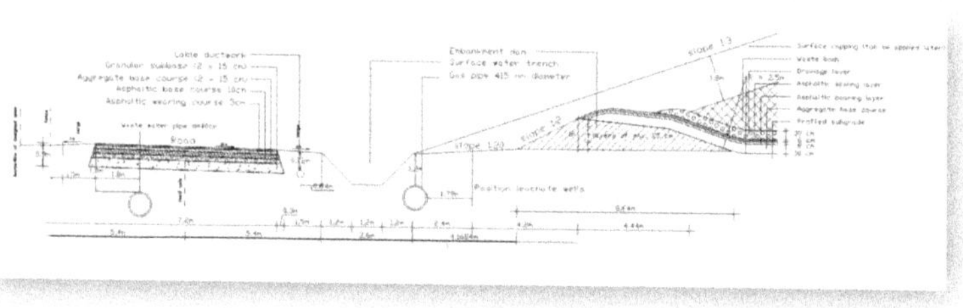

Figure 1: Example of a vertical section for a landfill in Muscat, Oman

- Landfill closure (landfill gas collection technology, leachate treatment, refurbishment of landfill drainage systems, site investigations and feasibility studies, monitoring etc.);
- Design and construction supervision of waste treatment plant (mechanical-biological pre-treatment, composting, fermentation, sorting and handling etc.); and
- Tender preparation and evaluation – technical and financial evaluation of offers submitted for solid waste management services, works and/ or supplies.

Contaminated site investigation and rehabilitation

- Contaminated site investigation (from historical records to technical sampling and ground water modelling);
- Clean up of contaminated sites (design and supervision of treatment methods such as Pump and Treat, Soil Vapor Extraction, Permeable Reactive Barrier, etc.);
- Securing of contaminated sites / stabilization of existing hazards (design and supervision of treatment measures such as surface sealing, slurry wall containment, etc.); and
- Monitoring systems.

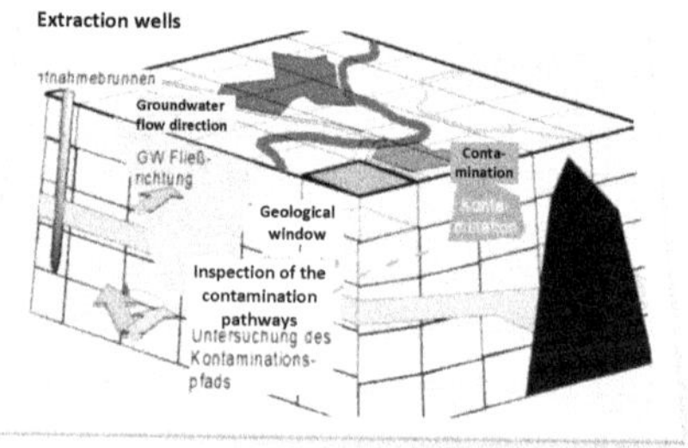

Renewable Energies

- Waste to energy (Biogas plants and RDF production plants);
- Landfill gas utilization; and
- Solar plants.

Planning of building decommissioning and demolition

- Site investigation / Groundwater management;
- Stability of deep excavations also in densely built areas;
- Controlled demolition;
- Investigation of hazardous materials;
- Measurement of emissions and imissions; and
- Surveys of current condition / Conservation of evidence.

Plant and working safety

- Health and safety plans;
- Health and safety coordination;
- Work place health and safety; and
- Plant safety.

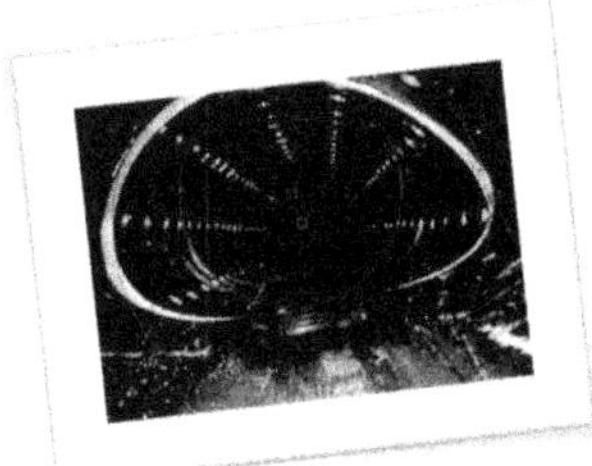

Environmental planning

- Environmental and social impact assessment;
- Risk assessment;
- Environmental expert advice; and
- Environmental damage assessment.

Geotechnical, soil mechanics and physical investigations

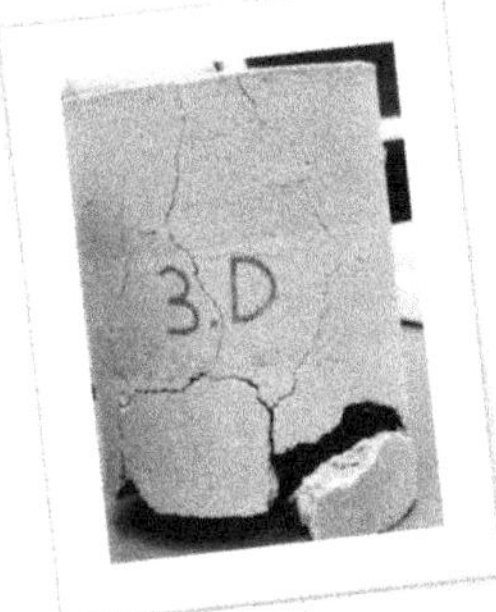

- Building assessment and foundation recommendations;
- Stability calculations;
- Settlement calculations;
- Execution of field tests and probing;
- Specification and supervision of boreholes;
- Disturbed and undisturbed sampling;
- Soil mechanical tests according to DIN ISO und ASTM;
- Soil physical tests according to DIN ISO und ASTM;
- Development of sealing materials from residues; and
- Testing of construction materials before being used as landfill sealing materials.

Geology und Hydrogeology

- Expert opinions;
- Ground water measurements;
- Drinking water protection areas;
- Execution of field tests and probing;
- Specification and supervision of boreholes; and
- Borehole probe, analysis and review.

Geodesy

- Terrestrial surveys;
- Settlement monitoring; and
- Inclinometer measurement.

Knowledge Transfer

- Organisation of symposiums and conferences;
- Training for professionals and local authorities; and
- Capacity building.

Main Fields of Activity

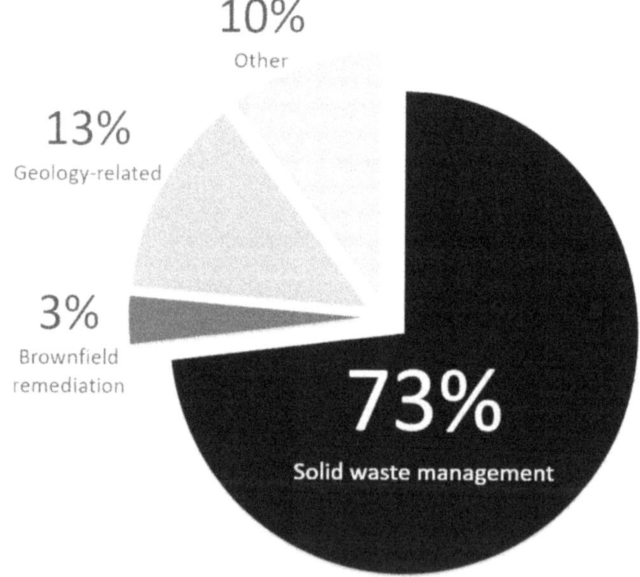

Figure 2: ICP's main fields of activity in the last five years

Structure

The companies that belong to **ICP mbH in Karlsruhe** or where **ICP mbH** has shares or owns a part of the social capital are the following:

STE Tunisia Société Tunisienne pour l'Environnement Immeuble SCI le Lac, Bureau N°32 6 rue du Lac TOBA, Les Berges du Lac, TN-1053 Tunis, Tunisia	**ICP Kaiserslautern** ICP Ingenieurgesellschaft Prof. Czurda und Partner mbH Am Tränkwald 27 D-67688 Rodenbach, Germany	**ICP Kempten** ICP Ingenieurgesellschaft Prof. Czurda und Partner mbH Illerstraße 12 D-87452 Altusried, Germany
ICP Leipzig ICP Ingenieurbüro Prof. Czurda Dr. Günther und Partner GmbH Fasanenweg 2 D-04420 Markranstädt, Germany	**ICP Rhein Main** ICP Rhein Main GmbH Berner Straße 107 D-60437 Frankfurt Germany	**ICP in Potsdam** ICP Braunschweig GmbH Geschwister-Scholl-Str. 76-77 D-14471 Potsdam, Germany
ICP Braunschweig ICP Braunschweig GmbH Berliner Str. 52j D-38104 Braunschweig, Germany	**ICP in Bitburg** ICP Ingenieurgesellschaft Prof. Czurda und Partner mbH Kopernikusstr. 1 D-54634 Bitburg, Germany	**ICP in Büren** ICP Braunschweig GmbH Briloner Str. 70 A D-33142 Büren, Germany
ICP Prüfungsgesellschaft mbH in Karlsruhe Auf der Breit 11 76227 Karlsruhe, Germany		

Employee Background

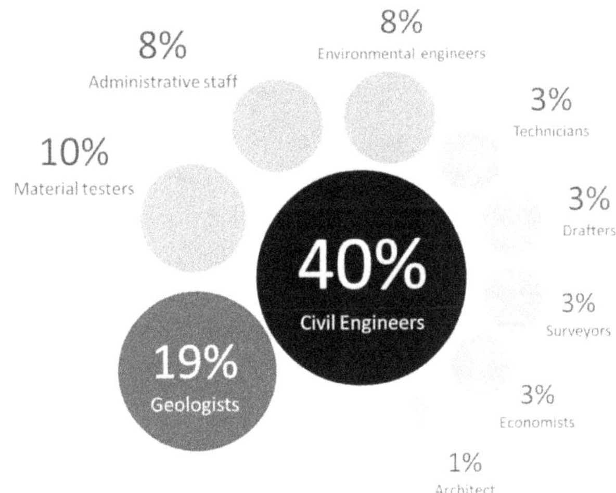

Figure 3: ICP Group's employee background

Annual turnover

The company turnover (ICP Group) is summarized in the following graph:

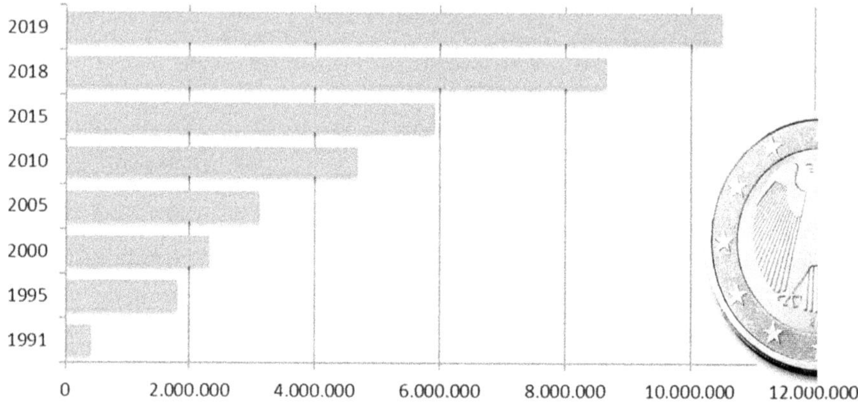

Figure 4: ICP Group's annual turnover growth 1991 – 2019 [EUR]

International Experience

ICP has experience in more than 50 countries.

Figure 5: ICP's experience in other countries

Content

I Waste Management and Zero Waste Strategies

Transitioning to a ZeroWaste Society 1
Peter S. Lobin

Enhancing Waste Reduction, Reuse and Recycling by Behaviour Change 5
Uttam Yadav

Zero Plastic Waste Cities – Sustainable Municipal Waste Management Through Social Business 23
Christina Jäger, Konstantin Münchau

II Practical Experience and New Facility Concepts

Paris XVII Material-from-Waste plant in Operation 35
Christophe Cord'homme, Frédéric Roux

III Processing and Recovery of Organic Waste Fractions

Biogas from OFMSW in Mexico by using BEKON Dry-Fermentation 51
Ignacio Benitez

Practical Implementation of Biological-Oxidative Desulfurization of Biogas. 63
Alejandra Lenis, Kristoffer Ooms, Cedric Thull

IV Mineral and Construction Waste

Management of Mineral Extraction Residues: Best Practices in Lombardy Region (Italy) 75
Alessandra Diotti, Alberto Clerici, Giovanni Plizzari, Sabrina Sorlini

Building Material Management: System Change in Minnesota and Beyond 87
Melissa Wenzel

The Effect of Granular Correction of Dredged Sediments by Other Wastes and the Mechanical and Environmental Properties of Established Mixtures 97
Zeinab Mkahal, Yannick Namindy-Pajany, Walid Maherzi, Nor-Edine Abriak

Remediation of Dredged Sediments by Electrokinetic Process Prior to Their Beneficial Reuse 111
Mathilde Betremieux, Yannick Mamindy-Pajany

V Waste management in Asia

Indonesia Waste Bank Transition and Transformational System: a Pathway of Circular Economy Development 129
Ade Brian Mustafa, Masahiko Haraguchi

A Technical, Economical, and Environmental Comparison of Composting and Anaerobic Digestion of Organic Waste Fraction of Municipal Solid Waste in Sri Lanka 143
Wijepala Abeysinghe Mudiyanselage Asanka Nuvansiri Illankoon, Sabrina Sorlini

Separate Collection and Recycling of Waste as an Approach to Combat Marine Litter – WWF Pilot Project in the Mekong Delta, Vietnam 157
Wolfgang Pfaff-Simoneit, Stefan Ziegler, Trinh Thi Long

VI Waste Management in Emerging and Developing Countries

Solid Waste Management Decision Making for Developing Countries through Life Cycle Assessment Tool: A State-of-the-Art Review 181
Ketan V. Shah, Melanie L. Sattler

Inclusive Circular Waste Management in the Refugee Camps – Strategy for Rohingya Refugee Camp 199
Piotr Barczak, Katharina Wiese, Enzo Favoino, Stephane Arditi

10 Years of National Solid Waste Policy in Brazil: The Legacy of Technological and Market Approach 221
Christiane Dias, Pereira

VII Concepts and Data on Waste Management

Green Deal – Tox-Free-Environment and SCIP Database 231
Beate Kummer

Less Waste, Less Landfilling, More Recycling 242
Peter Hoffmayer

More Recycling and More Economical Recycling by Optimising Collection 250
Clemens Pues

Material Flow Analysis of Materials for Reuse and Recycling in the Greater Dresden Area 259
André Rückert, Christina Dornack

Zero Waste Cardboard Technology to Support Product Stewardship 267
Aharon Arakel, Bithi Roy, Rubick Arakel, Warwick Madden

Take-Back of Waste Electrical and Electronic Equipment – A Successful Model? 277
Ralf Brüning, Julia Wolf, Stephan Löhle, Ute Schmiedel, Ines Oehme, Kerim Zaidi

VIII Waste Technology

Contribution of Biological Residual and Waste Materials to the Bioeconomy 283
Kyra Atessa Vogt, Iris Steinberg

How to Fail Successfully in Designing an Alternative Fuel Production 307
Hubert Baier

Organic Waste for the Production of Hard Carbons for Batteries – A Life Cycle Assessment Perspective 319
Huiting Liu, Claudia R. Tomasini, Linghan Lan, Jun Li, Xiang Zhang, Niklas von der Aßen, Marcel Weil

IX Circular Economy During the Corona Crisis

Waste Management Challenges in Romania During the COVID-19 Pandemic 335
Florin-Constantin Mihai, "Alexandru Ioan Cuza" University, Iasi, Romania

Using Computer Vision to Strengthen Resilience of Waste Sorting Infrastructure During the Pandemic 345
Victor Dewulf, Recycleye, London, United Kingdom

X Waste management in Western Asia, Middle East, Black Sea and Mediterranean countries

Households' Willingness to Pay for a Local Recycling Programm: A Case Study from Lebanon 357
Mary Abed Al Ahad, Ali Chalak, Souha Fares, Rima R. Habib

Sustainability Challenges in Solid Waste Management for Future Prospective I Gaza Strip: Rafah City Case Study 373
Samir Alnahhal, Samir Afifi

Waste Prevention and Recycling Strategy for Tbilisi City, Georgia 387
Virginie Herbst, Ludwig Streff, George Tavoularis

New Orientation of the Waste management Strategy in Durban, South Africa as Contribution to the City's Ambition to Achieve Carbon Neutrality 402
Wolfgang Pfaff-Simoneit, Logan Moodley, Federico di Penta

Plastic Wastes as a Resource: Lebanese Experience 415
Arwa Anouti ElZein

XI Slag recovery (Recovery of incinerator bottom ash)

Grate-for-Riddlings – Improved Fine Bottom Ash Treatment and Effects on the CO_2 Footprint of Waste Incineration Plant 431
Ivo Budde

Technical Assessment of a Bottom Ash Treatment Plant – a Case styudy 443
Kay Johnen, Alexander Feil

Process Development to Improve Volumetric Stability of Steel Slag 455
Mukund Manish, Praveen Kumar, D. Satish Kumar, Veeresh Bellatti, Balachandran G

XII Geomechanical Properties of MBT-Output

Settlement Characteristics of Saturated Mechanical Biological Treated Waste from Croatia 469
Nikola Kaniski, Nikola Hrncic, Igor Petrovic

Maximum and Minimum Void Ratio Characteristics of MBT Waste 483
Nikola Hrncic, Nikola Kaniski, Igor Petrovic

XIII Plastic Waste

Comparison of the Quality of Mechanically Recycled Plastics Made from Separately Collected and Mechanically Recovered Plastic Packaging Waste 499
Eggo U. Thoden van Velzen, Ingeborg Smeding, Marieke T. Brouwer, Evelien Maaskant-Reilink

Sorting of Film Mixtures to Increase the Recycling Rate 511
Maria Schäfer, Peter Clemenz, Heinz Schnettler

Conditioning of Plastic Fractions by Means of NIR-Sorting to Reduce Part of Plastics in Thermal Waste Treatment 523
Anett Kupka, Lukas Knauth

XIV Hydrogen Economy

Biomass Conversion and H_2-Production 533
Alfons Kuhles, Hubert Kohler

Hydrogen From Biogenic Sources with Negative CO_2 Balance for Fuel Cell Waste Collection Vehicles 545
Jens Hanke

XV Fire Protection in Waste Treatment and Recycling Plants 553

Why is Targeted Extinguishing Control the Better Solution? Prerequisites for Targeted Extinguishing Control Via Infrared Systems and the Technical Implementation in Practice
Mark Müller

XVI Recovery of Metals

Thermal Metal Recovery from Tertiary Waste 563
Kurt Bernegger, Helmut Lugmayr, Christian Mlinar

Spent Automotive Converters as a Secondary Resource of Platinum Group Metals 571
Martyna Rzelewska-Piekut, Pawel Krawczyk, Zuzanna Wiecka, Magdalena Regel-Rosocka

Hydrometallurgical Recovery of Metal Ions from Spent Catalytic Converters 581
Zuzanna Wiecka, Ewelina Lopinska, Martyna Rzelewska-Piekut, Magdalena Regel-Rosocka

XVII Separation and Processing of Waste Fractions

The Robust and Fast Way of Free-Fall Sorting of Ruble and Debris 591
Markus Eck

XVIII Advanced MBT Processes With Wet Mechanical Separation

Advanced Waste Treatment and Material Recovery by Wet Separation 601
Matthias Kuehle-Weidemeier

XIX Pyrolysis, Plasma, Gasification

Assessing Pyrolysis for the Recovery of the Composted Organic Fraction of the Municipal Solid Waste (OFMSW) 619
Jessica Graca, Marzena Kwapinska, Brian Murphy,J. J. Leahy, Brian Kelleher

Perspectives on Waste to Energy Conversion by Microwave Plasma Gasification 621
Melda Ozdinc Carpinlioglu

Note

The proceedings were arranged by ICP Ingenieurgesellschaft Prof. Czurda und Partner mbH with high accuracy. Nevertheless, errors cannot be fully excluded. ICP and the editors are not liable or responsible for the correctness of the information in this book. The authors take full responsibility for the content of their articles.

If a brand or trade name was used, there might be trademark rights valid, even if they were not explicitly stated.

The proceedings contain complicated technical terms. ICP do not take any liability for the correctness of the translation. Please check the plausibility of the content you are reading.

TRANSITIONING TO A ZEROWASTE SOCIETY

Peter S. Lobin

Seventh Generation Advisors, Santa Monica, California USA

Abstract: "The hardest part of introducing a disruptive idea is asking others to visualize the transformation and value it brings".

Few industries have changed as little in the past hundred years as the waste industry. A century ago, waste was picked up, put in a horse-drawn cart and taken away for disposal. Today, they have replaced the horse-drawn cart with a truck.

The pandemic offers us time to reflect upon aspects of our daily life we have taken for granted, or ignored, prior to COVID 19. Reimagining waste and recycling in a way that has long-term benefits for society opens the door to building a "ZeroWaste Society" (ZWS). The value proposition for a "ZeroWaste Society" is simple: maximize the value of local resources through coordination among stakeholders, gain economies-of-scale and improve operating efficiencies, resulting in greater economic activity, job creation and a reduced carbon footprint.

Keywords: *ZeroWaste Society, Garbologist, private-public-partnership (P3), stakeholders and beneficial reuse.*

Few industries have changed as little in the past hundred years as the waste industry. A century ago, waste was picked up, put in a horse-drawn cart and taken away for disposal. Today, they have replaced the horse-drawn cart with a truck.

Our current recycling system, built in a piecemeal fashion, started to take shape in the 1970s but has made only modest improvements over the past fifty years. Strangely the "Golden Rule" of recycling, *"Everything is recyclable, but not everything is economically recyclable"*, seems like it was predicting recycling's eventual death. If one wanted to add to recycling's obituary they would say, the lack of a uniform collection system (which causes confusion and contamination), the cost of transporting low value recyclables and no incentive for stakeholders to change were all causes of recycling's demise.

The pandemic offers us time to reflect upon aspects of our daily life we have taken for granted, or ignored, prior to COVID 19. Reimagining waste and recycling in a way that has long-term benefits for society opens the door to building a "ZeroWaste Society" (ZWS). The value proposition for a "ZeroWaste Society" is simple: maximize the value of local resources through coordination among stakeholders, gain economies-of-scale and improve operating efficiencies, resulting in greater economic activity, job creation and a reduced carbon footprint.

A ZWS can achieve economies-of-scale and operating efficiencies, by combining the waste and recyclable materials from multiple communities, lowering the cost of transportation, reducing processing costs and increasing market opportunities. Building a ZWS requires a public-private partnership (P3) providing private sector resources and investment local governments don't have.

The Recycling Partnership's recent report shows U.S. curbside recycling is currently capturing only 32 percent of the 37.4 million tons of commodities. This capture rate, while disappointing, doesn't include the largest opportunity for recycling organics, which on average represents close to 50% of municipal solid waste going to landfills.

A Zero Waste Society utilizes a hub-and-spoke infrastructure, with one or more hubs. A hub can be a specific campus or zone whose area can be defined roughly as falling within a 100-mile radius. The key to enacting a ZWS is for local governments controlling the waste and recycling stream to change from a line-item-expense model to a resource management approach. Coordination and communication among stakeholders will be key to designing, implementing, and maintaining a ZWS infrastructure.

Chicago is a perfect candidate to be a super hub in a "ZeroWaste Society". Chicago's industrial base combined with large population and corresponding volumes of waste, access to transportation, markets for recycled materials, universities, government laboratories and business community make it perfectly positioned to transition the region to a ZWS.

A multiple-hub infrastructure to optimize rural resources would work well by the states of Missouri, Iowa, Nebraska and Kansas, joining forces as the "Heartland" ZWS, which would encompass rural and urban areas. The Heartland ZWS has both the population, volume of waste, and industrial base to gain economies-of-scale and optimize these resources.

The State of California and it's progressive approach to waste and recycling, might be the perfect candidate to form a ZWS. The state government has enacted numerous programs and laws, diverting materials away from landfills while developing markets for value added materials. California's transition to a ZWS would be fast.

The large waste firms who own landfills will fight a transition to a ZWS. But these are for-profit firms and if change is inevitable, they will look at their assets and adjust. Not everything can be recycled, so the use of landfills in a ZWS will continue.

There isn't a roadmap, per se, for setting up a ZWS, but many policies, best practices and technologies are already in use that can be woven into initial plans. States are enacting legislation that reduces waste going to landfills while incentivizing stakeholders by offering grants and loans to achieve many of the goals of a ZWS. New technologies that reduce contamination and convert materials are coming to market. Corporations are investing in solutions for recycling materials they use or produce for packaging. The 10 states with bottle deposit laws consistently have higher rates of recycling, producing recyclables with less contamination and greater value.

Cities and local governments that control waste, must transition to a resource management approach to achieve a ZWS. A ZWS uses a flexible hub-and-spoke infrastructure that is adaptable to each region's unique geography, population, resources and stakeholders. Coordination among stakeholders and leveraging private sector resources through a P3 will be key to success. Building blocks that can be woven into a ZWS already exist in public policies, best practices, technologies, sustainable capital, etc.. Our current economic circumstances has started discussions about implementing a sustainable infrastructure bill, which is the perfect platform for transitioning to a ZeroWaste Society.

Enhancing Waste Reduction, Reuse and Recycling by Behaviour Change

Uttam Yadav

SEE, Abu Dhabi, United Arab Emirates

Abstract

Behaviour centric approach to Waste Reduction is an important dimension in devising the future strategies as the success rate of future strategies entirely depend on how well these are perceived, how effectively these invoke desired behaviour required for the future strategies to function and how effectively these are able to convince that these are beneficial to the Planet. The behaviours can be changed through Awareness on Key Information that influences behaviour dramatically, Sustainable Product Design that appeals the consumer without compromising the aesthetic and quality aspects, Strategic Alliances and Partnerships that can inspirationally foster involvement, innovation through Digital Platforms that materializes the intended behaviour change, improving access to the Recycling locations and Regulatory Regime that supports framework for waste reduction.

Keywords

Alternative Materials

Behaviour

Cultural / Paradigm Shift

Design for Behavioural Change / Sustainable Product Design

Extended Producer Responsibility

Persuasive Techniques

Product Lifecycle Responsibility

Pull Factor Framework

Stakeholder Management

Strategic Alliances and Partnerships

Sustainability Labels

1 Background

Everyone in this world would agree that waste generation is increasing at an alarming rate. The world has entered a significant phase of economic shift wherein

a) The product life cycles have shortened due to obsolescence, product design and upgradation of configuration (e.g. mobile phones, computers, tablets, etc.),

b) Rapid urbanization leading to redevelopment of existing infrastructure and communities in metros and cities (e.g. demolition of old buildings),

c) Increase in economic activities leading to over-pressure on ecosystems on account of higher demands leading to additional waste management issues, e.g. due to uncontrolled and illegal mining, etc.

d) Stringent health and hygiene standards, leading to single use (use and throw) packaging in most consumer items (e.g. water bottle)

e) Rise in income of middle and higher middle income category people boosting the so-called "throwaway culture".

f) Preference given equally by manufacturers and consumers to aesthetic appeal of the product leading to increased product packaging

g) Emergence of e-commerce business encompassing the lives to a significant extent leading to increase in single use packaging for delivery of articles, groceries, hot food, etc., resulting into quantum leap in waste generation

h) Increase in Emergency situations and Pandemics (such as COVID-19) leading to additional waste burden

With over 90% of waste openly dumped or burned in low-income countries [1], it is the poor and most vulnerable who are disproportionately affected. Poorly managed waste is contaminating the world's oceans, clogging drains and causing flooding, transmitting diseases, increasing respiratory problems from burning, harming animals that consume waste unknowingly, and affecting economic development, such as tourism.

In 2016 alone, the world [1] generated 242 million tons of plastic waste – equivalent to about 24 trillion 500-millimeter, 10-gram plastic bottles, and that's just 12% of the total waste generated each year. The global waste is projected to increase by 70% over the next 30 years – to a staggering 3.40 billion tons of waste generated annually.

A staggering 91 per cent of all plastic [2] is single-use. Since plastic became commonly used material almost six decades ago, it has cumulatively resulted in 8.3 billion metric tons of plastic pollution. The scariest part of this is that production of plastic is set to double over the next 20 years, despite increasing awareness of its detrimental impact on the environment. According to the United Nations Environment Program (UNEP), each year, an estimated eight million tons of plastic end up in the ocean – equivalent to a full garbage truck dumped into the sea every minute. Therefore, a novel approach for intercepting single-use plastic and removing it from the natural environment will be needed to ensure that it is placed back into the circular system.

Upper-middle and high-income countries provide nearly universal waste collection, and more than one-third of waste in high-income countries is recovered through recycling and composting. The low-income countries [1] collect about 48% of waste in cities, but only 26% in rural areas, and only 4% is recycled. Overall, 13.5% of global waste is recycled and 5.5% is composted.

Notwithstanding, such **sustainable waste management** requires a significant cost. In low-income countries, it comprises as high as 20% of municipal budgets. However, research suggests that it does make **economic sense** to invest in sustainable waste management. Uncollected waste and poorly disposed waste have significant health and environmental impacts. The cost of addressing these impacts is many times higher than the cost of developing and operating simple yet adequate waste management systems. The real solution to this problem, therefore, lies in the fact that how effectively all the associated stakeholders are able to convince and propagate this understanding amongst the masses to make it a **continual and sustainable movement**.

2 Introduction

There is only one Planet Earth, yet by 2050, the world will be consuming as if there were three planets [3]. As apparent from the statistics in the preceding section, a progressive, yet irreversible transition to a sustainable economic system (i.e., Sustainable consumption and production) is an indispensable part of the future strategies to

- accelerate the transition towards a **regenerative growth model** that gives back to the planet more than it takes,
- advance towards keeping its resource consumption within planetary boundaries to reduce its consumption footprint and
- significantly enhance the circular material use rate in the coming decade

It is therefore important that **resource consumption and circular material use** are scaled up into the mainstream for achieving climate neutrality by 2050 and decoupling economic growth from resource use, while ensuring inclusiveness of all masses.

Therefore, as the resource consumption and circular material use are at the **core** of every developmental decision taken by each entity across the globe, it is **critically important** that how **best** we can influence the people on this focus area to **drive the shift needed**.

3 Paradigm Shift

The human tendency, over a period of time, has changed to such an extent that we do not use **Resources** wisely, turning them into **Waste** and then invest heavily in infrastructure, equipment, energy, manpower, etc. to convert the Waste back to Resources. This, in no way, is sustainable consumption and production behaviour.

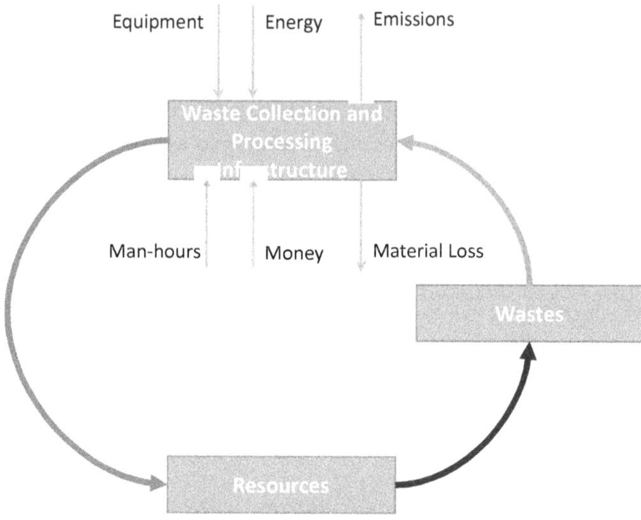

Figure 1 Waste Processing Cycle

Therefore, the planet is desperately calling for a **Paradigm Shift**, in first place, why to generate waste, and secondly, if generated how effectively it can be reduced, reused and recycled.

Addressing the root cause of this crisis, we need to first **Redefine** the existing model for **Economic Growth**, to be based not on **Consumption**, but by assigning a greater weightage on **Environment and Sustainability**.

Secondly, **human behaviour** towards the waste needs to be **radically changed**, which demands a **Cultural Shift** in our perspective towards accepting our wastes and the way they are generated and tackled. Technology can undoubtedly prevent certain quantum of waste generation from product use but not necessarily making a particular community, area or town **Waste Free.**

The cultural shift is not as simple as it looks as changing human behaviour is a herculean task. The greater challenge lies in garnering the attention of human beings from every walk of life, whom are unique by virtue of their culture, traditions and belief systems, their education, and their nature, attitude and behaviour, to the common goal of waste reduction.

The behaviour springs from and significantly influenced by awareness on underlying issues stemming from undesired or usual ignorant behaviour. Moreover, the behaviour can be significantly altered when there is a benefit attached to it, e.g., rewards on depositing recyclable waste, when innovatively highlighted, loud and clear.

It is also believed that **harmonizing the "Waste Segregation and Collection Systems & Protocols"** across the municipalities, cities, states and nation will aid in shaping up behaviour of people towards waste minimization by the way of better clarity, improved cognitive perception and gradual but consistent habituation.

In parallel, the Regulators also need to step-up roll out of **Smart Strategies** along with the **Regulation** that enhances synergies for transition into the circular economy as well as inspires the masses influencing their behaviour translating the best practices into the habits.

4 Finding a Fitting Solution

What lesson does the pandemic COVID-19 teach us? Primarily, the human behaviour (social distancing, wearing masks and washing hands frequently) was key to prevent the infection from further spreading. We also witnessed that this human behaviour was regulated through the penalties and fines on violation and there has also been plenty of innovative ways articulated to sustain the routine business.

Similarly, the effective and sustainable management of wastes relies on three pillars, namely, Innovation, Behavioural, and Regulatory. Let us get deeper into each of these pillars.

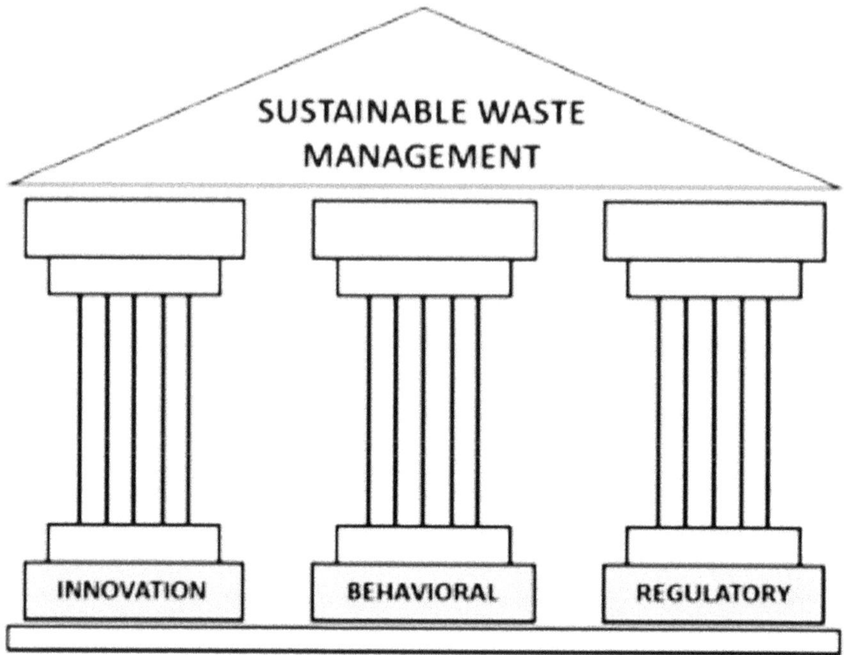

Figure 2 Pillars of Waste Management

4.1 INNOVATION:

The only way to discover the limits of the possible is to go beyond them into the impossible. Therefore, innovation continually evolves the way of conducting any activity.

I. **Redefining the Hierarchy of Waste Management**: The first step to begin our journey on embarking to this **Cultural Shift** is to revisit this governing criterion or the guiding principle on which the future strategies shall be built.

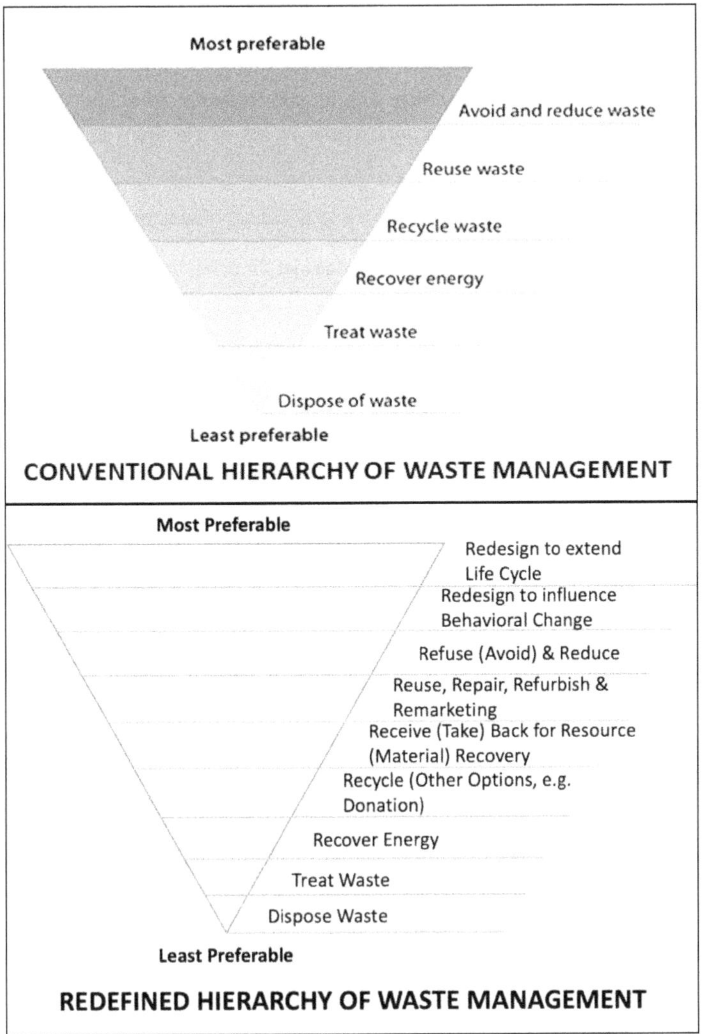

Figure 3 Re-Inventing the Hierarchy of Waste Management

- **Redesigning** not only includes all options that extends the life-cycle of wastes but more importantly, influence behavioural change, both intentionally and unintentionally.
- **Repair, Refurbish & Remarketing** included the options that enables remarketing of a repaired and refurbished product.

- **Receive, Take Back or Sell Back** includes the options that enables resource / component recovery, etc.
- **Recycle by Donation**: The functional items, meant for discarding, can be donated to such organizations which avails them to our underprivileged fraternity.

Finally, the target for Destruction and Disposal of wastes should be zero.

II. Calling for **Sustainable Product Design** as driver for creating a Circular Economy

- Eliminating or substituting the hazardous constituent in the product design
- Promote product designs that is modular, with separable components to allow replacement of damaged or faulty parts with new spare-parts, repair and reuse friendly, DIY repair features, etc. to restrict use and throw.
- Increasing recycled content in products without compromising quality, performance and safety
- Establish Standards and Regulations on Sustainable Product Design – Ensuring durability, long life span and avoiding premature obsolescence
- Incentivize sustainable product designs
- Use United Nations (UN) and similar Global Vehicles to make it popular through channels like United Nations Environment Program (UNEP), World Environment Day, WWF, Earth Hour, etc.

III. Promote the novel concept of **Design for Behavioural Change** to influence human behaviour: The design should seek approaches, like **persuasive techniques**, **user-centric designs**, etc. that seek to change human behaviour in terms of his/her cognition or attitude or in sense of his/her socially responsible actions.

IV. Incentivizing product-as-a-service where producers keep the ownership of the product or the responsibility for its performance throughout its lifecycle – a concept of **Product Life Cycle Responsibility**.

V. **Strategic Alliances and Partnerships** can create a symbiotic and synergetic impact wherein two specialized entities collaborate to provide solution to a multidimensional problem

- The Alliance to End Plastic Waste and China Petroleum and Chemical Industry Federation partnered to jointly tackle plastic waste in China.

- Massachusetts-based [4] **Preserve**, known for its reusable picnic ware and personal care products like shavers and toothbrushes made from recycled plastics, partnered with the **5 Gyres Institute** and **Renew Oceans** to launch a line of products made from recycled ocean plastic called POPi (Preserve Ocean Plastic Initiative). The POPi shavers and toothbrushes help remove plastic waste from shorelines and rivers to prevent it from entering the oceans and harming marine life.
- Connect sustainability lovers - those who deliver and those who admire: In USA [5], BBMG and Sustainable Brands joined with a group of global brand leaders including Procter & Gamble Company, Estée Lauder Companies, Happy Family Organics, Keurig Dr Pepper, Target, Johnson & Johnson Consumer Health, National Geographic, Heineken USA and Veocel conducted consumer research that revealed human need for premium, high-value product attributes with environmental and socially conscious design of goods, services and actions so consumers can shop consciously without feeling like there's a trade-off. This led to development of **Pull Factor Framework**, shown is figure below, a new way to build brands that create a pull factor by blending sustainability, human need and brand factor.

Figure 4 Pull Factor Framework

VI. Promote **Alternative Materials** for Packaging

- E-commerce industry being hugely packaging intensive; reusable delivery boxes can have a big impact on waste generation.
- Bio-based plastics, where the use of bio-based feedstock results in genuine environmental benefits, could be better alternative to conventional plastic packaging.
- Use of biodegradable or compostable plastics can be beneficial to the environment e.g., biodegradable and compostable material called **Flaxstic** [6] was used to create the first Pela case in 2011 for the iPhone 4.

VII. **Design New Communities** for greater waste sorting at a central location within the community for enhanced recycling, e.g., integrated resource recovery centre (IRRC) [7] that reduces community waste footprint

- improve access to the recycling point, e.g., used clothes donation boxes installed in each community at strategic locations [8].

Figure 5 Used Clothes and Shoes Bank

- Empower them to manage their waste footprints, e.g., they can assign some NGOs for collection from their community

VIII. Reducing the **Waste Management Footprint**

- Food waste: You can recycle plastic but what about food waste. The restaurants are now using new tools to capture, track, and categorize data on food waste. AI algorithms can forecast and predict sales, enabling restaurants, retailers, and other hospitality institutions to connect supply to demand more effectively, e.g., **Tenzo** AI software [9], Marketman Inventory Management Software [10].

- AI algorithms can help with food sorting during processing by analysing images and data from cameras, X-rays, lasers, and near-infrared spectroscopy. Some companies [11], such as Wasteless, are helping supermarkets and other retailers sell food before the expiration date by using AI-enabled tracking and dynamic pricing.

- Hydrogen fuelled waste collection vehicles, organic waste to biogas rather than composting (reduces CH4, a greenhouse gas, emissions), etc. are expected to augment reduction of overall waste footprint.

- Educational Reforms: Introducing the formal Diplomas, Degree and Masters Courses on Circular Economy is also expected to be important vehicle to bring the change.

4.2 BEHAVIOURAL

Behaviour is a distinct construct from mindset, knowledge, awareness, and concern. While people express concern about waste related issues, they continue to behave in opposite ways. Behaviour change is complex and it is possible to influence the behaviour by creating appropriate awareness on the issues attributed to the behaviour and motivating the right behaviour through acknowledgement, appreciation, awareness & education, enlightenment, incentives and rewards.

However, with time, our approach to unsustainable and wasteful resource use leading to explosion in waste generation has turned so casual that we need to **unlearn** many aspects that forms our current behavioural pattern and then begin learning to change our world. Various drivers for behaviour change are discussed below:

I. **Back to roots, your culture**: Since time immemorial, it was basic human instinct to use everything he gathered, nothing was left to waste. Reuse till the last trace was the Life Mantra. Truly, our culture in itself is an additional and main pillar of sustainability. We need to peep into our past, look at our local, Indigenous knowledge and ancestral practices, revisit them in terms of current perspective and strive to inculcate those traits in our world. A few examples are

- Living in harmony with nature; taking from nature only when needed and returning it to nature in the form easily broken down by the nature. No single use and throw away materials.
- Prefer buying loose grocery and vegetables rather than packaged in foam tray with plastic wrap. Carry cloth shopping bag to avoid carrying grocery and vegetables in plastic shopping bags
- Encourage reusable tableware, cutlery, etc. for serving food during social gatherings and ceremonies. Prefer restaurants with similar offerings
- Leading by example, e.g., by teachers, parents, managers, authority heads, etc. makes a big impact on behaviour of people.

II. **Nipping it in the Bud**: Children are the best ambassadors for implementing any change. Educating children brings about phenomenal change in behaviour towards waste and active participation in developmental activities. Teaching a primary school student has potential to change the approach of not only entire family but whole community. Making them aware of impacts of littering, chemical wastes, e-waste and unintentional release of micro-plastics on our environment, drinking water and foods could help the entire world nipping the issue in the bud.

III. **Governance**: The management of the businesses need to embrace **stakeholder management rather than shareholder management** to effectively deploy their capabilities, resources and infrastructure to address the key issues involving wastes from their products and services and the exact needs of all the stakeholders.

IV. **Community Outreach and Engagement**: To achieve behaviour change, we need to focus on community outreach by the way of engaging them, raising their level awareness and capacity-building campaigns amongst all stakeholders at large and community in particular.

- Social inclusion - If it's not inclusive, it's not sustainable. Exclusion can rob individuals of dignity, security, and the opportunity to lead a better life. Engagement with informal groups, e.g., waste pickers, through specific working arrangements, flexibility and incentives helps in maintaining a sense of autonomy and self-interest over and above livelihoods and opportunities. Social Inclusion provides an opportunity to education for the children of such groups of people, which has a potential to drive a major change.

- Community Outreach - Organizing special days and events within the community can enhance community outreach, e.g., Waste Free Day on the occasion of World Environment Day.

V. **Communication**: Beyond the logistics, waste management is also a communications business. Communication reminds people about the desired behaviour traits

- Simplifying information on any product consciously activates certain values and attitudes of individual, e.g., product label stating water saved while manufacturing
- Mobile applications, or apps, are also essential tools to enhance communication through information on nearest recycling centre, their timings, waste collection schedules, booking appointments, tracking, tracing and mapping of different wastes, etc. Moreover, spreading the word about the importance of recycling through social media is much more cost effective than other methods.
- Communicating at multiple avenues, e.g., school, community outreach, hoardings, TV advertisements, social media campaigns, adding a line below each Text message / SMS, package labelling, reminding them about culture, etc. would convince them about the severity of the impact,

VI. **Engagement and Empowerment**: The leadership and management at apex levels, sometimes, might struggle to find the right solutions wherein engaging and empowering representatives from different groups leads to a sustainable solution by the way of behavioural change

- Engaging people at different levels - one to one, group level, community level, organization level, gradually allows developing new habits and behaviours.
- Empowering your staff to find a generic but customized solution. For examples, allowing them to choose the type of "water bottle" that suits them to replace plastic single use water bottle. This would significantly add to desired sustainability goal.

VII. Striking the Right Chord on changing behaviour:

- Marketing and Advertising campaigns to spark change in consumer behaviour towards recycling, Regulatory and Advertising efforts to influence conscious minds of the people so that they practice it with unconscious mind.

- Waste Collection centres to offer Incentives and Rewards on recyclables to depositors when they deposit at collection centres but apply charges if collected from home. The manufacturers also can join the band-wagon wherein they charge for the packaging and reimburse when the packaging is deposited for reuse/recycling.
- Inspire to go for Durable and Recycled products: Concentrated and Sincere efforts need to be made to inspire people to adopt and stick to durable and recycle products.

4.3 REGULATORY

The regulatory regime, its broader application to cover broader waste related aspects, the clarity of its principles and accountability, the benefits from compliance, penal aspects of non-compliance, and the level and pace of its enforcement determines the behaviour towards effective waste management.

I. In order to promote the use of recycled products, **incentivize** the recycled products for rebate in taxes e.g., VAT, GST, etc. Incentivize the recycled products manufacturing industry, e.g., a lower tier for calculating taxes on profits.

II. Simultaneously, the Regulators need to mandate provision of sufficient alternatives to the consumers to opt, e.g., quality recycled products with identical specifications, product having equivalent specifications but with reduced environmental footprint, etc. This will shape the behaviour of the manufacturing industry towards products having reduced environmental footprint. Regulators need to work in this direction in parallel.

III. **Project Financing by Financial Bodies** – This process needs to include Community Consultation and Collaboration (CCC) at all stages of project thus making it an integral part of both Public and Private Sector Development Projects. The key decision makers should recognize this as a Value Enhancement Process. Empower community representatives (nominated on rotational basis, e.g., every year there is a different nominated Focus Group of community representatives to maximize the opportunity of participation and variety of ideas) at local level to participate in process of any development within their area. In addition to the clear scope of this mechanism, CCC to include mainly the following:

   - Consultation during Pre-Concept, Concept, Financing, Planning and Implementation and Post Development effectiveness and their potential mutual benefits.

- Participation during planning and implementation of any development project
- Empower them to monitor the project progress and discourses through community-driven approaches (e.g., social audits and community scorecards)
- 360-degree feedback throughout the project implementation via satisfaction surveys (e.g., Use of Social Media can aid the process)
- Complaint Redress Mechanisms that enable beneficiaries to present complaints
- Reporting and Recording - Transparent, regular and real-time on above process

IV. Mandatory **sustainability labels** (e.g., EU Ecolabel framework for energy-related products) and standards:

- Labelling Standards - to improve product awareness, e.g., indicate how much emissions will be reduced by recycling this packaging.
- Packaging Standard - More recyclable, reusable packaging e.g., jars, mugs, glasses, etc. All plastic packaging to be mandatorily recyclable to reduce the consumption of single-use plastics. Awareness on microplastics will help this for improving recycling behaviour.
- Common accessories for electrical or electronic equipment, e.g., common type of charger for mobile phones and tablets would not necessitate keeping multiple chargers, e.g., Apple [12] has recently launched iPhone-12 with no charging adaptor plug leading to packaging reduction by 70% plus e-waste reduction.

V. Grant **Industry status** to recycling, incentives and tax rebate. Recycling industry creates more jobs than disposal industry:

- A recent study (Cambridge Econometrics, Trinomics, and ICF – 2018) estimates [13] that applying circular economy principles across the EU economy has the potential to increase EU GDP by an additional 0.5% by 2030 creating around 700 000 new jobs.
- Notwithstanding, research by the University of Massachusetts [14] amid the 2008 recession revealed that, for $1 million in stimulus spending in the US, wind and solar technologies created approximately double the num-

ber of jobs compared with fossil fuel technologies, with solar creating 13.7 jobs versus only 6.9 jobs from coal.

VI. Accountability part to be appropriately addressed by regulatory regime

- **Extended producer responsibility (EPR)** - a policy approach that extends a producer's responsibility for a product to the post-use stage beyond point of purchase. The businesses are obligated to pay full net costs for managing packaging waste, including the collection, recycling, disposal and clean-up of litter.
- **1% of EBITDA for the Planet** to be a regulation for all businesses – mainly for restoring the ecosystem and sustainability initiatives directly under the concerned business supervision.

VII. **Other Regulatory Initiatives**:

- Planned and Progressive Phasing out production of single use products where alternatives exist, e.g., non-rechargeable batteries
- Introducing a ban on the destruction of unsold durable goods, except end-of-life goods upon material recovery.
- Material Recovery Targets in Construction and Demolition Wastes especially for Renovation and Redevelopment projects.
- Minimum mandatory green public procurement (GPP) and reporting for Government and Large organizations
- Strict enforcement of the "polluter pays" principle by the governments and local authorities
- Creating a fund from the penalties and fines collected through enforcement that shall be used only for development of Waste Reduction Systems

5 Conclusion

Although Innovation and Regulation can provide us technological solutions and administrative mechanisms, influencing the human behaviour seems to be the long-term and sustainable solution. Investing in influencing human behaviour should be the key criteria in developing further solutions for managing wastes and circular economy, as it significantly impacts the quantum and scale of solution. Creating and refreshing awareness, especially amongst primary school children, on impacts of littering, chemical wastes, e-waste and unintentional release of micro-plastics on our environment, drinking water and foods could help the entire world nipping the issue in the bud. Innovation and Regulation shall continue to shape-up the core behavioural change principles to bring about the desired change amongst the people. Design for Behavioural Change [15] could be the next game changer for reducing waste generation at source.

6 References:

[1] What a Waste 2.0: A Global Snapshot of Solid Waste Management to 2050 https://openknowledge.worldbank.org/handle/10986/30317

[2] Plastic Patrol https://www.outdoorjournal.com/featured/environment/plastic-patrol-app/

[3] United Nations Sustainable Development Goal #12 fact and figures https://www.un.org/sustainabledevelopment/sustainable-consumption-production/

[4] B Corps Create a Culture of Sustainable Living https://bthechange.com/b-corps-create-a-culture-of-sustainable-living-db9f2399e012

[5] Creating the Pull Factor Framework https://bthechange.com/b-corps-create-a-culture-of-sustainable-living-db9f2399e012

[6] Fuelling a Circular Economy through Sustainable Everyday Products https://bthechange.com/fueling-a-circular-economy-through-sustainable-everyday-products-c05c369eb4d2

[7] Introduction to Integrated Resource Recovery Centre (IRRC) Approach https://www.unescap.org/sites/default/files/Session%201_2_Waste%20Concern.pdf

[8] Donation Banks for Used Cloths and Shoes Installed at Aswaaq's Community Centres https://www.uaetoday.com/news_details.asp?newsid=31310

[9] Integrations https://www.gotenzo.com/integrations/

[10] Your Restaurant Inventory Management Software https://www.marketman.com/platform/restaurant-inventory-management-software/

[11] Sell more, waste less with dynamic pricing and smart markdowns https://www.wasteless.com

[12] iPhone 12 Pro https://www.apple.com/ae/iphone/

[13] Impacts of circular economy policies on the labour market https://circulareconomy.europa.eu/platform/sites/default/files/ec_2018_-_impacts_of_circular_economy_policies_on_the_labour_market.pdf

[14] GLOBAL ENERGY TRANSFORMATION https://www.irena.org/-/media/Files/IRENA/Agency/Publication/2018/Apr/IRENA_Report_GET_2018.pdf

[15] Design for Behaviour Change as a Driver for Sustainable Innovation http://www.ijdesign.org/index.php/IJDesign/article/view/2260/740

Author's address(es):

Uttam Yadav
Ruwais, Abu Dhabi
United Arab Emirates
Telephone +971 56 7941061
E-Mail pcuy@yahoo.com

7 Annexures

None

Zero Plastic Waste Cities
Sustainable Municipal Waste Management Through Social Business

Christina Jäger, Co-Founder & Managing Director

Konstantin Münchau, Research Analyst

Yunus Environment Hub GmbH, Wiesbaden Germany

Abstract

The world is facing a global plastic waste crisis. The rapid urbanization of many countries of the global south hereby constitutes one major factor for increased plastic waste generation and leakage into the environment. In many urban areas around the world, municipal solid waste management systems are overburdened, with the poorest communities carrying most of the burden. To tackle the manyfold root causes of plastic waste leakage, Yunus Environment Hub is implementing Zero Plastic Waste Cities around the world. Based on a modular approach and the concept of social business, Zero Plastic Waste Cities enhance municipal solid waste management systems with a particular focus on empowering the informal sector. This paper presents the methodology of the Zero Plastic Waste Cities program as well as best practice examples from different continents.

Keywords

Social business; plastic waste; municipal solid waste management; informal sector; waste pickers; circular economy

1. Yunus Environment Hub & Social Business

„We need to build a circular economy in order to achieve zero plastic pollution. Entrepreneurship and social business will go a long way in saving the planet"
Prof. Muhammad Yunus, Nobel Peace Prize Laureate & Co-Founder of YEH

Yunus Environment Hub (YEH) is the global social business network that creates solutions for the environmental crisis. Co-founded and led by Nobel Peace Prize Laureate Prof. Muhammad Yunus, YEH's vision is to create A World of Three Zeros: Zero Poverty, Zero Unemployment and Zero Net Carbon Emissions. To achieve this vision, we support, design and implement social business solutions related to any aspect of the current global environmental crisis and work towards scientifically established goals and recommendations to prevent its potentially catastrophic consequences. Tackling the issue from many different angles, YEH's programs focus on:

- Supporting the transformation from linear to circular economy
- Enhancing municipal solid waste management systems in developing countries

- Reducing global carbon emissions
- Preserving biodiversity by supporting local communities and indigenous people
- Promoting sustainable agriculture by empowering rural communities and farmers
- Providing clean energy
- Fostering solutions that provide access to clean water and sanitation

Within all our project activities, the social business concept constitutes the baseline method of our work. Following the definition of Nobel Peace Prize Laureate, Prof. Muhammad Yunus, a social business constitutes a unique form of business in the sense that it focuses exclusively on solving a particular social or environmental problem while ensuring financial self-sustainability. A social business hereby combines the best of two worlds: the social mission of a charitable organization with the acumen and creativity of traditional for-profit business.

In contrast to traditional for-profit businesses, its social mission enables the social business to form multi-stakeholder alliances with the private and the public sector to jointly follow a particular social objective. Unlike most non-profit organization, however, a social business is not depending on recurring donations or grants, but instead operates financially independent and self-sustaining. While it may require initial investments for the set-up of its business operations, these exclusively serve the achievement of the social mission. In the long run, the social business generates zero losses, while covering its operational expenditures through its revenue streams. A social business does not pay dividends to its investors or owners. Instead, any profits are reinvested into the expansion of the business and the achievement of its social mission.

Depending on the context of implementation, there exist two distinct forms of social businesses. A social business type I is a non-loss, non-dividend company that is owned by an investor or a group of investors with purely social motives. The social business solves a specific social or environmental problem and all profits generated through its operations are reinvested into expanding and improving the business. Over time, the investors may recoup their initial investments but do not earn any profits. In contrast, a social business type II is a profit-making company that – instead of investors – is exclusively owned by poor people, either through direct ownership or through a trust or cooperative. The profits of the social business always go into the alleviation of poverty, be it via reinvestments into the social business operations, or by providing decent living conditions to the bottom of the pyramid- owners of the social business.

In the context of YEH's Zero Plastic Waste Cities program, the social business approach thus allows to put a 100% target on the environmental impact while ensuring long-term financial sustainability. It enables the effective embedding of all project activities into the local context and the careful consideration of existing stakeholders. Additionally, by balancing the objective of fighting global plastic waste pollution with the requirement of financial stability, a social business enables the optimum combination of tangible results with a long-term impact on the targeted project regions.

2. The Global Plastic Waste Crisis

According to World Bank data, the world currently generates 0.74 kilograms of waste per capita per day with an expected increase of 40% by 2025 (KAZA ET AL., 2018). In Sub-Saharan Africa, South Asia and the Middle East, where population growth for decades has shown high figures, waste generation is expected to double by 2025 (KAZA ET AL., 2018). This trend is further enhanced as countries of the global south rapidly develop from low- to middle-income countries, a trend that is associated with high urbanization rates, a growing middle class and changing consumption patterns. Especially in the rapidly developing economies of the global south, high urbanization rates and increasing living standards go along with a significant increase in waste generation and a particular increase in plastic waste composition (LEBRETON & ANDRADY, 2019). Thus, while an already shocking amount of 12 million tons of plastic is leaked into the ocean annually (BOUCHER ET AL., 2020), this number is expected to double by more than 2.5 times by 2040 (PARKER, 2020). According to recent projections, the highest proportion of mismanaged plastic waste will hereby stem from Sub-Saharan Africa and Asia (LEBRETON & ANDRADY, 2019), where rapid urbanization and changing lifestyle patterns are complemented by an import of plastic waste streams from higher-income countries of the global north (KAZA ET AL., 2018).

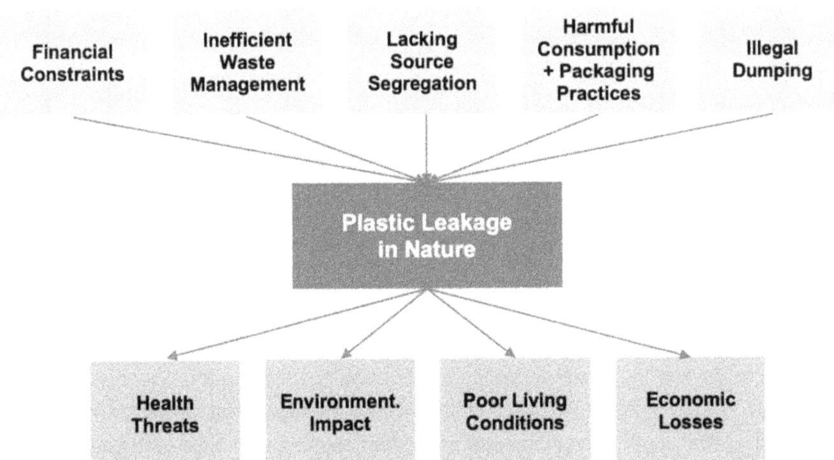

Figure 1: Origins and Consequences of Plastic Leakage in Nature in High-Leakage Areas

With alarming increases in plastic waste generation and leakage in the rapidly developing countries of the global south, the need for comprehensive waste management systems becomes more evident and urgent than ever before. However, in many of these countries, regional governments and city municipalities are overburdened with the rapid population growth of recent decades that has resulting in an influx of informal settlements and a significantly increased waste generation. While waste collection, sorting and recovery rates vary widely between high- and low-income countries – but also between high- and low-income areas within the same national context – waste disposal in most countries of the global south still heavily relies on open dumping and uncontrolled waste incineration. The lacking waste management capacities of regional and municipal governments hereby result in alarming effects on the environment, national economies as well as human health conditions.

With national, regional and municipal governments in countries of the global south being overwhelmed by the waste-related consequences of rapid population growth and urbanization rates, the informal sector plays a crucial part in municipal solid waste management. Especially along the lower end of the waste management value chain and in cities' lower-income areas and informal settlements, stakeholders from the informal sector carry out important waste management activities such as household collection or the recovery of valuable waste streams. For long, waste management experts have thus been emphasizing the importance of integrating informal sector activities into national waste management schemes and strategies. At the same time, national waste management agendas need to go beyond the mere acknowledgement of informal sector activities and instead need to make sure that safe working conditions and a sufficient income also reach into the informal sector.

It is against this backdrop of a global plastic waste crisis and overburdened municipal solid waste management systems that YEH has initiated the Zero Plastic Waste Cities program to end plastic waste leakage in nature and to enhance the waste management activities in urban areas of countries of the global south via a social business-based approach.

3. The Zero Plastic Waste Cities Methodology

The Zero Plastic Waste Cities program was initiated by YEH in late 2018 as a modular approach to developing social business solutions to sustainable municipal waste management in low- and middle-income countries. Since the rapid urbanization of most countries of the global south constitutes one major root cause for an increased plastic waste generation and leakage, the Zero Plastic Waste Cities program focuses explicitly on enhancing local waste management systems in cities and urban areas.

While plastic waste pollution and marine litter constitute issues of a global dimension, solutions need to be adapted locally, since realities and dynamics of waste management value chains vary widely across national contexts and even from city to city within one country. The Zero Plastic Waste Cities program is thus built around four modules that are independently applicable to local waste management realities and market dynamics. Based upon an extensive analysis of municipal waste management processes and major gaps in different high-leakage areas around the world, the four modules have been designed to offer solution propositions along different stages of the waste management value chain of waste collection, recovery, processing, recycling and disposal. The Zero Plastic Waste Cities program hereby constitutes YEH's general methodological framework to municipal waste management that is individually adapted to local conditions and requirements during each respective project assignment.

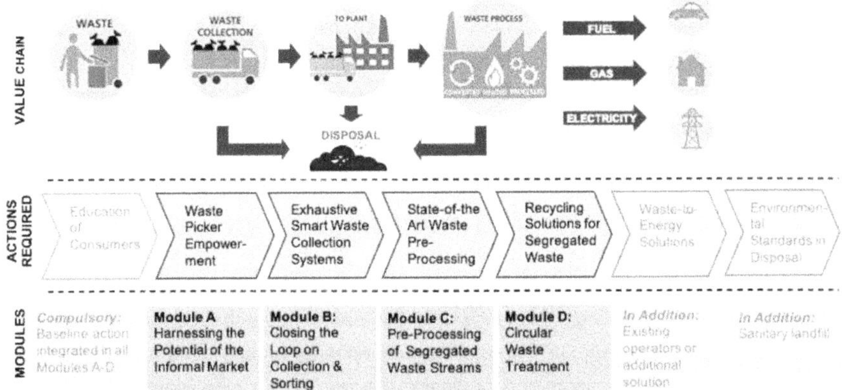

Figure 2: Modular Approach to Sustainable Waste Management

Preceded in any case by an extensive analysis of the local waste management system's dynamics and stakeholders as well as baseline awareness creation activities, the Zero Plastic Waste Cities program consists of the following four modules:

Module A – Harnessing the Potential of the Informal Market:

The first module analyses and strengthens the role of informal waste pickers in the respective solid waste management system. Operating at the lowest end of the waste value chain, informal waste pickers make their living by scavenging for valuable waste streams at cities' designated dumpsites as well as at smaller informal points of waste aggregation and along public roads. In many larger cities, several thousand waste pickers operate at the main dumpsites alone. While their everyday activities are characterized by a lack of protection equipment, hazardous working conditions and violence between rivaling groups, waste pickers in many countries are responsible for a majority of primary recovery of valuable recyclables such as plastic, glass, metal, textile or paper waste. Thus, the integration of informal waste pickers into solutions to enhancing municipal solid waste management systems is crucial.

Module B – Closing the Loop on Waste Collection and Sorting:

The second module aims at strengthening existing stakeholders and activities in order to establish exhaustive, segregated and smart waste collection systems in all income areas of a city. Especially in urban areas of countries of the global south, a wide gap exists between high-income areas on the one hand and low-income areas and informal settlements on the other with regard to the waste collection services available to local households as well as the waste collection rates achieved. While higher-income areas oftentimes achieve waste collection rates of 90% and higher, many low-income areas and informal settlements are underserved or not served by any means of formalized waste collection at all. Primary waste collection and sorting activities are thus widely carried out by informal to semi-formal stakeholders such as local community-based organizations, waste collection unions or youth groups. In order to enhance municipal solid waste management systems towards higher waste collection and sorting rates – and thus reduce waste leakage at the lower ends of the waste value chain – module B focuses on supporting and implementing local waste collection activities.

Module C – Pre-Processing of Segregated Waste Streams:

The third module focuses on the establishment of state-of-the-art waste pre-processing systems that are integrated into the existing municipal waste management value chain of collection, sorting, recycling and disposal. In the waste value chain, pre-processing systems and facilities constitute an important prerequisite for an adequate and efficient recycling of different waste streams and for closing the loop in municipal waste treatment. The establishment of social business-based material recovery facilities or sorting stations where different waste streams are adequately segregated and pre-processed thus forms a crucial element of enhancing municipal waste management systems.

Module D – Circular Waste Treatment:

Within the fourth module, social business-based recycling solutions for segregated waste streams are implemented to enhance municipal waste management systems towards circularity and high-quality recycling. The established recycling solutions may focus on recycled plastic products for the B2B and B2C market as well as on other recyclable waste streams such as organics, glass, paper or metal if existing recycling schemes are inadequate or inefficient.

This methodology of four individually or combined applicable modules enables the Zero Plastic Waste Cities program to effectively develop social business-based solutions for the specific needs of different urban waste management systems as well as according to the strategic objectives of project partners or funders. Avoiding the potential pitfall of a one-size-fits-all approach to the global plastic waste pollution, the modular approach allows for a financially flexible and efficient adaptation of the Zero Plastic Waste Cities program to different project locations and requirements around the world. Providing solution propositions along the entire waste management value chain, each specific Zero Plastic Waste Cities social business implementation is able to integrate existing local waste management stakeholders from the formal and informal sector into the project framework, e.g., as suppliers, customers, partners or employees of the social business.

Depending on the dynamics and local context of municipal waste management systems, three different implementation approaches to setting up a Zero Plastic Waste Cities programare possible as table 1 shows:

Table 1: Different Implementation Approaches of Zero Plastic Waste Cities

Initiation	**Scaling**	**Replication**
A new project or social business is initiated together with partners in the respective target region. An extensive pilot period is required to explore and analyze the waste management system on the ground as well as potential challenges to the proposed social business solution.	An existing solution that is already in place in the target country and successfully operated by a partner organization is taken as reference. Together with the partner, the project initiative or social business will be scaled to a target city. This approach allows to piggyback on a proven solution to create economies of scale.	An existing solution is taken (either from a different country or a model that is operated by another organization) and replicated in the target region together with local partners. The replication of successful social business solutions to other local contexts constitutes a long-term objective of any Zero Plastic Waste Cities projected rolled out by YEH.

4. Reference Projects

4.1. India & Vietnam: Recycling Low-Value Plastics

Since 2019, two Zero Plastic Waste Cities are piloted in Puducherry (India) and Tan An (Vietnam) with financial support of the Alliance to End Plastic Waste. Both cities were identified as project locations due to their close proximity to a river or the sea and their high plastic waste leakage paired with the existence of potential local project partners and dedicated municipal governments to address plastic in the environment. After an extensive analysis of each city's waste management system, the most urgent gaps for intervention were identified. While primary waste collection is conducted by private waste collection companies contracted by the municipality, the entire waste is broad to the local dumpsite. Bothformal and informal stakeholders extract recyclables mostly PET, metal and glass by purchasing from households and businesses before formal waste collection (informal stakeholders), during the waste collection (formal stakeholders) and at the dumpsite (informal stakeholders). In both cities, the on the ground analysis revealed that certain low-value plastic materials were not targeted either by the formal or the informal sector's recovery activities due to the low value of the material and the lack of segregation

Against this backdrop, a social business solution was developed that would focus on module D of the Zero Plastic Waste Cities program's modular approach and strengthen the local waste management systems by creating a trickle-down effect on collection, sorting, processing and recycling of previously unconsidered plastic waste fractions. In each city, a recycling facility has been established, where plastic pellets and high-density plastic panels are produced from locally recovered plastic material.

By creating a demand for low-value plastic material that had previously been unconsidered, the facilities increase the regional recovery and recycling rates for plastic waste. As social businesses,the recycling facilities in both cities are run by a local management team and offer direct employment opportunities to 20-30 waste workers employed at each facility. Furthermore, through the increased demand for different plastic waste fractions and direct material sourcing from the informal sector, the social businesses significantly increase the income opportunities for a great number of local waste pickers. The two Zero Plastic Waste Cities implemented in Puducherry and Tan An constitute an effective mixture of the program's modules A and D that allows to significantly increase recovery rates for plastic waste fractions and reduce leakage while simultaneously strengthening existing waste recovery activities carried out by informal waste pickers at the lowest end of the waste management value chain.

4.2. Kenya: Increasing Waste Recovery in Low-Income Areas

In 2020, YEH started a feasibility study to bring the Zero Plastic Waste Cities program to East Africa, starting with Nairobi and Mombasa as the two initial project locations in the region. To get an in-depth understanding of the municipal waste management system's dynamics and most urgent gaps for intervention in both cities, a five month-long mixed-methods analysis and feasibility study was conducted. Hereby, site visits and discussions with local waste management stakeholders revealed that waste collection from households and small shops in low-income areas and informal settlements is almost exclusively carried out by informal to semi-formally registered community-based organizations (CBOs). While these CBOs put high efforts into the primary collection services they carry out with limited capacities and means, the waste management system in both cities' low-income areas is characterized by an alarming lack of larger-scale waste sorting and pre-processing facilities. Existing recovery and sorting activities for recyclable waste components are merely carried out via a scattered landscape of small-scale informal waste collection CBOs, waste pickers andsmall aggregators, resulting in low efficiencies as well as very low profit margins for stakeholders from the informal sector. Furthermore, the lack of larger-scale waste sorting and pre-processing facilities results in a high degree of open waste dumping and littering with severe consequences for the environment as well as local communities.

Therefore, YEH is developing a social business-based solution that will enhance the municipal waste management systems of Nairobi and Mombasa with a special emphasis on modules B and C of the Zero Plastic Waste Cities program. In both cities, several material recovery facilities (MRFs) will be established in low-income areas with the aim of increasing the recovery rates for valuable waste streams (such as plastic, glass, paper or metal), and reducing open waste dumping and littering. Each MRF will consist of a concrete building equipped with conveyor belts and baling machines as well as running water and washrooms for local waste workers. With capacity building support from YEH, the MRFs will be run as a social business that areoperated by local waste management groupsthat will be empowered to generate revenues by selling sorted and pre-processed waste streams to existing recyclers in the country.

The MRFs will source their waste input streams from smaller transfer stations where local waste collection CBOs aggregate and pre-sort the household waste they have collected from the surrounding low-income areas. Offering secured storage as well as sanitary facilities for the CBOs' members, the transfer stations will further enhance the important waste collection activities carried out by primary waste collectors in low-income areas while serving as focal point for the subsequent waste sorting and pre-processing at the MRFs.

After the completion of the feasibility study including finalizing the business and financial plan for the Zero Plastic Waste Cities in Nairobi and Mombasa, YEH is aiming to implement first pilots by the end of the year.

4.3. Ethiopia: A Mobile Baling Solution to reduce transportation challenges

Starting in early 2021, YEH is conducting an analysis of the municipal solid waste management system of Addis Ababa, Ethiopia. In contrast to the dynamics outlined for Nairobi and Mombasa, Addis Ababa's solid waste management system is characterized by a high degree of formalization of primary waste collection efforts carried out by so-called collection unions. Furthermore, enhanced by Ethiopia's high demand for input materials, the informal sector achieves high recovery and recycling rates for a wide range of waste streams. The greatest need for intervention was thus identified at the level of lacking storage or compacting capacities at the level of waste collection unions and waste aggregatorsin the informal sector. This lack of pre-processing capabilities results in high inefficiencies and costs for storage and transport of aggregated quantities of recyclable waste streams for different stakeholders at the lower end of the waste management value chain. While the recovery rates of Addis Ababa's informal sector are remarkably high, the reigning inefficiencies in storage and transport thus result in low profit margins for most waste management stakeholders.

With the aim of improving the city's waste management system at the level of modules B and C of the Zero Plastic Waste Cities program, YEH is developing a mobile baling solution that will strengthen existing waste recovery activities and increase both efficiencies and revenues of waste collection unions and aggregators that trade with recovered recyclables. The mobile baling solution will consist of a fleet of trucks with two chamber-balers mounted on them. Serving the various waste collection unions, aggregators and waste traders in Addis Ababa, the mobile baling trucks offer the different stakeholders the option of significantly reducing the volume of their aggregated waste streams against a small baling fee. As a result, clients of the mobile baling solution will be able to store and transport three times the quantity of their waste streams, thereby significantly increasing their waste recovery efficiency and returns.

After completing the business plan for the mobile baling solution in the first half of 2021, YEH is aiming to create a MVP and test the social business idea in a pilot consisting of a first mobile baling truck serving waste traders and aggregators at Adddis Ababa's Merkato market area in the second half of 2021.

Table 2 summarizes the main characteristics of the three Zero Plastic Waste Cities projects.

Table 2: Summary of Zero Plastic Waste Cities Reference Projects

	Puducherry & Tan An	**Nairobi & Mombasa**	**Addis Ababa**
Identified Gaps	- Low recovery rates for low-value plastic - Income instability for informal waste pickers	- Low collection, sorting and pre-processing rates in low-income areas - Small-scale and inefficient recovery activities of the informal sector	- Inefficiencies and high costs for waste aggregation, storage and transport - Low profit margins for waste collectors and traders

Developed Social Business Solution	- Recycling facilities for low-value plastic that manufacture high-density plastic panels and pellets and source their input material from the informal sector	- Material recovery facilities and transfer stations set up in low-income areas run by umbrella organizations of local waste collection CBOs	- Mobile baling solution enabling waste collectorsand aggregators to increase their efficiency and overcome transporation challenges
Modules Targeted	- A & D	- B & C	- B & C
(Projected) Impact	- Increased recycling rate for low-value plastics - Decreased waste leakage - Direct job creation at the facilities + income stability for informal waste pickers	- Increased recovery and recycling rates for valuable waste streams - Reduction of open waste dumping and littering in low-income areas - Empowerment and professionalization of waste management CBOs into SMEs and social businesses	- Increased efficiency of waste recovery - Increased recovery and recycling rates for valuable waste streams - Increased revenues for waste collectorsand aggregators

5. Conclusion

The world is facing a global plastic waste crisis, that is expected to further unfold in upcoming decades. The rapid urbanization of many countries of the global south hereby constitutes one major factor for increased plastic waste generation and leakage into the environment and waterbodies around the world. As with most environmental challenges of today, it will predominantly be the world's poorest communities that will suffer the most from the consequences of plastic waste leakage and overburdened municipal solid waste management systems. Simultaneously, in most developing countries, stakeholders from the informal sector such as waste pickers, waste collection groupsand waste aggregators are keeping the lower end of the waste management value chain alive and running.

YEH's Zero Plastic Waste Cities program constitutes a modular approach to enhancing the waste management system of urban areas in developing countries through social business- based solutions. Due to the modular approach, the Zero Plastic Waste Cities program allows for a high adaptability to local needs for intervention as well as to the financial and strategic objectives of relevant project partners.

Building upon the existing experience in developing Zero Plastic Waste Cities in India, Vietnam, Kenya and Ethiopia, YEH is looking forward to implementing further social business solutions to municipal waste management in other locations in the upcoming years.

YEH is always looking to expand its network of partner organizations and bringing social business solutions where they are most needed. With still a long way ahead, we believe in the power of social business for enhancing municipal waste management systems around the world and for fostering a global circular economy to end plastic waste in the environment.

References

Boucher, J., Billard, G., Simeone, E., Sousa, J.	2020	*The Marine Plastic Footprint: Towards a Science-Based Metric for Measuring Marine Plastic Leakage and Increasing the Materiality and Circularity of Plastic.* Gland, Switzerland: IUCN, Global Marine and Polar Programme.
Kaza, S., Yao, L., Bhada-Tata, P., & Van Woerden, F.	2018	*What a Waste 2.0 – A Global Snapshot of Solid Waste Management to 2050.* Washington: The World Bank.
Lebreton, L., Andrady, A.	2019	*Future Scenarios of Global Plastic Waste Generation and Disposal*, in: Palgrace Communications 5(6).
Parker, L.	2020	*Plastic Trash Flowing into the Seas Will Nearly Triple by 2020 Without Drastic Action,* in: National Geographic. URL: https://www.nationalgeographic.com/science/article/plastic-trash-in-seas-will-nearly-triple-by-2040-if-nothing-done

Author's Address

Christina Jäger
Co-Founder & Managing Director
Yunus Environment Hub GmbH
Rheinstraße 109
65185 Wiesbaden
Christina.Jaeger@yunuseh.com

PARIS XVII MATERIAL-FROM-WASTE PLANT IN OPERATION

Christophe CORD'HOMME[1] **& Frédéric ROUX**[2]

[1]CNIM Group, PARIS – France - ccordhomme@cnim.com

[2]SYCTOM, Paris – France – roux@syctom-paris.fr

Abstract

Two years ago, the new Paris XVII sorting & recycling facility was delivered and the operation of this Material-from-Waste plant has started up.

SYCTOM (PARIS metropolitan agency for household waste treatment) is the public body responsible for the treatment and recycling of the waste produced by almost 6 million inhabitants of Paris and nearby suburbs, representing 10% of the population of France. Every year, some 2.3 million tons of municipal solid waste are processed. In the context of the increasing scarcity of raw materials and the energy transition, all this waste should be considered as resources, whether of material or of energy at least.

In 2015, SYCTOM awarded to CNIM consortium the contract to design, build and operate its new waste sorting facility in Paris (France). Located in the new Clichy-Batignolles quarter (17th arrondissement North of PARIS), this new large-capacity plant is processing the household packaging waste from more than 900,000 people. It is recovering packaging materials coming from the selective collection in preparation for recycling. This latest-generation sorting centre in Paris 17 is able to process up to 15 tonnes an hour of municipal waste. Thanks to its 13 optical sorting machines, it is super-efficient and fully automated. It helps achieve the target of recycling 75% of packaging as defined by legislation. Capable of sorting new plastics types, it represents a step forward for the recycling of household waste thanks to the implementation of the extension of waste sorting regulations.

The architects firm took also up the challenge of blending the installation's architecture into the urban environment of this new development zone. Situated on the northern edge of central Paris, this area is including the largest law courts complex in Europe with the new architectural attraction of the skyscraper "Palais de Justice" (Paris courthouse).

Keywords
Material-from-Waste, sorting, recycling, packaging, resources, household, circular economy…

1 Introduction

The SYCTOM, the PARIS metropolitan agency for household waste treatment, is the public body responsible for the processing and recycling of the municipal solid waste produced by the 6 million inhabitants in its eighty-four member towns (Paris and nearby suburbs), representing almost 10% of the population of France.

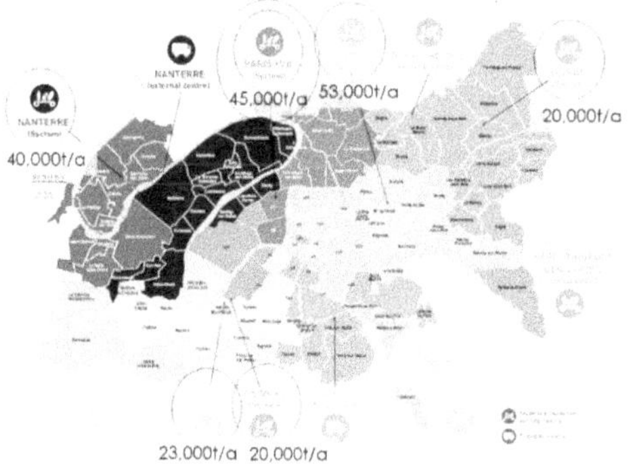

Figure 1 : Syctom sorting facilities

In 2019, SYCTOM processed 2,339,731 tons of municipal solid waste. In the context of the increasing scarcity of raw materials and the energy transition, all this waste should be considered as resources. In 2019, the 6 sorting centres of Syctom (including the new one in Paris 17) have treated 198,081 tons of household paper and packaging waste recovered by selective collection (yellow bin) (see figure 1).

This is a daily challenge for SYCTOM, which is constantly seeking innovations to optimize its facilities' performances (increasing energy efficiency, enhancing sorting and recycling processes) and to find optimal solutions for the processing of the various waste flows.

2 A new Material-from-Waste facility in Paris

2.1 New selective collection instruction

The fight against wastage not only urges us to consume better, but also to recycle the waste we produce more effectively. Syctom strives to achieve this objective by modernising its sorting centres and by organising a network of waste reception centres. These facilities are receiving the different flows coming from the selective collection of Parisian municipal waste and nearby suburbs (see figure 2). In 2022, plastic sorting will no longer be limited to bottles. It will cover all plastic packaging: jars, trays, films, etc. To comply with the Law for Energy Transition for Green Growth, which requires this expansion of the sorting instruction (see figure 3), Syctom is transforming its industrial tool.

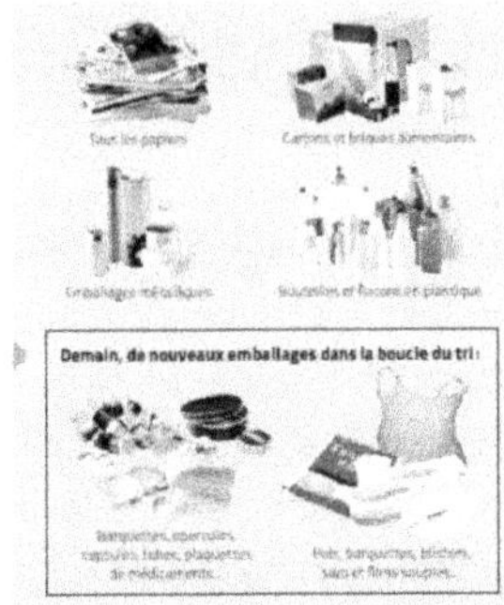

Figure 2: Evolution of "Yellow bin" content

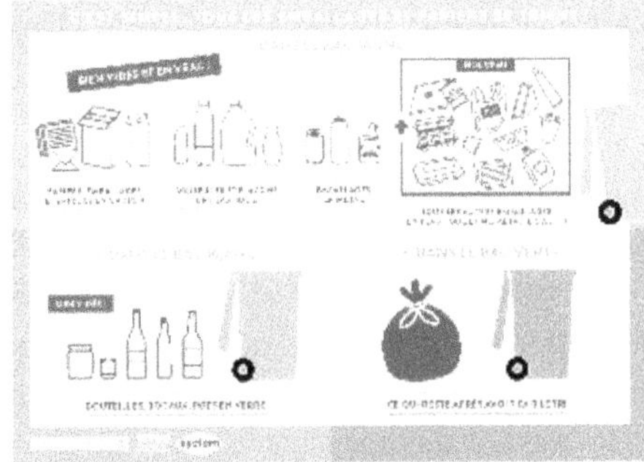

Figure 3: selective collection in Paris

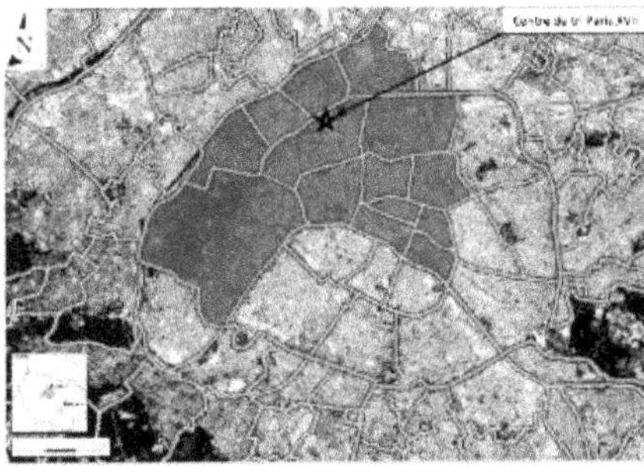

Figure 4: waste supply zone

Sort more and better, this is the objective set by Syctom in response to ever greater environmental challenges. The new Paris XVII sorting centre responds fully, thanks to its large capacity, efficient equipment and an optimized organization. It prepares for recycling household packaging waste from selective collection by 900,000 inhabitants of Paris and four neighbouring municipalities (see figure 4). It is the second one in Paris intra-muros. It gives Syctom a new generation tool capable of handling 45,000 tonnes of waste per year and implements the sorting instructions for all plastic and metal packaging.

2.2 A waste sorting centre in Paris intra muros

Situated on the northern edge of central Paris, the railway wasteland of the 17th arrondissement gave way to the new eco-district Clichy-Batignolles, transforming the image of the North-West of the French capital. At the heart of this 54-hectare “ZAC” (mixed development zone), this area is including the largest law courts complex in Europe with the new architectural attraction of the

Figure 5: Palais de Justice and Syctom sorting

skyscraper “Palais de Justice” (Paris courthouse), designed by star architect Renzo Piano.

In 2015, Syctom awarded to a consortium led by CNIM the contract to design, build turnkey and operate its new waste sorting facility in this zone. In addition to CNIM in charge of the design-and-build responsibility and of the operation and maintenance, this group includes Ateliers Monique Labbé, for the architecture, “Urbaine de travaux” for civil works, “Ar Val” for equipment, Ingerop and Segic for engineering. Established rue de Douaumont on a site of 1.1 hectares, the new Syctom sorting centre is well placed near the locations where wastes are produced.

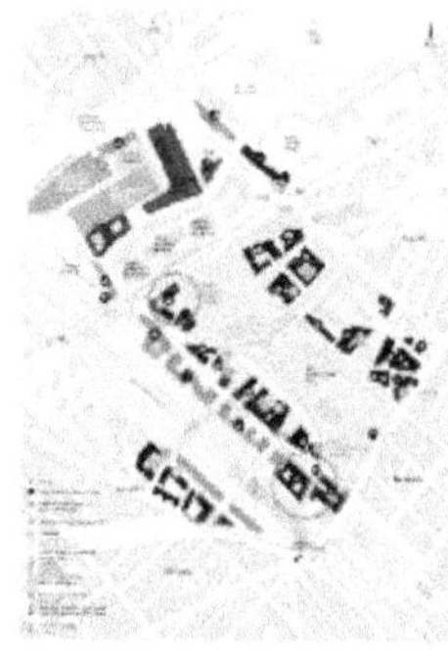

Figure 6: Clichy Batignolles development zone

It is a clear response to the challenges of the energy transition and the circular economy to improve recycling of different types of packaging waste.

3 A "new generation" centre adapted to new challenges

3.1 Principle

This latest-generation sorting centre in Paris 17 is able to process up to 15 tonnes an hour of municipal packaging waste. Capable of sorting new plastics types and so implementing the extension of waste sorting regulations, it represents a step forward for the recycling of household waste and help achieve the target of recycling 75% of packaging as defined by European legislation. It receives recyclable waste from selective collection trucks.

The received waste is the clean and dry conventional, packaging, paper, cardboard, plastics, metals, but also small waste electrical appliances and electronic equipment (WEEE). With this new sorting centre, the sorting instructions can be extended to food trays, polystyrene and plastic films, as it will be asked by 2022 for the sorting to the residents.

Figure 7: Paris XVII architect view

Thanks to its 13 optical sorting machines, it is super-efficient and highly automated to limit manual gestures and guide the activity of agents towards quality control; this is designed to improve working conditions and staff safety.

3.2 Input

As required by the 2015 French law on energy transition, the extension of sorting instructions for recycling is requiring by 2022, that residents will sort all plastic packaging and metal packaging at home. The Paris XVII sorting centre, with its high capacity and state-of-the-art equipment, is able to welcome and process these new flows in the best conditions.

The sorting guidelines apply to the following waste:

- Today :
    - All papers
    - Food cartons and bricks
    - Metallic packaging
    - Plastic bottles and flasks
- Tomorrow, new packaging in the sorting loop:
    - Trays, caps, capsules, tubes, medicine blisters
    - Pots, trays, blisters, bags and soft films.

3.3 The different steps of process for material sorting

The following block diagram illustrates the process principle of the sorting facility.

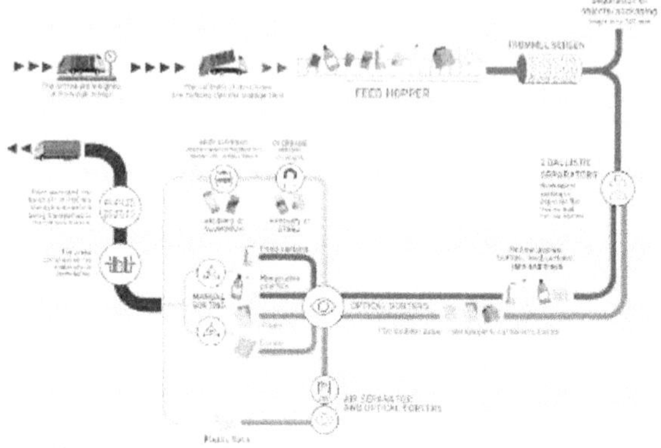

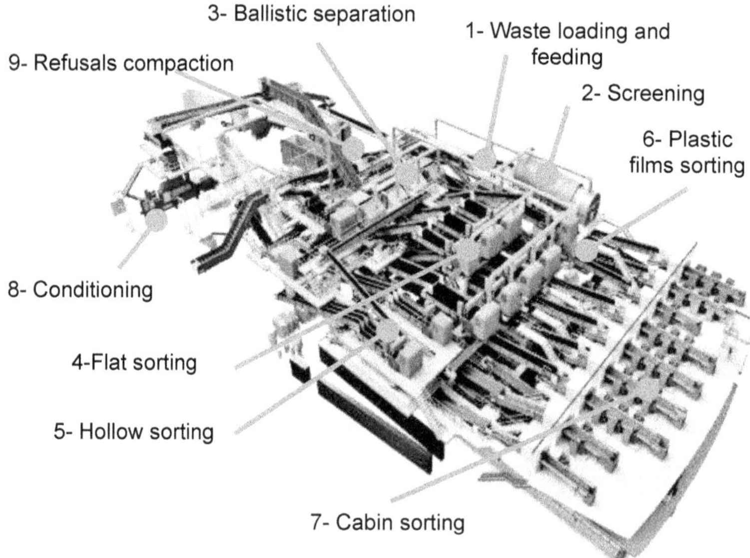

Figure 8: Paris XVII sorting process 3D view

With a waste reception 18 / 24h and an operation 6/7 days, the sorting horary capacity of 15 t/h allows treating 45,000 t / year with a recovery rate of 95%.

3.3.1 Waste loading and feeding

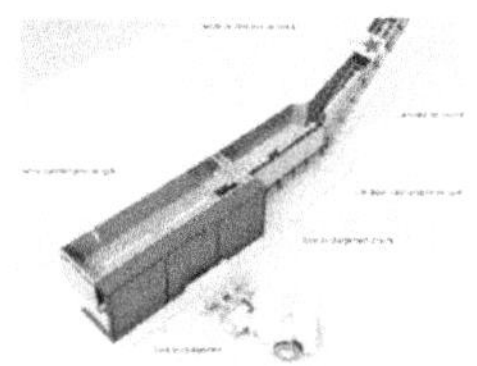

The wheel loader unloads the collected waste from the storage area into a hopper with an Alternating Moving Floor ("FMA") followed by the conveyor, which is feeding the sorting process.

3.3.2 Waste Screening

The trommel screen, a rotating cylinder with larger and larger holes, separates waste according to its size: below 90mm, over 350mm and in between (intermediate).

3.3.3 Ballistic separation

3D or "hollow" "heavy" bodies bounce on the ballistic separators and fall by gravity at the bottom. The 2D "flat" pieces, so-called "light" are blown by fans and evacuate on the top.

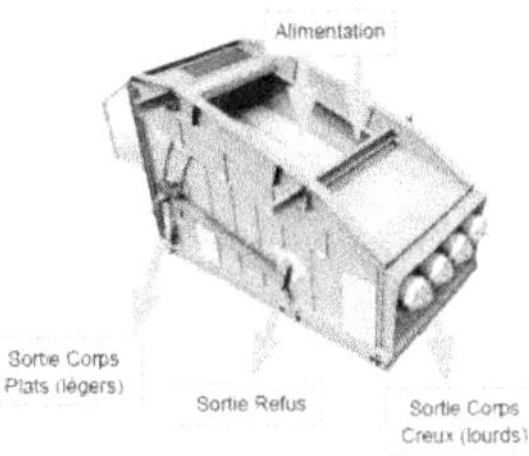

3.3.4 Flat bodies sorting

8 optical sorting machines (7 binary and 1 ternary) separate with positive and negative sorting Newspapers, Journals, Magazines (JRM), Recyclable Household Packaging (EMR), plastics and films.

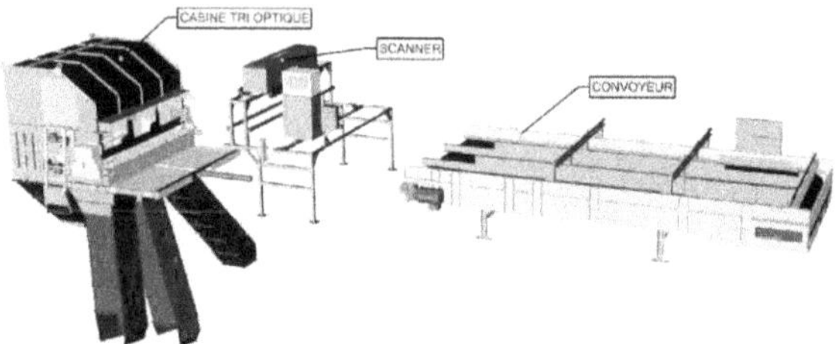

3.3.5 Hollow bodies sorting

4 optical sorting machines (1 binary and 3 ternary) separate PE-PP-PS (Polyethylene - Polypropylene - Polystyrene) + Packaging for Food Liquid (ELA), light and dark PET (Polyethylene terephthalate).

3.3.6 Plastic films sorting

1 ballistic separator and 1 optical separator allow isolating the films that feed the PEBD Films sorting table.

3.3.7 Cabin sorting

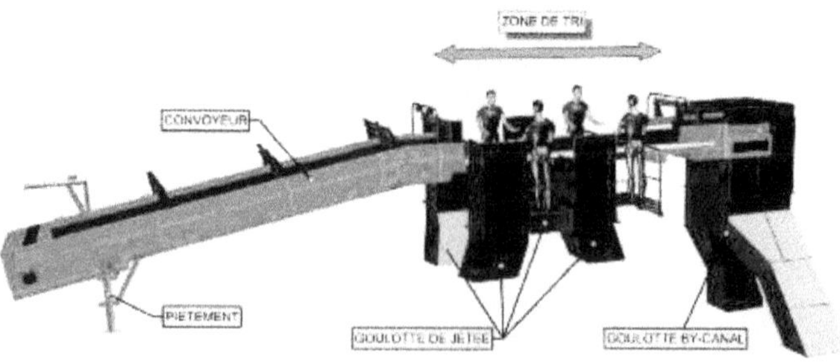

On 8 sorting conveyors, the sorting staff controls the quality of each product before baling:

- Dark PET
- Clear PET
- PP-PE-PS Polyethylene - Polypropylene - Polystyrene
- Food Fluid Packaging (ELA),
- Recyclable Household Packaging (EMR)
- Cartons
- Plastic films
- Newspapers, Journals, Magazines (JRM)
- Store cardboard (GDM)
- Aluminium
- Steel
- Small household appliances (PAM)
- Glass.

3.3.8 Conditioning

The sorted materials are packaged using two balers and a packer.

3.3.9 Compaction of process refusals

Alternating Moving Floor ("FMA") compactors are used for process refusals.

4 Lay out & Circulation

4.1 Lay out principle

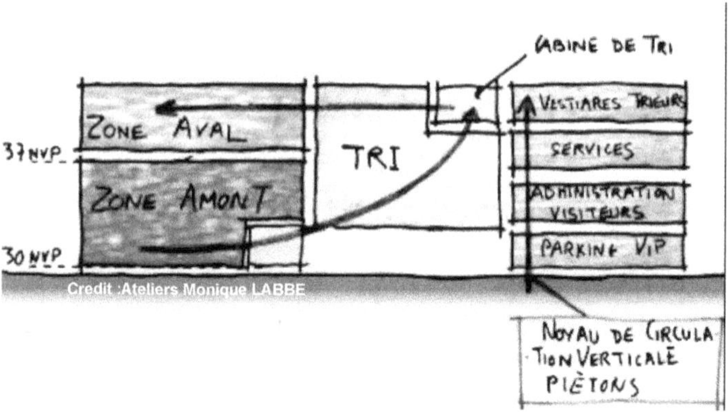

Figure 9: lay-out

A simple principle of functional layout has been adopted to achieve strict separation of functional areas and allow for the scalability of the sorting area.

Figure 10: sorting zone during erection of equipment

Figure 11: sorting equipment

4.2 Waste inlet

A traffic of 50 to 80 trucks a day is done for the loading of unsorted waste coming directly from selective collection.

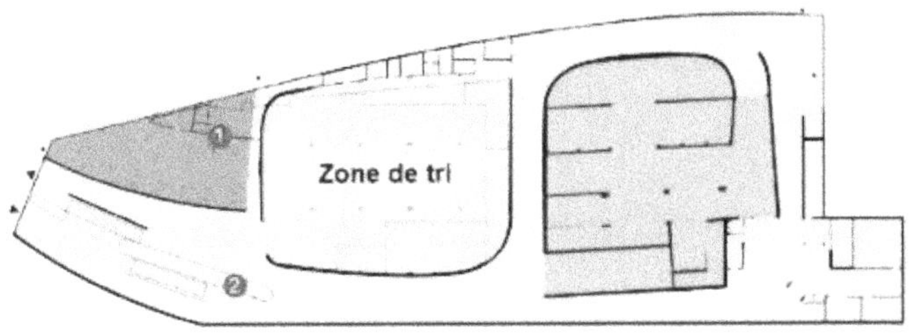

This waste flow arrives on the low level of the plant to be stored in the dedicated area for the waste reception on the right side after radioactive detection.

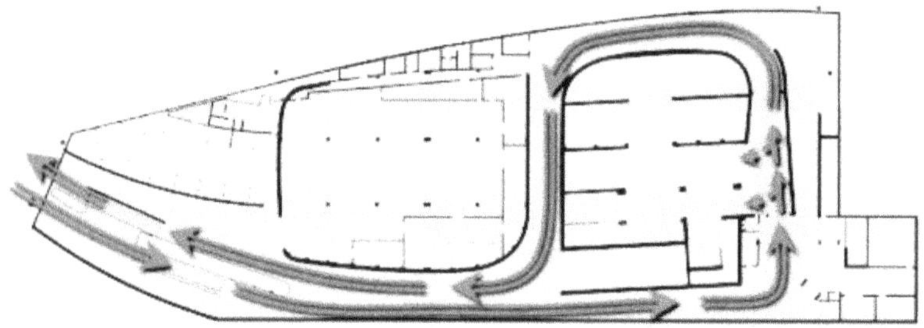

4.3 Material outlet

After sorting, recyclable materials leaves the plant from the dedicated area for storage of baled materials (on the right side) by a dedicated access on top level.

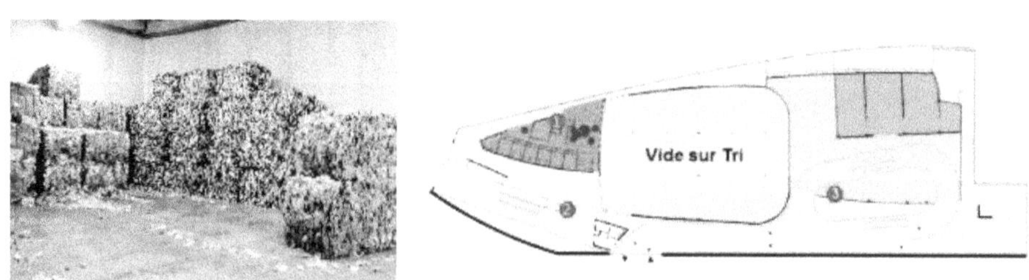

25 to 30 shipments are planned by road per week for plastics, metals and cardboard.

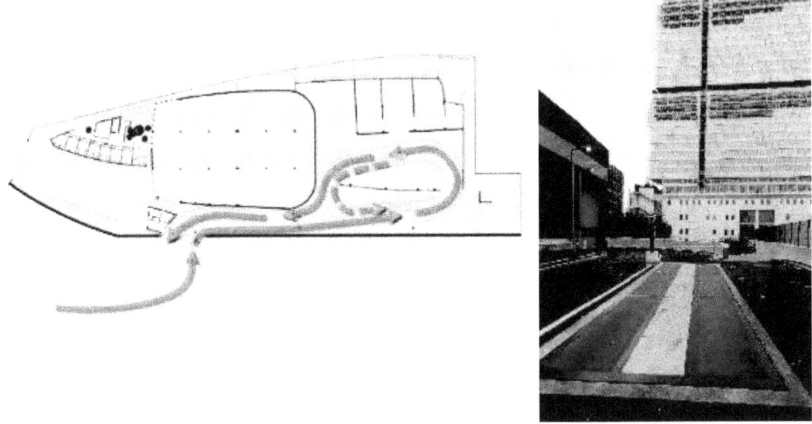

5 Exemplarity desire

Connected to the ring road and the railway, the sorting centre in Paris XVII is fully integrated into its environment. It took also up the challenge of blending the installation's architecture into the urban environment of this development zone and in the vicinity of the skyscraper "Palais de Justice", which is emblematic of the renewal of this district. The project is characterized by large vegetated areas, with suspended gardens and terraces, and a wide use of renewable materials. Equipped with photovoltaic panels, the roofs offer a landscape of streams using a wide variety of species, in favour of a bio-diversified environment.

Renewable material is used for the building facade and structure (wood). The North facade of the centre benefits from a specific treatment to limit the reverberation of traffic noise as shown on the following figures of noise mapping in day time.

Figure 12: Noise mapping

Status before plant construction (left) and after construction with absorbent facility facade (right)

With the help of careful insulation of the building, privileged natural lighting, luminaires with systematic presence detectors, the energy needs of the building is moderate. The building heating is provided by 85% renewable energy with the connection to the CPCU district heating. In the operating phase, the dust generated is directly sucked by an internal dust collection system.

The generated traffic by the plant operation is lower than 1% of neighbourhood traffic.

6 Construction

Started in 2017, the construction and start up was completed in 2019. The inaugural ceremony was held the 6th of June 2019 in presence of the French Vice-Minister of Environment, Mrs Brune Poirson.

7 Operation

The operation is assisted by a CAPM: computer assisted production management system, which goals are to measure performances & to help for operation.

Sensors are installed on sorting equipment and transport to weight each flow: upstream on inlet and downstream on material storage. This CAPM has been adapted from food industry and is linked to DCS (distributed control system).

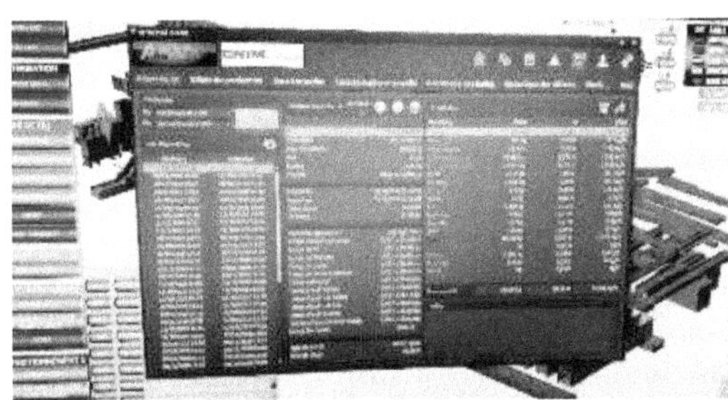

It gives the opportunity to produce immediate **reports:**

- daily
- by lot from one collection source
- by team
- by recycled material

It improves productivity for the automatic baling and continuous flow management by avoiding storage oversizing.

This system helps also for 3 D supervision by zones

- Inlet and refuses
- Optical sorting and manual checking
- Storage and baling

8 Credit and thanks

Credit photo and architecture Ateliers Monique LABBE

Credit SYCTOM and CNIM photos and documents

Web site : https://www.syctom-paris.fr/installations-et-projets/projets/paris-xvii/centre-de-tri.html

Video : https://www.youtube.com/watch?time_continue=2&v=fvxbvFGgx9Q

Authors' address

- Christophe CORD'HOMME

CNIM Environment & Energy
9 rue Francis de Pressensé, 93200 La Plaine Saint-Denis - France.
F: + 33 (0)1 44 31 11 00 / e-mail : ccordhomme@cnim.com

- Frédéric ROUX

SYCTOM
86 rue Régnault - 75013 Paris - France
F: + 33 (0)1 40 13 17 00 / e-mail : roux@syctom-paris.fr

Biogas from OFMSW in Mexico by using BEKON Dry-Fermentation

Ignacio Benitez

BEKON GmbH

Abstract

The BEKON process allows the production of biogas and compost from organic waste. Thanks to the robustness and simplicity of the BEKON process, it is possible to treat OFMSW in its fermenters without the need for sophisticated pre-treatment and/or post-treatment. This undoubtedly reduces operating and maintenance costs, as well as the energy balance between energy consumed and generated at the end of the process while producing biogas in a simple and effective way.

Keywords

Biogas, dry fermentation, batch fermenter, low maintenance and operating costs, OFMSW, biowaste, Carbon neutral

1 BEKON batch dry fermentation

The BEKON dry fermentation can be used for various types of organic waste. Both the organic fraction from residual waste (OFMSW), separately collected biowaste, green waste and a wide variety of agricultural or industrial organic waste can be utilised. The basic requirement for the use of the waste is a dry matter content of between 20% and 50%.

The BEKON process does not require any agitators or pumps that are in contact with the waste, so the process is completely independent of interfering materials (stones, sand, glass, wood, etc.). For this reason, no pre-treatment is necessary. However, in order to increase throughput and biogas yield, a screen (80-100 mm), a slow-speed shredder or a bag opener can be used depending on the waste properties. A massive shredding and/or addition of water is not necessary since the input does not have to be brought into a pumpable condition.

The BEKON dry fermentation is modular and consists of at least 4 independent fermenters. The capacity of a fermenter is between 3,000 and 5,000 t/ depending on its size and the retention time. A typical plant consists of 10 to 12 fermenters and has a capacity of 30,000 to 50,000 tonnes per year. At such a plant, 2 to 3 digesters are filled with waste per week.

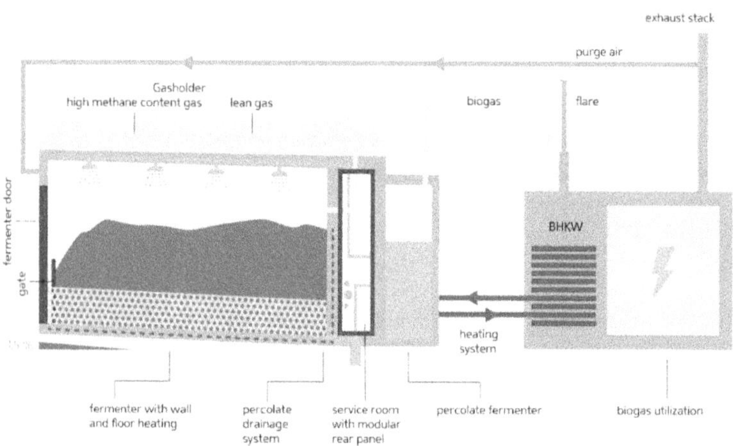

Figure 1: The BEKON process scheme

A wheel loader or automatic tunnel filling system fills the fermenter. The process starts either by inoculation with digestate or by an aerobic self-heating, followed by increased percolation with anaerobic process liquid. The optimum process temperature (mesophilic or thermophilic) is achieved through the large contact surfaces of the floor and wall heating system and enables a very high decomposition of the material. This optimize the living conditions of the microorganisms for biogas production in the fermenter without the need for further mixing of the biomass or the addition of further material.

Figure 2: Automatic tunnel filling system and wheel loader

The intensive inoculation with digestate and / or percolate leads to a rapid start of the anaerobic digestion process and therefore to a high biogas yield. The gas mixture accumulating within the early fermentation phase can directly be delivered to the gas storage unit, since all dry fermenters are operated in a batch process with a phased starting point. The large biogas storage unit buffers the discontinuous biogas production of the individual fermenters. To facilitate the fermentation process, the percolate is sprayed evenly over the input via a special nozzle system. The effective discharge of the percolate is ensured via the floor gradient to the back, drain channels and perforated drainage segments on the fermenter side walls. To optimize the drainage, the input can be flushed with biogas through the aerated spigot floor system. The percolate is collected in the heated percolate fermenter from where it is sprayed again over the input in the fermenter.

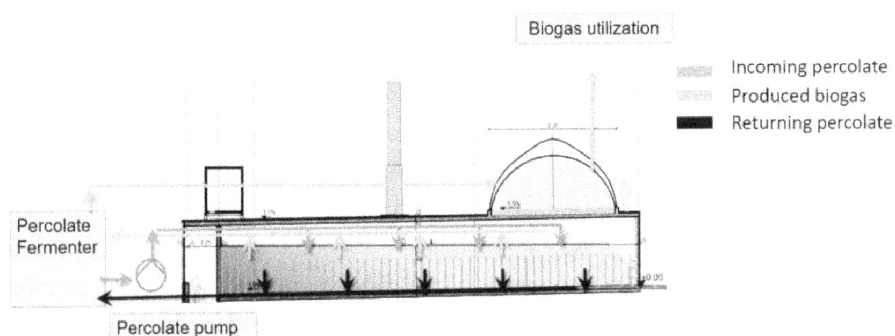

Figure 3: Percolate and biogas flow diagram

After three to four weeks, the biogas production in the fermenter diminishes. At this time, the percolation is stopped and the fermenter is aerated. The biogas-exhaust mixture is initially routed to the biogas storage until the methane content gets too low. The mixture then is flared by a lean-gas flaring system. Subsequently, the digestate is removed from the fermenter tunnels with a wheel loader. To ensure safe working conditions and emission protection, the fermenters are ventilated during the filling and emptying operations with a three-time air exchange rate. The exhaust air is treated by a biofilter.

The solid digestate is a well-drained residue in a stackable consistency, which can be further composted to a marketable product.

The generated biogas is normally used in a combined heat and power system (CHP) to produce electricity and heat. The continuous operation of the CHP is ensured by time-shifted filling and operation of the fermenters. The biogas is temporarily stored in a biogas storage for a few hours. Depending on the applicable laws, the generated electricity can be fed into the electricity network or can be directly marketed. Only a small amount of the waste heat generated is required for the self-consumption of the plant. Most of the thermal energy can be used externally, e.g. for feeding into a local or district heating grid or for drying materials.

As an alternative to electricity and heat generation, the biogas can be processed into biomethane and subsequently fed into the natural gas grid or used as a compressed natural gas fuel. The energy generated can thus be stored and used in a variety of ways.

Thanks to the support of the Eggersmann Group, BEKON (BEKON has belonged to the Eggersmann Group since 2016) can offer composting tunnels as well as separation plants. With these composting tunnels, the efficient conversion from anaerobic to aerobic conditions takes place. Depending on the waste, the composting technology enables the production of high-quality composts. The exhaust air from the composting tunnel is cleaned and deodorised by means of a scrubber and biofilter.

The first BEKON plant has been in operation since 2003, and since then 60 plants have been built worldwide and are being operated successfully. Through targeted internationalisation, BEKON has designed and provided the technology for 9 plants in North America, 1 in Asia and 1 in Australia.

2 Description on the project and its waste

Our project in Mexico is an OFMSW based project because there is no predominant source separation in Mexico that would allow direct access to a purer organic fraction.

Before the BEKON fermentation plant there is a separation plant, which basically consists of a bag opener, drum screen, manual separation and Fe separator. From this preliminary process, we obtain an organic fraction of 0-80 mm (OFMSW) which is used as substrate for our BEKON plant.

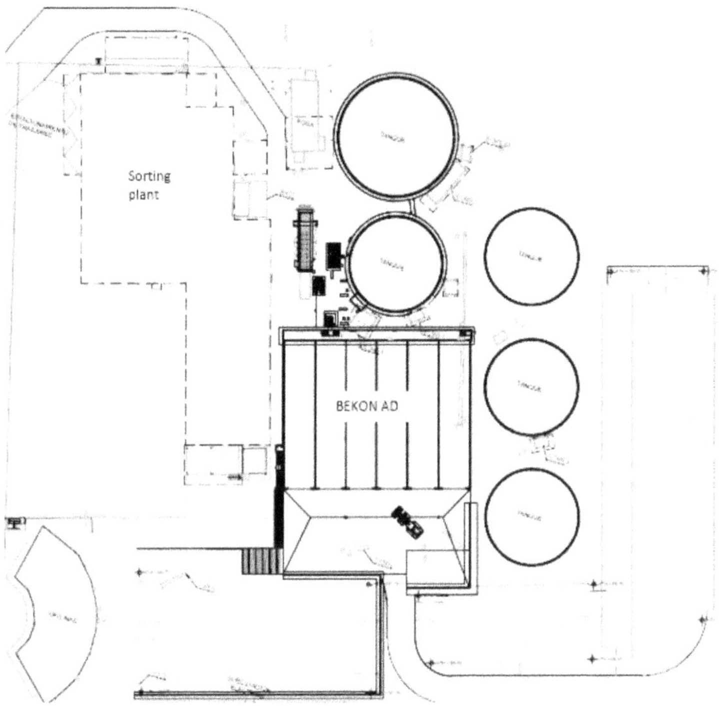

Figure 4: Site map Sorting and AD

The BEKON plant has an installed capacity of 633 kW and will generate 4,613,196 kWhel/a and 5,195,511 kWhter/a. It is designed for 21,000 t/a of OFMSW (0-80 mm) and has 6 fermenters of 31 meters long, 6.5 meters wide and 5 meters high.

Figure 5: OFMSW

In the past, our client tried to treat the OFMSW with a wet biodigestion plant but was unsuccessful due to the large number of impurities in the organic fraction.

The fermenters are filled and discharged periodically but not all at the same time to allow us to maintain constant biogas production.

Regarding the substrate it is, as mentioned above, MSW from the city of Linares and surrounding cities, which is composed as follows:

Elements	%
Organic	46,50%
Cartoon	16,63%
Clothing	11,95%
Crystal Pet	4,88%
High Density Bag	2,65%
Metals	2,63%
Low Density Crystalline Bag	2,48%
Multicolor Bag	2,44%
Dairy	1,72%
Polypropylene PP#5 (Hard)	1,64%
Tetrapack	1,37%
Low Density Bag Color	1,34%
Multicolor	0,97%
Disposable	0,83%
Footwear	0,45%
Glass	0,39%
Pet Green	0,34%
Aluminum	0,25%
Metal Wrap	0,21%
Electronic	0,18%
Exotic	0,14%
Total	**100%**

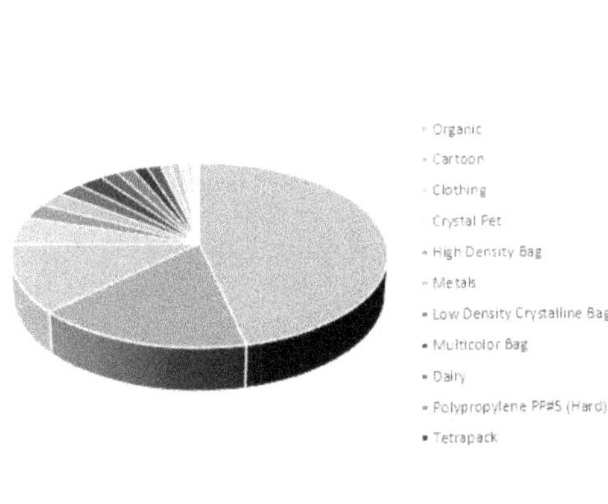

Figure 6: Waste composition

The biogas yield is 102 mN³/t according to laboratory tests performed to date. The substrate has a dry matter content of 58% and contains a lot of structure, which is quite advantageous for our system since our percolate can flow better through the material. For this process, we will use a combination between the percolate and digestate process in order to increase the biogas yield and reduce the operating costs.

To answer a question that you probably have about the CAPEX of the project I leave you with the following cost breakdown.

The total costs of the project can be divided into two categories, which are as follows:

- Engineering and equipment costs: 3,359,600.00 €.
- Civil works costs: 1,031,900.00 €.

It is important to note that the civil works were carried out by a local company using local materials. These costs vary from country to country.

3 Key project factors in Mexico

Within the BEKON project in Mexico we have different factors that we can mention and that contributed to BEKON finally being the technology used by the client considering the geographic conditions, simple separation plant and the type of waste to be treated, which contains a high degree of impurity that hinders conventional biodigestion processes.

Figure 7: Prefabricated rear wall

These factors include the following:

Operating costs

Since our process is dry biodigestion, we do not need large amounts of water to dilute the substrate and pump it, which saves us a substantial amount of water considering that this is a natural resource that is becoming increasingly scarce. Adding large quantities of water to treat the substrate would be counterproductive to the concept of generating energy from waste since water is a limited and valuable resource, not a waste. On the other hand, the material leaves the biodigester quite dry, which facilitates the composting process without the need for an additional process between biodigestion and composting. As for the percolate produced by the material itself, it is used to constantly percolate the substrate.

Having such a compact process, as the BEKON operation without agitators and with pumps that do not come into contact with the waste, results in low energy consumption. In addition, both pre-treatment and post-treatment do not have to be as intensive, which is undoubtedly reflected in the operating costs of a biogas plant. An additional factor is that the material to be pumped is water/percolate and not substrate or dense material that could damage or hinder the work of the pumps. As for the digestate, it should not be pressed in order to remove the water from the digestate, but the digestate should be dry enough to be composted. Thanks to these advantages, our plant has a self-consumption of approximately 8% of the electricity produced. As for the heat required, this is 15%, i.e. xxx kW.

The cost of personnel is also relatively low because the simplicity of the process does not require more attention from the operator. For this purpose, the operator will make use of 2 workers but not full time which allows him to make better use of the working staff. The remaining time will be invested in the separation plant.

Maintenance costs

In the case of BEKON the maintenance costs are quite low since the biodigesters have no moving or fragile components that can be damaged by contact with the material or the biogas such as agitators or pumps. In a project abroad and far away from Germany like Mexico, this is a great help since finding spare parts in the local market or having to travel there for maintenance generates high costs.

Expected maintenance costs are 35,000.00 € per year, which corresponds to less than 1% per year of the CAPEX cost of the biogas plant.

Simplicity of the process

In the case of Mexico, the simplicity of BEKON's process played an important role considering different factors such as distance, time difference, language, etc. which make the execution of a project difficult. The simpler the process and way of operating the plant the better for all parties. BEKON offers a very simple and robust process thanks to the lack of specific components that are difficult to understand and maintain. On the other hand, there is a saving of both time and money thanks to different ready-to-assemble prefabricated components such as rear walls. This undoubtedly lightens the operation of a plant on another continent and in another culture.

4 Biogas Yield

According to first laboratory studies of the OFMWS the biogas yield of the project in Mexico is 102 mN^3/t. Based on previous projects and BEKON's experience, our company guarantees 80% of the biogas yield achieved in laboratories.

From the generated biogas both electricity and heat will be produced as usual by a 633 kW CHP. The electricity will be sold publicly while a part of the heat produced will be used to heat the fermenters and in the future other processes of our customer's plant.

5 Compost

After the anaerobic process of 4 weeks, the digestate is extracted by wheel loader and subjected to a windrow process in order to stabilize and dry the material. For the composting process there will be different windrows as you can see in the picture below. The material to be composted will spend about 8 weeks in the windrows, during which time it will be stabilized and dried. It is expected that approximately 10,950 t/a of the 20,450 t/a will come out of the windrows. This is a great help for the operator as he has to move less waste.

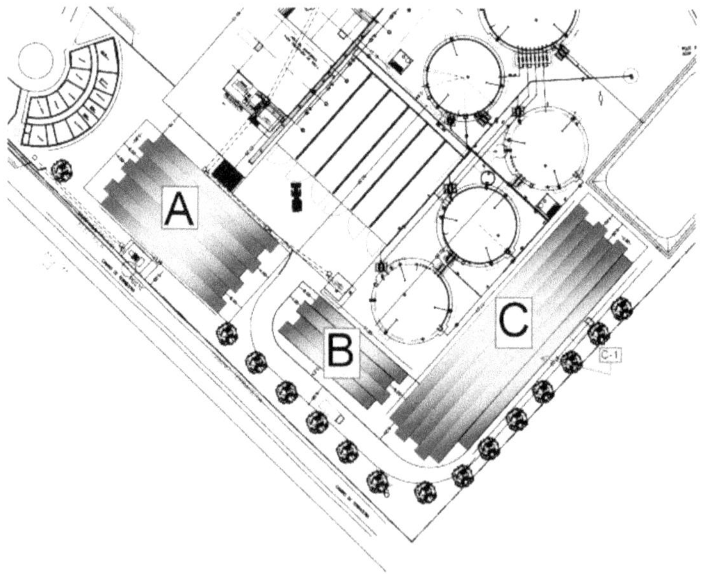

Figure 8: Composting and Screening process

After this and once the material is drier, the compost is screened at 20 mm to remove impurities that could not be removed in previous processes. This way the compost is ready to be marketed as soil for different uses such as construction, filler material, etc. thus avoiding the generation of greenhouse gases such as methane gas. This generates an additional income through the sale of the compost and contributes to the fact that the organic fraction of different cities in the region of Nueva Leon is treated according to high quality standards.

Figure 9: Composting and Screening process

Author's address

Ignacio Benitez
BEKON GmbH
Feringastraße 7
D-85774 Unterföhring
Telefon: +49 89 907795923
Email: i.benitez@f-e.de
Website: www.bekon.eu

Practical implementation of biological-oxidative desulfurization of biogas

Dipl.-Ing. Alejandra Lenis, Dr.-Ing. Kristoffer Ooms, Cedric Thull, M. Sc.

Research Institute for water and waste management at the RWTH Aachen University (FiW), Aachen, Germany

Abstract

The testing and systematic investigation of the biological-oxidative desulphurization of biogas under anoxic conditions was carried out. The hydrogen sulphide contained in the biogas was absorbed into the washing liquid in a biotrickling filter. The washing liquid loaded with hydrogen sulfide was subsequently fed into a separate reactor, where it was microbiologically decomposed into sulfate and sulfur with added nitrate. The process was tested on site at an agricultural fermentation plant where biogas is produced. Hence, a practical investigation of the process under realistic conditions was possible. The investigation was carried out at a pilot scale plant designed and built for this project. Operating conditions, such as trickling liquid velocity (TLV) and hydraulic retention time (HRT) and nitrogen to sulphur ratio (N/S-ratio) were varied to determine their influence on H_2S elimination capacity.

Keywords hydrogen sulphide, anoxic, biotrickling filter, biogas, biodesulfurization, sulfur oxidizing bacteria

1 Introduction

The most commonly used biogas desulphurization processes can be classified into three categories: chemical, physical and biological. Chemical and physical processes include the adsorption and absorption of hydrogen sulfide (H_2S) from the biogas. This is often carried out in a countercurrent absorption column or scrubber, where mostly caustic solutions are implemented for scrubbing (Schobert, 2013). Other methods, such as the addition of iron salts into the digester to prevent the formation of H_2S, are also widely used. Among the most common biological methods for biogas desulphurization are the aerobic desulphurization processes. These include biotrickling filters, bioscrubbers and direct biological desulfurization in the fermenter. For this purpose, oxygen (O_2) is required, which is added to the gas space of the fermenter or to a separate process unit by air introduction. The microbiological oxidation reaction of H_2S with O_2 allows almost a complete removal of H_2S and at the same time the formation of pure sulfur. As a disadvantage of the introduction of air into the gas space of the fermenter is the deterioration of the combustion properties of the biogas due to the mixing with the additional nitrogen (N_2) from the air. This practice also proposes a danger since an explosive atmosphere can be built by mixing biogas with air. Biotrickling filters and bioscrubbers are designed for indirect air entry so that a deterioration of the biogas quality can be decreased, however the danger of an explosive atmosphere still remains. Since the anoxic desulfurization of biogas does not require oxygen, neither an explosive atmosphere nor deterioration of biogas combustion properties occur (González-Cortés, 2020).

2 Material and methods

2.1 *Experimental set-up*

Within the framework of the Nitro-SX project, which was funded by the German federal ministry of economics and energy (BMWi), the Nitro-SX pilot plant was designed and built. It consists of two main process units, shown in Figure 1. These are a biotrickling filter (BTF), where the absorption of H_2S from the biogas into the washing liquid (agricultural digestate) takes place and a moving-bed biofilm reactor (MBBR), where the actual microbiological oxidation of dissolved H_2S is monitored. The pilot plant was installed at an agricultural fermentation plant and it was fed biogas produced in one of the anaerobic reactors.

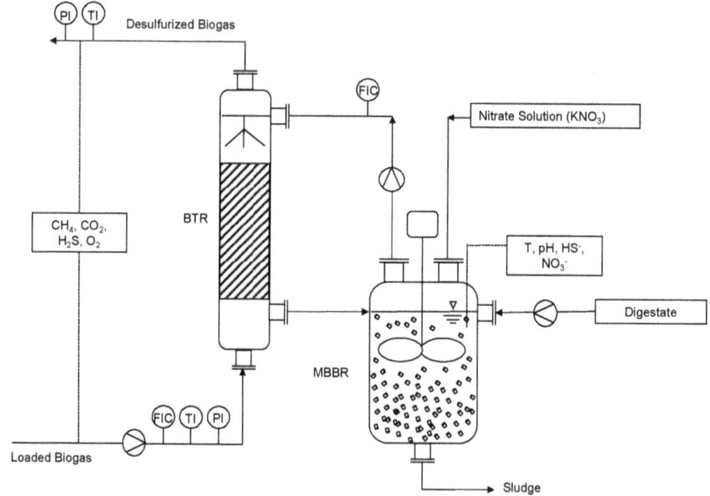

Figure 1: Process flow diagram of the Nitro-SX pilot plant

The BTF is made of semi-transparent polyvinyl chloride (PVC) and is filled with a carrier material, EvU Pearls®, which increases the exchange surface between gas and scrubbing liquid (approximate surface 755 m^2/m^3) and also as a biomass carrier. The carriers are made of recycled PVC and have a density close to that of water (1 kg/L). The BTF has a total volume of 0.16 m^3, with a packed bed volume of 0.088 m^3 (Height to Diameter ratio H/D = 1.75) and is divided into three layers. Dried raw biogas is uniformly introduced into the sump of the BTF through a nozzle. The raw biogas is pumped from the agricultural fermenter into the Nitro-SX plant by a knf Neuberger GmbH gas pump (N816 KT.29 DC-B-M). The biogas flow is controlled by flow meter (model H250 M40, KROHNE, Germany). In the BTF, the pressure is kept constant at max. 50 mbar overpreassure with the help of an overpressure flow valve (Type BS15i.D14ShJ.SSCP.TT, Bormann & Neupert GmbH, Germany). Additionally, the system pressure (Siemens, Sitrans P210, Germany) and biogas temperature (electrotherm GmbH, Germany) are measured at the inlet and outlet of the gas system.

The loaded biogas flows through the BTF from bottom to top and the digestate flows in countercurrent from top to bottom. During the process, the H_2S dissolves in the digestate, which has fully wetted the carrier material.

The sump of the BTF communicates hydraulically with the MBBR through a pipe. Therefore, the digestate level in the sump corresponds to the digestate level in the MBBR. This construction prevents biogas from flowing from the BTF into the MBBR.

The digestate from the MBBR is fed into the upper part of the BTF through a nozzle, which evenly distributes it onto the packing material. The digestate is recirculated by a controllable pump (Type BN 1-6L, Seepex GmbH, Germany). A level measurement (Type KQ6005, ifm GmbH, Germany) installed in the upper part of the BTF monitors a possible damming up of digestate.

2.2 Analytical Methods

The gas analyzer type SWG100 from the company MRU GmbH was used to measure the gas composition of the biogas at two different sampling points: on the raw gas side before entering the BTF and on the clean gas side after leaving the BTF. The gas analyzer measures methane (CH_4), carbon dioxide (CO_2), hydrogen sulfide (H_2S) and oxygen (O_2). The H_2S measurement time is user-dependent and was set to carry out a measurement every 10 minutes.

In the MBBR, nitrate as well as three other nitrogen compounds (NO_3^--N, NO_x, NO_x-N), the Total Dissolved Sulphur Analyzed (TDSA), temperature and pH were measured. The spectrophotometers OPUS and TpH-D from the company TriOS Mess- und Datentechnik GmbH were used for this purpose.

Four nitrogen compounds were measured photometrically using a xenon flash lamp as light source. The xenon flash lamp emits light which passes through the medium in the optical path and is partially adsorbed. The measurement method is based on the physical effect of light absorption. The absorption measure is substance-specific. Accordingly, nitrate nitrogen (NO_3^--N) can be detected in the concentration range between 0.5 and 160 $mg_{N\text{-}NO3}$/L. The measuring interval was set to 10 s.

The second sensor built into the spectrophotometer is the TpH-D, which is based on the electrochemical measurement method for determining pH and temperature using a salt bridge to protect the electrolyte. The sensor takes into account the temperature dependence of the pH value. The measuring interval was set to every 2 s.

2.3 Pilot plant operation

This study encompasses the results of 13 experiments carried out during the experimental phase of the project. Both the BTF and the MBBR were analysed separately, in ordert to determine the influence of the different operating conditions on the performance of the plant. The studied conditions were biogas and digestate temperature (which were not controlled), BTF overpressure, which was varied between 50 and 10 mbar, biogas hydraulic retention time (HRT) in the BTF and digestate trickling liquid velocity (TLV) in the BTF. With the help of the analytical measurements, the nitrate consumption rate was estimated. The performance of the BTF was described by the following operating characteristics (Dumont, 2015):

Inlet Load IL = $(\dot{V}_{biogas}/V_{packed\ bed})*C_{S\text{-}H2S,in}$ $[g*m^{-3}*h^{-1}]$ (1)

Elimination Capacity EC = $((\dot{V}_{biogas}/V_{packed\ bed})(C_{S\text{-}H2S,in} - C_{S\text{-}H2S,out})$ $[g*m^{-3}*h^{-1}]$ (2)

Removal Efficiency RE = $100*(C_{S\text{-}H2S,in} - C_{S\text{-}H2S,out})/C_{S\text{-}H2S,in}$ [%] (3)

Hydraulic Retention Time HRT = $V_{packed\ bed}/\dot{V}_{Biogas}$ [h] (4)

Trickling Liquid Velocity TLV = $Q_{digestate}/A_{BTF}$ [m/s] (5)

Where $\dot{V}_{biogas}$ corresponds to the gas flow rate (m^3/h), $V_{packed\ bed}$ is the fixed packed bed volume (m^3), $C_{H2S,in}$ and $C_{H2S,out}$ are the H_2S inlet and outlet concentrations in the biogas, $Q_{digestate}$ (m^3/h) corresponds to the liquid flow rate and A_{BTF} (m^2) is the BTF cross sectional area.

The IL and EC describe the H_2S concentration flow into the BTF and out of the biogas respectively and the RE indicates in percent how much H_2S was eliminated in the BTF. The HRT gives the time the biogas spends in the BTF, while the TLV states how fast the digestate flows through the BTF.

As a first step, the suitability of the OPUS spectrophotometer for use in this application was analysed. It was found that the digestate is too turbid for the measurement technique. For this reason, a dilution of the pure digestate had to be performed. The dilution ratio of digestate volume to clean water volume to allow measurement was determined practically and is 0.11.

According to the stoichiometry of the equations taking place, the N/S-molar ratio influences the outcome of the reaction process into producing more sulfate or more sulfur (González-Cortés, 2020).

$$5H_2S + 2NO_3^- \rightarrow 5S + N_2 + 4H_2O + 2OH^- \quad (6)$$

$$5H_2S + 8NO_3^- \rightarrow 5SO_4^{2-} + 4N_2 + 4H_2O + 2H^+ \quad (7)$$

The N/S-molar ratio was calculated as follows:

$$N/S = (C_{NO3-}/M_{NO3-})/[(C_{HS-}/M_{HS-}) + (C_{H2S}/M_{H2S})] \quad [mol_N/mol_S] \quad (8)$$

Since the OPUS spectrophotometer can only measure up to a maximum NO_3-N concentration of 160 mg/L, the N/S-ratios were always lower than 0.47. In order to estimate the nitrate consumption rate, nitrate dosing was carried out manually according to the measured H_2S and HS^- value in the digestate. The nitrate consumption rate is given by the following equation.

$$r_{NO3-} = \Delta c_{NO3-}/\Delta t \quad [mg_{NO3-}/(L*d)] \quad (9)$$

An overview of the experimental conditions are given in the following ***Table 1***:

Table 1: Operating conditions in BTF and MBBR during experimental phase

BTF					
Temperature [°C]	**Overpressure [mbar]**		**HRT [s]**	**TLV [m/h]**	**IL [g*m⁻³*h⁻¹]**
Biogas: 16 – 21 Digestate: 21 – 25	50/ 40/ 30/ 20		333.3/ 422.23/ 575.76/ 753.9	1.75/ 2.78/ 3.58/ 4.77	0.33 – 1.43
BTF + MBBR					
Temperature [°C]	**Initial Nitrate concentration [mg/L]**	**N/S-ratio [-]**	**HRT [s]**	**TLV [m/h]**	**IL [g*m⁻³*h⁻¹]**
Biogas: 16.1 – 27.4 Digestate: 21.47	6.94 – 121.15	0.04 – 0.47	437 – 364	3.44	5.05 – 18.9

3 Results and discussion

3.1 Dependency of removal efficiency and elimination capacity on process pressure

The results presented in this chapter correspond to the measured daily average values. In the following ***Table 2***, the operating conditions for two sets of respectively four experiments are given.

Table 2: Operating conditions in BTF for both experiment sets – pressure variation

Temperature [°C]	Overpressure [mbar]	HRT [s]	TLV [m/h]
Biogas: 15 – 26 Digestate: 21 – 25	50/ 40/ 30/ 20	422.23	1.68
Biogas: 16 – 21 Digestate: 21.3 – 23.9	50/ 40/ 30/ 20	440.7	1.75

The results of the experiments are displayed in the following Figure 1 to illustrate the dependency of EC and RE on IL for 4 overpressure values. The diagonal line cutting through the diagram represents the theoretical removal efficiency (RE) of 100 % (Dumont, 2015). Hence, the closer the experimental values are to the diagonal, the higher the RE.

For seven out of the eight tests the RE was above 90%, which is recognizable by the proximity of the measured values to the diagonal. The highest RE was recorded under 50 mbar for both experimental sets and were 98.88 % (HRT = 422.2 s, TLV = 1.68 m/h) and 98.75 % (HRT = 440.7 s, TLV = 1.75 m/h). The lowest RE was recorded under 40 mbar and was 47.61 % (HRT = 422,2, TLV = 1,68 m/h). However, the RE measured for the corresponding experiment under similar conditions was 97.53 % (HRT = 440.7, TLV = 1.75 m/h). A possible explanation for the discrepancy can be the fact that the pilot plant was not operated the day prior to the experiment, which might have led to a drying of the carrier material, therefore decreasing the RE. Without considering the experiment at 40 mbar with the RE of 47.61 %, the lowest RE was recorded under 20 mbar and was 93.16 %. Also to be noted is that although the digestate and biogas temperatures showed considerable variations between 21.3 and 27.9°C, the removal efficiency was more strongly influenced by pressure.

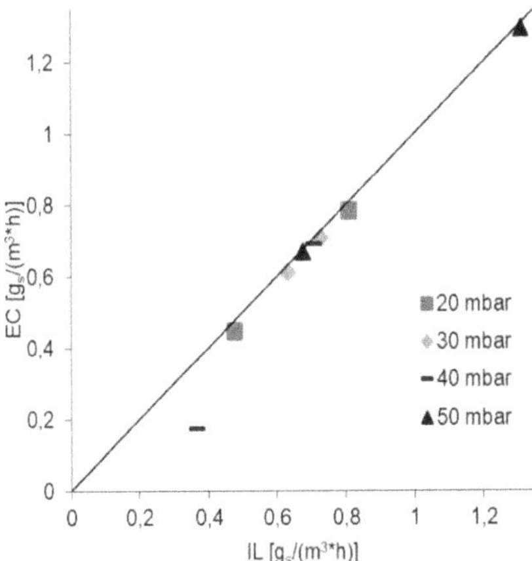

Figure 2: Dependency of Elimination Capacity (EC) on Inlet Load (IL) for four different experiments

3.2 Dependency of removal efficiency and elimination capacity on trickling liquid velocity

The results presented in this chapter correspond to hourly average values. The influence of four different TLV on RE and EC was studied under 50 mbar. A summary of the test conditions can be taken from the following ***Table 3***.

Table 3: Operation conditions for experiment set – TLV variation

Temperature [°C]	Overpressure [mbar]	HRT [s]	TLV [m/h]
Biogas: 17.1 – 26.1 Digestate: 22.6 – 24.8	50	443.4	1.75/ 2.78/ 3.58/ 4.77

In ***Figure 3*** below a tendency can be observed in which lower TLVs lead to a lower deviation or distance from the diagonal. This distancing behavior was quantified by calculating the mean squared error (MSE) between the experimental results and the theoretical value. The MSE was highest for the highest TLV and was $(560\ \%)^2$ and lowest for the TLV 2.78 m/h with a MSE of $(0.3\ \%)^2$. This scattering behavior can be interpreted as a higher process stability for lower TLVs. This observation is also consistent with mean removal efficiency. The highest mean RE was obtained at a TLV of 2.78 m/h and was 99.5 %, the lowest mean RE was 81.4 % at a TLV of 4.77 m/h. The second highest mean RE was obtained at a TLV of 3.5 m/h and was 99.2 %. For the lowest TLV the mean RE was 98.7 %.

The fact the highest TLV value leads to lower RE and EC can be explained by a short residence time of the digestate in the BTF, which leads to insufficient mixing between the gas and liquid phase and consequently dissolution of H_2S is incomplete. Therefore lower trickling velocities allow for a more complete dissolution process of H_2S and higher removal efficiencies.

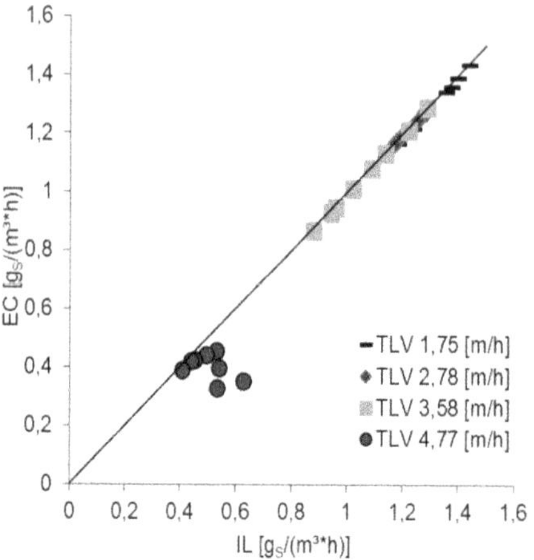

Figure 3: Dependency of Elimination Capacity (EC) on Inlet Load (IL) for four trickling liquid velocities (TLV)

3.3 Dependency of removal efficiency and elimination capacity on hydraulic retention time

The results presented in this chapter correspond to hourly average values. The influence of four different HRT on RE and EC was studied under 50 mbar. All test conditions can be taken from the following ***Table 3***.

Table 4: Operation conditions for experiment set – HRT variation

Temperature [°C]	Overpressure [mbar]	TLV [m/h]	HRT [s]
Biogas: 17.1 – 26.1 Digestate: 22.6 – 24.8	50	1.68	753.9/ 575.8/ 422.23/ 333,3

The following ***Figure 4*** showcases the course of the EC as a function of the loading rate IL. Here it can be observed that all experimental results do not deviate significantly from the theoretical course regardless of the H_2S inlet loading IL. On average the RE at HRT of 575.8 s was the lowest with a mean value of 95.86 % and a MSE of $(23.75\ \%)^2$. The highest RE value was observed at a HRT of 422.2 s and was 98.8 % with a MSE of (1.98 %)^2. The RE for HRTs 753.9 and 333.3 were 96.3 % and 97.9 % respectively. These results indicate that the variation of HRT have a lower effect on EC and RE than TLV, as the RE were higher on average and the MSE were lower than the previously discussed experiments under variation of TLV. Also as in the previous chapter, shorter residence times of the biogas in the BTF have a negative effect on the elimination efficiency.

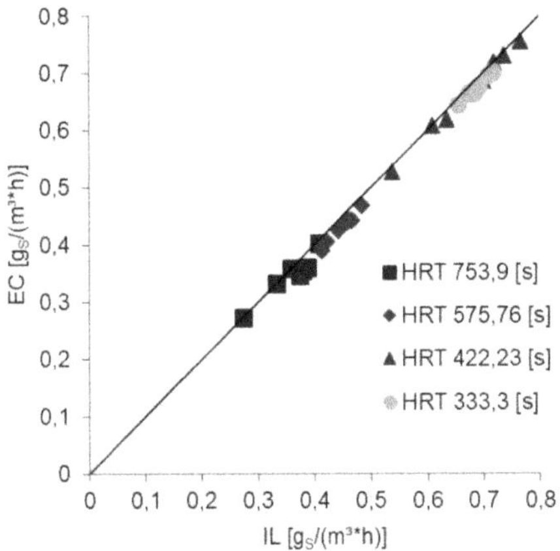

Figure 4: Dependency of Elimination Capacity (EC) on Inlet Load (IL) for four hydraulic retention times (HRT)

3.4 Effect of N/S-ratio on H_2S elimination

During the last experimental phase, the removal efficiency as well as the nitrate consumption rate in the Nitro-SX plant was investigated at higher IL values for an experiment period of 10 hours. For this purpose, a biogas mixture was produced from the digester of the biogas plant and from a biogas cylinder. The composition of the biogas from the digester consisted on average of 63.47 Vol.-% CH_4, 35.51 Vol.-% CO_2, and 20 ppm H_2S. The composition of the biogas from the cylinder was 61.59 Mol% CH_4, 38.41 Mol% CO_2, and 1.892 mol-ppm H_2S. The experiment was performed under 50 mbar. In accordance with the results from the previous chapters, the HRT was between 437 s and 364 s and the TLV was constant at 3.44 m/h over the test period. The following ***Figure 5*** shows the average hourly EC values as a function of IL over the entire test period. During the first four hours of the experiment, an IL between 5.05 and 5.68 $g_S/(m^3{*}h)$ was kept constant to allow bacteria to adapt to the conditions.

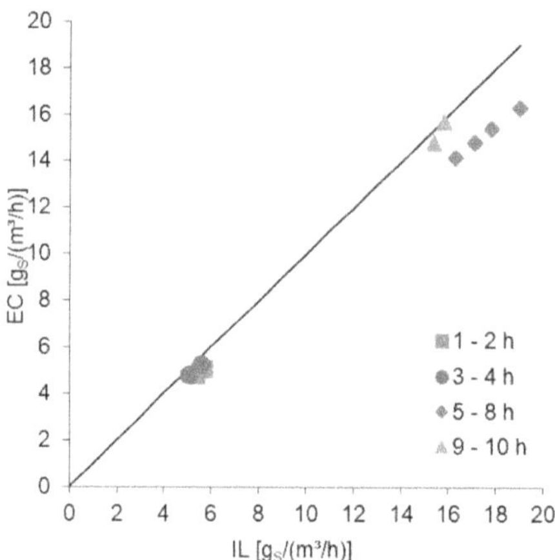

Figure 5: Dependency of Elimination Capacity (EC) on Inlet Load (IL) for four experiment intervalls

The RE during this adaptation time was 89,5 % the first two hours and increased to 95.2 % in the following two hours. After the fourth hour of the experiment, a jump in the RE from 95.15 % to 85.72% is seen, which is consistent with the increase in IL from 5.56 to 18.9 $g_S/(m^{3*}h)$, the latter being the highest IL value during this experiment. ***Figure 5*** also shows that although the IL decreased steadily from the 5th hour of the experiment, the RE remained constant from hour 5 to hour 8 (RE_{mean} = 84.4 %, σ = 0.51%). From hour 9, RE increased to the maximum value of 99.14% and during the following hour it decreased again to 96.48%. This alternating behavior can be attributed to a possible adaptation of the microorganisms in the digestate and in the BTF. It can be seen that both the IL as well as adaptation time have a significant influence on the RE and hence EC.

The influence of N/S-ratio in the liquid phase on RE was analysed. In the following ***Figure 6***, the courses of both N/S-ratio and RE over the experiment time period can be seen. It can be noted that both RE and N/S-ratio courses tend to showcase opposite behaviors, with N/S-ratio going down as RE increases during the experiment. This behavior is corroborated by estimating the correlation coefficient, which is – 0,7. A negative correlation coefficient corresponds to a negative linear relationship between the two quantities (Papula, 1999). This means that the smaller the N/S-ratio is, the greater the RE. Thus, the N/S-ratio in this process configuration could be used as a parameter for predicting the RE in the gas phase, as well as a control parameter of the process.

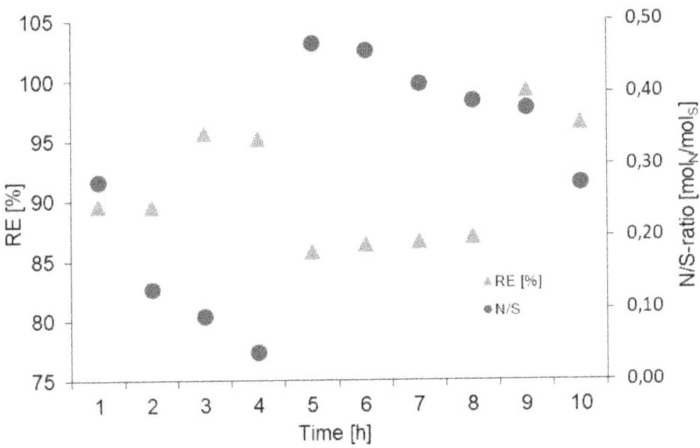

Figure 6: RE and N/S-ratio courses over experiment time

Nitrate consumption rate was also estimated. During the first four hours of experiment, the nitrate consumption rate in the digestate was 4.45 mg/(L*h). This rate was lower than for the following four hours of experiment, which was on average 9.05 mg/(L*h). During the last two hours of experiment the nitrate consumption rate was 12.97 mg/(L*h). The different consumption rates can be explained by the different initial nitrate and H_2S concentrations going into the system. This can also be explained by Le Chatelier's principle. This principle states that equilibrium reactions in which an increase in reactant concentration takes place cause a shift of the equilibrium towards the product side (Mortimer, 2003). In the following ***Table 5*** a summary of consumption rates for this and two previous studies is shown.

Table 5: Summary of estimated consumption rates

	Nitro-SX	(Fernandez, 2014)	(Almenglo, 2015)
NO_3-N-consumption rate [mg/(L*d)]	193.03*	956.2	308
N/S-ratio	0.47	0,77**	0,5
Initial NO_3-N-concentration [mg/L]	121.2	430	1500**

*projected value for a 24 h period of mean consumption rate between the 5th and 8th hour of experiment

**calculated average value

If the highest consumption rate in this study is assumed for a period of 24 h, then the nitrate consumption rate would be in agreement with literature values. It can be identified in Table 5 that initial nitrate concentration has a more significant influence on consumption than N/S-ratio.

4 Literature

Schobert, H. H.	2013	Chemistry of fossil fuels and Biofuels. Cambridge University
González-Cortés, J.J. et al.	2020	Simulateneous removal of ammonium from landfill leachate and hydrogen sulphide from biogas using a novel two-stage oxic-anoxic system. Science of the total environment. Elsevier
Dumont, E.	2015	H_2S removal from biogas using bioreactors: a review. International Journal of Energy and Environment
Papula, L.	1999	Mathematik für Ingenieure und Naturwissenschaftler Band 3. Vieweg Verlag
Fernandez, M. et al.	2014	Biogas biodesulfurlzation in an anoxic biotrickling filter packed with open-pore polyurethane foam. Journal of Hazardous Materials.
Almenglo, F. et al.	2015	Operational conditions for start-up and nitrate-feeding in anoxic biotrickling filtration process at pilot scale. Chemical Engineering Journal
Mortimer, C.E.	2010	Chemie. Georg Thieme Verlag KG.

Author's address(es)

Dipl.-Ing. Alejandra Lenis
Forschungsinstitut für Wasser- und Abfallwirtschaft an der RWTH (FiW) e.V.
Kackertstraße 15 - 17
52072 Aachen
Telefon +49 241 80 26842
E-Mail
lenis@fiw.rwth-aachen.de

Dr.-Ing. Kristoffer Ooms
Forschungsinstitut für Wasser- und Abfallwirtschaft an der RWTH (FiW) e.V.
Kackertstraße 15 - 17
52072 Aachen
Telefon +49 241 80 26822
E-Mail
ooms@fiw.rwth-aachen.de

Cedric Thull, M. Sc.
RWTH Aachen University

Management of mineral extraction residues: best practices in Lombardy Region (Italy)

Alessandra Diotti[1], Alberto Clerici[1], Giovanni Plizzari[1] and Sabrina Sorlini[1]

[1] Department of Civil Engineering, Architecture, Land, Environment and Mathematics, University of Brescia - Via Branze 43, 25123 Brescia, Italy

Abstract

Wastes from extraction and processing of mineral resources represent one of the main waste streams in Europe. These residues involve different types of materials that show a great potential for recovery, including the fractions that must be removed for the extraction of mineral resources (e.g. topsoil), as well as the scraps directly obtained from the phases of excavation (e.g. crushed rocks, stone powder) and processing (e.g. residual sludge) of stones and aggregates. In this context, this research work aims to study the valorisation of these residues and presents the results of a study developed on several companies located in the provinces of Brescia and Bergamo (Lombardy Region – northern Italy). The results demonstrated that all the generated residual materials can be recycled in several sectors, both industrial and construction. Moreover, the provinces analysed have proved to be virtuous realities, which, over the years, have built a solid collaborations network to achieve the maximum recovery of these mineral residues.

Keywords

Mineral resources, extraction, processing, crushed rocks, stone powder, residual sludge.

1 Introduction

The construction sector is the major cause of demand for building and construction materials. These materials (e.g. sand, gravel, clay) are generally used in the construction of engineering works (e.g. building, schools, bridges etc.), roads and other infrastructures, such as airways or railways (EUROMINES, 2021). In this context, the need for suitable mineral raw materials makes quarrying activities strategically significant in the European economy. At the same time, although mining activity is essential for the economy development, it involves a severe depletion of non-renewable natural resources due to intensive mining and a critical environmental degradation. Additionally, quarrying and processing activities generate significant amounts of residual wastes. According to the Waste Framework Directive 2008/98/EC, which established ambitious targets for EU waste legislation in terms of reducing the amount of waste and maximizing recycling and reuse operations, a key contribution to a more circular economy may come from the extractive wastes (BLENGINI ET AL., 2019). In 2016, 2,590 million tonnes of natural resources were extracted in Europe (EU-28) by means of quarrying and mining activities. Italy, with 167.8 million tons, is in fifth place after Germany (572 million tonnes), France (331 million tonnes), Poland (269 million tonnes) and the United Kingdom (259 million tonnes), confirming itself as one of the EU countries traditionally most representative of the sector. In reference to these quantities, the most extracted materials in Italy are limestone with about 65 million tonnes (39% of the total extracted) and sand and gravel with 54.9 million tonnes (32% of the total extracted) (ISTAT, 2019). In this context, the Lombardy region (northern Italy) is the first Italian region for quantities of quarrying minerals extracted (over 22.3 million tonnes, equal to 13.3% of the national total). In the same reference year, mining and quarrying activities contributed to 25% (640 million tonnes) of the total wastes produced in EU (28) (EUROSTAT, 2018). Italy, with a production of about 0.75 million tonnes of mining and quarrying wastes (0.45% of the related amount of minerals extracted) is in fifteenth place in EU. These data highlight that Italy, in this sector, is already oriented towards a circular economy model and is able to implement, and promote on the national territory, the recycling and reuse of the residues produced by stone and aggregate extraction and processing activities. So, the main goal of this work was to study the possibilities of valorising these residues and, at the same time, to promote the best practices in their management adopted on the regional territory. In particular, in the present study only the quarrying residues will be analysed (from stone and aggregate extraction activities). On the other hand, residues from mining activities will not be analysed.

The amount of residues produced depends on the type of material extracted upstream of the process (i.e. stones or aggregates). In stones quarries, generally dedicated to marble extraction (commercial name for a wide family of calcareous rocks), only 30% of the material extracted is considered suitable for commercial purposes. The remaining 70% are precious residues of the industrial process mainly composed of stone fragments (crushed rocks of different sizes) which still have an economic value but, sometimes, are still destined for disposal. (AUKOUR & AL-QINNA, 2008; PAPANTONOPOULOS ET AL., 2007). Conversely, by their nature, aggregate quarries generate much lower quantities and types of residues (e.g. washing aggregate sludge). The reason is that, as mentioned above, natural stones such as marble, limestone, granite, and more are quarried for commercial purposes and, therefore, must comply with certain shape and size requirements to be economically attractive. In practice, the extracted stone material (100%) is processed and transformed into blocks, slabs and more (30%) and, consequently, generates large quantities of residues (70%) that must be properly managed. In this context, the stone residue classification depends on the procedures adopted during the industrial process. In particular, the process is divided into two main sub-processes: the quarrying phase (directly carried out in quarry) and the processing phase of the extracted material (carried out in laboratory). During the quarrying phase, the solid mountainous rock is cut into large unfinished rough blocks which are subsequently sent to the processing laboratories and transformed into slabs and other smaller final products (processing phase). Each of these two processes generates different types of residues including topsoil and residual soil, crushed stone, stone dust, sludge, and others. A significant part of these residues shows a great potential for recovery and can be considered a valuable resource to be reintroduced into the economic market. However, in Italy, sometimes it is not adequately exploited due to unclear legislation and inefficient reuse or recovery practices (DINO ET AL., 2017).

In the literature, several authors have studied the possibility of valorising these residues both in the construction and other sectors, such as paper industry, plastics, paints, pharmaceutical and food fields, and so on (FURCAS, 2015). This is mainly due to their chemical composition as these materials have the same composition as the original stones. In particular, depending on the quarrying basin, the extraction of marble, limestone or travertine generally originates residues with a high content of calcium carbonate ($CaCO_3$) for which there is a continuous market demand not only from the construction sector.

Concerning waste stone reuse, the authors Sudarshan and Vyas, 2019 and Ulubeyli et al. (2016) evaluated the feasibility of use crushed stone as coarse aggregate in concrete production. The results demonstrated that the water permeability of the concrete mixture decreases as the crushed stone content increases due to the poor adhesion between the coarse material and the cement paste. Moreover, the resistance to chloride penetration and sulphates attack was significantly higher in concrete made with crushed stone aggregates than in the reference concrete. Arel, 2016 also highlighted improved workability and mechanical properties. Other authors (DINO ET AL., 2017) identified further potential recovery fields, for example in the production of asphalt, cement, gypsum, rubber, fillers, paper, paints, plastics, or as filling material and cliff boulders. Regarding the stone dust, generated by both the extraction and dry processing of ornamental stone, several studies analysed the possibility of using this residue as a substitute for cement (in the case of marble or limestone dust) (TAJI ET AL., 2019; AREL, 2016) or fine aggregates (SILVA ET AL. 2013; SOYKAN & ÖZEL, 2012) in concrete production. The results demonstrated that the replacement of cement with stone powder improve the concrete compressive strength by 25% and 10-20% is the optimal replacement ratio. The same was found with the fine aggregate substitution. The results showed that a substitution of up to 60% by weight leads to improvements in the concrete mechanical properties. Other interesting opportunities have been studied by other authors. Sadek et al., 2016 demonstrated the feasibility of using the stone powder as a mineral additive in the production of self-compacting concrete (SCC). Bilgin et al., 2012 investigated the possibility of using marble powder (with an addition of up to 10%) as a fine material in the production of industrial bricks. Differently, Awad & Abdellatif, 2019 studied the possibility of reusing marble dust as a reinforcement in the low-density polyethylene (LDPE) matrix. In particular, the results showed that the mechanical properties (i.e. flexural strength, compressive strength and hardness) of LDPE composites are improved by adding up to 15% by weight of marble dust to the composite matrix. Finally, Soydal et al., 2018 demonstrated the possibility of use as a filler in epoxy-based composites. At the same time, also the stone extraction and processing sludge (SEPS) and the washing aggregate sludge (WAS) can find different application in the industrial sector. In particular, SEPS can be used as Portland cement substitute (up to 10%) in concrete production (RANA ET AL., 2015) or as a filler in the autoclaved cellular concrete production (FORMIA ET AL., 2015). Sala et al., 2014 also demonstrated the feasibility of using wet SEPS as a component of bonding agents utilised for laying stone material or as an addition to lime mortars. Finally, Marras & Careddu, 2018 evaluated the possibility of using SEPS as an additive for tire production, as a substitute for calcium carbonate. Differently, Chamorro-Trenado et al., 2016 and Sorlini et al., 2017 highlighted the possibility of using WAS to produce porous fired clay bricks and unfired bricks, respectively. Despite these several reuse opportunities highlighted by the literature and the great recovery potential of these residues, they are still often destined for disposal (LUODES ET AL., 2012; LUCIANO ET AL., 2020) or destined for environmental restoration applications and disused quarries reclamation. Additionally, in recent times, the Italian stone sector has been facing severe economic difficulties linked to the increasing presence on the national and international markets of products and producers from foreign countries such as China, India, and Brazil. In this context, for a new upswing in the sector, the full use of the extracted resources and the valorisation of all the residues produced is fundamental. Within this framework, the present study aims to study the valorisation of the residues generated by stone and aggregate extraction and processing activities and to promote the best practices in their management adopted on the regional territory. In particular, the paper presents the results of a study developed in collaboration with Confindustria Brescia and Confindustria Bergamo on several companies located in the provinces of Brescia and Bergamo (the Lombardy Region). By means of technical visits and questionnaires submission, the different types of mineral residues produced by the companies and their best management practices were analysed. Specifically, in this paper only the results related to some companies will be shown.

2 Materials and method

2.1 Materials

To best represent the stone and aggregate extraction and processing chain present in the Lombardy region (with reference to the provinces of Brescia and Bergamo), several technical visits were carried out at representative companies of the entire supply chain. In particular, these companies are representative of:

- stone quarries (6 companies);
- natural aggregate quarry (1 company);
- stone processing laboratories (2 companies);
- residue treatment and end-use activities (3 companies).

As mentioned above, in this paper the results related to 1 stone quarry and 1 aggregate quarry will be analysed.

2.2 Methodology

In this section, in order to technically outline the current stone and aggregate excavation and processing chain, primary information and operational-management data were collected during technical visits performed at the selected companies. In particular, during the visits the following data were collected through questionnaires submission: (1) technologies and procedures currently adopted in the mineral extraction and processing chain, (2) types and quantities of residues produced and (3) final applications of the valorised residues.

3 Stone and aggregate extraction and processing chain

3.1 Stones

Figure 1 shows the stone quarrying and processing chain adopted in the provinces of Brescia and Bergamo and the related residues produced. In particular, in this section only the stone quarrying activities will be analysed taking one of the most representative ornamental stone quarries of the Brescia extraction area as a reference. Generally, the quarrying phase is carried out in open-air using dry mechanical techniques such as diamond wire, chainsaw machines or hydraulic drilling machine. In particular, during the cutting phase, the diamond wire is made to slide near holes previously made by a drilling machine in order to make a precise cut on the sides of the block. Sometimes, when the block presents flaws or impurities, the quarrying phase is performed by explosive (detonating cord). In this case, on the base of the block, coplanar holes are made at 25 cm from each other using a compressed air drill and, in these holes, explosive charges are inserted and detonated. The increase in volume and the pressure caused by the detonation leads to the breaking of the block along the perforation lines (dynamic splitting) and its sliding forward. Subsequently, by means of an excavator or a wheel loader, the extracted block is sent to the squaring station and the squaring phase is performed using a diamond wire and water, as a coolant and fine dust suppression liquid. The squared blocks represent approximately 20% of the solid stone extracted. The remaining 80% is classified as residual material. Generally, the residues produced by the stone quarrying activity are:

- topsoil and residual soil: material consisting of vegetation and soil that covers the mineral deposit. It also includes fractured rock and/or soil present within the deposit.
- stone dust: powdery residue produced by dry cutting and crushing of the ornamental stone.
- crushed stone: granular aggregate consisting of sharp-edged rock fragments resulting from the cutting or crushing of natural stone.
- Stone Extraction and Processing Sludge (SEPS): liquid-solid residue obtained from the processing and wet cutting of stone blocks. It represents the finest component produced by the entire excavation chain.

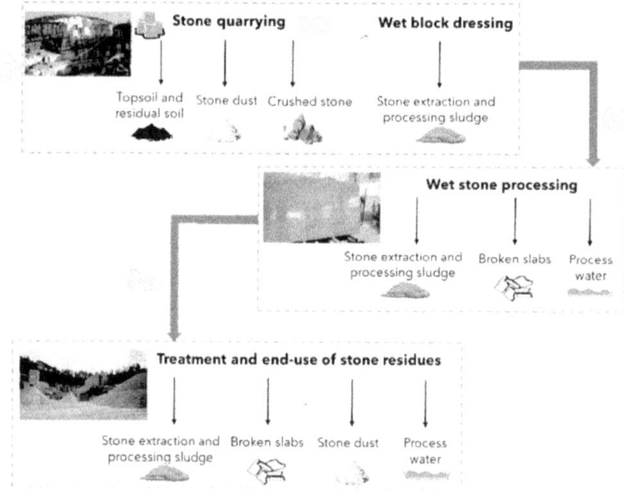

Figure 1 Stone extraction and processing chain

3.2 Natural aggregates

Figure 2 shows the aggregate quarrying and processing chain adopted in the Brescia province and the related residues produced. In particular, the results related to 1 natural aggregate quarrying activity will be presented. Generally, the quarrying phase is carried out in open-air using mechanical techniques such as dredger and excavator. The excavated material is transported and unloaded in the aggregate processing plant. Through conveyor belts the material is sent to a sieve that separates the earthy impurities and selects the aggregates with a grain size higher than 30 mm (destined to the crushed aggregate production line), from those with a grain size lower than 30 mm (destined to the granular aggregate production line). In particular, the aggregates with a particle size lower than 30 mm are washed and sieved to produced different particle sizes: 0-4 mm, 4-8 mm, 8-20 mm, 20-30 mm. These materials are then stored in heaps and subsequently sold as natural aggregates for concrete production. Conversely, the material with a grain size higher than 30 mm is stored in heaps and then crushed with a mechanical mill. After this process, the aggregates are washed and sieved to produced different particle sizes: 0-6 mm, 6-9 mm, 11-16 mm, 16-22 mm and 22-32 mm. These crushed materials are then stored in heaps and subsequently sold for industrial use.

Generally, the residues produced by the natural aggregate quarrying activity are:

- topsoil: material consisting of vegetation and soil that covers the mineral deposit.
- washing aggregate sludge (WAS): liquid-solid residue derived from the treatment of natural aggregate washing water.
- process water: liquid water used for natural aggregate washing or for mineral stone processing.

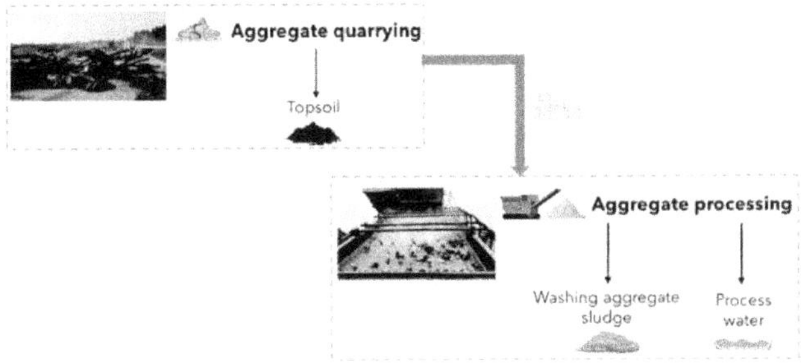

Figure 2 Natural aggregate extraction and processing chain

4 Residue's valorisation

The residues produced by the stones and aggregates extraction activities (limited to the companies analysed), are fully valorised in the construction sector or, in some cases, directly in the production site. Table 1 shows the final applications of the analysed residues.

4.1 Crushed stone

Crushed stone, which represents about 55-88% of all residues generated by stone quarries, presents several reuse solutions. Since this material has the same chemical composition as the original stone, it is generally converted into products with a renewed economic value. In particular, the marble extracted from the Brescia area generates residues with a high calcium carbonate content ($CaCO_3$) which allows them to be re-used in different applications. In the analysed case, the crushed stone is generally re-used as cliff boulders, aggregate for concrete production or for riverbank's restoration. Specifically, this residue is never destined for disposal.

4.2 Residual sludge

4.2.1 Stone extraction and processing sludge (SEPS)

The quality of the stone extraction and processing sludge (SEPS) produced both during the quarrying activity and the on-site stone cutting strongly depends on the quality of the original stone. As for crushed stone, the high calcium carbonate contained in the original stone (excavated in the Brescia area) allows the realization of ancillary works directly in the production site, as well as the use for remodelling and environmental restoration works. In some cases it is also used for cement production.

4.2.2 Washing aggregate sludge (WAS)

In the analysed case, WAS is currently used for ancillary works directly in the production site and for remodelling and environmental restoration works.

4.3 Stone powder and residual soil

In the analysed case, as for SEPS, the stone powder and the residual soil are generally used for ancillary works directly in the production site and for remodelling and environmental restoration works. In some cases, the residual soil derived from aggregate quarrying activity is used in the gardening sector and for the production of stabilized granular material.

4.4 Process water

The process water used for natural aggregates washing is conveyed to a settling vessel and a flocculant is added to settle the suspended solids. After that, the clarified water is reintroduced into the manufacturing process.

Table 1 Final application of the analysed residues

Residue	Final application
Crushed stone	Cliff boulders, aggregate for concrete production, riverbank's restoration.
Stone extraction and processing sludge (SEPS)	Ancillary works directly in the production site, remodelling and environmental restoration works, cement production.
Washing aggregate sludge (WAS)	Ancillary works directly in the production site, remodelling and environmental restoration works.
Stone powder	Ancillary works directly in the production site, remodelling and environmental restoration works.
Residual soil	Ancillary works directly in the production site, remodelling and environmental restoration works, gardening works and stabilized granular material production.
Process water	Treatment and re-entry into the plant for a new aggregate washing process.

5 Conclusions

In this study, the operational-management data of several companies' representative of the stone and aggregate extraction and processing chain of the provinces of Brescia and Bergamo (Lombardy Region – northern Italy) were analysed. The aim of the study was to evaluate, by analysing real application experiences, the valorisation of the residues produced by these activities in the construction sector or in other final applications. In particular, in the present paper, only one stone quarrying activity and one natural aggregate quarrying activity were analysed. The results demonstrated that all the generated residual materials can be recycled in several sectors, both industrial and construction. Moreover, the provinces analysed have proved to be virtuous realities, which, over the years, have built a solid collaborations network in order to achieve the maximum recovery of these mineral residues.

Literature

Arel H.S.	2016	Recyclability of waste marble in concrete production 2016. Journal of Cleaner Production 131179-188. http://dx.doi.org/10.1016/j.jclepro.2016.05.052
Aukour, F. J., & Al-Qinna, M. I.	2008	Marble Production and Environmental Constrains: Case Study from Zarqa Governorate, Jordan 2208. Jordan Journal

		of Earth and Environmental Sciences, 1(1).
Awad A.H., Abdellatif M.H.	2019	Assessment of mechanical and physical properties of LDPE reinforced with murble dust 2019. Composites Part B 173 106948. https://doi.org/10.1016/j.compositesb.2019.106948
Bilgin N., Yeprem H.A., Arslan S., Bilgin A., Gunay E., Marsoglu M.	2012	Use of waste marble powder in brick industry 2012. Construction and Building Materials 23 449-457. doi:10.1016/j.conbuildmat.2011.10.011
Blengini, G.A., Mathieux, F., Mancini, L., Nyberg, M., Viegas, H.M.	2019	Recovery of critical and other raw materials from mining waste and landfills State of play on existing practices 2019. https://doi.org/10.2760/600775
Chamorro-Trenado M.A., Pareta-Marjanedas M.M., Berthelsen-Molist B.E., Janer-Adrian F.X.	2016	The exploitation of sludge from aggregate plants in the manufacture of porous fired clay bricks 2016. Materiales de Construccion. Vol. 66, Issue 323. http://dx.doi.org/10.3989/mc.2016.03915
Dino, G. A., Chiappino, C., Rossetti, P. G., Franchi, A., & Baccioli, G.	2017	Extractive waste management: a new territorial and industrial approach in Carrara quarry basin. Italian Journal of Engineering Geology and Environment, (2). https://doi.org/10.4408/IJEGE.2017-02.S-05
Euromines	2021	Mining in Europe 2021. http://www.euromines.org/mining-europe
EUROSTAT	2018	Statistics Database 2019. https://ec.europa.eu/eurostat/web/environment/waste/database
Formia A., Tulliani J., Zerbinatti M., Palmero P.	2015	Tecnologia brevettata per materiali da costruzione a partire da fanghi di segagione 2015. Politecnico di Torino.
Furcas, C.	2015	Sustainable management of natural stone waste. A proposed re-use for the production of mortars and the assessment of their potential CO2 sequestration capacity.
Istituto nazionale di statistica (ISTAT)	2019	Le attività estrattive da cave e miniere 2019.
Luciano, A., Reale, P., Cutaia, L., Carletti, R., Pentassuglia, R., Elmo, G., & Mancini, G.	2020	Resources Optimization and Sustainable Waste Management in Construction Chain in Italy: Toward a Resource Efficiency Plan 2020. Waste and Biomass Valorization, 11(10), 5405–5417. https://doi.org/10.1007/s12649-018-0533-1

Luodes, H., Kauppila, P. M., Luodes, N., Aatos, S., Kallioinen, J., Luukkanen, S., Aalto, J.	2012	Characteristics and the environmental acceptability of the natural stone quarrying waste rocks. Bulletin of Engineering Geology and the Environment, 71(2), 257–261. https://doi.org/10.1007/s10064-011-0398-z
Marras G., Careddu N.	2018	Sustainable reuse of marble sludge in tyre mixtures 2018. Resources Policy 59, 77-84. https://doi.org/10.1016/j.resourpol.2017.11.009
Papantonopoulos, G., Ntua, M. T., Bonito, G. N., Adam, P. K., & Christodoulou, I.	2007	A study on best available techniques for the management of stone wastes 2007. 3rd International Conference on Sustainable Development Indicators in the Minerals Industry.
Rana A., Kalla P., Csetenyi L.J.	2015	Sustainable use of marble slurry in concrete 2015. Journal of Cleaner Production 94 (2015) 304-311. http://dx.doi.org/10.1016/j.jclepro.2015.01.053
Sadek D.M., El-Attar M.M., Ali H.A.	2016	Reusing of marble and granite powders in self-compacting concrete for sustainable development 2016. Journal of Cleaner Production 121 19-32. http://dx.doi.org/10.1016/j.jclepro.2016.02.044
Sala E., Clerici A., Cominoli L., Giustina I.	2014	Proposta di riutilizzo degli scarti delle lavorazioni in laboratorio del Marmo di Botticino: la sostenibilità come risparmio ambientale e economico 2014. Atti Convegno di Studi “Quale sostenibilità per il restauro?”, Bressanone, Scienza e Beni Culturali XXX. 2014, pp. 585- 595, ISBN 978-88-95409-18-4, Edizioni Arcadia Ricerche
Silva D., Gameiro F., de Brito J.	2013	Mechanical properties of structural concrete containing fine aggregates from waste generated by the marble quarrying industry 2013. J. Mater. Civ. Eng. (ASCE) 26 (6), 04014008. https://doi.org/10.1061/(ASCE)MT.1943-5533.0000948
Sorlini S., Rondi L., Bettini N., Collivignarelli C., Plizzari G.A.	2017	Technical and environmental characterisation of recycled aggregate for reuse in bricks 2017. MATEC Web of Conferences 120, 03015. DOI: 10.1051/matecconf/20171200.
Soydal U., Kocaman S., Marti M.E., Ahmetli G.	2018	Study on the reuse of marble and andesite waste in epoxy-based composites 2018. Wiley Online Library. DOI 10.1002/pc.24313
Soykan O., Özel C.	2012	Effects to polymer concrete properties of particle size of marble dust. In: Int. Constr. Congress, ICONC2012-SDÜ.
Sudarshan K.D., Vyas A.K.	2019	Impact of fire on mechanical properties of concrete containing marble waste 2019. Journal of King Saud University – Engineering Sciences 31 (2019) 42-51.

		http://dx.doi.org/10.1016/j.jksues.2017.03.007
Taji I., Ghorbani S., de Brito J., Tam V.W.Y., Sharifi S., Davoodi A., Tavakkolizadeh M.,	2019	Application of statistical analysis to evaluate the corrosion resistance of steel rebars embedded in concrete with marble and granite waste dust 2019. Journal of Cleaner Production 210, 837-846. https://doi.org/10.1016/j.jclepro.2018.11.091
Ulubeyli G.C., Bilir T., Artir R.	2016	Durability Properties of Concrete Produced by Marble Waste as Aggregate or Mineral Additives 2016. Procedia Engineering 161, 546-548.

Author's addresses

Eng. Alessandra Diotti
Department of Civil Engineering, Architecture, Land, Environment and Mathematics, University of Brescia
Via Branze 43
25123 Brescia, Italy
a.diotti@unibs.it

Prof. Alberto Clerici
Department of Civil Engineering, Architecture, Land, Environment and Mathematics, University of Brescia
Via Branze 43
25123 Brescia, Italy
alberto.clerici@unibs.it

Prof. Giovanni Plizzari
Department of Civil Engineering, Architecture, Land, Environment and Mathematics, University of Brescia
Via Branze 43
25123 Brescia, Italy
giovanni.plizzari@unibs.it

Prof. Sabrina Sorlini
Department of Civil Engineering, Architecture, Land, Environment and Mathematics, University of Brescia
Via Branze 43
25123 Brescia, Italy
sabrina.sorlini@unibs.it

Building Material Management: System change in Minnesota and beyond

Melissa Wenzel

Minnesota Pollution Control Agency, Saint Paul, MN USA

Abstract

Some construction and demolition (C&D) landfills in Minnesota are contaminating our state's groundwater. Only 20% of Minnesota's C&D materials are recycled; despite MN having a material management hierarchy that focuses on prevention and reuse over landfilling and even recycling. How do we change a system? Minnesota is still trying to figure that out, but has spent a lot of time and energy working with stakeholders on how to move to a more sustainable building material management system. Learn about the latest developments from a US state that typically boasts high recycling rates, while admitting that we're behind the curve in this large and largely ignored issue. Starting slowly and engaging many seems to be the right path, especially during and after our once-a-century pandemic. But is it enough? Are we doing enough to make a lasting difference? We hope so; it's definitely a work-in-progress development.

Keywords: *building material management, waste, construction, demolition, reuse, recycling, preservation, reuse, climate*

Introduction

Construction and Demolition (C&D) landfills in Minnesota have largely been managed in the same way since the late 1980s, when rules for the design, construction, and operation of these landfills were promulgated. The regulations for C&D landfills have remained unchanged since 1988; however, the Minnesota Pollution Control Agency (MPCA) developed the 2005 Demolition Landfill Guidance Document to create more consistency in the siting, design, monitoring, and operation of C&D landfills, including specifications for three classes of landfills depending on the waste accepted. At the time, C&D waste was assumed to be inert, and therefore accepted at Class I and Class II unlined landfills. Only Class III C&D landfills, accepting both C&D and industrial waste, were required to construct a liner and collect leachate. The release of the MPCA's 2005 Demolition Landfill Guidance Document led to the installation of monitoring wells at more C&D landfills to collect groundwater data. With this additional data from the C&D sites, permit hydrologists noted the presence of Arsenic (As), Boron (B), and Manganese (Mn) at unlined C&D landfills across the system, with concentrations exceeding the drinking water health-based values set by the Minnesota Department of Health. These findings raised questions about the previous assumption that C&D materials are inert, and led the agency to revisit the protections and monitoring needed when this waste is managed at end-of-life through land disposal. A complete summary of the data is available in the MPCA Groundwater Impacts of Unlined Construction and Demolition Debris Landfilling report.

In an effort to identify opportunities for reducing the environmental impacts of the State of Minnesota's building sector, the Minnesota Pollution Control Agency (MPCA) conducted a study on the generation and composition of construction and demolition (C&D) materials originating in Minnesota and disposed of at permitted solid waste facilities with C&D disposal areas, typically C&D landfills. The goal of this report is to gain a stronger understanding of the material categories and prevalence of those materials within the C&D waste stream. 1

The results and analysis completed in this study help the MPCA look at how different materials are being wasted, which is useful in determining how to improve building material management strategies for recycling and reuse. For instance, are there replacement materials that would have a smaller environmental footprint from a material life cycle perspective? Are there materials that would be more durable and have a longer life before disposal? What reuse and recycling markets are strong in Minnesota or surrounding states? What markets need support and expansion, and how can Minnesota diversify markets to make them more stable? Answering these questions will allow the MPCA to define program strategies and identify priorities to most effectively prevent building material waste and identify alternative management methods to keep materials in use as long as possible before disposal.

Figure 1. Aggregate statewide C&D waste composition, 2019

2

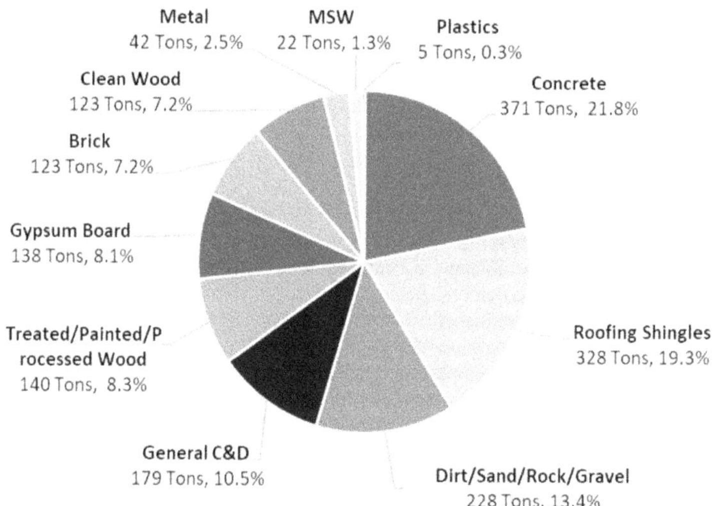

In addition to mitigating the environmental impacts of landfilling, there are clear opportunities for Minnesota to improve sustainability in earlier stages of material design, manufacturing, and reuse, and building design, maintenance/preservation, and removal. The state's low recycling rate for C&D material and its growing number of Metro region home tear-downs resulted in the MPCA's decision to examine the environmental impacts of the entire C&D and building system. The state's building sector contributes significant greenhouse gas emissions, and as a result it was important for the MPCA to consider the broader climate impacts of buildings, not just end-of-life management.

The U.S. Environmental Protection Agency (EPA) estimated that 600 million tons of C&D debris were generated in the United States in 2018, which is more than twice the amount of municipal solid waste (MSW) generated in the same year (EPA, 2020). In order to determine the amount of C&D debris generated in Minnesota, the MPCA used annual reports from permitted solid waste facilities which includes the types and amounts of waste managed. While the MPCA additionally collects detailed MSW generation data from counties, the same level of granularity isn't required for C&D data. Furthermore, while the state of Minnesota's efforts to reduce MSW generation and recycle more residential waste have grown, the same prioritization hasn't historically been given to the reuse and recycling of C&D materials. The MPCA estimates that about 20% of C&D material is currently recycled, despite many constituents in the C&D waste stream having more potential for reuse and recycling. At this point, the MPCA has limited formal data on C&D material handled outside the solid waste stream.

In 2018, the MPCA held four stakeholder meetings for organizations and individuals involved in the C&D sector across the state to provide feedback on opportunities and barriers within the current material management system. Approximately 100 stakeholders participated in one or more of four stakeholder sessions in Crookston, Duluth, Mankato, and St. Paul. At each session, participants discussed a set of questions in small, facilitated groups. The MPCA staff documented and coded 750 discrete comments, identifying emerging themes.

Workgroup overview

As a follow-up to the four initial sessions held in 2018, the MPCA convened two stakeholder groups in the fall of 2019, to gather input on approaches for improving the environmental protections and sustainability of the state's building sector. The Rule Advisory Panel (RAP) efforts focused on end-of-life management, with the ultimate goal of rewriting the statewide Rules for C&D landfills. In contrast, the Sustainable Building Group (SBG) worked to develop recommendations for reducing the environmental impacts of the full building system, prioritizing activities and strategies that aim to extend the useful life of existing buildings and materials.

Sustainable building group and subgroups

The role of the SBG was to identify system gaps and opportunities for improving the sustainability of new construction, building preservation and maintenance/renovation, building removal, and materials management across Minnesota. The stakeholder workgroup represented a wide range of industry players, including local government staff, architects, building preservation specialists, product manufacturers, reuse retailers, material recyclers, deconstruction and demolition managers, and landfill operators. Members represented their organization and similar organizations across the state, acting as a liaison between the workgroup and external parties interested in the recommendation development. Members were selected based on their expertise, commitment to communicating within their sector, geographic representation, and ability to work on a team.

The SBG full workgroup met every 4-6 weeks throughout the year, and five topic-specific subgroups met as needed between the full workgroup times. The subgroup topic areas included:

1. **Upstream building design** - This subgroup recommended strategies to reduce the environmental impacts through building design and material choices.
2. **Material value** - This subgroup recommended strategies for creating market infrastructure and demand for various reused materials.
3. **Regulation for collaboration** - This subgroup recommended new statewide or local regulations, policies, or sample ordinance, that support collaboration practices in material reuse.
4. **Material processing and recycling efficiencies** - This subgroup identified the processing steps necessary to ensure an efficient assessment, deconstruction, and distribution of materials, and provided recommendations to ensure effective storage and movement of reuse, recycling, and waste materials.
5. **New financial system** - The current financial system was not created to incentivize and support the capture of construction and demolition materials for future use (reuse, recycling, etc.). This subgroup met at the start of the year, but decided to not meet separately from the broader workgroup as the recommendations coming from the other subgroups were likely to have a large influence on the financial ideas put forward.

Process & recommendation development, stakeholder involvement

The year-long stakeholder process involved five main phases:

1. Research and goal-setting

MPCA staff and external presenters provided relevant information and facilitated workgroup discussions about sustainable building and materials management strategies, limitations of the state's building system, and existing frameworks and tools available to support transitioning to lower impact approaches. Workgroup members also identified priorities and goals they hoped to address with the SBG recommendations.

2. Brainstorm subgroup ideas

Using their specific subgroup charters and the overarching workgroup goals as guides, the subgroup teams brainstormed 65 unique concepts that would encourage more sustainable building processes, material selections, and material management methods within Minnesota. This initial list of 65 unique concepts is available in Appendix C.

3. Evaluate subgroup ideas based on goals and key values

In order to narrow the list of 65 unique workgroup concepts, subgroup members evaluated their specific list based on the following criteria:

- Reduces greenhouse gas emissions
- Creates resiliency to climate change
- Reduces toxicity to the ecosystem
- Follows the Waste Management Hierarchy, prioritizing reduction strategies first, then reuse and repair strategies, and finally recycling

- Promotes equitable systems
- Adds sustainable, trained, local jobs
- Applies and is accessible to both the Metro and Greater Minnesota regions
- Touches on all building types (commercial, residential – single family, residential – multi-family, etc.)

Using the scores from the evaluation, subgroups identified their top three to five concepts. This generated a new list of 22 ideas.

4. **Vote and narrow to top ideas**

   The SBG workgroup members individually ranked the 22 ideas produced by the subgroups. The list of 22 ideas and their ranking from the SBG workgroup is available in Appendix D. This final exercise led to the selection of the final five SBG ideas.

5. **Refine and finalize recommendations**

   During the final stage of development, the subgroups dissolved and workgroup members formed new teams for each of the five final ideas. The idea-specific groups added detail and refined the recommendations before presenting them to all workgroup members, MPCA leadership, members of the RAP, and members of the public that chose to attend.

In order to create an inclusive and transparent stakeholder process, all SBG meetings were open to the public and external input was encouraged.

Final recommendations

The SBG workgroup reached consensus on five final ideas. In cases where the idea was more conceptual, additional research and discussion by subgroup members and MPCA staff helped solidify concepts into more specific, tangible actions. SBG workgroup members concluded their involvement in the stakeholder process by advising statewide implementation of five recommendations and supporting activities in order to improve the sustainability of Minnesota's building sector.

Idea	Final recommendation
Establish a statewide, state-funded deconstruction training program that accompanies deconstruction and demolition licensing	Establish a statewide, state-funded deconstruction training program that accompanies deconstruction and demolition licensing
Create a financial incentive for buildings to be preserved, while still meeting energy efficiency standards	Create a statewide grant program for building preservation projects
Create a deconstruction ordinance template that cities can adopt, establishing consistent standards across the state	Create three tiers of deconstruction ordinance templates that cities/counties can select from and adopt
Create, implement, and enforce a statewide diversion requirement of C&D waste (including reuse and recycling) for new construction, demolition, additions, and/or renovation projects	Implement a statewide diversion requirement for C&D waste for new construction, additions, renovations, and building removal
Create a financial incentive that encourages the use of reused, more repairable, and/or reusable materials in new building construction and renovation projects	Create a statewide rebate program for reused building materials in new building construction and renovation projects

Recommendation interconnections and dependencies

Each of the five final SBG recommendations aim to improve different aspects of the building system, but they depend strongly on the implementation and success of the other recommendations.

Deconstruction training program

- **Reuse and recycling markets**: Markets are needed for materials that come from deconstruction.
- **Partnerships**: Partnerships are needed with existing programs that focus on deconstruction training and skills building.
- **Deconstruction and demolition certifications**: Certification requirements need to be established for both deconstruction *and* demolition to help level the playing field and move towards a more cost-comparative system.
- **Deconstruction ordinance**: Passing deconstruction ordinances that require certified teams to perform the services helps incentivize and drive demand for this type of trained work.

Building preservation grant

- **Reuse markets**: Markets are needed with reused materials to incorporate in preservation projects.
- **Partnerships**: Partnerships are needed with existing programs that focus on the operational efficiencies of buildings.

- Reused material rebate: Joint promotion with the Reused Material Rebate reinforces the hierarchy, by encouraging use of grant funds first for repair, then replacement with reused materials, then replacement with new materials.

Deconstruction ordinance templates

- **Reuse and recycling markets**: Markets are needed for materials that come from deconstruction and diversion.
- Deconstruction and demolition certifications: Requiring certifications for demolition and deconstruction contractors in ordinances, ensures consistent and more easily tracked services.
- Diversion targets: The lowest tier option for deconstruction ordinance templates can connect details from the Statewide Diversion Requirement.

Diversion requirement for C&D waste

- **Recycling markets**: Markets are needed for materials to meet diversion requirements.
- Deconstruction and demolition certifications: Requiring certifications for demolition and deconstruction contractors alongside a statewide diversion requirement ensures consistent and more easily tracked services.

Incentive for reused material rebate

- **Reuse markets**: Markets are needed with salvaged materials to be sold for reuse.
- **Deconstruction ordinance**: Passing deconstruction ordinances helps create material supply to support this program.

Upon completion of the C&D Composition Study, the following recommendations were made to help further C&D data sets as well as progress towards state sustainable building material management goals. They include:

Routinely update Minnesota statewide C&D composition research: A comprehensive analysis of Minnesota's C&D materials in the disposal stream provides insight to the overall management of materials in this sector. The MPCA has historically performed statewide MMSW characterization studies, and this study expands the dataset to include C&D materials. In order to document change over time and track progress through different initiatives, C&D composition studies should be conducted at least as frequently as MMSW characterization studies. Given the homogenous loads that can be delivered to C&D landfills, such as roll-offs of roofing shingles or concrete, it is important to do routine composition studies so these individual loads of specific materials do not skew composition results, especially when loads may be tied to either seasonally-specific construction and demolition projects with seasonally-driven material composition.

Increase facility participation in future studies: Should the MPCA conduct future waste characterization studies, it is recommended that additional C&D disposal facilities in the Metro region of the state be recruited to participate in field data collection, ensuring a more accurate representation of the statewide waste stream. This study by design focused on Class I C&D landfills. Future studies should also look at capturing waste characterization data from additional Class II and Class III C&D sites, as well as processing areas.

Conduct statewide C&D facility “at the gate” surveys: While Minnesota Rules require facilities to report C&D material quantities, a gap exists with providing more granular details on the generation of C&D materials, as indicated by some loads containing wastes not classified as C&D waste. Future studies would benefit from “at the gate” surveys at a large subset of the state’s C&D disposal facilities to document the breakdown of Construction, Demolition, Renovation, and Mixed/Non-C&D loads. Gate surveys are conducted by identifying all vehicles entering a facility throughout the day, noting the type of materials, the weight or volume, and the originating generator class. This information captures all load types (C&D, MMSW, special waste, etc.) and gives perspective on the overall statewide waste stream.

Create a robust and thorough contractor, hauler, and retailer survey: Based on current practices, it appears there are notable reuse and recycling efforts already occurring in Minnesota. Surveying is needed to verify this assumption further, and strengthen the statewide data quantifying the amount of materials being reused and recycled, identifying which markets are the strongest and weakest, and confirming what support is needed to scale these efforts.

Implement statewide reuse or recycling requirements: Recognition is growing across the United States about the environmental benefits of deconstructing buildings instead of demolishing them in order to extend the usable life of existing materials. Implementing statewide reuse and recycling requirements and/or deconstruction ordinances at a local level can help create consistency and establish necessary support systems for this transition. Requirements could reduce the barriers that currently make it difficult to deconstruct or separate materials for recycling at the site. As part of an ordinance, a contractor would have to allow for extra time to deconstruct and capture materials, leveling the playing field for this more sustainable practice. Examples of other factors needed to shift building and material management practices include, but aren’t limited to, creating deconstruction and demolition certifications, providing incentives for material reuse and recycling, and securing adequate material sorting and storage spaces on job sites.

Next steps

The MPCA will continue circulating the final recommendations to gather additional and diverse feedback, refining the details in the proposed programs and legislative requests. This follow-up work will focus in large part on defining specific metrics and targets for each of the recommendations, ensuring realistic, yet ambitious goals. The development and implementation of the SBG recommendations will heavily depend on the support and statewide involvement of numerous partners. In some cases, the proposed programs from this stakeholder process will be better initiated and managed by local units of government or industry experts.

The MPCA will continue to develop elements of its work to support sustainable building efforts as part of the agency’s Strategic Plan goal to prevent risks associated with construction and demolition materials, and especially the agency’s commitment to advancing sustainable materials management and reducing the associated environmental and human health impacts.

References:

Minnesota Pollution Control Agency	2005	Demolition Landfill Guidance Document. https://www.pca.state.mn.us/sites/default/files/w-sw5-04.pdf
Minnesota Pollution Control Agency	2019	Groundwater Impacts of Unlined Construction and Demolition Debris Landfilling report. https://www.pca.state.mn.us/sites/default/files/w-sw5-54a.pdf
United States Environmental Protection Agency	2021	Facts and Figures about Materials, Waste and Recycling; Construction and Demolition: Material-Specific Data.: https://www.epa.gov/facts-and-figures-about-materials-waste-and-recycling/construction-and-demolition-debris-material

Author's address:

Melissa Wenzel
Minnesota Pollution Control Agency
520 Lafayette Road North
Saint Paul, MN 55155
Phone: 651-757-2251
Email: melissa.wenzel@state.mn.us

The effect of granular correction of dredged sediments by other wastes on the mechanical and environmental properties of established mixtures

Zeinab MKAHAL, Yannick MAMINDY-PAJANY, Walid MAHERZI, Nor-Edine ABRIAK

Univ. Lille, Univ. Artois, IMT Lille Douai, JUNIA, ULR 4515 – LGCgE, Laboratoire de Génie Civil et géo-Environnement, F-59000 Lille, France.

Abstract

Due to scarcity of natural aggregates, dredged sediments and other solid wastes can be used as secondary raw materials in civil engineering. This study investigates the advantage of adding municipal solid wastes from incinerator (MSWI) and construction and demolition wastes (CDW) to the waterways sediments (SED) for their use in road construction. In this sense, granular mixtures were prepared and characterized according to geotechnical guidelines for sub-base road materials conception. California bearing index were measured for dredged sediments and mixtures including sand and or CDW and MSWI. Results reveal the improvement of mechanical performance for certain mixtures especially those containing CDW. Furthermore, the leaching behavior of trace elements in granular mixtures demonstrates the low mobility of contaminants.

Keywords

Dredged Sediments, Technical Road, Granular Corrector, Municipal Solid Waste From Incinerator, Construction and Demolition Waste, Batch Equilibrium Leaching Test, Trace Elements.

1 Introduction

In France, in the fluvial domain, an average of 6 million m^3 of sediments is dredged annually from the 525,000 km of national rivers and canals (Mouvet & Piantone, 2008). According to national and international environmental regulations, traditional solutions such as waterway resuspension and inshore storage are becoming more restricted for dredged sediments. After dredging, sediments acquire the status of waste, which poses a number of environmental issues related to their management in landfills or their reuse in beneficial ways.

In the last few decades, various research and development projects have been developed to explore the beneficial reuse of dredged sediments in civil engineering such as in technical roads, landscaping, development of industrial platforms, and shoreline erosion control (Achour et al., 2014; M Boutouil, 1998; Colin, 2003; Damas, 2015; Dubois et al., 2009; KASMI, 2014; Maherzi et al., 2017; Mamindy-Pajany et al., 2019; Samara, 2007; Scordia, 2008; Ulbricht, 2002; Zentar et al., 2008). In the road construction sector, around 200 million tons of natural aggregates are required each year in France (SETRA, 2012b), which is a great opportunity for the valorization of dredged sediments.

In the two last decades, numerous European countries have developed their national environmental guidelines to allow reusing mineral solid wastes in the field of civil engineering (Suer et al., 2014). In France, a methodology has been established for assessing the environmental characteristics of waste materials prior to their reuse in technical roads taking into account the protection of human health and groundwater quality in specific scenarios (SETRA, 2011). Among the treatments, the addition of mineral granular corrector is usually implemented to increase the mechanical performance of sediments, which was largely reported in previous studies (Colin, 2003; Zentar et al., 2008; Mamindy-Pajany, 2018; Boutouil & Saussaye, 2011; Kasmi et al., 2016). In this sense, this paper focuses on the use of granular correction technique to formulate road materials.

Until now, dredged sediments are being mixed with natural sand for road applications to be treated after by lime and or hydraulic binder. Based on literature, some studies used solid wastes to improve the granular skeleton of conventional aggregates. For example, cullet from glass bottles (Kou & Poon, 2013), construction and demolition wastes (Kwan et al., 2012), bottom ash from coal combustion power plants (Siddique, 2010; García Arenas et al., 2011), and bottom ash from municipal solid waste incinerators; Dabo et al., 2009; Ginés et al., 2009) were used for various applications in civil engineering. Recently, Wang et al. (Wang et al., 2015) have proved the possibility of producing non-load-bearing concrete masonry units, including solid mineral wastes, as coarse aggregate to strengthen the granular skeleton of stabilized/solidified sediments. However, this work has focused only on the construction materials (concrete masonry blocks) and no results have been reported for technical road applications.

The objective of this paper is to test construction and demolition waste (CDW) and municipal solid waste from incinerator (MSWI) as new granular correctors for dredged sediments prior to their reuse in technical roads as backfill materials. For this purpose, granular correction strategy has been defined for various mixtures using Talbot-Fuller-Thompson curves to optimize their compactness. Each mixture was characterized by its geotechnical and mechanical properties to verify the ability of its application as a backfill material. Finally, batch equilibrium leaching tests (NF EN 12457-2) were performed to check the environmental behavior of materials and their mixtures.

2 Materials and methods

2.1 Materials

Waterway sediments were collected from a deposit site of "Voies Navigables de France" at Noyelles-Sous-Lens. After collection, the dredged sediments have been quartered and stored in barrels at the laboratory to obtain homogeneous and representative samples. It must be noted that sediment samples are classified as non-hazardous wastes, which allow studying their valorization.

To provide a suitable compactness of dredged sediments and to expand their reuse in road construction, mineral granular correctors were added. Therefore, each type of sediments is mixed with natural sand (S), municipal solid waste from incinerator (MSWI), and construction and demolition waste (CDW). All these materials were collected from the "Hauts-de-France" region: the sand (0-6 mm) from Lille city, the CDW (0-4 mm) from a local construction waste deposit in Douai city, and the MSWI (0-6 mm) from RMN (Recyclage De Materiaux Du Nord) company in Fretin city.

2.2 Methods

2.2.1 Mixtures design:

To optimize the mixtures of sediments with different granular correctors (S, CDW, and MSWI), the methodology used is based on improving the grain size distribution of the mixtures. The Talbot-Fuller-Thompson method was adopted to obtain proportion of each constituent in order to ensure a well-graded mixture. This semi-empirical method makes it possible to define a spindle formed from reference granular curves ensuring the optimal arrangement of grains with minimum porosity and maximum compactness.
The SED is constituted of 75.6% of fine fraction, 13.6% of sand, and 10.8% of gravel. Natural sand (S) is composed of 16% of fine fraction, 35.7% of sand, and 48% of fine gravel. Construction and demolition waste is made up of 5% fine particles (<63 µm), 74% sand, and 21% fine gravel. MSWI is composed of 3.6% of fine particle (<63 µm), 57.2% of sand, and 39.2% of fine gravel. The granular size distribution of sediment, S, CDW, and MSWI confirm the possibility to constitute mixtures with optimized granular skeleton. To define the composition of mixtures (n = 4), the compactness was optimized with binary formulation including dredged sediments and natural sand. After that, some mixtures were defined by partially or totally replacing the sand fraction by CDW and /or MSWI.

2.2.2 Geotechnical analyses

Raw sediment and road materials were prepared and characterized as recommended by the French technical guide of backfill and subgrade conception for soils (SETRA-LCPC, 2000).

Grain size distribution

For each material, the grain size distribution (XP P94-041) is determined by washing of the sample on a 63 µm sieve. Then, the dried refusal at 105 °C is sieved with an analytical sieve with several sieves ranging from 10 to 0.063 mm. The sieving is carried out by vibration at amplitude of 1 mm for 7 minutes.

Methylene blue value

The methylene blue value, which represents the adsorption capacity of methylene blue on the particle surfaces, is calculated in order to estimate the sensitivity of materials to water according to the European standard NF P94-068.

Atterberg limits

According to the European standard NF P94-051, the liquid (W_L) and plastic (W_P) limits are determined by the percussion cup and rolling thread methods, respectively. The plasticity index (I_p), defined as the difference between the liquid and the plastic limits, reflects the clay content of the soil.

Loss of ignition and water content

Organic matter contents are determined according to the French test standard XP P 94 047 by the loss of ignition method at 550 °C for three hours.

The initial water content of materials is determined using oven drying at 105°C until a constant solid mass is obtained.

2.2.3 Mechanical characterization

The main mechanical characteristic that defines the ability of the material to be used in backfill is the California bearing index (CBR). Thus, mechanical performances of raw sediments and road materials were evaluated through Normal Proctor and CBR tests according to EN 13286-2 and NF P 94-093, respectively.

The CBR is often used to measure the bearing capacity of compacted material. The test consists of measuring the forces to be applied to a piston to introduce it at constant speed into compacted material. This test is preceded by the compaction of samples according to Normal Proctor tests, which are required to investigate the optimal water content, allowing the highest level of compaction at a field site.

Normal Proctor test consists of compaction of a sample, after humidification at several water contents, according to a conventional process and at a specific energy (0.6 MJ/m^3). The corresponding dry density of the material is defined for each water content. From the Proctor curve (curve of variations in density as a function of water content), the compaction characteristics are determined and they correspond to the material maximum dry density value obtained for an optimal value of the water content.

2.2.4 Environmental assessment

Batch equilibrium leaching tests

Road materials as well as dredged sediments (SED), CDW, and MSWI were submitted for environmental analysis. In this study, French limit thresholds were used for inorganic components (As, Ba, Cd, Cr, Cu, Mo, Ni, Pb, Sb, Se, and Zn, as well as chlorides, fluorides, and sulfates) (SETRA, 2012b) (SETRA, 2012a). In the case of dredged sediments, no guideline values have yet been established, and the most restrictive thresholds that are already available for CDW were applied.

Batch equilibrium-leaching tests were conducted on all mixtures, SED, CDW, and MSWI to check their environmental conformity (NF EN 12457-2). The compacted materials i.e. dredged sediments and their mixtures displaying water contents close to optimum proctor were selected. Then, the materials were placed in an oven at 40°C maximum. After drying, samples were systematically ground to obtain 95% of the sieve residues of less than 1 mm. Two samples were evaluated for each material. The leaching tests were conducted with a liquid-to-solid ratio of 10 (L/Kg). Under specific test conditions, the sample with added ultrapure water (18.2 MΩ.cm) was agitated for 24 h. The liquid phase was separated from the solid by decantation and filtration through cellulose acetate membranes (0.45 µm). The filtrate was analyzed for As, Ba, Cd, Cr, Cu, Mo, Ni, Pb, Sb, Se, and Zn by inductively coupled plasma optical emission spectrometry ICP-OES and for chlorides, fluorides, and sulfates by ionic chromatography

3 Results and discussion:

3.1. Granular correction of dredged sediments

Natural sand (S), CDW, and MSWI were used as granular correctors to optimize the granular skeleton of dredged sediments (SED). The compactness of the mixtures is related to the granular size distribution of each material and it was optimized by the granular spindle of Talbot-Fuller-Thompson.

Sand (70%) was used to improve the granular size distribution of SED and the sand fraction was then substituted by 57% of CDW or MSWI (Table 1). The well graded distribution curves and the proportions of each grain size play an important role in the optimization of the compactness. Thus, it can be noted that sediments are fine which justify the need to add granular corrector to fit the T-F-T spindle. Despite their discrepancies in terms of initial characteristics, there are few differences in the proportion of added corrector materials (70%). The advantage of granular correctors is revealed by the improvement of the compactness of all mixtures (60-66%) compared to sediments (55%) alone (Table 2).

Table 1 Mixtures design made up of (SED) mixed with S alone or with wastes (CDW, MSWI)

Mix design					
	SED				
	100 SED	30 SED(70 S)	30 SED(30 S, 40 CDW)	30 SED(30 S, 40 MSWI)	20 SED(30 CDW, 50 MSWI)
Sediment	100%	30%	30%	30%	20%
Sand	0%	70%	40%	40%	0%
CDW	0%	0%	30%	0%	30%
MSWI	0%	0%	0%	30%	50%

Several studies have reported the beneficial use of granular correction methods to improve the geotechnical and mechanical properties of dredged sediments applied in road construction (Colin, 2003; Zentar et al., 2008; Boutouil & Saussaye, 2011; Kasmi et al., 2016). They have used the uniformity and curvature coefficients and compressible packing model to define the proportion of granular correctors. Overall, in studied dredged sediment samples, the sand was used as a granular corrector and it reached

70% in the designed mixtures. In this research, additional mixtures without sand have also demonstrated their ability to formulate road materials. Indeed, mixture containing only SED, CDW, and MSWI has been developed (Table 1). Wang et al.,(2015) have also reported the feasibility of dredged sediments correction by granular wastes i.e. glass and MSWI. However, the treatment applied to dredged sediments was used to manufacture construction products like non-load-bearing concrete. These findings highlight the relevance of our approach aiming to optimize the replacement of sand based granular correctors by mineral wastes such as CDW and MSWI for various valorization scenarios in civil engineering.

3.2. Geotechnical classification of granular mixtures

After optimizing, four mixtures (Table 1) were selected and characterized for their geotechnical properties according to the French technical guide of backfill and subgrade conception for soils.

Sediment samples were assimilated to a soil of class A1. These fine soils are not very plastic soils, which suddenly changes their consistency for small variations in water content, in particular when their natural water contents are close to optimum moisture contents. Similar class was found by Kasmi (2014) where fluvial sediments were classified as A2, A4, and B5. These classifications confirm that dredged sediments do not have a suitable geotechnical properties to be used alone as road materials. This is mainly due to their high sensitivity to water (Zentar et al., 2008; Achour et al., 2014)

The mixture designed with SED was also classified as B5 (Table 2). The proportion of fines and the low plasticity of B5 soil bring their behavior very close to that of A1 soils. Hence, all of the selected mixtures were sensitive to water.

Table 2 Geotechnical classification of dredged sediments as well as their mixtures with Sand (S), Construction and demolition waste (CDW) and municipal solid waste from incinerator (MSWI).

	Mixtures

	100 SED	30 SED(70 S)	30 SED(30 S, 40 CDW)	30 SED(30 S, 40 MSWI)	20 SED(30 CDW, 50 MSWI)
Compactness (%)	55	66	61	63	60
GTR class	A1	B5	B5	B5	B5
Measured CBR %	3.2	15.9	20.4	7.5	28
CBR required	8	12	12	12	12

3.3. CBR index

Based on French technical guide of backfill and subgrade conception for soils, to quantitatively characterize, at the project study stage, the conditions defining a fine and wet soil, it is necessary to refer mainly to the CBR of the soil measured at its natural water content on a compacted test specimen at "Normal Proctor" energy.

The CBR of sediment was 3.2 and 5 (Table 2). All the mixtures reach the requested thresholds with the exception of the mixture "30% SED (30% S, 40% MSWI)" (7.5 instead of 12) (Table 2).

Indeed, the CBR index obtained for mixtures containing CDW exceeds those obtained for mixtures of sediments with natural sand alone. However, partial sand replacement by CDW in mixtures allows the CBR to be increased by about 1.2 compared to the mixture of SED with 70% of sand. On the contrary, the same rate of sand replacement by MSWI in mixtures reduces the CBR compared to the mixture of SED with sand alone.

Replacing the sand with CDW seems to be the best option. Indeed, all the mixtures containing this solid waste meet the required threshold and exceed the results obtained for the mixtures containing MSWI at the same percentage of replacement. This result is confirmed by the comparison made between CDWs and non-weathered MSWIs for their use in road construction. The various parameters studied for the CDW are much more suitable than those of the MSWI and in particular the CBR (Vegas et al., 2008).

Nevertheless, the CBR obtained for mixtures with CDW or MSWI allows the alternative materials to be used also as backfill without treatment for all the mixtures. These results are consistent with those obtained in the literature review (Cardoso et al., 2016), which emphasizes the effects of the incorporation of recycled aggregates on the properties of the compacted granular material. In fact, the CBR measured immediately or after immersion indicated that most recycled aggregate are comparable to high quality natural aggregate for use in unbound pavement layers or in other applications requiring compaction (Alnedawi & Rahman, 2021; Barbudo et al., 2012; Leite et al., 2011; Vegas et al., 2011). As far, Forteza (2004) demonstrated that MSWI is suitable for use as backfill when the CBR index and other parameters meet regulatory requirements.

The mixture of sediments with only CDW and MSWI allows sand replacement up to 100%. This is asserted by obtaining mixtures of sediments which respect the required thresholds without resorting to additional treatments (hydraulic binders). In addition, the highest value of CBR is obtained when 20% of SED is mixed with 30% CDW and 50% MSWI without sand.

3.4. Environmental assessment

3.4.1 Batch equilibrium leaching tests (NF EN 12457-2)

Threshold are defined for (i) type 1 usages for underlying courses of capped pavement or shoulder sublayers, (ii) type 2 usages for covered engineering embankments associated with road infrastructure (e.g. phonic protection) or for capped shoulders, and (iii) type 3 usages for some unpaved or uncapped road construction usages. In this study, all leaching parameters (As, Ba, Cd, Cr, Cu, Hg, Mo, Ni, Pb, Sb, Se, and Zn, as well as chlorides, fluorides, and sulfates) were measured with only Ba, Cr, Cu, Sb, Zn, chloride, fluorides, and sulfates exceeding at least one of the thresholds for authorized usages (Table 3).

The release of inorganic constituents from a material is controlled by a combination of physical and chemical processes, which depend on the properties of the solid material, the environmental exposure conditions or scenario, and the specific contents of inorganic pollutants in the material.

Only elements that can be quantified by ICP-OES were reported in (Table 3).

Antimony is the only element that exceeds the required limits for type-3 use (0.08 mg/kg) where its leaching concentration for SED samples is 0.12 ± 0.01 (mg/kg) (Table 4). In addition, SED samples has a high leaching concentration of Zn (5.33 ± 0.07 mg/kg); therefore, they it cannot be applied alone for any type of use (threshold of 5 mg/kg for all types of use). Likewise, SED samples releases 14,690 mg/kg of sulfates, which exceeded the type 3 of use threshold (1,300 mg/kg).

The MSWI respects the limits for all type of use (SETRA, 2012a). For CDW, the leachability of Cr (1.2 ± 0.02 mg/kg) exceeds the acceptance criteria for type 3 usage (0.6 mg/Kg) (Table 5).

For mixtures, no tested samples exhibited Ba values over the limits for any type of use which is in accordance with the leaching behavior of sediment, CDW, and MSWI.

Although the CDW showed values above the Cr-leach limit for type 3 of use, all the mixtures were below leaching limits, which can be explained by the dilution effect of other granular constituents by 70%.

The mixtures of SED that contain CDW (pH=11.6) exceeded the acceptance criteria for their use in type 2 (5mg/kg). The leaching of Cu from mixture 20 SED (30 CDW, 50 MSWI), exceeds by 1.8 mg/kg the threshold for type 3 of use however, the mixture 30 SED (30 S, 40 CDW), with 7.3 ±0.2 (mg/kg) of Cu leaching, does not respect the threshold for type 2 of use (Table 6).

Nevertheless, all of these mixtures have a basic pH between 10.5 and 10.6. Hence, Cu leaching is controlled by pH (Chatain et al., 2012)(Chatain et al., 2012). In SED mixtures with pH > 10, the mixture "20 SED (30 CDW, 50 MSWI) with 20 % of SED and high amount of MSWI, leaches 4.8 ±0.2 (mg/kg),which is less than leaching of Cu from the mixture with 30% of sediments 30 SED (30 S, 40 CDW) that releases 7.3 ±0.2 (mg/kg) of Cu.

The leaching concentrations of Sb from all road materials are equal to or greater than the threshold of type 3 of use (0.08 mg/kg) (Table 7). For Zn, it can be noted that the leaching of this element is pH-dependent as in alkaline conditions the concentration of Zn in all mixtures is below the quantification limit (Dijkstra et al., 2008; Engelsen et al., 2010; Cappuyns & Rudy, 2008)

As reported in (Table 8), regarding the leachability of anions (fluorides, chlorides and sulfates), only the sulfate exhibited values over the limits. In all road materials, sulfate concentrations exceed the threshold required to be use in type 3 (1,300 mg/kg). Contrariwise, the leaching of sulfates from 30 SED (30 S, 40 CDW) and 30 SED (30 S, 40 MSWI) mixtures was 5,270 ± 183 and 6,560 ± 212 (mg/kg) respectively, which exceeds the limit required for type 2 of use (5,000 mg/kg).

4 Conclusion

Results highlighted that the partial substitution of natural sand by CDW and MSWI seems to be the best option to increase the mechanical properties of backfill road materials. In addition, the sand fraction replaced by ternary mixtures of SED, CDW, and MSWI allowed obtaining acceptable road materials without resorting to any treatments with hydraulic binders.

From an environmental point of view, road materials have acceptable release levels for most inorganic pollutant, which makes it possible to consider their use in at least one of the use families listed in the application guide related to the environmental acceptability of sediments in road construction.

Table 9 Concentration of trace metal cations (Ba; Cu and Zn), oxyanions (Cr and Sb), and anions (Cl−, SO4 2− and F−) in the leachate from dredged sediments and their mixtures. Threshold for different types of use for CDW and MSWI were reported to validate the environmental conformity

								Threshold				
Inorganic pollutants Concentration (mg/Kg)	100 SED	CDW	MSWI	30 SED2 (70 S)	30 SED (30 S, 40 CDW)	30 SED (30 S, 40 MSWI)	20 SED (30 CDW, 50 MSWI)	Type 1 use for CDW	Type 1 use for MSWI	Type 2 use for CDW	Type 2 use FOR MSWI	Type 3 use for CDW
Ba	0.28	0.31±0.01	0.12±0.004	0.27±0.01	0.18±0.01	0.23±0.01	0.11±0.01	36	56	25	28	25
Cr	<0.007	1.17 ±0.02	0.03 ±0.003	<0.007	0,41±0.01	0.02±0.001	0.27±0.01	4	2	2	1	0.6
Cu	1.72±0.03	0.22±0.001	0.3±0.01	0.57±0.01	7.3±0.19	0.94±0.01	4.8±0.15	10	50	5	50	3
Sb	0.12 ±0.01	<0.056	0.44±0.004	0.08±0.003	0.08 ±0.01	0.19±0.01	0.2±0.01	0.6	0.7	0.3	0.6	0.08
Zn	5.33±0.07	< 0.05	< 0.05	< 0.05	< 0.05	< 0.05	< 0.05	5	50	5	50	5
Fluorides	3±0.01	9.7 ±0.07	2.9±0.01	4.6±0.07	12±0.01	2.5±0.07	9.3±0.14	60		30		13
Chlorides	31±0.01	188 ±0.7	406±19	21 ±0.7	107±1.4	168±3.5	268±2.1	10,000		5,000		1,000
Sulfates	14,690 ± 1	1,245 ±49	1,835±77	4,865±35	5,270±183	6,560±212	4,545 ±7	10,000		5,000		1,300
pH	7	11.6	9.6	7.8	10.6	8.3	10.5					
Conductivity(µs.cm)	2,270	1,405	640	1,008	1,143	1274	1,053					

5 Litterature:

Achour, R., Abriak, N. E., Zentar, R., Rivard, P., & Gregoire, P. (2014). Valorization of unauthorized sea disposal dredged sediments as a road foundation material. *Environmental Technology (United Kingdom)*, *35*(16), 1997–2007. https://doi.org/10.1080/09593330.2014.889758

Alnedawi, A., & Rahman, M. A. (2021). Recycled Concrete Aggregate as Alternative Pavement Materials: Experimental and Parametric Study. *Journal of Transportation Engineering, Part B: Pavements*, *147*(1), 04020076. https://doi.org/10.1061/jpeodx.0000231

Barbudo, A., Agrela, F., Ayuso, J., Jiménez, J. R., & Poon, C. S. (2012). Statistical analysis of recycled aggregates derived from different sources for sub-base applications. *Construction and Building Materials*, *28*(1), 129–138. https://doi.org/10.1016/j.conbuildmat.2011.07.035

Boutouil, M. (1998). *Traitement des vases de dragage par stabilisation/solidification à base de ciment et additifs (treatment of dredging muds by stabilization/ solidification with cement and additives).* ,. University of Le Havre, France.

Boutouil, Mohamed, & Saussaye, L. (2011). Influence de l'ajout d'un correcteur granulométrique sur les propriétés des sédiments traités aux liants hydrauliques. *European Journal of Environmental and Civil Engineering*, *15*(2), 229–238. https://doi.org/10.3166/EJECE.15.229-238

Cappuyns, V., & Rudy, S. (2008). *The application of pHstat leaching tests to assess the pH-dependent release of trace metals from soils, sediments and waste materials.*

Cardoso, R., Silva, R. V., Brito, de J., & Dhir, R. (2016). Use of recycled aggregates from construction and demolition waste in geotechnical applications: A literature review. *Waste Management*, *49*, 131–145. https://doi.org/10.1016/j.wasman.2015.12.021

Chatain, V., Benzaazoua, M., Loustau Cazalet, M., Bouzahzah, H., Delolme, C., Gautier, M., Blanc, D., & de Brauer, C. (2012). Mineralogical study and leaching behavior of a stabilized harbor sediment with hydraulic binder. *Environmental Science and Pollution Research*, *20*(1). https://doi.org/10.1007/s11356-012-1141-4

Colin, D. (2003). Valorisation de sédiments fins de dragage en technique routière. *Thèse de Doctorat, Université de Caen/Basse Normandie*, 147.

Dabo, D., Badreddine, R., De Windt, L., & Drouadaine, I. (2009). Ten-year chemical evolution of leachate and municipal solid waste incineration bottom ash used in a test road site. *Journal of Hazardous Materials*, *172*(2–3), 904–913. https://doi.org/10.1016/j.jhazmat.2009.07.083

Damas, O. (2015). *Programme SITERRE : procédé de construction de Sols à partir de matériaux Innovants en substitution à la TERRE végétale et aux granulats de carrière.*

Dijkstra, J. J., Meeussen, J. C. L., Van der Sloot, H. A., & Comans, R. N. J. (2008). A consistent geochemical modelling approach for the leaching and reactive transport of major and trace elements in MSWI bottom ash. In *Applied Geochemistry* (Vol. 23, Issue 6, pp. 1544–1562). https://doi.org/10.1016/j.apgeochem.2007.12.032

Dubois, V., Abriak, N. E., Zentar, R., & Ballivy, G. (2009). The use of marine sediments as a pavement base material. *Waste Management*, *29*(2), 774–782. https://doi.org/10.1016/j.wasman.2008.05.004

Engelsen, C. J., Van Der Sloot, H. A., Wibetoe, G., Justnes, H., Lund, W., & Stoltenberg-Hansson, E. (2010). Leaching characterisation and geochemical modelling of minor and trace elements released from recycled concrete aggregates. *Cement and Concrete Research*, *40*(12), 1639–1649. https://doi.org/10.1016/j.cemconres.2010.08.001

Forteza, R., Far, M., Seguí, C., & Cerdá, V. (2004). Characterization of bottom ash in municipal solid waste incinerators for its use in road base. In *Waste Management* (Vol. 24, Issue 9, pp. 899–909). https://doi.org/10.1016/j.wasman.2004.07.004

García Arenas, C., Marrero, M., Leiva, C., Solís-Guzmán, J., & Vilches Arenas, L. F. (2011). High fire resistance in blocks containing coal combustion fly ashes and bottom ash. *Waste Management*, *31*(8), 1783–1789. https://doi.org/10.1016/j.wasman.2011.03.017

Ginés, O., Chimenos, J. M., Vizcarro, A., Formosa, J., & Rosell, J. R. (2009). Combined use of MSWI bottom ash and fly ash as aggregate in concrete formulation: Environmental and mechanical considerations. *Journal of Hazardous Materials*, *169*(1–3), 643–650. https://doi.org/10.1016/j.jhazmat.2009.03.141

KASMI, A. (2014). *Prétraitement et traitement des sédiments fluviaux en vue Prétraitement et traitement des sédiments fluviaux en vue d'une valorisation en technique routière*.

Kasmi, A., Abriak, N. E., Benzerzour, M., & Azrar, H. (2016). Environmental impact and mechanical behavior study of experimental road made with river sediments: recycling of river sediments in road construction. *Journal of Material Cycles and Waste Management*, *19*(4), 1405–1414. https://doi.org/10.1007/s10163-016-0529-5

Kou, S. C., & Poon, C. S. (2013). A novel polymer concrete made with recycled glass aggregates, fly ash and metakaolin. *Construction and Building Materials*, *41*, 146–151. https://doi.org/10.1016/j.conbuildmat.2012.11.083

Kwan, W. H., Ramli, M., Kam, K. J., & Sulieman, M. Z. (2012). Influence of the amount of recycled coarse aggregate in concrete design and durability properties. *Construction and Building Materials*, *26*(1), 565–573. https://doi.org/10.1016/j.conbuildmat.2011.06.059

Leite, F. D. C., Motta, R. D. S., Vasconcelos, K. L., & Bernucci, L. (2011). Laboratory evaluation of recycled construction and demolition waste for pavements. *Construction and Building Materials*, *25*(6), 2972–2979. https://doi.org/10.1016/j.conbuildmat.2010.11.105

Maherzi, W., Benzerzour, M., Mamindy-Pajany, Y., van Veen, E., Boutouil, M., & Abriak, N. E. (2017). Beneficial reuse of Brest-Harbor (France)-dredged sediment as alternative material in road building: laboratory investigations. *Environmental Technology (United Kingdom)*. https://doi.org/10.1080/09593330.2017.1308440

Mamindy-Pajany, yannick, Priez, C., & Nadah, J. (2019). *Valorisation de sédiments de dragage non immergeables, avancées technologiques dans le domaine de la consrtuction routiere*.

Mamindy-Pajany, Y. (2018). *Contribution à la gestion environnementale des sédiments marins non-immergeables provenant de sites portuaires : Approche couplée (Géo) -chimie - Ecotoxicologie - Génie Civil,HDR*.

Mouvet, C., & Piantone, P. (2008). *Sédiments, groupe de réflexion usage et gestion, Rapport final, BRGM/RP-59859-FR, 360 p., 2 fig., 6 ann.*

Samara, M. (2007). Valorisation des sédiments fluviaux pollués après inertage dans la brique cuite. *Thèse de Doctorat, Ecole Centrale de Lille*, 139.

Scordia, P. (2008). *Caractérisation et valorisation de sédiments fluviaux pollués et traités dans les matériaux routiers Pierre-Yves*.

SETRA-LCPC. (2000). guide technique: Réalisation des remblais et des couches de forme - Guide Technique - Fascicule I Principes généraux. In *Setra Lcpc*.

SETRA. (2011). *Acceptabilité de matériaux alternatifs en technique routière - Évaluation environnementale* (SETRA). Guide méthodologique.

SETRA. (2012a). *Acceptabilité environnementale de matériaux alternatifs en technique routière Les mâchefers d'incinération de déchets non dangereux (MIDND).*

SETRA. (2012b). *Guide d'application: Acceptabilité environnementale de matériaux alternatifs en technique routière Les mâchefers d'incinération de déchets non dangereux (MIDND).*

Siddique, R. (2010). Utilization of coal combustion by-products in sustainable construction materials. *Resources, Conservation and Recycling*, *54*(12), 1060–1066. https://doi.org/10.1016/j.resconrec.2010.06.011

Suer, P., Wik, O., & Erlandsson, M. (2014). Reuse and recycle - Considering the soil below constructions. *Science of the Total Environment*, *485–486*(1), 792–797. https://doi.org/10.1016/j.scitotenv.2014.03.044

Ulbricht, J. . (2002). Contaminated sediments: raw materials for bricks. *Symposium Dragage*.

Vegas, I., Ibañez, J. A., Lisbona, A., Sáez De Cortazar, A., & Frías, M. (2011). Pre-normative research on the use of mixed recycled aggregates in unbound road sections. *Construction and Building Materials*, *25*(5), 2674–2682. https://doi.org/10.1016/j.conbuildmat.2010.12.018

Vegas, I., Ibañez, J. A., San José, J. T., & Urzelai, A. (2008). Construction demolition wastes, Waelz slag and MSWI bottom ash: A comparative technical analysis as material for road construction. In *Waste Management* (Vol. 28, Issue 3, pp. 565–574). https://doi.org/10.1016/j.wasman.2007.01.016

Wang, L., Tsang, D. C. W., & Poon, C. S. (2015). Green remediation and recycling of contaminated sediment by waste-incorporated stabilization/solidification. *Chemosphere*, *122*, 257–264. https://doi.org/10.1016/j.chemosphere.2014.11.071

Zentar, R., Dubois, V., & Abriak, N. E. (2008). Mechanical behaviour and environmental impacts of a test road built with marine dredged sediments. *Resources, Conservation and Recycling*, *52*(6), 947–954. https://doi.org/10.1016/j.resconrec.2008.02.002

Author's address:

MKAHAL Zeinab

PhD student in civil engineering at IMT-LILLE-DOUAI
France, 59500 Douai
+33 0659741912
Zeinab.mkahal@imt-lille-douai.fr

Remediation of Dredged Sediments by Electrokinetic Process Prior to Their Beneficial Reuse

Mathilde BETREMIEUX *PhD Student*

Yannick MAMINDY-PAJANY *Dr HDR*

Univ. Lille, Univ. Artois, IMT Lille Douai, JUNIA, ULR 4515 – LGCgE,

Laboratoire de Génie Civil et géo-Environnement, F-59000 Lille, France.

Abstract

The electrokinetic (EK) remediation method is based on the introduction of a low-intensity direct current into a contaminated sediment (or soil) in order to mobilise the metal pollutants within the solid matrix. In this study, electrokinetic tests were carried out on polluted river sediments to study the mobility of trace metals (As, Ba, Cd, Cr, Cu, Mo, Ni, Pb, Sb, Se and Zn). Two reagents were confronted: demineralised water (DW) and a newly developed BioSurfactant (BS) obtained from microbial sources. The results showed that BS effectively decreases in particular As (81.3%), Ba (80%), Cr (97.3%), Cu (82%) and Zn (94.5%) in the liquid phase. Although this technique is very effective, there seems to be a need to extend the operating time of the EK and/or to use more acidic reagents. This would allow controlling the pH of the matrix, and improving the removal rates in the solid phase.

Keywords

River sediment, BioSurfactant, Electrokinetic, Inorganic pollution, Remediation, Solid phase, Liquid phase

1 Introduction

Waterways maintenance produces several tons of sedimentary waste every year whose storage arise increasing harbouring and environmental management issues. Meteoritic events facilitate indeed, the runoff of both metallic and hydrocarbon pollutions into fluvial and aquifer systems, involving potential ecological outcomes. Therefore, dredging sediments treatment becomes a major step before perennial land storage or valorisation in civil engineering pathways. Different techniques (separation techniques, biological and chemical processes, incineration, flotation etc.), that can be used alone or in combination, have been lately developed in this purpose (Amrate, 2005). The choice of adequate treatment depends notably on sediments nature and the typology of associated contaminants (Kribi, 2005).

For use in civil engineering, highly effective pollutant extraction techniques are usually implemented such as ElectroKinetic (EK) treatments (Benamar et al. 2007, Ammami, 2013, Song et al. 2014). The method leans on ionic contaminants transportation by electromigration (Yang et al. 2014), and interstitial non-ionic contaminants movement through electro-osmosis (Acar & Alshawabkeh, 1993, Acar et al. 1993, Shapiro & Probstein, 1993, Pamukcu & Wittle, 1993, Acar et al. 1995, Cameselle, 2015).

The EK process produce water electrolysis, the generated H^+ and OH^- ions involve pH gradients between the anode and cathode of EK devices (Acar et al. 1995). Such pH variation may favour inorganic pollutants desorption (Tang et al. 2018). These variations are usually delayed during EK treatment by sediments' high acid/base buffering capacity (Yeung & Hsu 2005, Masi et al. 2017). Overall, pH variation appears to be the most de-termining factor regarding inorganic elements' mobility according to the majority of studies (Reddy & Chinthamreddy 2003, Ottosen et al. 2005). A decrease in pH promotes trace elements mobility, through the dissolution of bearing phases and/or the alteration of metals' redox states (Gidarakos & Giannis, 2006). Conversely, high pH are detrimental to trace elements migration and favour rather their adsorption onto minerals phases (e.g. iron and manganese oxides) via cation/H^+ exchange (Kennou et al. 2015).

Classic chemical reagents used as electrolytes in EK devices may induce secondary soil and groundwater pollution (Gonzini et al. 2010). In this study an environment friendly BioSurfactant (BS) was chosen as electrolyte on the basis of its biodegradability, and its economic viability. Here we chose to test an anionic cyclic lipopeptid produced by Bacillus subtilis, recently developed by HTS Bio® Laboratories and not yet commercialized. This type of anionic BioSurfactant lowers the surface tension of water to low concentrations and has good biodegradability rates (Chen et al. 2015). Micellar BioSurfactants remove trace metals through electrostatic adsorptions or, ionic exchanges. They also allow trace elements complexation thanks to chelating agents. As polar groups bind with the metals, the neo formed ensemble (BioSurfactant molecule/pollutant) becomes more soluble in water (Aşçi et al. 2010).

The main objective here was to identify the contribution of this newly developed BioSurfactant during the EK-removal of 11 frequent trace contaminants (As, Ba, Cd, Cr, Cu, Mo, Ni, Pb, Sb, Se and Zn). Its performances were compared to the yields of Deionised Water (DW) in order to highlight its advantages in both liquid and solid phases. This comparison had also the objective to identify the points that would require compositional adjustments for future investigations. The electric current, redox potential and pH parameters were monitored in pore waters as well as the electrolytic chambers to identify the effect of BS on one hand, and to stress out the potential influence on these key metrics regarding trace metals' removal on the other hand.

2 Materials and methods

2.1 Sediment origin and EK cell design

Sediments were collected from the Scarpe River, located in Saint-Laurent-Blangy (Haut-de France, France), using an excavator, and conserved in containers (25L) at 5°C. In this study, a glass-made system was used as EK cell (**Figure 1**). It consiste of three parts which were: (i) A central compartment for the sediments with three cross sections (S1, S2 and S3) for zoned monitoring and localised analyses after treatment; (ii) Two electrolyte compartments on both extremities as depicted in **Figure 1**.

The dimensions (L × W × H) are 200 mm x 100 mm x 160 mm for the sediment chamber and 100 mm x 100 mm x 160 mm for the electrolytes chambers. Graphite plates (100 mm x 10 mm x 200 mm) were implemented as electrodes and two perforated grids (100 mm x 2 mm x 160 mm) were disposed to separate the sediment compartment from the electrolyte chambers.

After each experiment, the EK cell was cleaned with demineralized water, and nitric acid (65%). The complete experimental dispositive comprised a DC power supply, a solution tank, two membrane pumps for reagents circulation (60ml.min^{-1}), a multimeter for electric intensity measurement, and two overflow tanks receiving the excess of solution. In order to avoid the contamination of the anode and cathode chambers, filter papers were sided to the perforated grids. On a chronologic viewpoint, the central chamber was first filled with 4kg of sediment, and then the electrolytic reagents were added in the two lateral chambers before 72h of rest time with the aim to optimise sediments saturation. The graphite electrodes were consecutively inserted into the anode and cathode chambers, and connected to the DC power supply. The HTS Bio® BioSurfactant implemented in this study bases on surfactin from natural fermentation process.

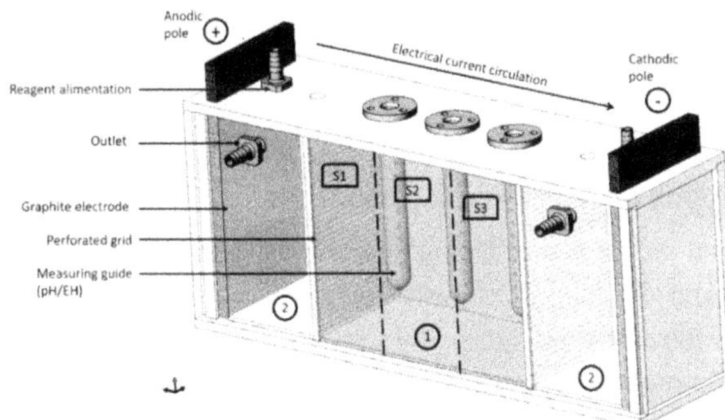

Figure 1 Schematic representation of the experimental device (Electrokinetic cell). S1, S2 and S3 correspond to the cross sections for zoned monitoring.

2.2 Setting during the Electrokinetic tests

Table 1 presents the experimental conditions. The cell was covered (with aluminium foil) to prevent microorganism contaminations and BS photodegradation. Electrical current, pH and redox, were measured twice per day (morning and evening) in triplicates using a calibrated multi-probe device (pH/redox), and copper wires connecting the sediment to a multi-meter were implemented for current intensity measurements.

Table 1 EK experimental conditions (DW: Deionized Water; BS: BioSurfactant)

	Current (V)	Duration (Days)	Anolyte and catholyte
EXP.A	14	7	DW
EXP.B			BS (0.6 g.l^{-1})

At the end of each EK-run, the sediment sample was divided into 3 equal sections from anode to cathode (S1, S2 and S3 as depicted in **Figure 1**) in order to verify the preferential migration of studied trace metals. The samples were dried at 40°C before quantifying inorganic elements' concentrations.

The efficiency of inorganic elements removal was calculated as following Equation (1).

$$\%R = \frac{C_i - C_f}{C_i} \times 100\% \qquad \text{Eq (1)}$$

With :

$\%R$: The removal, expressing the depollution of a given trace element;

C_i : The initial concentration of a given trace element;

C_f : Its final concentration (after EK treatment);

3 Results and discussion

3.1 Raw sediments' physico-chemical properties

X ray analyses performed on the raw sediment underlined the strong occurrences of quartz, Muscovite albite and kaolinite, which were very likely the erosive outcome of the Scarpe river watershed. Simultaneous ignition loss suggested indeed, the strong occurrence (11.91 ± 0.005%) of organic matter (OM) and carbonates. Raw sediments' chemical content determined through X-Ray fluorescence (**Table 2**) has shown the predominance of SiO2 due to the strong representation of mineral phases shown at X ray. The organo-mineral ensemble presented a pH of 7.48 ± 0.02 before EK treatments.

Table 2 Trace element (ICP-OES) and major elements (XRF) assessed in the raw sediment

	Trace elements		**Major elements (Oxides)**	
	Leaching concentration (mg.kg-1)	**Solid phase concentration (mg.kg-1)**	**Oxides**	**Concentration (%)**
As	0.19 ±0.01	6.78 ±0.27	Na_2O	1.1
Ba	1.21 ±0.008	406.31±1.40	MgO	0.7
Cd	0.003 ±0.00	0.67 ±0.05	Al_2O_3	8.6
Cr	0.07 ±0.002	44.17 ±0.80	SiO_2	75.0
Cu	0.60 ±0.003	41.75 ±0.69	P_2O_5	0.6
Mo	0.24 ±0.005	1.30 ±0.00	SO_3	0.1
Ni	0.22 ±0.003	19.33 ±0.23	K_2O	2.1
Pb	0.02 ±0.00	56.19 ±1.82	CaO	7.3
Sb	0.12 ±0.012	2.75 ±0.15	TiO_2	0.8
Se	0.14 ±0.017	1.85 ±0.00	Fe_2O_3	3.3
Zn	0.96 ±0.015	333.24 ±10.60	ZrO_2	0.1

3.2 Background parameters (pH / redox), and electric currents' evolution during the treatment

3.2.1 Electric current's evolution

The electric charges of total chemical species in movement generate an electric intensity which varies as a function of the entire system's resistances (Gao et al. 2013). The circulation of the H^+ and OH^- ions, as well as inorganic elements ions, is linked to the magnitude of electric intensity. Its evolution during the treatment with DW (EXP.A) as displayed in **Figure 2a** exhibited high values in the first 48h before reaching a more stable state around 8-9mA similar to the results of Maturi & Reddy (2006) and Colacicco et al. (2010) while the experiment with BS (EXP.B) showed an opposite trend. An examination per sections showed that S1 and S2 had close behaviours with a rise until 72h, and a strong decrease at the end of the treatment. S3 on the other hand, displayed a more pronounced lowering compared to the other sections with average values around 6.5mA. As shown in **Figure 2b**, electric current in EXP.A was lower than in EXP.B, suggesting that the use of BS improved electric current circulation within the sedimentary matrix, probably as an effect of micelles formation and emulsification process.

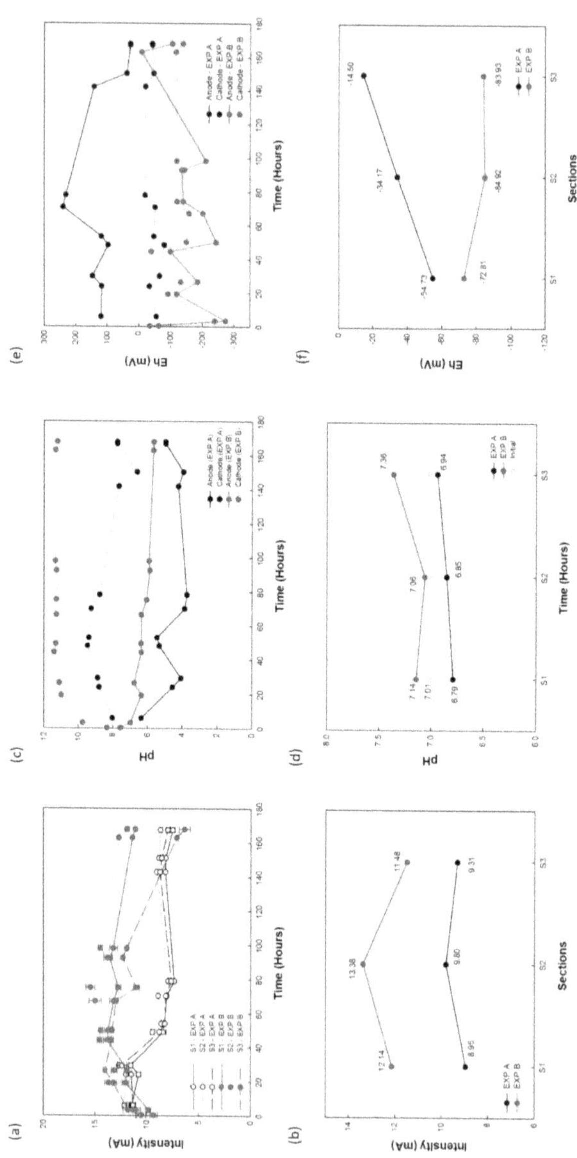

Figure 2 (a) Electrical conductivity evolution during the EK experiments conducted with EXP.A and EXP.B (S1, S2 and S3 are cross sections realized from the anode to the cathode); (b) Average electrical intensity in the three sections (S1, S2 and S3) during the EK treatment with both electrolyte; (c) pH evolution within the reagents chambers (anode and cathode) during the EK experiments; (d) Average of pore waters' pH measured within the 3 sedimentary sections (S1, S2 and S3) during EK treatment; (e) Redox evolution at both anodic and cathodic poles during the EK treatment; (f) Average pore water redox values within sediment sections (S1,S2 and S3) during the treatment.

For both experiments, the section S2 presented higher current intensity compared to the other sections, suggesting mobile ions electromigration from the anode towards the centre of the EK cell. The electric intensity lowering at the cathode was very likely favoured by localised ions precipitation as shown by Ammami et al. (2014) and Bahemmat & Farahbakhsh (2015).

3.2.2 Evolution of pH

The monitoring of pH magnitudes in pore water and the two electrolyte compartments (anodic and cathodic) is recapitulated in **Figure 2c and 2d** In the anodic compartment, it decreased during the treatment with both electrolytes (**Figure 2c**). The use of DW as electrolyte (EXP.A) induced values ranging between 6.35 and 4 while BS (EXP.B) induced oscillations between 7.44 and 5.64, suggesting plausible oxidations at the anode (Mosavat et al. 2012, Moghadam et al. 2016). For the cathodic compartment, the pH increased from 8.01 to 9.47 during the first 72h as a probable effect of OH^- income from water electrolysis (Gidarakos & Giannis, 2006), and decreased to 7.74 at the end of the EXP.A. For EXP.B, it varied in an alkalisation trend with values ranging between 8.33 and 11.22. Regarding pore water pH, a slight decrease was observed during EXP.A (**Figure 2d**), whereas EXP.B exhibited increasing values, especially near to the cathode. Such alkaline pHs usually prevent inorganic elements solubilisation and migration (Virkutyte et al. 2002). Alkaline conditions promote carbonates precipitation and contribute therefore to the buffering capacity of the sediment (Singh et al. 1998). In EK systems, OH^- ions relative lack of mobility compared to H^+ induces the gradual appearance of an acid front (Mosavat et al. 2012). However here, the pore water pH measured in both experiments was rather neutral, which suggested an absence of acidification and carbonates solubilisation during the treatment. This may have noteworthy implications regarding the fate of the studied trace elements, since carbonates strongly constrain their desorption by promoting background alkalinisation (Hlavackova, 2005).

3.2.3 Redox potential evolution

BS (EXP.B) allowed an important redox reduction in both anodic and cathodic poles (**Figure 2e**). In the sedimentary matrix, DW involved an oxidizing medium near to the anodic pole and a reducing environment close to the cathode compartment, whereas BS inducted a more homogeneous reduction in pore waters nearing both poles. Indeed, with BS as electrolyte (EXP.B), redox monitoring results (**Figure 2f**) indicated reducing properties for pore water with pronounced values in the sections near to the cathode (-84mV). Demineralised water (EXP.A) did not allow such reduction of the medium, and showed a more pronounced redox delta between the three sections (**Figure 2f**). Besides, more reducing conditions were observed near to the anode (-54.7mV), rendering this zone more favourable to the occurrence of reduced mineralogical phases i.e. sulphides (Singh et al. 1998).

Previous studies suggested that redox potential increases in the anode regions and decreases at the cathode as an effect of water electrolysis (Ammami, 2013). Low pH combined with high redox potential increases inorganic elements desorption from sediments and improves electrokinetic yields, (Singh et al. 1998, Nystroem et al. 2005). Although pore water did not change significantly here, the depollution rates remained quite satisfactory within the sedimentary matrix (Cf. elements evolution described in the next sections). Conversely, a decrease in the redox potential promotes minerals precipitation and notably sulphides formation (Tack et al. 1997). Oxidizing conditions favour therefore the neoformation of secondary phases, and contaminants co-precipitation or adsorption (Gäbler 1997, Tack et al. 1997, Lions et al. 2007, Lions et al. 2010, Isaure 2012). Secondary phases such as amorphous iron and aluminium (hydro-oxides), have large specific surface areas and play an important role in metals and metalloids retention processes (Meima & Comans 1998, Dijkstra et al. 2004) but are usually less stable than primary phases which tends to eventually increase the global background solubility (Calmano et al. 1993, Morse & Luther 1999, Achard 2013).

3.3 Trace elements evolution in the solid phase

3.3.1 Behaviour with Deionized Water as electrolyte (EXP.A)

The studied trace elements (As, Ba, Cd, Cr, Cu, Mo, Ni, Pb, Sb, Se and Zn) and removal rate (%R) from the solid phase, calculated via equation 1, are presented in **Table 3** as a function of sediments cross sections. Regarding Zn, poor extractions were observed using DW as electrolyte (EXP.A). Similar trends were reported by Song et al. (2014) which suggested that Zn behaviour was driven by high precipitations in the section nearing the anodic pole. Here, Zn first seemed difficult to extract from the solid phase, and exhibited greater enrichment tendencies near to the anode (-12.9%). For Cu, Ni, Pb and Zn, precipitation/complexation phenomena of free forms were very likely involved in the enrichments observed after treatment as suggested by Isaure (2012). Note that the relatively high %R values for Cu may result from the corrosion of copper wires implemented for electrical intensity measurements. Thus, these Cu removal yields were plausibly doped by this potential experimental bias. As, Cu, Cr, Ni, Pb and Zn had low removal yields and presented punctual enrichments within the solid phase. Eh and pH spatial evolutions as a function of distance from the polar extremities drove very likely these particular behaviours

Table 3 Metals removal efficiencies in the three sections ($\%R_{S1}$, $\%R_{S2}$ and $\%R_{S3}$), in solid phase and liquid phase after treatments, for both electrolyte

	Solid phase efficiency (total content test)						**Liquid phase efficiency (leaching test)**					
	EXP.A			EXP.B			EXP.A			EXP.B		
	$\%R_{S1}$	$\%R_{S2}$	$\%R_{S3}$	$\%R_{S1}$	$\%R_{S2}$	$\%R_{S3}$	$\%R_{S1}$	$\%R_{S2}$	$\%R_{S3}$	$\%R_{S1}$	$\%R_{S2}$	$\%R_{S3}$
As	4.6	4.9	9.7	19.0	22.5	17.7	52.8	79.2	37.7	81.3	57.9	54.6
Ba	69.3	69.5	69.3	70.9	71.2	71.0	75.1	75.1	74.7	80.0	78.1	77.5
Cd	34.0	45.4	33.2	77.5	77.5	77.5	16.7	16.7	16.7	*-66.7*	*-66.7*	*-66.7*
Cr	4.8	7.2	7.5	11.2	10.9	10.6	93.2	93.2	93.2	97.3	97.3	97.3
Cu	*-30.3*	*-30.5*	*-32.2*	*-30.8*	*-26.9*	*-28.4*	75.0	68.4	70.9	82.0	76.1	79.2
Mo	53.8	53.8	53.8	7.7	7.7	7.7	15.4	16.9	19.6	13.8	13.0	11.2
Ni	*-11.0*	*-5.9*	*-4.8*	*-2.6*	*-3.4*	*-2.7*	59.1	46.4	60.4	60.7	56.6	62.7
Pb	*-0.4*	*-0.5*	1.3	4.4	6.9	6.4	32.5	32.5	32.5	*-40.0*	*-40.0*	*-40.0*
Sb	56.4	56.4	56.4	40.0	40.0	40.0	*-4.5*	5.0	0.8	*-8.7*	8.2	9.3
Se	16.2	16.2	16.2	0.0	0.0	0.0	66.5	66.5	66.5	66.8	31.1	19.5
Zn	*-12.9*	*-7.9*	*-6.9*	*-1.1*	*-1.4*	*-1.1*	89.4	85.5	87.6	92.1	94.2	94.5

Cu enrichments were higher in the section S3, consistently with cations global migration towards the cathode (Ammami et al. 2014). Ni and Zn on the other hand, were enriched in section S1 i.e. closer to the anode. According to the general directions of cation migration in EK systems (Ammami et al. 2014), they should have migrated to the cathode, and provoked therefore more enrichment in the section S3 instead of S1. A plausible explanation for this incoherency was that the drying phase (40°C) inherent to our analytical design involved these enrichments.

3.3.2 Elements behaviour in the solid phase using the BioSurfactant as electrolyte (EXP.B)

Although trace elements removals were globally higher with the BioSurfactant (EXP.B) than Deionized Water (**Table 3**), data underlined a few disparities following elements nature. Ba and Cd for instance, displayed respectively 71.2% (S2) and 77.5% (S1, S2 and S3) of abatement, while Se concentrations seemed barely affected.

The micellar properties of BioSurfactants are acknowledged to enhance desorption/solubilisation mechanisms (Ammami et al. 2015, Mao et al. 2015). For Mo, Sb and

Se on the other hand, the use of the BioSurfactant did not improve significantly the removal yields (**Table 3**), probably as a result of the oxyanion forms of molybdenum, selenium and antimony unsuited for micellar interactions. Indeed, the anionic BS has rather a good affinity with cationic pollutants, that may explain the good yields observed for cations such as Cd (77.5%). BS efficiency per element gave the hierarchy Cd>Ba>Sb>As>Cr>Co>Mo>Pb>Zn>Ni>Cu, which was probably governed by elements predisposition to interact with the micellar compound. In contrast to EXP.A, %R in the different sections did not vary significantly which suggested that elements mobilization here was homogeneous and independent from the distance to the polar compartments.

Ammami (2013) observed Cd, Cu, Pb and Zn strong accumulations in the central section of the treated matrix due to basic pH effects on cations migration. In his approach, the most important abatements were found under the configurations with the highest current intensities, which was consistent with our study. Indeed, the use of BioSurfactant (EXP.B) led to higher electrical intensity (**Figure 2a and 2b**), and higher elemental abatements compared to the experimental design implementing DW as electrolyte (EXP.A). As previously observed in EXP.A, here also significant Cu enrichments were noted after treatment. The enrichment magnitudes were quasi similar in the two experiments and were probably both doped by the copper wires implemented for intensity measurements.

3.4 Trace elements evolution in the liquid phase

3.4.1 Metals evolution in the liquid phase with DW as electrolyte (EXP.A)

After treatment with DW as EK electrolyte, leachates pH values were 7.12, 7.21 and 7.22 in S1, S2 and S3 respectively, which were quasi equal to the raw sediment pH (7.48). The same measuring spots exhibited redox magnitudes of 189, 197 and 195mV respectively. The examination of trace elements behaviour underlined better removal yields in the sections near to electrolyte compartments (S1 and S3). Despite relatively basic pH conditions, high depollution rates with the following order Cr>Zn>Ba>Cu>Ni>As>Se>Pb were observed, and poorer yields for Mo, Cd, Sb under the DW experimental configuration. Solubilisation phenomena appeared predominant although pore water pH remained globally neutral (pH ≈ 6, 7)

Metals removal from the liquid phase depends particularly on background pH. Ammami et al. (2015), showed for instance, that Zn precipitates under hydroxide species at pH>7, while the precipitation of elements such as Cr, Cu and Pb requires more acidic conditions (pH>4-5). Our pH conditions (**Figure 2c and 2d**) tended to favour Zn cationic state, which is admitted more labile. For, Cr, Pb and Cu, the acido-bacic conditions here were theoretically unfavourable to mobile forms and fitted rather with sedimentary immobilization. However, with an average interstitial water pH of 6.86, good removal rates were observed for these elements: Cr (93.2%); Cu (75%); Pb (32.5%).

Generally speaking, acidic and reducing conditions are the most favourable configurations for metal solubilisation, with higher effects of pH compared to the influence of redox potential (Chuan et al. 1996). Chuan et al. (1996) for instance showed that metal solubility increases with lowering redox. Removal results here (**Table 3**) under evolving conditions suggested that both redox and pH have significant influence. In addition to pH and redox, the difference of electric potential is also evoked to explain ionic species migration. Indeed, electrophoresis is known to confer surface charges onto solubilized colloidal particles (Idivan & Nunes, 2018). The particle in this case is not immobilized in an electric field, but carried towards the electrodes. However, it can be slowed by accumulating compensator ions.

3.4.2 Evolution in the liquid phase with the BioSurfactant as electrolyte (EXP B)

Leachates pH in S1, S2 and S3 after treatment with BS were 7.27, 7.37, 7.26 respectively, which corresponded to the raw sediment pH (7.48). In the same way, the redox potential were 206, 205 and 205mV, which also corresponded to the redox value of the raw sediment (205mV). Despite non-acidic interstitial water (pH=7.19), most elements (As, Ba, Cu, Mo, Se) were depolluted in greater quantities in S1, and a few (Cd, Pb and Sb) exhibited enrichment trends. As previously mentioned, cations preferentially migrate towards cathodes (Ammami et al. 2014), but here micellar interactions with BS allow the formation of anionic groups that can migrate towards the anodic pole. Data showed that the BioSurfactant caused elements migration mainly in S1 and S3 (**Table 3**). Except for As, Ba, Cu, Mo, Se, which were mainly removed in S1, elements as Ni, Sb and Zn had higher yields in S3, with respective %R values of 62.7, 9.3 and 94.5% (**Table 3**).

Cd, Pb and Sb enrichments (respectively 67, 40 and 9% enrichments) probably as an effect of Eh increase. Sulphides oxidation and secondary phases dissolution are indeed known to involve aqueous concentrations' rise (Meima & Comans 1998, Lions et al. 2007). Achard (2013) evoked similar phenomena to explain the increase K, As, Mo, Ni and Zn, in pore water concentrations after EK treatment through probable dissolution and desorption. Cappuyns & Swennen (2005) also reported that elements as As can be released into pore waters through sediments oxidation. Such oxidation mechanism involves Fe-mono-sulphides and pyrite whose alteration and leads to the release of associated trace metals. On the other hand, oxidation tends to decrease pH values, which also may affect trace elements' mobility.

The lowering trend observed here for Zn, Cu, and Ni were also stressed out by Mulligan (2009) using similar anionic electrolytes (Saponin, and other BioSurfactants) on contaminated sediment. Batch washing steps included in Mulligan (2009) study drove very likely the difference in magnitude between their data and the trends observed here with BS. Their highest removal for Zn (33%) and Pb (24%) were achieved with $30g.l^{-1}$ of saponin at pH 5. Similarly, their highest Cu removal (84%) was achieved with 2% rhamnolipids (pH 6.5). The BioSurfactant here combined with the EK method was more effective at lower concentrations. 94.5% of Zn was indeed, removed at $0.6g.l^{-1}$ of BS. The removal of Cu (82 %) is close to the rate measured by Mulligan (2009).

4 Conclusion

BioSurfactants as rhamnolipids, saponin, and sophorolipid, have already shown good yields for the remediation of contaminated matrices (soil/sediment/water), by improving contaminants solubilisation and emulsification. In this study, the emphasis was put on a newly developed BioSurfactant (HTS Bio) with good biodegradability and low toxicity in order to expand the common database.

BS has been beneficial in removing most trace elements from the mobile aqueous phase. Indeed, 81.3% of As, 80% of Ba, 97.3% of Cr, 82% of Cu, 94.5% of Zn, 13.8% of Mo, 62.7% of Ni, 66.8% of Se and 9.3% of Sb were removed when BS was implemented as electrolyte. The new electrolyte was nevertheless, less satisfactory regarding Mo, Cd, and Pb for which its contribution did not exceed classical Deionized Water's yields. These last elements (Cd and Pb) exhibited aqueous enrichments trends probably as a result of pH and redox new configurations.

BS presented more contrasted results regarding the remediation of the solid matrix. Although removal values were quite satisfactory for elements as Ba (71.2%) and Cd (77.5%), performances were relatively poor for Sb (40%), As (22.5%), Cr (11.2%), Mo (7.7%) and Pb (6.9%). Regarding Mo and Se, which can display oxyanionic states, BS did not perform better than DW. This lack of affinity with oxyanion was very likely driven by BS's polarity, more suited for cations attraction. Overall, BS tended to depollute more efficiently in the section close to the anodic pole compared to the rest of the EK Cell, probably as a result of anionic micelles' formation in this area of the experimental device.

This new electrolyte was promising and future investigations can be considered with BS-like surfactants with a special focus on the mechanisms of inorganic elements solubilisation in presence of this type of foamy compound. A main enhancement point would require the use of more acidic reagents in order to promote the digesting mechanisms of pollutants carrier phases.

5 References

Acar, Y. B., & Alshawab-keh, A. N.	1993	*Principles of Electrokinetic Remediation*. Environmental Science and Technology, 27(13), 2638–2647. https://doi.org/10.1021/es00049a002
Acar, Y. B., Alshawabkeh, A. N., & Gale, R. J.	1993	*Fundamentals of extracting species from soils by electrokinetics*. Waste Management, 13(2), 141–151. https://doi.org/10.1016/0956-053X(93)90006-I
Acar, Y. B., Gale, R. J., Alshawabkeh, A. N., Marks, R. E., Puppala, S., Bricka, M., & Parker, R.	1995	*Electrokinetic remediation: Basics and technology status*. Journal of Hazardous Materials, 40(2), 117–137. https://doi.org/10.1016/0304-3894(94)00066-P
Achard, R.	2013	*Dynamique des contaminants inorganiques dans les sédiments de dragage : rôle spécifique de la matière organique naturelle* (Université de Toulon). Retrieved from https://tel.archives-ouvertes.fr/tel-00874422
Ammami, M. T.	2013	*Contribution à l'étude des processus électrocinétiques appliqués aux sédiments de dragage Soutenue* (Université du Havre). Retrieved from http://www.theses.fr/2013LEHA0007
Ammami, M. T., Benamar, A., Wang, H., Bailleul, C., Legras, M., Le Derf, F., & Portet-Koltalo, F.	2014	*Simultaneous electrokinetic removal of polycyclic aromatic hydrocarbons and metals from a sediment using mixed enhancing agents*. International Journal of Environmental Science and Technology, 11(7), 1801–1816. https://doi.org/10.1007/s13762-013-0395-9
Ammami, M. T., Portet-Koltalo, F., Benamar, A., Duclairoir-Poc, C., Wang, H., & Le Derf, F.	2015	*Application of biosurfactants and periodic voltage gradient for enhanced electrokinetic remediation of metals and PAHs in dredged marine sediments*. Chemosphere, 125, 1–8. https://doi.org/10.1016/j.chemosphere.2014.12.087
Amrate, S.	2005	*Etude du transport d'ions dans des milieux poreux sous l'action d'un champ électrique. Application à la décontamination de sols pollués*. Université des sciences et de la technologie Houari Boumediene.
Aşçi, Y., Nurbaş, M., & Sağ Açikel, Y.	2010	*Investigation of sorption/desorption equilibria of heavy metal ions on/from quartz using rhamnolipid biosurfactant*. Journal of Environmental Management, 91(3), 724–731. https://doi.org/10.1016/j.jenvman.2009.09.036
Bahemmat, M., & Farahbakhsh, M.	2015	*Catholyte-conditioning enhanced electrokinetic remediation of co and Pb polluted soil*. Environmental Engineering and Management Journal, 14(1), 89–96.

		https://doi.org/10.30638/eemj.2015.011
Benamar, A., Baraud, F., & Alem, A.	2007	Traitement des sédiments de dragage : un enjeu du développement durable. *25e Rencontres de l'AUGC, 7. 23-25 mai, Bordeaux.*
Calmano, W., Hong, J., & Forstner, U.	1993	*Binding and mobilization of Heavy Metals in contaminated sediments affected by pH and redox potential.* Wal. Sci. Tech, 28(8), 223–235.
Cameselle, C.	2015	*Enhancement of Electro-Osmotic Flow during the Electrokinetic Treatment of A Contaminated Soil.* Electrochimica Acta, 181, 31–38. https://doi.org/10.1016/j.electacta.2015.02.191
Cappuyns, V., & Swennen, R.	2005	*Kinetics of element release during combined oxidation and pHstat leaching of anoxic river sediments.* Applied Geochemistry, 20(6), 1169–1179. https://doi.org/10.1016/j.apgeochem.2005.02.004
Chen, W. C., Juang, R. S., & Wei, Y. H.	2015	*Applications of a lipopeptide biosurfactant, surfactin, produced by microorganisms.* Biochemical Engineering Journal, 103, 158–169. https://doi.org/10.1016/j.bej.2015.07.009
Chuan, M. C., Shu, G. Y., & Liu, J. C.	1996	*Solubility of heavy metals in a contaminated soil: Effects of redox potential and pH.* Water, Air, and Soil Pollution, 90(3–4), 543–556. https://doi.org/10.1007/BF00282668
Colacicco, A., De Gioannis, G., Muntoni, A., Pettinao, E., Polettini, A., & Pomi, R.	2010	*Enhanced electrokinetic treatment of marine sediments contaminated by heavy metals and PAHs.* Chemosphere, 81(1), 46–56. https://doi.org/10.1016/j.chemosphere.2010.07.004
Dijkstra, J. J., Meeussen, J. C. L., & Comans, R. N. J.	2004	*Leaching of heavy metals from contaminated soils: An experimental and modeling study.* Environmental Science and Technology, 38(16), 4390–4395. https://doi.org/10.1021/es049885v
Gäbler, H. E.	1997	*Mobility of heavy metals as a function of pH of samples from an overbank sediment profile contaminated by mining activities.* Journal of Geochemical Exploration, 58(2–3), 185–194. https://doi.org/10.1016/S0375-6742(96)00061-1
Gao, J., Luo, Q., Zhang, C., Li, B., & Meng, L.	2013	*Enhanced electrokinetic removal of cadmium from sludge using a coupled catholyte circulation system with multilayer of anion exchange resin.* Chemical Engineering Journal, 234, 1–8.

		https://doi.org/10.1016/j.cej.2013.08.019
Gidarakos, E., & Giannis, A.	2006	*Chelate agents enhanced electrokinetic remediation for removal cadmium and zinc by conditioning catholyte pH.* Water, Air, and Soil Pollution, 172(1–4), 295–312. https://doi.org/10.1007/s11270-006-9080-7
Gonzini, O., Plaza, A., Di Palma, L., & Lobo, M. C.	2010	*Electrokinetic remediation of gasoil contaminated soil enhanced by rhamnolipid.* Journal of Applied Electrochemistry, 40(6), 1239–1248. https://doi.org/10.1007/s10800-010-0095-9
Hlavackova, P.	2005	*Evaluation du comportement du cuivre et du zinc dans une matrice de type sol à l'aide de différentes méthodologies.* L'Institut National des Sciences Appliquées de Lyon.
Idivan, A., & Nunes, V.	2018	*Transport d'ions sous l'effet d'un champ électrique en milieu poreux : application à la séparation de terres rares par électrophorèse à focalisation.* Institut National Polytechnique de Lorraine. Retrieved from https://tel.archives-ouvertes.fr/tel-00011215v2
Isaure, M.	2012	*Spéciation et transfert du zinc dans un dépôt de sédiment de curage contaminé : évolution le long du profil pédologique.* Université Joseph-Fourier. Retrieved from https://tel.archives-ouvertes.fr/tel-00717491
Kennou, B., El Meray, M., Romane, A., & Arjouni, Y.	2015	*Assessment of heavy metal availability (Pb, Cu, Cr, Cd, Zn) and speciation in contaminated soils and sediment of discharge by sequential extraction.* Environmental Earth Sciences, 74(7), 5849–5858. https://doi.org/10.1007/s12665-015-4609-y
Kribi, S.	2005	*Décomposition des matières organiques et stabilisation des métaux lourds dans les sédiments de dragage Introduction générale.* Institut National des Sciences Appliquées de Lyon. Retrieved from http://theses.insa-lyon.fr/publication/2005ISAL0064/these.pdf
Lions, J., Guérin, V., Bataillard, P., Van Der Lee, J., & Laboudigue, A.	2010	*Metal availability in a highly contaminated, dredged-sediment disposal site: Field measurements and geochemical modeling.* Environmental Pollution, 158(9), 2857–2864. https://doi.org/10.1016/j.envpol.2010.06.011
Lions, J., Lee, J. van der, Guéren, V., Bataillard, P., & Laboudigue, A.	2007	*Zinc and cadmium mobility in a 5-year-old dredged sediment deposit.* Experiments and modelling. Journal of Soils and Sediments, 7(4), 207–215.

		https://doi.org/10.1065/jss2007.05.226
Mao, X., Jiang, R., Xiao, W., & Yu, J.	2015	*Use of surfactants for the remediation of contaminated soils: A review.* Journal of Hazardous Materials, 285, 419–435. https://doi.org/10.1016/j.jhazmat.2014.12.009
Masi, M., Ceccarini, A., & Iannelli, R.	2017	*Model-based optimization of field-scale electrokinetic treatment of dredged sediments.* Chemical Engineering Journal, 328, 87–97. https://doi.org/10.1016/j.cej.2017.07.004
Maturi, K., & Reddy, K. R.	2006	*Simultaneous removal of organic compounds and heavy metals from soils by electrokinetic remediation with a modified cyclodextrin.* Chemosphere, 63(6), 1022–1031. https://doi.org/10.1016/j.chemosphere.2005.08.037
Meima, J. A., & Comans, R. N. J.	1998	*Application of surface complexation/precipitation modeling to contaminant leaching from weathered municipal solid waste incinerator bottom ash.* Environmental Science and Technology, 32(5), 688–693. https://doi.org/10.1021/es9701624
Moghadam, M. J., Moayedi, H., Sadeghi, M. M., & Hajiannia, A.	2016	*A review of combinations of electrokinetic applications.* Environmental Geochemistry and Health, 38(6), 1217–1227. https://doi.org/10.1007/s10653-016-9795-3
Morse, J. W., & Luther, G. W.	1999	*Chemical influences on trace metal-sulfide interactions in anoxic sediments.* Geochimica et Cosmochimica Acta, 63(19–20), 3373–3378. https://doi.org/10.1016/S0016-7037(99)00258-6
Mosavat, N., Oh, E., & Chai, G.	2012	*A review of electrokinetic treatment technique for improving the engineering characteristics of low permeable problematic soils.* International Journal of GEOMATE, 2(2), 266–272. https://doi.org/10.21660/2012.4.3i
Mulligan, C. N.	2009	*Recent advances in the environmental applications of biosurfactants.* Current Opinion in Colloid and Interface Science, 14(5), 372–378. https://doi.org/10.1016/j.cocis.2009.06.005
Nystroem, G. M., Ottosen, L. M., & Villumsen, A.	2005	*Electrodialytic removal of Cu, Zn, Pb, and Cd from harbor sediment: Influence of changing experimental conditions.* Environmental Science and Technology, 39(8), 2906–2911. https://doi.org/10.1021/es048930w
Ottosen, L. M., Pedersen, A. J., Ribeiro, A. B., & Hansen, H. K.	2005	*Case study on the strategy and application of enhancement solutions to improve remediation of soils contaminated with Cu, Pb and Zn by means of electrodialysis.* Engi-

		neering Geology, 77(3-4 SPEC. ISS.), 317–329. https://doi.org/10.1016/j.enggeo.2004.07.021
Pamukcu, S., & Wittle, J. K.	1993	*Electrokinetic treatment of contaminated soils, sludges, and lagoons. Final report.* https://doi.org/10.2172/10185835
Reddy, K. R., & Chinthamreddy, S.	2003	*Sequentially Enhanced Electrokinetic Remediation of Heavy Metals in Low Buffering Clayey Soils.* Journal of Geotechnical and Geoenvironmental Engineering, 129(3), 263–277. https://doi.org/10.1061/(asce)1090-0241(2003)129:3(263)
Shapiro, A. P., & Probstein, R. F.	1993	*Removal of Contaminants from Saturated Clay by Electroosmosis.* Environmental Science and Technology, 27(2), 283–291. https://doi.org/10.1021/es00039a007
Singh, S. P., Tack, F. M., & Verloo, M. G.	1998	*Heavy metal fractionation and extractability in dredged sediment derived surface soils.* Water, Air, and Soil Pollution, 102(3–4), 313–328. https://doi.org/10.1023/A:1004916632457
Song, Y., Benamar, A., Mezazigh, S., & Wang, H.	2014	*Extraction de métaux lourds des sédiments par méthode électrocinétique.* 1055–1062. https://doi.org/10.5150/jngcgc.2014.116
Tack, F. M., Lapauw, F., & Verloo, M. G.	1997	*Determination and fractionation of sulphur in a contaminated dredged sediment. Talanta, 44(12), 2185–2192. https://doi.org/10.1016/S0039-9140(97)00035-0*
Tang, J., He, J., Xin, X., Hu, H., & Liu, T.	2018	*Biosurfactants enhanced heavy metals removal from sludge in the electrokinetic treatment.* Chemical Engineering Journal, 334. https://doi.org/10.1016/j.cej.2017.12.010
Virkutyte, J., Sillanpää, M., & Latostenmaa, P.	2002	*Electrokinetic soil remediation - Critical overview.* Science of the Total Environment, 289(1–3), 97–121. https://doi.org/10.1016/S0048-9697(01)01027-0
Yang, J. S., Kwon, M. J., Choi, J., Baek, K., & O'Loughlin, E. J.	2014	*The transport behavior of As, Cu, Pb, and Zn during electrokinetic remediation of a contaminated soil using electrolyte conditioning.* Chemosphere, 117(1), 79–86. https://doi.org/10.1016/j.chemosphere.2014.05.079
Yeung, A. T., & Hsu, C.	2005	*Electrokinetic Remediation of Cadmium-Contaminated Clay.* Journal of Environmental Engineering, 131(2), 298–304. https://doi.org/10.1061/(asce)0733-9372(2005)131:2(298)

Author's address(es) :

PhD Student. Mathilde BETREMIEUX - Dr HDR. Yannick MAMINDY-PAJANY
Centre de recherche IMT Douai - 764 Boulevard Lahure, 59500 Douai ;

E-Mail :

mathilde.betremieux@imt-lille-douai.fr

yannick.mamindy@imt-lille-douai.fr

Indonesia Waste Bank Transition and Transformational System: a Pathway of Circular Economy Development

Ade Brian Mustafa [a], **Masahiko Haraguchi** [b]

[a] *School of Environmental Science and Engineering, Shanghai Jiao Tong University, Shanghai, 200240, PR China*

[b] *Research Institute for Humanity and Nature, Kyoto, Japan*

Abstract

Community-based waste management has been established in Indonesia since 2008, namely *Bank Sampah* (Waste Bank). Waste banks are functionally based on rewards that households can obtain when collecting, sorting, and depositing waste into waste banks. By 2020, there exist 11330 units of waste banks across 34 provinces in Indonesia. Since its inception, waste banks have contributed to raising public awareness, increasing recycling rates, and reducing wastes in Indonesia landfills. However, additional innovation must be carried out to improve waste bank systems. In particular, it is vital to integrate waste bank systems into the circular economy mechanisms. Here, we present an overview of the latest development and waste bank database, structural regulation of the waste bank, highlighting the key drivers of organizations associated with the waste bank, and supports from the Indonesian government. Furthermore, we review practices and literature in a circular economy to be attained in the waste bank systems. Several essential aspects will be elaborated, such as mainstreaming the waste bank circularity (materials) and industrial ecology, waste bank and environmental footprints, waste bank and social-catalyst accelerators, and waste bank on natural regeneration systems (interrelationship within nexus). The paper concludes that a sustainable waste bank can be achieved while integrating a community-based activities economy.

Keywords

Waste Bank, Circular Economy, Sustainable Development, Community-Institutional Capacity Building, Indonesia

1 Introduction

The concept of waste banks (WB) is to collect and manage recyclable waste from society, allowing them to earn a profit by depositing waste (Maryati et al., 2018). Since its inception in Bantul, Yogyakarta in 2008, WB began to develop in various cities in Indonesia (Rohmawati, 2015). Indonesia WB plays an essential role in the recycling mechanism, and this is frequently called """"""""social engineering"""""""" practice that leads to improving waste segregation and sorting at source. This platform is usually set up in neighborhood settings. Lubis (2018) stated that households deposit their trash on a regular basis, which is weighed and received a monetary value based on the price set by waste collectors. WB is a site at which the WB administrator provides the waste depositor with service activities. The waste management (WM) mechanism resembles a conventional commercial bank. The working arrangement for WB comprises of sorting, delivering, weighing, selling, and documenting and sharing waste transactions between depositor and aggregator in the saving book (Lubis, 2018).

To some extent, the recyclables price is regulated by whether the waste is retrieved by the WB service or delivered directly by households themselves. The WB operates within the community, principally in subdistrict and district level, so they could use to deposit their waste and generate income from the waste (Wulandari et al., 2017). WB management generally depends on the recycling sector's involvement and cooperation in the local community where a WB is located (Priyo et al., 2018). While the role of an online-based technology system (through apps usage) for the WB institution becomes an important aspect (Wulandari and Fajar, 2018). Besides, the WB may generate economic growth and opportunities for the household because when they exchange their waste, they will acquire a reward (Widiyanto et al., 2017) and provide value creation of materials by means of recycling and upcycling (Bappenas, 2021).

Gamaralalage et al. (2016) reported that WB has the potential to be identified as a successful community-based program in Indonesia for encouraging the transition from solid waste to resource management principles and practices at the grassroots level. In conformity with Satori et al. (2020), WB's activity following the collection of recyclable waste is the sale of materials to collectors with informal business networks. On a regular basis, the WB exchanges recyclable materials with collectors, junk dealers, or agents. At the level of a junk dealer, recycled waste classification is usually more specific, and plastics have even been listed. Equivalently, it is marketed to the recycling manufacturing as a compound of raw materials for a specific industry.

Sulami et al. (2018) noted The Republic of Indonesia's Ministry of the Environment and Forestry (MoEF) enforced the realizations of the 3R concept by employing community-based approaches through WB and proposed integration of extended producer responsibility (EPR) into the WB. Such a shift in consumption trends needs a fundamental shift, especially in business models. EPR should also be adopted to enforce a policy solution whereby manufacturers are held liable for the end-of-life of their goods beyond the sale. Even producers are held liable for environmental harm caused by their products (Maitre-Ekern, 2020). Furthermore, Azwir (2019) stated that under EPR provisions, the WB could also act as a collection/dropping point. From a retailer standpoint, this collection/dropping point is meant as a starting point for the replacement of goods and/or packages that are subject to EPR requirements.

WB is an alternative solution for reducing the ever-increasing volume of waste (in the landfill). Besides, developing and inculcating environmental awareness through the WB in a social collaboration frame is very important (Khair, 2019). At present, the circular economy (CE) concept is establishing in any part of the world. While in Indonesia, since the last decade, WB shows a considerable increase and engagement. However, this community-driven waste management (CDWM) are not fully implementing the CE. The WB is only operating in their current conventional way, but WB can achieve more profits with CE's framework and provide more benefits. WB and CE will be fostering integrated WM systems in Indonesia through various values, initiatives, practices and in-line with sustainabilities.

This research paper would reveal the critical understanding of fundamental WB and CE through investigated the following questions: how to establish a more profitable CE framework within the developing Indonesia waste banks institution through database observation, interconnected key-drivers while integrating by its community-based activities economy? Respectively in this paper, the approach and framework of study will be described first, following by a comprehensive analysis of WB development. Additionally, the interrelation and implication of WB and CE will be expanded as well as policy implication and conclusion of the study.

2 Method

Indonesia WB baseline information is obtained from Mr Putra Fajar Alam, MSi, the founder of smash.id (digital platform) which is an integrated platform database that provides information on Indonesia WB. The smash.id platform is chosen through initial communication with the Directorate General of Solid Waste, Waste, Toxic and Hazardous Waste Management, Ministry of Environment and Forestry of the Republic of Indonesia (MoEF), Dr Ir Novrizal Tahar in 2020, and obtained the database with the permission. Thus far, this platform is the only source of WB management system information in Indonesia. Several data aspects that are used and processed in this research is the unit number of the WB in Indonesia provinces (per October 2020), community contributors involved, proportion of waste processed/managed categories, provincial data of WB financial turnover per month (recapitulated per October 2020), and the performance comparison between registered and active WB.

Moreover, literature studies of WB development in Indonesia were conducted using the searching tools of scientific publication platforms (keywords consist of: 'Indonesia waste bank' for English-based literature and 'Bank Sampah Indonesia' for Indonesian language literature). However, the authors only able to identify the size of Indonesia WB since 2012 only. Despite the inception of WB in 2008, the existing and available literature emerged since 2012. Previous statistical data of WB is unavailable. An in-depth analysis throughout the national database (from government or institutional report) is administered to obtain the economical pricing of waste materials on a 2019 basis. While the average quantity of waste fraction managed in the WB is obtained from literature in several cities for representativeness. For creative recycling products, the average prices are obtained from various online resources after evaluating trusted sources through WB forums (Bank Sampah Nasional) communication, which is involved by numerous WB associations across Indonesia's provinces. Whereas analysis of registered WB is conducted, which includes all listed WB in the database, and active WB is the number of WB with confirmed active status (per October 2020). Both values are compared in terms of financial turnover, the economic potential from general recyclables waste, and the unit scenario of creative recycling products.

Furthermore, the WB and CE development's interconnection is explained through an in-depth analysis of Indonesia's WB and its potential for circular works from several perspectives. To support the analysis, the author was actively involved in WM, WB, and CE discussions, including online webinars and focus group discussions. Thus, an integrative review approach is applied in this research, following Snyder (2019) that described new and evolving topics or mature subjects through integrative literature reviews. The purpose of using an integrative review approach in the context of mature topics is to analyze the base of knowledge, critically review and possibly re-conceptualize, synthesis and build on the theoretical basis of the particular topic as it evolves. A more innovative selection of data is also needed for this form of analysis. The objective is usually to include viewpoints and observations from various fields, rather than to cover all papers ever published on the subject. Eventually, the authors highlighted the Indonesia WB potential to be more circular by linkage the mechanism with the example of CE practice in different locality, which in conformity with typical Indonesia WB, customs, and communities.

3 Analysis of the Indonesia Waste Bank Development

The latest database obtained from the Government of Indonesia (GoI), under the support of the Directorate General of Waste Management per October 2020, revealed that 11330 WB units are established in Indonesia. Based on the database provided in Fig. 2, East Java province holds the highest WB number, reached almost 3000 units, while Jakarta and West Java accounted for around 2000 WB, respectively. However, Central Java province has only around 1000 WB, while another remaining province has a mere quantity of below 600 units. Only East Nusa Tenggara, Bangka Belitung, and North Borneo have a WB below ten units (the smallest amount).

Fig. 2 shows that the density of the WB units, which is predominantly located in Java Island. The Java region consists of six provinces, namely DKI Jakarta, West Java, East Java, Central Java Banten, and Yogyakarta cumulatively have 8439 WB. This quantity shares for 74.48% of WB distribution nationally. Notably, the escalating number is correlated with the province's population, as the Java region shares 57.49% of Indonesia's population. The waste amount is relatively higher than other Indonesia regions, and the proximity of demand-institution (agent) or availability of private company that processed the WB waste materials is primarily positioned in the Java region. This is related as the WB should have a channel (parties) that willing to purchase the waste they collect, for example, a plastic pellet factory that requested plastic waste as its raw materials.

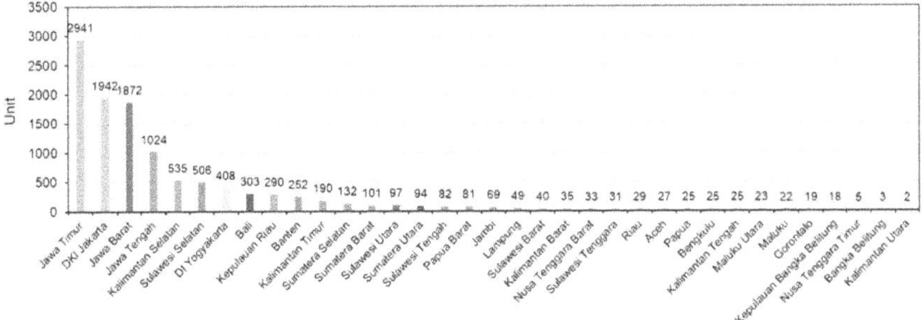

Fig. 2. WB unit in 34 Indonesia Provinces, which majority located in Java Island (data obtained from Directorate General of Waste Management, MoEF and Smash.id digital survey in 2020)

The progression is correlated with the presence of the Indonesian Plastic Recycling Association (Asosiasi Daur Ulang Plastik Indonesia/ ADUPI) since 2015. From this standpoint, ADUPI promotes the standardization of recycled raw materials, which provides business certainty for WB, collectors, and informal scavengers on the front lines of waste collection (ADUPI, 2020; Agrofarm, 2020). Following this, the EPR systems proposed by PRAISE (Packaging and Recycling Association for Indonesia's Sustainable Environment) conducted their projects (CSR scheme) mostly in the Java region. Another reason for existing and growing WB entities is supported by public literacy in environmental aspects. As more people comprehend that waste-as-a-resource could enhance the economy, they are initiated to find a network (channel) and organize the WB.

It is noticeable from these outcomes that the WB is relatively less prevalent in Celebes and Moluccas, Borneo, Small Sunda region, and Sumatera. This condition indeed transpired due to minimal participation and engagement from the community itself to manage the waste, less community-based involvement practice in the areas, and the minimal number of institutional channels that are probably willing to accept the segregated valuable waste (less demand). Furthermore, the waste quantity are relatively lesser compare to the Java region.

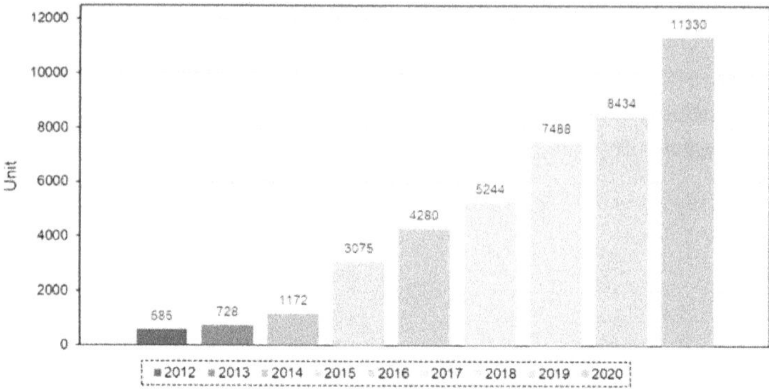

Fig. 3. Development of Indonesia WB since 2012 until 2020, data obtained from: Satuharapan (2012); Temesirecycling (2013), MoEF (2014), Tempo (2015; 2016), MoEF (2017), Media Indonesia (2018), Samudera Nesia (2019); and MoEF with Smash.id report (2020)

Fig. 3 presents the compilation dataset of WB development in Indonesia from 2012 to 2020, which stemmed from the government report, non-governmental organization report, and former national mass media. The depicted graph reveals that by 2020 there were 11330 WB in Indonesia. The growth number is relatively substantial as from 2014 to 2015 and 2017 to 2018 around 2000 WB were inaugurated successively. A significant increment is highlighted during 2019-2020; approximately 2900 WB were established in Indonesia.

This national database provides evidence that the number of WB increases significantly during eight years of observation by the year unit. The initial time provided in this chart is 2012 because the previous year's data is undocumented. This reflected at that time, and the WB documentary efforts were minimal. The escalating trend of the WB is primarily due to its benefit to the community, economy, and environment. The local government and enterprise sector support are also taken into consideration.

4 Discussion

In the review of the literature, WM still has profound and complex environmental challenges in Indonesia. As a consequence of increasing landfill heaps, while landfills' space and capacity is severely limited. On the other hand, industrial stakeholders face challenges in terms of raw material availability and production efficiency. The application of a CE is considered to be effective in responding to these problems. To further explore the potential and opportunities available, it is necessary to analyze the WB's existence to achieve a more integrated and sustainable CE in Indonesia. Accenture (2015) stated that the essence of the CE revolution is excluding the notion of """""""waste""""""" and recognizing that everything has value, shifting from efficiency into effectiveness in handling inputs and outputs, and developing a much closer relationship with consumers by the use of a sustainable business model.

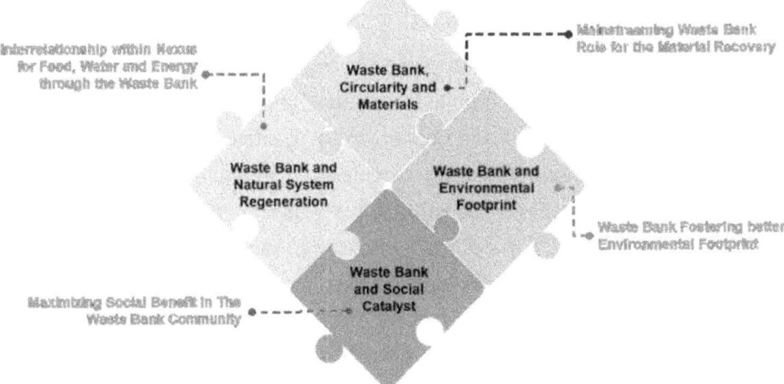

Fig. 12. The framework of Indonesia waste bank and circular thinking systems, divided into four relevant aspects. These categorize is selected as the recyclables and organic waste materials categories.

Fig. 12 displays four concepts of interrelationship from the transition of Indonesia WB for a more CE. The conception in this framework will be thoroughly explained in the following discussions. It is prominent to highlight that moving towards a CE would make a crucial contribution to environmental conservation and alleviate the climate crisis. According to Jakarta Globe (2020), Indonesia has already made some progress towards sustainable development, yet the government is trying to expand an integrated ecosystem for a CE. WRI (2020) stressed that organizations embracing circular business models would have a viable business added benefit as the GoI promoting recycling-based policies.

4.1 Waste Bank, Circularity and Materials

By leveraging its fundamental principles to gain a competitive advantage, the CE contributes to greater resource efficiency and sustainable development (Barros et al., 2020). Through MoEF (2019), the GoI campaign for separating waste at-a-source since 2019 was a strategic way to be implemented. Furthermore, the MoEF with Media Indonesia (2019) stated that the National Movement for Sorting Waste from Home seeks to increase the recycling rate, composting rate, and energy recovery rate. Another goal is to realize producers' obligations to reduce waste and encourage a CE in waste management. To improve waste processing capabilities, tangible actions are needed to change mindsets and lifestyles to achieve preferable WM and maintain environmental sustainability. Other efforts related to aspects that support a CE include providing fiscal incentives, policies to reduce scrap imports to increase the potential for domestic waste as raw material for the recycling industry, standardization of recycling products, and policies on using recycling products. Occasionally, if no industry becomes a permanent off-taker, accumulating recyclables at WB poses a risk. As a result, forming a partnership between recycling sector and WB is fundamental.

The waste sorting movement is a continuation of the waste-minimal movement that has been seen as massive in the community to ensure that waste can be sorted, collected, and transported to the processing and final processing sites. This process of waste sorting will enable factories and/or recycling industries to achieve more significant investments. Sorting household waste helps to drive the waste bank's operation. According to the Ministry of Industry (2020), the plastics recycling industry is currently developing in Indonesia, especially for PET and PP plastic types. The recycling rate for both is above 50%. The CE model has economic value for society and can support the industry for waste processing, where WM in Indonesia itself has been regulated in Law No. 18 of 2008. This applied CE replaces the linear economy that has been implemented in plastics WM.

The CE is supposed to achieve sustainability objectives through the effective and circular use of resources. In summary, recycling is indeed a key part of the CE. To maintain more waste materials more circular through Indonesia WB, the strategies proposed are to establish resource efficiency and CE holistic regulation of materials and elaborate with community-driven materials recovery. This approach involves local WB to strengthening the domestic demand for raw materials of products and recycling corporation/industry and providing transparent and mutual benefit EPR scheme of certain products associated in the WB. Besides, promoting an entire producer liability system will convince manufacturers to switch to the Performance Economy for economic reasons. WB could implement the return logistics loops where the business model arranged to back to the manufacturers of goods and/or back to the materials producers (Stahel and EMF, 2019). While the government incentives and awards for recovery products from the WB could enhance the producers industry's active participation to collaborate with the WB. Besides, communicating the CE within the WB and local society resulting in more circular flows.

4.2 Waste Bank and Environmental Footprint

Recycling is one strategy that can provide environmental benefits. Waste type mostly favorable sent to the WB is recyclable. The handling process of these relatively simple than that of organic or biowaste. In light of the CE's shift to food waste or organic-based waste, the WB's role in producing high-quality compost can be expanded. However, marketing this product is somewhat challenging as the public in urban settings is not interested in agriculture. Therefore peri-urban zone could be the best option to market the compost. While the government campaigned for urban agriculture and verticulture, some compost proportion may be purchased by urban society. Whereas a particular institution like the Ministry of Agriculture should facilitate the composting product initiative from the WB and accommodate them towards critical soils or infertile land in the provinces or country, the government can also directly purchase the compost in the WB association with reasonable prices.

Conclusively, the environmental benefit could be obtained by utilizing WB as local CE platforms with their various creative products, and the linkage with the recycling industries would foster the recovery rate of materials, thereby reducing waste in the dumpsite or river. However, not every WB, particularly in their early establishment, is ready to provide innovative environmental-friendly approach systems. As a result, industries or governments must provide training to WB organizers to improve their capacity and awareness. A post-supervision program to evaluate progress could be included in the training. They should also be aware of and concerned about those WB who are financially disadvantaged, and they should help them survive. The benchmark outlined above is critical for the WB to endure more and contribute to CE and sustainability.

4.3 Waste Bank and Social Catalyst

Residents' involvement determines the WB adaptation in each community and the WB program's long-term viability. As a result, management at the community level must be considered. The establishment of a WB will assist local authorities in empowering and managing community-based WM responsibly. Organizing the WB properly, comprehensively, and mutually beneficial society is a strong criterion because cooperation and partnership are required. Naim (2016) deliberated that efforts to fabricating tolerance and awareness could be conducted through dialogue, discussion, meetings, and cooperation among the community. It is obvious that the WB program is a collaborative action that is effectively and continuously implemented through the use of a model of collaboration centered on society (Amalia and Pramusinto, 2017). As a matter of fact, WB administrator has a significant and dominant role in building communication between the actors to build trust and mutual understanding. This generates legitimacy among the actors and dedication. Process transparency and inclusiveness participatory are the variables that determine how the society-centered collaboration model is implemented.

Collaboration between government, society, and producers is the WB program's goal to have a significant multiplier effect, both in WM and with a beneficial impact on society. It is expected that producers and the community's participation will stimulate the value of sorted waste and open a revenue stream for sorting waste. Participation effort is essential for carrying out community empowerment activities (Maryati et al., 2018). As a result, the central and local governments' efforts, particularly the village government, are required to increase community participation. One of these is the WB program's socialization of the surrounding community. When a high-level of involvement of society occurs, transfer-knowledge and development for CE could flourish accordingly. Information and communication are required to raise producer and consumer awareness of their responsibilities for materials during their service lives. The greater the amount of recycled material used in a product, the more prominently this performance should be advertised and publicized. This accelerates more engagements between WB and society and shaping a community-based activities economy. Each WB has different programs, depending on local wisdom and program plans regionally. Hence, the WB should be acknowledged by WB association or government, which maximizes mutual benefit and economic sharing.

4.4 Waste Bank and Natural System Regeneration

As a regenerative mechanism, the CE can have many positive consequences that enhance the quality of life, the environment, and the community. The CE avoids the use of non-renewable resources while retaining or improving renewable resources, such as returning indispensable nutrients to the soil to promote regeneration or by using renewable energy instead of non-renewable resources. As a CDWM platform, the WB has the potential to regenerate natural systems in its respective region. It is crucial to encourage more WB, and organic-based waste volume processed, contributing more towards nature and soil ecology. Besides composting and agricultural organic fertilizer support to the soils, the local WB could enable themselves through innovative agriculture infrastructure derived from recycled plastic materials. Moreover, some interlinkage with urban farming should be established. In rural settings, the community-waste bank-nature relationship can be built closely because farmers certainly need compost for their plants.

One of the larger issues emerging is that WB could extend their framework into comprehensive water management, as numerous WB cleaned the plastic (bottles) with water. Thereafter, tap water (after cleaning usage) should be filtered and safe to be given back to the outlet. Another innovation is protecting the water bodies from plastic pollution. More WB should soundly campaign to society about recycling and sorting plastics to the WB institution; thus, protecting our oceans and the livings underwater from any marine debris threats. WB works can also be developed in terms of energy sustainability to optimize the manure waste (where WB location near the cattle-farming territory) to produce biogas. In rural areas, by means of WB's existence through investment in the biogas installation, the harvested biogas can be utilized by the local people. These approaches eventually reduce people's dependency on non-renewable natural gas. In another circumstance, for the WB located nearby the paddy field areas, the WB can also exploit the abundance of rice husk waste as a stable energy source or combine with certain organic materials to produce preliminary products for RDF (refused-derived fuels), which may be transferred towards RDF station for input materials.

5 Conclusion

The establishment of WB in Indonesia shows the emerging trend during the past decade. Reflecting the more robust WM systems in the country. WB contributes in capturing recyclables materials which important as the input for manufacturing. Integration and implementation of more CE values in the WB could uplift more benefits, likewise the transition for waste-as-resources and renewable resources in the industrial sector through the recycling/upcycling process. The expansion under the four branches framework provided in this research can be used to fostering WB works in more circular pathways. The key to success is cooperation among the central and local governments, society, business sector, and academicians. The potential of WB units and its contributor is a strength for circular collaboration and sustainable development in the future. Indonesia WB should be coping with strategies that are in line with CE practices, life-cycle thinking and regenerative thinking. Some aspects, such as material circularity and sustainability, environmental footprint, social catalyst, and harmony, as well as natural system restoration and regeneration, are promising with the WB prominent role and social cooperation. These frameworks could be expanded through long-lasting CDWM and connectivity with manufacturers or related stakeholders, with a circular business model that eventually fosters resource efficiency.

6 Literature

Accenture.	2015	Creating Advantage in a Circular Economy - Waste to Wealth. Accenture Strategy; Accenture. accenture.com/_acnmedia/Accenture/Conversion-Assets/Dot-Com/Documents/Global/PDF/Strategy_7/Accenture-Waste-Wealth-Infographic.pdf
Agrofarm.	2020	Bank Sampah Salah Satu Penggerak Perekonomian Tanah Air. Agrofarm. https://www.agrofarm.co.id/2020/06/24710/
Asosiasi Daur Ulang Plastik Indonesia (ADUPI).	2020	Bank Sampah: Standarisasi Bahan Baku Beri Kepastian Usaha. ADUPI. http://adupi.org/berita/bank-sampah-standarisasi-bahan-baku-beri-kepastian-usaha/

Azwir, A.	2019	Analisis Pengelolaan Bank Sampah Mandiri Berbasis Partisipasi Masyarakat (Studi Kasus Desa Mororejo, Kecamatan Kaliwungu, Kabupaten Kendal). eprints.undip.ac.id. http://eprints.undip.ac.id/75287/
Barros, M. V., Salvador, R., Francisco do Prado, G., Carlos de Francisco, A., & Piekarski, C. M.	2020	Circular economy as a driver to sustainable businesses. Cleaner Environmental Systems, 100006. https://doi.org/10.1016/j.cesys.2020.100006
Gamaralalage, D., Premakumara, J., Soedjono, E., Kataoka, Y., & Fitriani, N.	2016	Central Bringing Excellence in Open Access Transition from Waste Management to Resource Management: A Potential of Waste Bank Program in Indonesian Cities. https://www.jscimedcentral.com/EnvironmentalScience/environmentalscience-4-1037.pdf
Jakarta Globe	2020	Indonesia Launches Circular Economy Initiative With Denmark, UNDP. Jakarta Globe. https://jakartaglobe.id/business/indonesia-launches-circular-economy-initiative-with-denmark-undp/#:~:text=Indonesia%20is%20preparing%20to%20create
Khair, H.	2019	Study on Waste Bank Activities in Indonesia Towards Sustainable Municipal Solid Waste Management (pp. 1–102). Graduate School of Environmental Engineering, The University of Kitakyushu. https://core.ac.uk/download/pdf/298622458.pdf
Lubis, R.	2018	Managing Ecopreneurship: the Waste Bank Way with Bank Sampah Bersinar (BSB) in Bandung City, Indonesia. in International Journal of Multidisciplinary Thought. http://www.universitypublications.net/ijmt/0703/pdf/P8RS37.pdf
Maitre-Ekern, E.	2020	Re-Thinking Producer Responsibility for a Sustainable Circular Economy From extended producer responsibility to pre-market producer responsibility. Journal of Cleaner Production, 286(1), 125454. https://doi.org/10.1016/j.jclepro.2020.125454

Maryati, S., Arifiani, N. F., Humaira, A. N. S., & Putri, H. T.	2018	Factors influencing household participation in solid waste management (Case study: Waste Bank Malang). IOP Conference Series: Earth and Environmental Science, 124, 012015. https://doi.org/10.1088/1755-1315/124/1/012015
Media Indonesia	2019	Gerakan Pilah Sampah dari Rumah dan Circular Economy. https://mediaindonesia.com/; media group. https://mediaindonesia.com/humaniora/273759/gerakan-pilah-sampah-dari-rumah-dan-circular-economy
Ministry of Environment and Forestry (Kementerian Lingkungan Hidup dan Kehutanan).	2018	Bank Sampah Tumbuhkan Sirkular Ekonomi Masyarakat. ppid.menlhk.go.id. http://ppid.menlhk.go.id/siaran_pers/browse/1667
Ministry of Industry (Kementerian Perindustrian).	2020	Sistem Pengelolaan Sampah Plastik Melalui Circular Economi. http://bbkk.kemenperin.go.id/page/index.php. http://bbkk.kemenperin.go.id/page/bacaartikel.php?id=hunjTGvOaFtgvh4M2_dZMHXRk-pioBnitbGvLR_AH6_w
Ministry of National Planning and Development Indonesia (Bappenas).	2021	The Economic, Social, and Environmental Benefits of a Circular Economy in Indonesia https://www.id.undp.org/content/indonesia/en/home/library/the-economic-social-and-environmental.html
Naim, N.	2016	Membangun Kerukunan Masyarakat Multikultural. Jurnal Multikultural & Multireligius, 15(1). http://jurnalharmoni.kemenag.go.id/index.php/harmoni/article/download/218/181
Rohmawati, D.	2015	Kewiralembagaan dalam Pengelolaan Sampah Berbasis Masyarakat di Bank Sampah Gemah Ripah, Badegan, Bantul. Jurnal Studi Pemuda, 4(2), 297–314. https://doi.org/10.22146/studipemudaugm.36814
Satori, M., Amaranti, R., & Srirejeki, Y.	2020	Sustainability of waste bank and contribution of waste management. IOP Conference Series: Materials Science and Engineering, 830, 032077. https://doi.org/10.1088/1757-899x/830/3/032077
Snyder, H.	2019	Literature review as a research methodology: An overview and guidelines. Journal of Business Research,

		104(104), 333–339. https://doi.org/10.1016/j.jbusres.2019.07.039
Stahel, W. R., & Ellen Macarthur Foundation.	2019	The circular economy : a user's guide. Routledge.
Sulami, A. P. N., Murayama, T., & Nishikizawa, S.	2018	Current Issues and Situation of Producer Responsibility in Waste Management in Indonesia. Environment and Natural Resources Journal, 16(1), 70–81. https://ph02.tci-thaijo.org/index.php/ennrj/article/view/106711
Widiyanto, A. F., & Rahab, R.	2017	Community participation in bank of garbage: Explorative case study in Banyumas regency. Masyarakat, Kebudayaan Dan Politik, 30(4), 367. https://doi.org/10.20473/mkp.v30i42017.367-376
WRI Indonesia.	2020	How to Build a Circular Economy | WRI Indonesia. wri-indonesia.org. https://wri-indonesia.org/en/blog/how-build-circular-economy
Wulandari, D., Utomo, S. H., & Narmaditya, B. S.	2017	Waste Bank: Waste Management Model in Improving Local Economy. International Journal of Energy Economics and Policy, 7(3), 36–41. https://www.econjournals.com/index.php/ijeep/article/view/4496
Wulandari, S., & Fajar Alam, P.	2018	The Use of Online Waste Management System on Bank Sampah Induk Bantul. Ecotrophic: Jurnal Ilmu Lingkungan (Journal of Environmental Science), 12(2), 186. https://doi.org/10.24843/ejes.2018.v12.i02.p08

A technical, economical, and environmental comparison of composting and anaerobic digestion of organic waste fraction of municipal solid waste in Sri Lanka.

Wijepala Abeysinghe Mudiyanselage Asanka Nuvansiri Illankoon, Sabrina Sorlini

Civil, Environmental, International Cooperation and Mathematical Engineering, University of Brescia - Via Branze 43, 25123 Brescia, Italy.

Abstract

Due to the expansion of urbanization and industrial development in developing countries, a huge quantity of municipal solid wastes (MSW) is generated every day. Sri Lankan MSW consists of a large fraction of organic waste and still, there is no well-developed MSW management system in Sri Lanka. Composting and anaerobic digestion (AD) have been identified as the more promising options to treat organic waste valorizing its energy and nutrient content. Both options vary in different aspects. The goal of the present study is to identify how both technologies are beneficial for people and the environment by doing a cost-benefit analysis and greenhouse gas emissions comparison for both scenarios. The present study proved that AD is much environmentally friendly than composting. Also, in economic and technical aspects composting is at the spearhead than AD.

Keywords

waste management, MSW, OFMSW, composting, AD, Sri Lanka

1 Introduction

The rate of municipal solid waste (MSW) generation is increasing rapidly worldwide due to population growth. This impact is high for many Asian countries due to their unplanned urbanization, and centralized economic activities that affect to increase urban population irregularly. Therefore, proper management of MSW is a momentous duty of the local authorities (LA) and other related bodies to ensure the protection of human health and environmental standards. However, facilities related to MSW management are far from the satisfactory level in many Asian counties (MENIKPURA, GHEEWALA AND BONNET, 2012; VISVANATHAN ET AL., 2004). As a developing country in South Asia, Sri Lanka (SL) also faces severe difficulties with regard to MSW management in many urban and suburban areas. According to the nearest data and statistics from Central Environmental Authority (CEA) in SL, approximately 11,163 tons of MSW was generated per day in 2019 and per capita, waste generation was 0.5 kg per day. KORAI, MAHAR AND UQAILI, 2016 have derived an equation (i) to estimate future waste generation.

$$Eq_{(MSW)} = Gr_{(MSW)} * Pq_{(MSW)} * n + Pq_{(MSW)} \qquad (i)$$

Where Eq $_{(MSW)}$, Gr $_{(MSW)}$, Pq $_{(MSW)}$, and n denoted as estimated quantity, growth rate (growth rate was considered as 0.45% according to the CEA database), present quantity, and during n years respectively. Figure 1 represents the waste generation each year. In 2025 the quantity of organic waste generation will reach nearly 25000 tons per day. The rapid increase of MSW generation is a challenging issue in SL due to an unsustainable management system and poor disposal methods, which create stringent environmental issues.

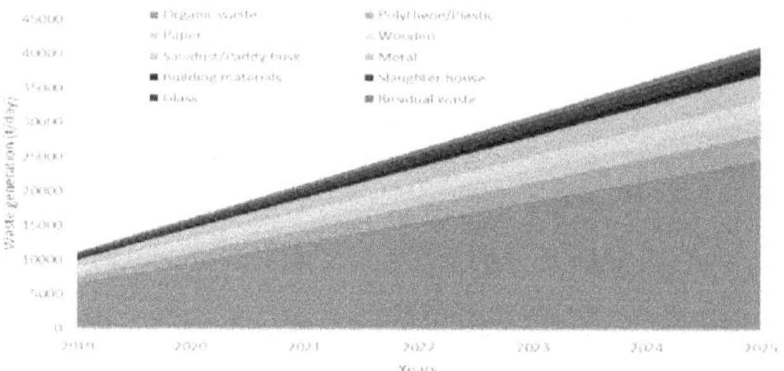

Figure 1 Projected MSW generation in Sri Lanka

The organic waste fraction of municipal solid waste (OFMSW) mainly consists of food and garden waste of the residences of the municipality. Waste collection methods and quantity of collection vary in each municipality (Figure 2). LAs currently collect only a portion of the waste stream, due to scarcity of resources and financial problems. Some LAs practice source separation and bring the wastes into the sites and sort into the material recovery. Also, composting and anaerobic digestion (AD) are being used by the LAs to manage OFMSW but on a smaller scale (KOTTE HEWA, 2018). These treatment methods have not been very successful in the country due to various reasons KOTTE HEWA, 2018). Therefore, this study evaluated two biological waste management scenarios, namely, composting and AD, to manage OFMSW in SL from an environmental sustainability perspective.

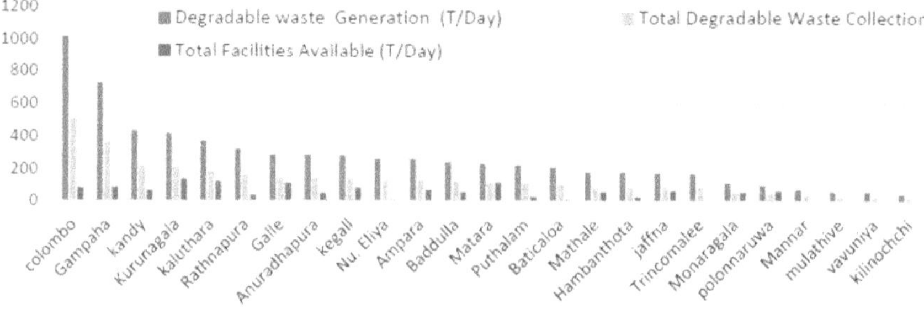

Figure 2 Comparison of Current composting facilities, current generation, and collection

2 Technical and Economic analysis

2.1 Technical and economic aspects of composting

Five different sizes of composting plants were analyzed for this study. Following data were collected from each of the plants through interviews of different shareholders and studying relevant environmental impact assessment reports. Such as capacity of the plant, input and output quantities, diesel usages, type of waste, and sale price. The technologies associated with composting are less compared to other waste management techniques. However, due to these fewer technologies, it can be easily adopted into low-income countries and their waste management systems.

Current Composting Technique in Sri Lanka

Pre-processing

Before starting composting, the waste usually has to undergo some pre-processing stages. The first stage is sorting – removing unnecessary materials from the MSW that are unable to compost. The second stage is size reduction – reducing the particle size of the input materials and the third stage is treating feedstock. Most LAs are responsible for maintaining composting sites in their region, according to their facilities from small-scale static pile compost to large-scale composting. However, the windrow composting technique is widely adopted around the country. The technique of composting is changed according to the type of waste and other environmental and socio-economic factors because, waste like household food waste and high moisture content organic waste cause some odor problems, some hygiene problems, and attraction of animals. Therefore, an outdoor windrow system is used to minimize the above problems. As well as indoor windrow technique is suitable for gardening and mixed MSW. However, the static piles' technique is not popular in Sri Lanka. Although, according to the information that we collected, most of the compost sites conduct their process in the following way (Figure 3).

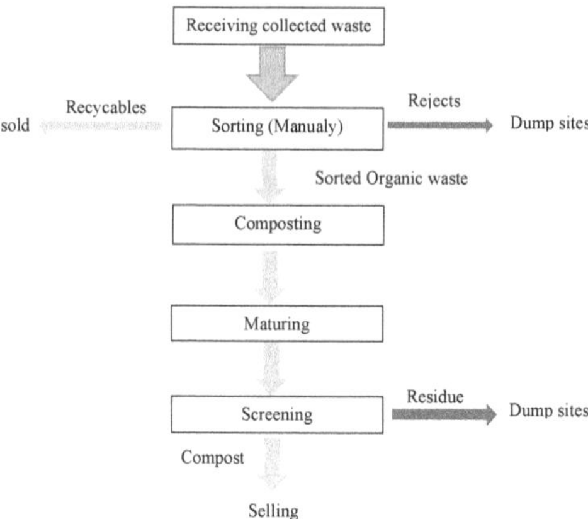

Figure 3 Process flow diagram windrow composting in Sri Lanka for mixed waste

Cost-Benefit Analysis for Composting

Financial analysis was conducted by accessing the data obtained from the Kaduwela Municipal Council in order to identify the tremendous benefits attaining from composting. For the analysis following cost categories such as initial capital for the building of the site, tools and other equipment required for composting, administrative cost (wages of workers, delivery, loading, unloading, sorting, pilling, electricity and water bills, etc.) the expenses for the open dumping landfilling were subtracted to find the real composting cost, compost production (expenses for adding nutrients, packaging costs) and expenses for the operating of the recycling products marketing center were considered. Also, the following benefits like revenue gaining from selling composting and revenue gaining from recycling products were considered. During the analysis, some fundamental assumptions were considered. Such as the lifetime of the composting site is assumed to be 25 years. The rate of waste production does not change within the considered timeframe, the cost of the land for both composting and dumping site is assumed the same and hence the cost of acquisition of the land is neglected for the analysis, government rules, regulations, and principals do not change throughout the project, and the other environmental benefits of the composting practice do not take into account due to lack of data. In addition to that other economic, environmental, and social benefits can be categorized as follows (Table 1).

Table 1 Considered costs criteria (Source: Kaduwela Municipal Council, Sri Lanka, 2018)

Parameters	Value (LKR)
Initial capital	17 million
Administrative cost	516000 per month
Compost production	
The capacity of compost production per day	1.33 MT
Packaging cost (per month)	30 LKR per 50 kg = 24000LKR
Price of composting and revenue (per month)	15 LKR per kg = 598500 LKR
Recycling	
Revenue from recycling products	60000 LKR per month
Sales center maintenance cost	5000 per month
Total income (Compost + recycling)	658500 per month
Simple Pay-back period considering waste management, composting, and recycling production	27 months

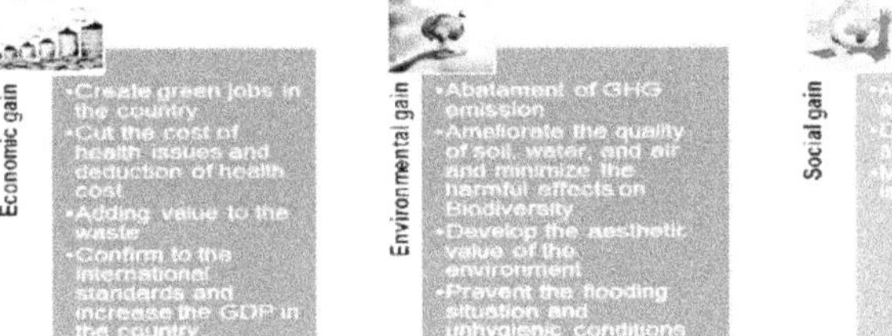

Figure 4 Economic, environmental, and social benefits of composting

2.2 Technical and economic aspects of anaerobic digestion

Following data were collected from each of the plant: capacity of the plant, input and output quantities, the ratio of water and waste, type of waste, and ownership. According to the gathered information and data, the main reason for adapting AD technology for waste management can be summarized as; avoiding the problem of improper waste assembling for household waste, prevention of the issue of bad odor caused due to earlier dung handling, escaping the problems of collecting of firewood as energy derived from AD can be used, lower the expenses of LP gas, and the slurry can be used for agricultural purposes.

Current AD Techniques in Sri Lanka

According to the intervied and literature data, Sri Lanka is being used four types of AD plants technologies, such as, Chinese fixed dome, plug flows, floating drum and Sri Lankan tunnel type. However, these technologies are depending on institutional and personal preferences. Five different sizes of AD plants were analyzed for this study. The capacity of the plant depends on the input waste quantities, in Sri Lanka AD plants are popular in livestock farms and it is still not common in general household practices. Even though AD plants have been established in other places according to the anlysed LA's data 74% of plants are established in livestock farms, 17% from the household, 4% from temples and other religious places, 3% of AD plants are established for commercial purposes and another 2% are established swine farm. However, their average capacity is varied from 8 m^3 to 20 m^3

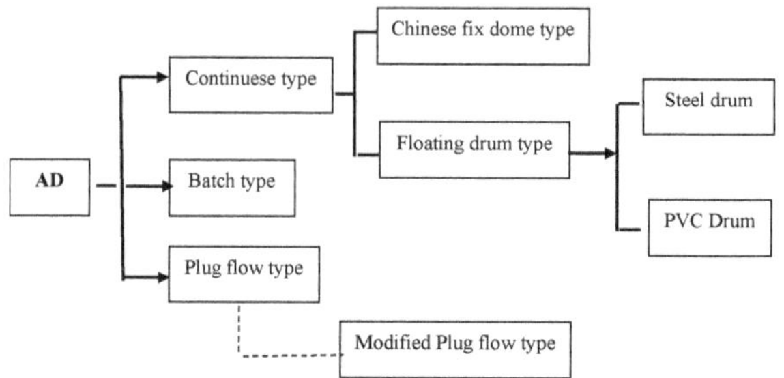

Figure 5 AD Systems practicing in Sri Lanka

Operational steps

As concerns waste and water mix in the inlet of the AD system, water and waste ratio is determined by experience, rather than measuring. Then the floating particles like glass and hay are eliminated. Generally, household waste is fed directly to the digester agitating, in order toprevent the formation of scum. Stirring is done using a rod. Rod is let in from various points from the inlet. However, practically stirring is not done by users even though the stirring rate is applied once per fifteen days. Monitoring the issues of the system enquiring gas pressure is necessary to save reactors from high gas pressure. High inner air pressure will make damages to the reactor walls and finally, it leads to occur leaks. Maintenance of frozen water inside the gas pipe is an issue, although water can be returned towards the reactor if the pipe is kept in a slope scum. Accumulation can be prevented by doing agitating. However, when deeply considered some site data and interviewed information, there is no clear relationship between the type of waste and the capacity of the system. Likewise, there is no definite relationship between the volume of waste and capacity. The likely reason could be AD digester capacity was not chosen concerning the category of waste and volume of waste. Most of the operators who worked in AD plants have declared that biogas leakage is a common problem and can be detected from the digester top. Leakage most likely resulted because of extreme interior air pressure or digester construction failure. During the data collection, we were able to find that few AD systems are not still generating energy, though feeding was done within the earlier three months. This may be caused as a consequence of insufficient microbial reaction because of inadequate introductory feeding. As well as customer complaints have arisen sometimes and have asked for repair regarding leakages of AD systems and AD systems that do not generate biogas. Although still could not give the service for repairing which indicating aftercare service is insufficient.

Cost-Benefit Analysis for AD

The cost-benefit analysis was carried out based on the information and data gathered from the interviews from the AD plant stakeholders who related to the AD process in Sri Lanka. According to survey results, AD plant cost is mainly dependent on the place, type of AD plant (technology), funding agent (Government, Non-Government Organization, or personal), and supply and demand.

Capital costs: The AD plant is able to manage any type of short-term organic waste in a particular area. According to the interviewed data and information, 40% of the cost goes to battery limits, buildings, and services. However, in Sri Lankan scenario land acquisition cost can be easily omitted from the list because all the biogas plants are established on organizational and personal perception. Therefore if it is a personally funded system, the AD plant can be established from their own land and if it is a government-funded program, government land can be used for the project.

Operating costs: Usually, each AD plant should assign 20% of its income for the operation costs. However, this portion should be enough to cover maintenances and repairs. These costs are also divided into two parts like fixed costs and variable costs. Fixed costs category consists of labor costs, routine equipment restitution costs, and maintenance costs. Furthermore, if the AD plant is at household-level operational costs are negligible. Because family members can conduct the process and not need to assign any amount of salary.

Benefits of AD Plant

The main economic benefit regarding AD is the cost of the fuel it replaces. Also, the benefits are depending on which products it generated and which fuel or which material can be replaced by the output of an AD plant. There is some significant benefits associate with a well-functioning AD plant. Such as benefit as fertilizer, benefit as a sustainable waste management practice, and maintaining circular economic concepts, Minimize the greenhouse gas impact. Also, there are some drawbacks are connected with the AD plant due to poor maintaining and low technological practice. Such as Methane Leakage and Water contamination.

Table 2 Considered costs criteria (Source: "Sirilak Umanga" biogas unit, Sri Lanka, 2019)

Description	**Value**
The total cost of the biogas unit	LKR. 14,250,000.00
Quantity of biogas production per month (if total input is around	4500 m^3

3000/kgs/day)	
This value is equivalent to 1800 kg of LP gas and taking the price of one kg of LP gas as LKR. 120.00 it has an LP equivalent value of 216000 per month.	
If the Gas is used to run a 50 kW generator and sold to the grid at LKR 20 per unit the monthly value is (4500 units per month X LKR 20)	LKR. 90,000.00
Simple Pay-back period considering electricity generation from the unit	158 months
monthly biogas slurry production 1500 liter x 25 days	37500 liters
Value of liquid fertilizer (Considering liquid fertilizer as LKR. 3 per liter) minimum	LKR. 112,500.00
Total income per month (electricity sale + slurry)	202500.00
Simple Pay-back period considering waste management, fertilizer, and biogas production	70 months

LP gas- Liquified petrolium gas, LKR-Sri Lankan Rupees

*** The value of CO_2 saving & reduction in free Methane emission is not accounted here, The existing cost of waste management is also not accounted in the calculations.*

2.3 Composting versus anaerobic digestion

Based on current practicess for both technologies,the better technique for handle collected OFMSW by LAs is composting when it is compared with AD as shown in Table 3. Specially it needs low initial capital and easy to handle.

Table 3 Comparison between composting vs AD of collected OFMSW.

	Composting		**Anaerobic digestion**	
Area	Need large area	cons	Need small area	pros
Technique	Simple	pros	Complex	cons
Product	Customer demand is high	pros	Customer demand is high	pros
	Handling is easy because of solid form	pros	Handling is difficult due to gas formation	cons
Labor power	Need more labor	cons	Need less labor	pros
Expense	Need low initial capital and maintenance cost	pros	Need high initial capital and maintenance cost	cons
Odor	Unpleasant	cons	Odorless	pros

Currently las do not collect all the amount of generated waste. Since, some parts are remaining in residence's houses. Therefore, household level AD is much more ideal waste management technique for onsite waste handling. Because, it saves money of purchasing energy and the slurry can be used as a biofertilizer (Table 4).

Table 4 Comparison between composting vs AD of non-collected OFMSW.

	Household composting		**Household anaerobic digestion**	
Output	Compost enriches soil condition and minimizes the requirement for chemical fertilizer	pros	Biogas can be used as a fuel for cooking, lighting, etc. and replace firewood	Pros
			The slurry can be used as a biofertilizer	Pros
Odor	Bad odor due to open process	cons	The odor is not an issue due to the closed operational process	Pros
Area	The low area is enough	pros	Need more space to construct the reactor	Cons
Process	Simple	pros	complex	Cons
Expense	Low initial capital for composting bin or barrel	pros	High initial capital for the construction of reactor, gas pipes, and other materials	Cons
	Save money on chemical fertilizer	pros	Save money on chemical fertilizer (due to bio-slurry)	Pros
			Save money of purchasing energy	Pros
Additional benefits	No extra benefits	cons	Minimize deforestation due to lower the demand for firewood	Pros

3 Greenhouse gas emission analysis

"Do-nothing scenario"

Murphy, McKeogh and Kiely, 2004 have defined that the do-nothing scenario for Greenhouse Gas (GHG) generation is landfilling of OFMSW. Around 65% of volatile is decomposed over time due to its slow digestion process. Table 6 is emphasized the GHG generation in the "do-nothing" scenario. It is assumed that the generated landfill gas dissipates into the atmosphere. Also, Murphy and McKeogh, 2004 have evinced that the generation of 9.2kgCO_2 equivalent per m^3 landfill gas is emitted due to this manner.

3.1 GHG emission analysis of composting

Complete aerobic transformation of OFMSW has defined as in the equation i (Tchobanoglous, Peavy and Rowe, 1985). The changes in the elemental composition during the decomposition process and final composition of the OFMSW are highly dependent on the process conditions (Liwarska-Bizukojc and Ledakowicz, 2003). As well, the generation of GHG quantity dependent on some physical factors such as moisture content, volatile solids, and destruction of volatile solid over time (Table 5).

$$C_aH_bO_cN_d + 1/4(4a + b - 2c - 3d)\ O_2 \rightarrow aCO_2 + 1/2(b - 3d)\ H_2O + dNH_3 \quad \text{(ii)}$$

Table 5 GHG generation during the decomposition process (Source: Data obtained from several EIA reports issued by CEA Sri Lanka)

Dry solids	**400 kg/t OFMSW (60% moisture content)**
Volatile solids	260 kg/t OFMSW (65%)
Decomposition of volatiles over time	169 kg/t OFMSW (65%)
Generation of GHG	169 m^3/t OFMSW
Generation of GHG	1,555 $kgCO_2$/t OFMSW

Typical municipal organic waste consists of Carbon, Hydrogen, Oxygen, Nitrogen, and Sulphur with the value of 48%, 6.4%, 3.76%, 2.6%, and 0.4% respectively. Also, 5% of ash is in their dry basis (TCHOBANOGLOUS, PEAVY, AND ROWE, 1985). Based on the data obtained from the CEA of Sri Lanka, OFMSW contains 60% of moisture. Therefore, the content of wet OFMSW can be defined as $C_{21.6}H_{125}O_{57.7}N$. Furthermore, the plenary composition of OFMSW described as;

$$C_{21.6}H_{125}O_{57.7}N + 23.25\ O_2 \rightarrow 21.6\ CO_2 + 61\ H_2O + 1\ NH_3 \quad \text{(iii)}$$

Based on equation 2, molecular weight can be balanced as follow.

$$1321.4 + 744 = 950.4 + 1098 + 17$$
$$2065.4 = 2065.4$$

It is summarized as 1t of OFMSW is equivalent to 719kg of CO_2.

However, complete decomposition does not take place at the compost plant and about 50% of the volatile decomposition takes place. Therefore, around 360kg of CO_2 is released into the environment by 1t of OFMSW at the compost plant. As well, CO_2 is released into the environment due to the utilization of diesel (2.69 $kgCO_2$/L) for numerous purposes in the composting plant. According to the data and information received from the different composting plants in Sri Lanka, they use only BobCat to mix waste and spend 30min to 1 hour per day. The volume of diesel consumption is varied from 0.2-0.5 liters per ton and the average value is taken as 0.35L per ton. Therefore, another 0.95~1 $kgCO_2$/t is added to the final value. Also, diesel consumption for the transportation of waste is excluded from the analysis. The comparison of analysis is represented in table 6.

3.2 GHG emission analysis of AD

According to the information gained from different AD plants stakeholders around 130m^3 of biogas generated from the 1t of OFMSW. Mainly, GHG emissions in the AD plant are from the combustion process. MURPHY, MCKEOGH AND KIELY, 2004, have indicated that this value as 1.96 $kgCO_2/m^3$ biogas production. As well as, degradation of compost in the land application was ignored for ease of comparison. Moreover, emission of biogas from the AD plant is considered as negligible with the well maintenance of the plant. In point of fact, this is a simplified GHG analysis, to allow comparison of AD and composting process in SL.

3.3 GHG analysis: Comparison of Composting and AD

Table 6 represents that an AD creates a positive impact on the environment than composting by replacing fossil fuel in the electricity generation sector. An additional benefit of using biogas as a cooking energy source is, it can replace LPG used for cooking. In general, biogas contains 55-70% Methane and 30-45% Carbon dioxide, and a small quantity of some other gases. Its flame temperature is between 870^{0}C-1500^{0}C and the calorific value is approximately 20 MJ/m^3. Whereas LPG has 47 MJ/kg which gives a replacement ratio of 1 LPG: 2.35 Biogas m^3. As well as, a well-maintaining AD plant gives multiple benefits rather than waste management like it provides a clean and low-cost fuel for cooking, generating electricity and have a higher potential to sell to the national grid and production of high-quality organic fertilizer.

Table 6 GHG emissions comparison of composting and AD

Description	**Composting CO_2 kg/t**	**AD CO_2 kg/t (electricity generation)**
Direction GHG	361	255
GHG generation in "do-nothing" scenario dissipation of landfill to the atmosphere	1555	1555
Net greenhouse gas production	-1194	-1300
GHG savings from displaced electricity[1]	0	103
Net GHG savings	1194	1403

[1] An average of 0.734tCO_2 produce in the generation of 1MW$_e$h; 140 kWeh is the net production per tonne of OFMSW.

4 Conclusion

The progress of both analyzed composting and AD plants looks to be dependent on adequate political engagement from relevant politicians as well as own commitment of the officers connected to the project. Progressive political assistance is crucial to achieving success in all tasks such as getting funds for initial capital, acquiring land for plants, and selling the end product. Lack of cooperation between officers in composting plants and the officers in the Ministry of Agriculture in regard to the quality of the produced compost. Most of the local authorities do not use their total organic fraction of MSW as the feedstock to their composting plants because still, they are able to sell only a small quantity of compost as there is no well-developed marketplace. This condition leads to send a large volume of MSW into landfills. Community participation in the segregation of waste sources is very low in some locations and can be significantly developed by arranging public awareness programs. Also, public awareness about the benefits of both technologies is not sufficient.

In 2013, Central Environmental Authority has launched a project named "PILISARU". Pilisaru gives financial supports directly to local authorities for building composting plants, buying specific instruments, and training employees. Although the government has invested large capital for these projects, government involvement is not sufficient for developing its standard for using agricultural purposes. Further, there is no well-arranged marketplace for selling the products. According to interviewed data, the inner and outer pipes of most AD plants are narrow and consequently, it leads to creating blocks of the gas supply. Many AD plants are suffered from issues related to sulfur dioxide and other sulfur products. This quickens the corrosion of cookers and also households have to keep cookers outside due to the unpleasant smell of biogas. Mainly, there is no proper storage system for the extra production of biogas in some areas.

4.1 Recommendation

- Enhance political responsibility as the main aspect of the favorable operation of both AD and composting projects.
- Both composting and AD plants should be constructed near the earlier waste dumping sites.
- It is necessary to give competent training for project officers and relevant employees regarding the technology (operation, maintenance, etc.) when building AD and composting plants.
- The value of composting and biogas should be promoted among people through arranging awareness programs. Stakeholders should be informed regarding the benefits of using these products and the risk of releasing methane into the atmosphere. Thus, create a market for these products.

5 Literature

Visvanathan, C., Trankler, J., Joseph, K., Chiemchaisri, C., Basnayake, B. and Gongming, Z., 2004. Municipal solid waste management in Asia. *Asian regional research program on environmental technology (ARRPET). Asian Institute of Technology publications. ISBN 974-417-258-1*, 974(417-258).

Korai, M., Mahar, R. and Uqaili, M., 2016. Optimization of waste to energy routes through biochemical and thermochemical treatment options of municipal solid waste in Hyderabad,

Menikpura, S., Gheewala, S. and Bonnet, S., 2012. Sustainability assessment of municipal solid waste management in Sri Lanka: problems and prospects. *Journal of Material Cycles and Waste Management*, 14(3), pp.181-192.

Kotte Hewa, P., 2018. *Study of organic solid waste management in Sri Lanka using centralized com-posting and household scale anaerobic digstion.* Doctorate. Hokkaido University, Japan.

Pakistan. *Energy Conversion and Management*, 124, pp.333-343.

Murphy, J. and McKeogh, E., 2004. Technical, economic and environmental analysis of energy production from municipal solid waste. *Renewable Energy*, 29(7), pp.1043-1057.

Murphy, J., McKeogh, E. and Kiely, G., 2004. Technical/economic/environmental analysis of biogas utilisation. *Applied Energy*, 77(4), pp.407-427.

Tchobanoglous, G., Peavy, H. and Rowe, D., 1985. *Environmental engineering*. Estados Unidos: McGraw-Hill Interamericana.

Author's address

Wijepala Abeysinghe Mudiyanselage Asanka Nuvansiri Illankoon
Civil, Environmental, International Cooperation, and Mathematical Engineering
University of Brescia,
Via Branze 43, 25123 Brescia (IT)
Telephone +393808929470
E-Mail: a.wijepalaabeysi@unibs.it

Separate Collection and Recycling of Waste as an Approach to Combat Marine Litter
WWF Pilot Project in the Mekong Delta, Vietnam.

Wolfgang Pfaff-Simoneit

WPS Consult UG (LLC)

Stefan Ziegler

WWF Germany

Ms Trinh Thi Long

WWF Vietnam

Abstract

The main causes of land-based marine litter are insufficient waste management systems inland in combination with increasing sales of single-use plastic products, mainly packaging. As part of its global commitment to the conservation of natural resources and biodiversity, WWF has been strongly involved in the international negotiations on marine plastic litter and played an active role in shaping the policy environment. WWF also wanted to demonstrate in a pilot project carried out in the Mekong Delta in Vietnam that a high financial contribution for covering the costs of the waste management can be generated by comprehensive recycling and utilisation of waste. Segregation at source is essential to ensure quality. However, experience shows, that this is not respected by all households to the desired extent. For this reason, the collection concept to be tested envisaged the immediate check of the sorting quality and post-sorting of the collected fractions when required. The pilot project has proven that a labour-intensive separate collection with immediate post-sorting is very suitable for private households and small businesses. The separately collected and directly post-sorted fractions have a very high-grade purity and quality. Although the costs of the labour-intensive collection are considerably higher, the total costs of waste management are significantly reduced due to the significant reduction of residual waste and thus the gate fees to be paid for their disposal.

Keywords

Labour-intensive separate collection; combatting marine litter; performance data; cost-efficiency; waste recovery effectiveness

1 Background and Objectives

Marine Litter, and especially plastic waste, has been identified as a major environmental issue during the last years. More than 80% of the plastics entering the sea is stemming from land-based sources. Rivers and waterways are the most important entry paths as they transport waste thrown or flushed into the rivers and dumped at their shores into the oceans.

The main causes of land-based marine litter are insufficient waste management systems inland in combination with increasing sales of single-use plastic products, mainly packaging. As part of its global commitment to the conservation of natural resources and biodiversity, WWF has been strongly involved in the international negotiations on marine plastic litter and played an active role in shaping the policy environment. In order to demonstrate decision-makers how the causes of marine littering can be combated in less developed countries, WWF launched a pilot project to improve the waste management services in Vietnam in 2018. The key to change the situation lies in improving the financial sustainability and the introduction of efficient waste management systems. In addition to these measures at local level, actions to prevent the generation of waste are required. The producers and distributors of plastic products have to reduce the placing of plastic into the economic cycle and to ensure orderly and reliable collection after use at their own cost.

A significant amount of plastic waste entering the oceans is stemming from only a few countries in South-East-Asia, of which Vietnam is one. The Mekong River is amongst the most important entry paths for marine litter worldwide. WWF has therefore selected the Province of Long An in the Mekong Delta as a model area. Long An is a typical province for this region with predominantly rural structures, some urban agglomerations and its capital Tan An. More than 80% of the 1.5 Million inhabitants live in rural areas, about 10% live in the capital Tan An.

To prepare the project, WWF has commissioned a feasibility study in which the present waste management system was analysed and various alternatives for the improvement of the waste management in Long An Province were developed and assessed. Key findings were:

- The central causes for the input of waste, especially plastic waste, into water bodies were identified to be poor or even absence of collection systems on the one hand, and the inadequate disposal of waste in unsecured dumps on the other. This is especially the case in rural areas. Only about two thirds of the population in the Long An Province are connected to a regular waste collection system.
- There is a high demand for all types of recyclable materials contained in the waste including for compost and compost products.
- At the local level, the main challenge is to ensure regular waste collection at reasonable cost. At the provincial level, environmentally sound disposal facilities are to be realised and the environmental standard of existing facilities needs to be improved, respectively.

The extensive recovery and recycling of waste proved to be the most suitable option in terms of both ecological and economic criteria. Segregation at source and separate collection are essential for such a strategy. WWF therefore wanted to test a collection concept in a pilot project that allows the highest possible added value to be generated from the waste in order to make a high financial contribution for covering the costs of the waste management system.

2 Collection Concept

The waste collection concept is the decisive component in a strategy aiming at resource recovery and high waste utilisation rates. Sustainable demand for secondary raw materials and products derived from waste and high revenues can only be achieved if high product quality is guaranteed. This is why waste should ideally be segregated at source by the waste producers to prevent the contamination of recyclables and organic waste with pollutants and impurities. However, experience in many countries shows, that the desired segregation at source is often not complied with by all participating households and waste generators, at least not from the beginning. It must be anticipated that a more or less large proportion of the population will only gradually understand and comply with the requirements of proper source segregation of waste. While the separation of recyclable materials is quite common for large parts of the population, the separation of organic waste and the knowledge, that it also represents a valuable raw material, is not very widespread.

The collection concept must take these weaknesses into account. For this reason, the collection concept to be tested envisaged the immediate check of the sorting quality and post-sorting of the collected fractions in the event that the households / waste producers have not sorted the materials well or not at all.

Post-sorting needs to take place before the waste is compressed in the collection truck. If the waste is compacted, organic waste and liquids are squeezed into the voids and contaminate the recyclable components. Especially sensitive materials such as paper and cardboard, but also plastics, packaging material and textiles, lose considerable market value as a result and can become even unmarketable. This is the very reason why the materials obtained out of compacted mixed waste are of significantly lower quality, which leads to sales problems and low or even no revenues. Organic waste comes into contact with contaminating substances such as glass splinters, cigarette butts, cotton swabs, etc. and with harmful pollutants. Especially heavy metals, which enter the waste via batteries, fluorescent tubes, electrical waste, etc., are easily dissolved due to the acidic decomposition phase, which organic waste undergoes very quickly, and thus contaminate the waste and lead to high heavy metal loads in the compost. Moreover, liquids containing pollutants such as waste oil, solvents, paints, varnishes or outdated medicines, cosmetics, etc. become mixed with the organic waste and contaminate the substrate if they are not collected separately.

By sorting waste before it is compacted, a high degree of purity of the raw materials is ensured. Thus, a high quality of the resulting products, which is particularly decisive to achieve high quality compost and high quality secondary raw materials. The quality of the materials is highest when they are kept separate. Therefore, segregation at source is the most effective measure to ensure quality. Post-sorting is a quality assurance step to prevent unsorted waste from being mixed with the pure materials.

Post-sorting can be either carried out by the waste collector directly during collection, or in a centralised sorting facility to which the waste is transported without being compacted. In Tan An the administration opted for 'sorting at cart'-model. The waste collectors were provided with trolleys equipped with a table on which waste receptacles provided by the households for collection can be emptied and waste can be sorted. Figure 1 shows a collection cart.

Figure 1: Type of collection cart used in the pilot project

The waste was collected in an alternating mode, i.e. the different fractions were collected on different days of the week. Alternating collection imposes a greater obligation on waste producers to pre-sort waste because only one fraction is collected at a time. Simultaneous collection, in comparison, offers greater flexibility and is easier for the residents to handle. They can make their waste available for collection as needed at any time. In summary, it can be concluded that the alternating collection is technically simpler but organizationally more demanding. The cost are more or less in the same range.

The decision on the collection concept – alternating or simultaneous collection - was taken based of practical considerations, since for an alternating collection the trolleys do not need to be separated in several compartments. However, the trolleys have to be equipped with extra bags to store the impurities sorted out by the collector from the main load.

3 Objectives of the Pilot Project

The objective of the pilot project is to test the practicability, effectivity and efficiency as well as social acceptance of the separate collection concept in practice. The planning assumptions were to be verified, strengths and weaknesses of the approach identified in order to optimise the concept and adapt best to the local framework conditions. To this end the following parameters were monitored and evaluated:

- Collection performance: The quantities of waste collected per collector and time unit are a decisive parameter for collection cost and route planning. The collection quantities was measured and evaluated for each collection tour, taking into account the settlement structure and the origin of the waste.
- Composition of the collected waste: The separately collected fractions were weighed separately and the composition of the fractions analysed at the beginning of the pilot project. Another waste composition analysis will be carried out at the end of the project.
- Willingness of the population to cooperate. The evaluation of this parameter helps to develop strategies for improving environmental awareness and the willingness of waste producers to cooperate.
- Interaction of the various actors and processes: The processes and roles of the different actors were to be tested and optimised, in particular cooperation with Women's Union or Neighbourhood associations who supported public relations and motivation of the waste producers to cooperate.
- Possibilities for integration and willingness of the informal sector to participate in this concept.
- Market situation for recycled materials and compost: Even if a high demand for secondary raw materials and compost was found in the Feasibility Study, the actual acceptance of the market for such products had to be tested. Measures to create and develop markets can be developed on this basis.
- Cost and potential revenues: The cost of the collection system represent the core criterion to decide on the roll-out of the collection concept. Potential revenues were to be taken into account in the cost assessment.

The pilot project forms the central basis for optimising the collection concept and its roll-out to the other quarters of the city and the throughout the Province of Long An.

The project was launched on 1 August 2020 with an official ceremony and public meeting. The first collection day was the 2 August 2020. This paper represents an interim report as the pilot project is ongoing till June 2021. The paper mainly reports on technical and economic parameter like collection performance, recovery rates and costs. The experiences on the socio-economic aspects of the project approach will be reported at a later stage.

4 Execution and Evaluation of the Pilot Project

4.1 Pilot Area

The pilot project is to carry out both in an urban and a rural area. It was decided to start with the urban area since it seemed to be easier to start a new collection model in an area where a waste collection system is already in place and people are used to provide waste at a certain date for collection. The quarter of Bin Dong 2 in Ward 3 of Tan An City was selected for the first part of the pilot project. The Number of participating households is 425, which corresponds to 1,855 persons.

4.2 Quantities and composition of household waste

In order to gain the basic data on the waste quantities generated and the waste composition, a one week waste audit campaign was carried out between 19 and 25 October 2020 in the Bin Dong 2 quarter. The waste generated from a total of 117 households with 434 people living in them was collected and analysed. This corresponds to about 25% of the total number of households / people living in Bin Dong 2.

The waste audit yielded the following findings:

Waste quantities	Specific waste quantity per household		1.7 kg/hh/d
	Specific household waste quantity per person		0.46 kg/cap/d
	Waste from business	20% /	0.12 kg/cap/d
Waste composition	Recyclable waste		20%
	thereof fully recyclable	*3%*	
	potentially recyclable[1)]	*15%*	
	not recyclable[2)]	*2%*	
	Organic waste		62%
	Residual waste		18%

1) Dirty 'nylon bags' (PE), milk and beverage boxes requiring specialised recycling technologies
2) Other plastics, dirty carton & paper

Figure 2 shows the composition of the total waste. Kitchen waste has the highest share with around 46%. Surprisingly (or indeed not surprisingly because waste is predominantly provided in plastic bags for collection) the second important component is so-called 'nylon bags', which consist mostly of PE, with almost 12% by weight. At third place come the coconut shells with almost 10% by weight. 7.5% consists of garden waste, mainly tree and bush leaves, branches and barks etc.

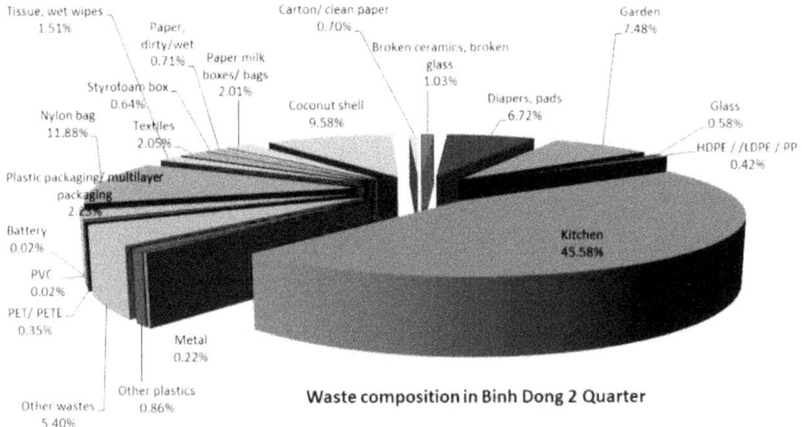

Figure 2: Household waste composition in quarter Binh Dong 2

4.3 Effectiveness of the separate collection concept

Table 1 shows the collection quantities and the recovery rate for the recyclables and organic waste. In total 60,201 kg of waste have been collected during this first eight weeks in the quarter Bin Dong 2, thereof 26,444 kg organic waste, 258 kg recyclables and 33,499 kg residual waste. The comparatively low quantities of separately collected organic waste are mainly due to the fact that the waste producers were instructed to throw coconut shells into the residual waste, as the operator of the composting plant does not have a suitable shredder for shredding the hard coconut shells.

The figures in Table 1 show that the waste quantities which would otherwise have to be disposed of at the landfill could be reduced by almost 45% from the very beginning. This so-called 'diversion from landfill rate' could have been even higher if households were instructed to put the coconut shells to the organic waste fraction. In this case it can be assumed that the diversion from landfill would achieve more than 50%. If also the dirty plastic bags and other potentially recyclables were recovered during post-sorting, the diversion from landfill rate would lie even at around 65%, as Table 2 shows.

Table 1 Waste generation and recovery rate Binh Dong 2

Waste generation and recovery rate Binh Dong 2					
Generated waste quantities			**Collected waste quantities**		**Recovery rate***
	[%]	[kg/d]	[kg/d]	[%]	
Total	100%	1.075	1.075	100%	100%
Organic waste	53%	570	472	43,9%	83%
Coconut shells	10%	108	-	0,0%	
Fully recyclables	3%	32	5	0,5%	16%
Potentially recyclables	15%	161	-	-	
Residual waste	19%	204	598	55,6%	

Table 2 Corrected waste generation and recovery rate Binh Dong 2

Corrected waste generation and recovery rate					
Generated waste quantities			**Assumed Recovery rate***	**Theoretically recovered waste quantities**	
	[%]	[kg/d]		[kg/d]	[%]
Total	100%	1.075	100%	1.075	111%
Organic waste	53%	570	83%	473	44,0%
Coconut shells	10%	108	83%	89	18,9%
Fully recyclables	3%	32	16%	5	0,5%
Potentially recyclables	15%	161	80%	129	12,0%
Residual waste	19%	204		379	35,2%

The utilisation of coconut shells in the composting process is basically possible and has very probably a positive influence on the composting process, as it forms structural material that allows better aeration of the compost heaps. Although the coconut shells are not completely degraded in one composting cycle, this is not detrimental as the mature compost is screened at the end, and the shells that have not fully decomposed are recovered and returned back to a new composting cycle. In this way, the share of the scarce structuring material is increased. If necessary, the coconut shells could first undergo a separate treatment to wash out tannins that could hinder biodegradation. However, this needs to be investigated in the course of composting trials to develop the optimal procedure.

Although the so-called 'nylon bags' (which is mostly PE) are very dirty, as they are mainly used to package the waste, they could in principle be recycled if the processing plant is equipped with an appropriate washing system.

Milk and beverage carton boxes are basically also recyclable, provided the recycling plant is equipped with respective process steps. However, the extent to which such plants are available in Vietnam or the materials are purchased from traders needs to be clarified in further investigations.

The very high recovery rate of more than 80% of the organic waste generated proves that the approach chosen to collect these materials separately is very effective. In contrast, the poor recovery rate of only 16 % for fully recyclables is absolutely unsatisfactory. The reasons for this can presently only be presumed. But even if the surveys conducted cannot conclusively explain this low value, is highly likely that households have maintained their habits and sold the recyclables to itinerant collectors from the informal sector or leave it to them for social reasons.

In view of the very low quantities of separately collected recyclables and the weak recovery effectiveness it was decided, to reduce the number of separately collected fractions to two – organic waste and residual waste.

4.4 Collection performance

As a basis for the economic evaluation of the collection concept, the collection performance was examined. For this purpose, the waste collectors were accompanied over a period of one week and the time spent for the individual work steps was stopped and recorded. The workflow comprises the following steps:

Collection

- Picking up of waste receptacles
- Emptying on sorting platform, checking content and sorting
- Bringing receptacles back
- Recording / Taking notes
- Maybe instruction of households
- Going to next household

Once the trolley is filled:

Transport and recording

- Going to transfer point
- Weighing of trolley, recording of collection quantity
- Going back to collection area

Table 3 shows the average time needed for the collection processes separately for organic and residual waste. The time required for the collection of organic waste is about 50% higher as for the collection of residual waste. This is essentially due to the more time consuming post-sorting of organic waste.

Table 3: Collection time per household

Waste type	**Collection time [s] per household**					
	Pick up	**Empty-check-sort**	**Bins back**	**Record**	**Instructions**	**Total**
Organic	13.05	35.33	0.30	10.89	1.91	60.15
Residual	15.10	11.47	-	9.53	3.37	39.47

The time needed to get to the collection area, to transport the full trolleys to the transfer point and back to the collection area depends on the walking distances and thus on the housing density, as well as on the location of the transfer points. A general planning value for the time required for these work steps can therefore not be given. In the pilot project about 1/3 of the daily working time was spent for going to the collection area, transport, weighing and recording of quantities while about 2/3 of the daily working time were used for collection activities.

Table 4 lists the preliminary planning parameters which can be derived from the time and performance measurements carried out in Bin Dong 2.

Table 4: Preliminary planning parameter

Parameter	**Organic waste**	**Residual waste**
Collection time per household	60 seconds	40 seconds
Walking speed with empty trolley	1.13 meter/second / 4 km/h	
Walking speed with full trolley	0.95 meter/second / 3.5 km/h	
Collection performance net (related to pure collection time)	1.85 kg/min/person 110 kg/h/person	7.0 kg/min/person 420 kg/h/person

The average walking distance from and to the collection area was 306 m, the average transport distance was 286 m. The average velocity for walking to the collection area and pushing the empty trolley was 1.15 m/s, when transporting the full trolley the average velocity was 0.96 m/s. The net collection performance, i.e. the quantities collected per pure collection time, was 1.85 kg/min for organic waste and 7.09 kg/min for residual waste. The brut collection performance, i.e. the collected quantities per total working time, was 1.64 kg/min for organic waste and 3.39 kg/min for residual waste.

The average collection quantity per collection round was 125 kg for organic waste and 90 kg for residual waste. Together with the tare weight of the trolley of about 30 kg, the collectors had to push a weight of up to 150 kg and more. It becomes obvious that this is only possible in flat areas and with paved paths and alleys. In areas with uphill and downhill gradients, as well as in poor path conditions, hand-pushed collection trolleys are not suitable. In such situations motorized trolleys or small pick-ups are imperative. At the same time, motorising the collection vehicles would increase collection performance, as the time required to drive distances would be significantly reduced. In addition, larger quantities could be loaded per tour, which would further improve collection performance.

4.5 Cost of separate waste collection and overall cost balance

The measured collection performance allows to determine the costs of separate waste collection with the currently applied system. The calculation is based on the following cost data:

- Average salary of a collector: 10,000 kVND/cap/month / 370 EUR/cap/month
  120,000 kVND/cap/year / 4,450 EUR/cap/year
- Specific personnel cost: 343 kVND/cap/day / 12.70 EUR/cap/day
  43 kVND/cap/hour/ 1.60 EUR/cap/hour
- Equipment: Investment cost: 5,500 kVND / 204 EUR
  Depreciation, maintenance: 1,397 kVND/a / 51.70 EUR/a

In Table 5 the costs for collection, transport and disposal for the current system and the new collection concept are compared. The cost of the present system were provided by Urenco Tan An, which is the municipal waste management company. For the cost comparison different scenarios are considered:

- Pre-test data: Cost calculation based on the collection quantities and recovery rates measured during pre-test in quarter Binh Dong 2
- Scenario 1: The theoretical collection quantities and recovery rates based on corrected waste fractions in Table 2 are used
- Scenario 2: A slight increase of the recovery rate for organic waste is assumed
- Scenario 3: New transport trucks with a loading capacity of 20 m³ are utilised for transport to the disposal facility

Table 5: Cost comparison with current collection system

Cost comparison [kVND]		Share	Primary collection [kVND/t]	Transport to disposal facility [kVND/t]	Disposal / Gate fee [kVND/t]	Cost per ton [kVND/t]	Total cost [kVND/t]
Current system*	Residual waste	100%	258	452	400	1,110	**1,110**
Pre-test data	Organic waste	44%	454	452	0	906	**979**
	Residual waste	56%	184	452	400	1,036	
Scenario 1	Organic waste	55%	454	452	0	906	**965**
	Residual waste	45%	184	452	400	1,036	
Scenario 2	Organic waste	60%	454	452	0	906	**906**
	Residual waste	40%	184	452	400	1,036	
Scenario 3	Organic waste	60%	454	350	0	804	**868**
	Residual waste	40%	184	380	400	964	

* Cost data provided by Urenco Tan An

The shaded boxes mark the parameters changed in the respective scenarios. In all scenarios it is assumed that no gate fee has to be paid for the organic waste, as it is a raw material that serves as the basis for the production of a saleable product - namely compost. The cost analysis for the composting process shows that the production of compost is highly profitable even when assuming high production cost and low revenues. This means that the operator of the composting plant not only does not need a gate fee for organic waste to operate the composting plant profitably, he could even pay for the raw material.

The transport costs in Scenario 3 are calculated under the assumption that Urenco Tan An invests in the acquisition of new, efficient transport vehicles. Although the investment costs are very high for a waste compaction truck for transporting residual waste and a Rotopress truck for organic waste, the specific transport costs are significantly lower than for the current transport, which is mostly carried out with small trucks having only low loading volumes of 10 m³.

5 Preliminary Conclusions

The experience with the pilot project and the measurements carried out so far allow the following preliminary conclusions:

- The approach of labour-intensive separate collection with immediate post-sorting is very suitable for private households and small businesses. High collection and recovery rates of over 80% are achieved in relation to the potential contained in the waste.
- The separately collected and directly post-sorted fractions have a very high-grade purity and quality.
- Although the costs of the separate collection combined with immediate post-sorting are considerably higher compared to the currently applied collection system, the total costs are significantly reduced in comparison to the currently applied system due to the significant reduction of gate fees to be paid for disposal. Proceeds from the sale of the separately collected materials are not yet included in the cost calculation.
- The approach is not suitable for larger waste sources such as schools, hotels, large restaurants, administrations, hospitals and the like. If the amount of waste generated by a location exceeds about 10 times the volume of a household (household equivalent), the use of handcarts with direct post-sorting is not practicable and efficient. For such large sources of waste, a specific collection is to be organised with larger, motorised vehicles, but without compaction, e.g. small pick-up vehicles. The post-sorting is to be carried out in a nearby small sorting plant.
- Handcarts are only suitable in flat areas and on paved roads and for short distances. Motorisation of the trolleys is generally desirable due to the high load weights. In rural areas, the use of motorised collection vehicles is indispensable.

6 Literature

WWF Vietnam / Anh, T.G., Hang, N.G. | 2020 | Waste Audit Report Bin Dong 2, Ward 3, Tan An City, Long An Province (not published)

Infrastruktur & Umwelt | 2019 | Development of a Waste Management Concept for-Long An Province / Vietnam, Final Report (not published) available on request from the authors

Pfaff-Simoneit, W. | 2010 | Sectoral Approaches in Solid Waste management to link development and climate change mitigation;

		ISWA World Congress 2010; http://www.iswa.org
Pfaff-Simoneit, W.	2013	Development of a sectoral approach to establish sustainable SWM systems in developing and emerging countries in light of climate change and increasing shortage of resources' (edited in German language). Doctoral Thesis. ISBN 978-3-86009-203-3
Pfaff-Simoneit, W.	2017	Adapted selective waste collection concepts for developing and emerging countries, Waste-to-Resources 2017

Author's addresses

Dr.-Ing. Wolfgang Pfaff-Simoneit
WPS Consult UG (LLC)
Darmstadt | DE
Tel. +49 (0)6151 9698 185
Mob. +49 (0)160 8369 408
Email: Wolfgang.pfaff-simoneit@t-online.de

Dr. Trinh Thi Long
Wetland coordinator and Fresheater Practic technical advisor
HCMC Office, WWF-Vietnam
Tel: +84 2839321874
Fax: +84 2839321847
Mobile: 0903198528
Email: long.trinhthi@wwf.org.vn

Dr. Stefan Ziegler
Senior Referent Asien
Senior Conservation Advisor Asia
WWF Deutschland
Frankfurt/Main | DE
Tel.: +49 69 79144-290
Mobile: +49 151 18854993
Email: stefan.ziegler@wwf.de

Solid Waste Management Decision Making for Developing Countries through Life Cycle Assessment Tool: A State-of-the-Art Review

Ketan V. Shah[1], Melanie L. Sattler[2]
University of Texas at Arlington

Abstract
Solid waste management (SWM) in developing countries is an exigent and emerging problem. Improper decision making has led to great loss of human lives, financial budget crunches and environmental impacts. To avoid such problems, several researchers have attempted to develop solid waste management assessment methodologies frameworks, and life cycle assessment models to support tactical and strategic decision making.
These methodologies can help decision-makers choose among collection systems, waste processing technologies, and disposal options, in terms of cost and environmental impact.
Therefore, the objective of this state-of-the art review is to assess SWM decision support methodologies/models for potential application in developing countries. Published papers from 2012 through 2018, as well as the most common LCA models for SWM, were identified and analyzed.
Out of twenty-six independent tools studied, six tools were short-listed for a more detailed comparison. Despite various tools available globally, for life cycle assessment for SWM, none of them contains waste processing technologies, default values, and other critical parameters for developing countries. A tool is needed, for example, which allows less than 100% waste collection efficiency, vermi-composting as one of the major processing technologies, and a regional landfill approach.
Keywords: Solid Waste Management, Life Cycle Assessment, Decision-Support tools, Decision making Model.

1. Introduction

Urbanization has increased in speed and scale in recent decades, with more than half of the world's population now living in urban areas [1]. By 2050, urban dwellers probably will account for 86% of the population in developed countries and 64% of the population in developing countries [2]. Rapid urban population growth has resulted in several land-use and infrastructure challenges, including municipal solid-waste (MSW) management. The World Bank in 2012 estimated that 340 million metric tons of MSW were landfilled annually, and this may grow to 570 million metric tons by 2025 due to increases in population, urbanization, and economic development. National and municipal governments often have insufficient capacity to meet the growing demand for solid-waste management services [1]. Solid-waste management is already the single largest budget item for many cities ([3], [3,a]).

Not only will cities globally need to find ways deal with increasing volumes of waste in the coming decades, but cities in developing countries will need to find methods that are more protective of human health and the environment. In developing countries, open dumpsites are the most common method of disposing of waste. 540 million metric tons of solid waste in developing countries was dumped in the open annually in 2018; this may increase by 70% over the next 30 years. Dumpsites have been linked to many harmful health effects, including skin and eye infections, respiratory problems, vector-borne diseases such as diarrhoea, dysentery, typhoid, hepatitis, cholera, malaria and yellow fever, high blood lead levels and exposure to heavy-metal poisoning [6,a].

Decision-making regarding SWM in developing countries is made more difficult due to the fact that data on waste generation and composition are largely unreliable and insufficient; rarely does the data contains system losses or any informal sector activities [7]. Without proper data, it is difficult to design sound strategies or to make wise budget decisions on waste management ([6, a],[8]).

As developing countries attempt in the coming decades to overcome the challenges of growing volumes of MSW, unsustainable open dumping practices, and lack of waste-management data, decision-support tools can be valuable.

Therefore, the objective of this review is to assess SWM decision support methodologies/models for potential application in developing countries. Published papers for SWM- LCA models over a period from 2012 through 2018, as well as the most common LCA models for SWM, were identified and analyzed. In addition, the effect of critical variables affecting solid waste management will be evaluated. Areas of contention and uncertainties will be discussed and recommendation for future research are provided.

2. Methodology

2.1 Review of SWM LCA studies

LCA studies of MSW, published from 2001-2018, were collected from various databases such as Elsevier, Science direct, ResearchGate, ASCE Database, and Google Scholar. Important variables influencing environmental impacts of SWM were identified.

As shown in Fig 1 below, it was found that around 64% of the studies worldwide use LCA models to simplify MSW management and for the calculation of environmental emissions, benefits and burdens. SimaPro was used in the highest number of studies. A model's suitability depends on its goal, availability, accessibility, availability of documentation, clarity and transparency in working procedures, language of the study, computer cost, database availability, system language, research aim, and user preference. 56 studies used equations rather than models for LCA evaluations.

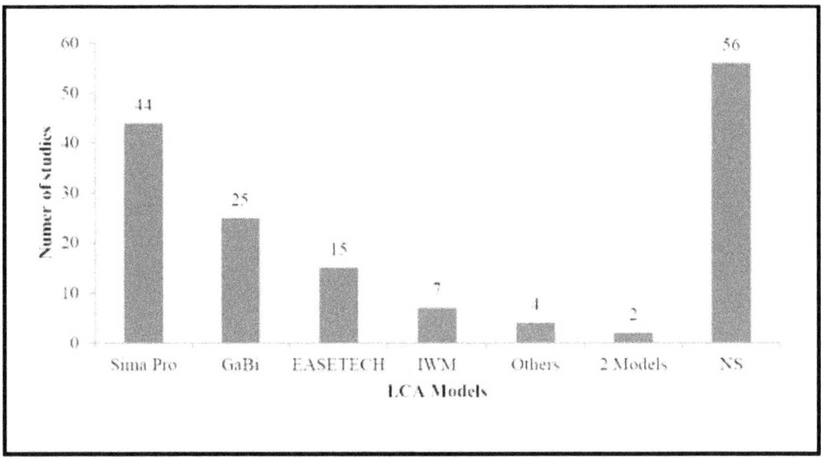

Fig 1. *MSW- LCA models worldwide used in studies 2012-2018.*

Source: (Yadav and Samadder, 2018a)

Most of the LCAs conducted for Asia used computer models. In the studies conducted by Aye and Widjaya (2006) and Buttol et al. (2007), the clear guidance for selecting a particular LCA model is available. It was observed that 24% of the LCAs in Asian countries used SimaPro as shown in Figure 2 below.

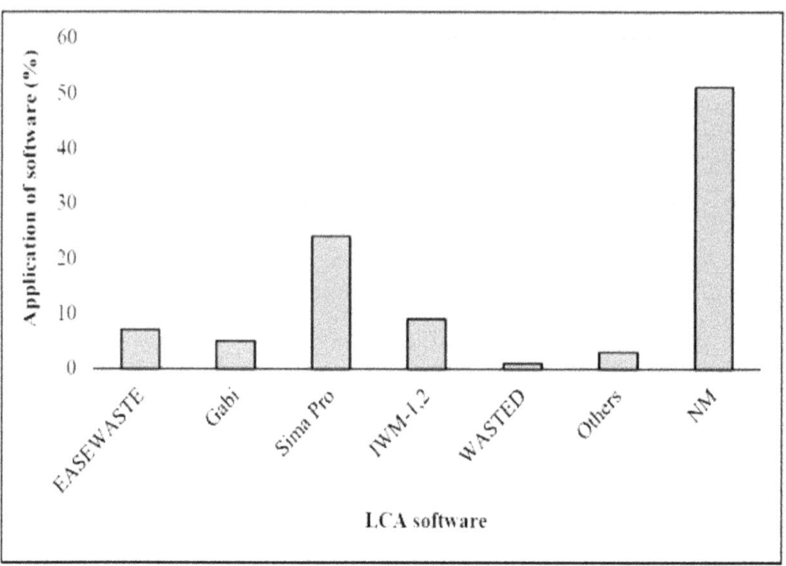

Fig 2. *Distribution of the type of computer model (software) used (%) in LCA studies on SWM in Asian countries. NM: Not mentioned, case -study based.*

Source: (Yadav, 2018)

Indian Scenario: Only 5 studies on LCA of MSWM since 2011 have been conducted in the biggest democratic country of the world, India. There was no study record before 2011. A detailed comparison of the studies is shown in Table 1 below.

Table 1. *Detailed comparative analysis for 5 LCA studies for India*

Sr. No.	Reference	Software (SW) & Methods	Scenarios (S) or Options	Impact Assessment of Parameters	Results
1	**Bohra et al., 2012**	FU- 1t of MSW SW- NM Method-NM	S1. 22% of MSW to aerobic composting and remaining 78% to sanitary landfill S2. 10% to aerobic composting facility, 12% of MSW to RDF plant and disposal of remaining 78% to sanitary landfill S3. 22% of MSW to biomethanation and remaining 78% to sanitary landfill S4. 12% of MSW to RDF, 10% to biomethanation and disposal of remaining 78% to sanitary landfill S5. 100% to sanitary landfill S6. 33% of MSW to biomethanation and remaining 67% to sanitary landfill S7. 18% of MSW to biomethanation, 15% to RDF and disposal of remaining to sanitary landfill S8. 22% of MSW to mass incineration remaining 78% to sanitary landfill S9. 22% of MSW to biomethanation, 17% of MSW to composting and remaining MSW to sanitary landfill S10. 11% of MSW to RDF palletization, 11% to aerobic composting, 11% to biomethanation and disposal of remaining MSW to sanitary landfill S11. 16% of MSW to RDF palletization, 10% to aerobic composting, 16% to biomethanation and	GWP	The results showed that the scenario with maximum diversion from sanitary landfill had least GWP.

			remaining MSW to sanitary landfill		
2	**Babu et al. (2014)**	FU- NM SW- NM Method-NM	S1. Open dumping S2. Landfill without LFG S3. Landfill with LFG S4. Bioreactor landfill system	GWP AP POP EP EC	The result found that bioreactor landfill option was better option among the four scenarios.
3	**(Sharma and Chandel, 2017)**	FU- 1 ton SW- Gabi6.0 Method- IPCC	S1. Open dumping and bioreactor landfill S2. MRF and sanitary landfill S3. MRF, composting and sanitary landfill S4. MRF, AD and sanitary landfill S5. MRF, composting, AD and sanitary landfill S6. MRF, composting and incineration S7. MRF and incineration	GWP EP AP HTP	The study found that S5 (MRF, composting, anaerobic digestion and sanitary landfill) had lowest environmental impact
4	**Yadav and Samadder (2017b)**	FU- 1 ton SW- Simapro Method- CML 2 baseline 2000	S1. Collection and transportation S2. Baseline scenario S3. Composting and landfilling S4. Recycling, composting and landfilling	GWP MAE FAE EP AD (fossil fuel) AD AP ODP POP HTP TE	The study found that S4 was the best scenario for the study area due to low environmental impacts.
5	**Harshit Khandelwal and Sunil Kumar et al., (2018)**	FU- 1 ton SW- Gabi 8.5.0.79 model	S1. composting combined with landfilling S2. MRF & composting combined with landfilling	GWP EP AP ADP POCP	The study concluded that S2 was the best scenario in

		Method: CML-1A impact characterization	S3. MRF & anaerobic digestion (AD) combined with landfilling S4. MRF, AD & composting combined with landfilling	HTP	terms of least environmental impact

AD- Anaerobic Digestion; ADP-Abiotic Depletion Potential; AP- Acidification Potential; EC -Energy Consumption; EI- Environmental Impact; EP- Eutrophication Potential.

FWE-Fresh Water Aquatic Ecotoxicity; GWP- Global Warming Potential; HTP- Human Toxicity Potentials; MAE- Marine Aquatic Ecotoxicity; NM- Not Mentioned; ODP- Ozone Layer Depletion Potential; POP; Photochemical Oxidation Potential, (POCP) Photochemical ozone creation potential.

2.2 Review of LCA tools for SWM:

2.2.1 *Collection of LCA SWM tools and narrowing to a short-list*

An extensive literature search was carried out for the 26 LCA software tools for SWM (Appendix A, Table B). Most of these tools include all four phases of the life cycle (raw material acquisition, manufacturing/construction, use, and end-of-life).Since waste management deals with end-of-life, a few specialized tools were also identified, which focus only on this phase (e.g., WRATE, MSW-DST).

Of the 26 SWM LCA tools identified, tools which give life cycle inventory numbers (LCI, e.g. tons of pollutant emitted) as the final output (like EPIC-CSR, MIMES, WASTED), without including an impact analysis phase, were eliminated from further evaluation. Databases used for LCIs (e.g., Impact 2002+) were by themselves eliminated from further evaluation. Also, tools like SSWMSS, MSWI, MIMES, WISARD, which do not have enough information for vetting and analysis (because of dearth of data available and lack of updates since development), were eliminated from further screening steps. The following eight tools were thus selected for a second level of evaluation: WRATE, MSW-DST, EASETECH/ EASEWASTE, HOLIWAST, WARM, WAMPS, SWOLF, and SWEET.

2.2.2 *Detailed evaluation of the short-listed tools*

The LCA tools identified for screening as described above were considered for detailed evaluation. The identified tools were scrutinized based on the year of development, documentation available, pollutants evaluated, updated emission factors, technical support available, and applicability to various geographical locations. The selection criteria are listed in Table A, of the Appendix.

The eight tools listed were further evaluated to select up to six tools for more detailed evaluation. Waste fraction can be changed in most of the tools selected for evaluation, so that it can be applied to different locations. The US tools developed with this criterion were selected viz: (WARM, MSW-DST, SWOLF). From the remaining five tools, EASETECH, WRATE,(European models) and SWEET (US model) were determined to provide the most flexibility and application variability (e.g., collection, processing, material recovery, recycling, etc.), while the remaining two tools HOLIWAST and WAMPS did not possess such flexibility in application (i.e.., user cannot chose the recycled fraction of the waste, or the waste percent recycled) therefore, SWEET, EASETECH and WRATE, were selected for further screening.

2.2.3 *Comparative Analysis of LCA Models/Software Tools*

Table 2 compares major features of the 6 short-listed MSW-LCA tools, including material categories, outputs, and impact categories. The four US models are available free; the two European models cost considerable amounts. All six have built-in data defaults (e.g. emission rates and capital/operating costs) based on US or European values. WARM is the only model that includes all 3 categories of MSW, construction and demolition debris, and electronic waste. EASETECH includes the most impact categories, and MSW-DST includes the second most.

Table 2. *General information about MSW-LCA short-listed tools*

Consideration	WARM [17]	MSW-DST	SWOLF	EASETECH	WRATE	SWEET
Procurement Cost	Free	Free	Free to non-commercial use. Educational version available.	€ 5,000 (Approx. $5,533)	£1,400/yr. (Approx. $1,807/year)	Free
			Cost for commercial use is not yet determined.	(Approx. $5,700)	(Approx. $1,800/year)	

Version and Year[4]	13 (2015)	1.0 (2002)	1st (2015)	2.0.0 (2014)	3.0.1.5 (2014)	1.0 (2017)
Country/Region of Materials and Management Strategies Modeled	**U.S.**	**U.S.**	**U.S.**	**Europe**	**Europe**	**U.S.**
Material Categories						
MSW	✓	✓	✓	✓	✓	✓
CDD	✓	-	-	-	-	-
Electronic Waste	✓	-	✓	-	✓	-
Tool Outputs						
Simultaneous Comparison of Multiple Scenarios	-	-	-	-	✓	✓
Process-specific Emissions		✓	✓	✓	✓	✓
Construction, Operating, and Maintenance Cost Estimates	-	✓	✓	-	-	-
Impact Categories						
Global Warming	✓	✓	✓	✓	✓	✓
Ozone Depletion	-	-	-	✓	-	-
Human Toxicity—General	-	-	-	-	✓	-

Human Toxicity—Carcinogenic	-	√	-	√	-	-
Human Toxicity—Non-Carcinogenic	-	√	-	√	-	-
Ionizing Radiation	-	-	-	√	-	-
Smog Formation	-	√	√	√	-	-
Eutrophication	-	√	√	-	√	-
Freshwater Eutrophication	-	-	-	√	-	-
Marine Eutrophication	-	-	-	√	-	-
Ecotoxicity	-	√	-	√	-	-
Freshwater Aquatic Ecotoxicity	-	-	-	-	√	-
Depletion of Abiotic Fossil	-	-	√	√	-	-
Developing country scenario	-	-	-	-	-	□

Source: Tolaymat (2018) and literature review by the author in this paper[21].

Table 3 compares the different models in terms of waste collection, transportation, and treatment/disposal options. Only SWOLF, WRATE and SWEET have a source segregation step for organics, which is important for the collection routes and cost concerned with transportation of the material. For the other models, all the waste has to go to the same place (e.g. composting), which is a significant limitation. Only WRATE has multiple transport vehicle options. Only WRATE and SWOLF have single- and dual-stream MRF facilities.

Table 3 *LCA tools comparison on material processing technologies*

Consideration	WARM	MSW-DST	SWOLF	EASETECH	WRATE	SWEET
Materials Collection						
Bin/cart options	-	√	√	-	√	-
Drop-off	-	√	√	-	√	-
CDD Collection	-	-	-	√	√	-
Source Separated Organics	-	-	√	-	√	√
Materials Transport						
Multiple Fuel Options	-	√	√	√	√	√
Multiple Vehicle Options	-	-	-	-	√	-
Multiple Modes (e.g., Rail, Ship)	-	√	√	√	√	-
Multiple Road Options	-	-	√	√	√	√
Transfer Station	-	√	√	-	√	-
Material Recovery Facility (MRF)						
Single Stream	-	√	√	√	√	√
Dual Stream	-	-	√	-	√	-
Mixed EOL materials	-	√	√	-	√	-
Landfill						
MSW Landfill	√	√	√	√	√	√
Ash Landfill	-	√	√	√	√	-
Carbon Storage	√	√	√	√	-	-
Leachate	-	√	√	√	√	-
Landfill Gas (LFG)—Generation	√	√	√	√	√	√
Rate Adjustable						
LFG—Flaring	√	√	√	√	√	√
LFG-to-Electricity	√	√	√	√	√	√
LFG—Direct Beneficial Use	-	-	-	√	-	√
Emerging Technologies						
Gasification	-	-	√[1]	-	√	-

Pyrolysis	-	-	-	-	✓	-
Anaerobic Digester	-	-	✓	✓	✓	✓
Incineration						
Mass Burn	✓	✓	✓	✓	✓	✓
Refuse-Derived-Fuel	-	✓	✓	-	✓	-
Incineration without Energy Recovery	-	-	✓	✓	-	-
Composting						
Windrow Composting	✓	✓	✓	✓	✓	✓
In-vessel Composting	-	✓	✓	✓	✓	-
Backyard Composting	-	-	-	-	✓	-

1 *Planned but not included in version evaluated.*

2 *SWOLF has an editable impact assessment method section that could add or remove categories.*
EASETECH has multiple impact assessment methods available.

3 *The EASETECH license is provided free to the registered user and the training cost (provided by DTU) for consultants, authorities, and developers to become registered users is € 5,000 (Approximately $5,533).*

4 *The latest version available at the time of the study was evaluated.*

5 *"✓" and "- "represent presence and absence, respectively, of the associated consideration/feature.*

6 *SWEET data was retrieved from the SWEET software in this research work.*

Source: Tolaymat (2018) and the literature review by the author in this paper[21].

Table 4 compares the six LCA tools based on waste processing options. SWEET can modeled collection vehicle type while it cannot model Transfer station as compared to other tools.

Table 4 *LCA tools comparison based on waste processing options*

Processing Options	**Options**	**WARM**	**MSW-DST**	**SWOLF**	**EASETECH**	**WRATE**	**SWEET**
Landfill Gas Treatment Options	None, Landfill with no LFG treatment	✓	✓	✓	✓	✓	✓
	Flaring	✓	✓	✓	✓	✓	✓
	LFG-to-electricity	✓	✓	✓	✓	✓	✓
	LFG-to-electricity	✓	✓	✓	✓	-	-

	with bioreactor						
Source-Separated Organics Processing	Collection and Composting	✓	-	✓	✓	✓	✓
	Collection and anaerobic digestion with gas-to-electricity	-	-	√a	✓	✓	-
Backyard Composting	Decreased organics collection due to home composting	✓	✓	✓	✓	✓	✓
Materials Recovery	Single stream MRF	✓	✓	✓	✓	✓	✓
	Dual stream MRF	✓	-	√a	✓	✓	-
	Mixed waste MRF	✓	✓	✓	✓	✓	-
MRF Automation	Various levels of manual vs automated work	-	-	√a	-	-	-
Recycling Plastics vs Recycling Glass		✓	✓	✓	✓	✓	-
Pay-as-You-Throw		✓	✓	✓	✓	✓	✓
CDD Recycling	Landfilling of CDD	✓	-	-	-	-	-
	Recycling of CDD	✓	-	-	-	-	-
E-waste Collection and Recycling	Landfilling of e-waste	✓	-	√a	-	✓	-
	Recycling of e-waste	✓	-	√a	-	✓	-

Collection Vehicle Fuels	Diesel	✓	✓	✓	✓	✓	✓
	CNG			✓a		✓	✓
	Biodiesel					✓	-
Collection Vehicle Types	Vehicles with different mechanisms for waste collection	-	-	-	-	-	✓
Transfer Station	Adding a centrally located transfer station	✓	✓	✓	✓	✓	-
Thermal Treatment Options	Mass burn WTE	✓	✓	✓	✓	✓	✓
	Gasification	-	-	✓a	-	✓	-
	Pyrolysis	-	-	-	-	✓	-
Plastic Incineration	Plastic incineration	✓	✓	✓	✓	✓	✓
Plastic Recycling	Plastic recycling	✓	✓	✓	✓	✓	-
RDF Recovery Before Landfilling	RDF production from fresh MSW	-	✓	✓	-	✓	-
RDF Recovery After Landfilling	RDF production from landfill mining	-	-	-	-	-	-

The latest version available at the time of study was evaluated. a Not in the version evaluated but is expected to be included in future versions. "✓" indicates that the listed scenario can be modeled using the tool.

Source: Tolaymat (2018) and the literature review by the author in this paper

All tools, except WARM, provide a process-specific emissions breakdown. The tools evaluate a wide range of impacts categories; nevertheless, global warming (GHG) emissions, traditional air pollutants, energy consumption, cost etc.) are the specific classification of impact ([18],[19],[20]), that can be used to compare the results of the tools. Only MSW-DST and SWOLF analyze and provide cost information as an output.

The decision-support tools were compared based on flexibility (Table 5).

Flexibility with tool parameters	WARM	MSW-DST	SWOLF	EASETECH	WRATE	SWEET
Allows user to adjust material composition and generation rate (e.g., 1,000 tons/day)	□		□	□	□	□
Allows modification of material properties	-	□	□	□	-	□
Allows input of additional material types (in addition to those provided by tool)	-	-	-	□	-	-
Allows addition/modification of processes (aside from adjusting default values)	-	-	□	□	-	-

Flexibility of the tool is particularly important for users who try to adjust the input parameters to reflect an alternative geographic area than the one used to develop default conditions of the model. While WRATE and EASETECH are designed to represent conditions in Europe, these tools have the flexibility to simulate other conditions by modifying the input parameters (e.g. composition of materials).

Table 5 *Comparison of Tool Flexibility for Key LCA Parameter Categories*

Table 6 compares the ability of the tools to accommodate material-specific physical, biological, and chemical properties defined by the user. It is evident that in SWOLF, EASETECH and MSW-DST, users have the most flexibility to modify material properties.

Table 6 *Comparison of Tool Flexibility for Key Material-Specific Properties*

Material Properties		WARM	MSW-DST	SWOLF	EASETECH	WRATE	SWEET
Physical	Moisture content	-	□	□	□	-	□
	Energy content	-	□	✓	□	-	-
	Ash content	-	□	□	□	-	-
	Material density	-	□	□	-	-	-
	Combustion efficiency	-	□	□	✓	-	□
Biological	Decay rate constant (k)	□	□	□	□	-	□
	Methane generation potential	-	-	□	-	-	□
Chemical	Elemental constitution	-	-	□	□	-	-

Generation sectors shown in Table 7 shows that, only SWEET considers the developing countries sectors for analysis. In addition, only EASETECH considers the industrial sector.

Table 7 *Comparison of Material Generation Sectors Considered by the Tools*

Generation Sectors	WARM	MSW-DST	SWOLF	EASETECH	WRATE	SWEET
Residential (single-family)	□	□	□	□	□	□
Multi-family	-	□	□	□	-	□
Commercial	-	□	□	-	□	-
Industrial	-	-	-	□	-	-

Sectors in Developing country	-	-	-	-	-	□

Table 8 below summarizes planned updates for the 6 short-listed tools.

Table 8 *Future planned updates of the six shortlisted tools.*

Tools	Future planned updates
WARM_v14	No known updates
MSW-DST	Note that the forthcoming next version of the MSW-DST will include AD as well. It will use the same process models as SWOLF. Adding new processes and direct import of background process (will be possible in future version of the tool) Next generation of the MSW-DST is currently under development with the US EPA and a test version is expected to be completed soon. The next generation MSW-DST is a version of SWOLF. It is expected that adding new processes and direct import of background process will be possible in future version of the tool.
SWOLF	No known updates
EASETECH	A complex structure calculating chemical and physical interaction of each step of the waste treatment technology is under development. This will allow for estimation of emissions and resource consumptions at each step rather than redesigning these interactions for each case study
WRATE	No known updates
SWEET v3	Impact of targeting different sources of open burning pollution.

Source: Eldbjørg et al. (2018). SWEET was compared using the literature material.

The only MSW management decision support tool designed for developing countries is Solid Waste Emissions Estimation Tool (SWEET). SWEET is designed as a screening tool for reducing short-lived climate pollutants, not for detailed decision-making. In addition, it does not estimate costs of waste management alternatives and has significant shortcomings as follows:

2.2.4 *SWEET: Solid Waste Emissions Estimation Tool (shortcomings and limitations):*

Year: SWEET was developed on behalf of the US EPA for the Climate and Clean Air Coalition (CCAC) in 2017 while ver 3.1 released in Sept 2020.

1) Waste management practices vary not only city by city, but business by business and residence by residence. These variations are not accounted in the tool.
2) Local factors such as geography and economic development may impact the waste stream, so that the average waste composition and generation rate data may not reflect real local conditions.
3) Using this tool gives a good overall picture (as the defaults are based on US) of what might be expected but does not provide detailed estimates of what is going on at the local level.
4) The tool is built on the assumption that businesses of a certain type produce similar wastes at similar rates, regardless of the location or size of the business, and that generators are generally the same throughout the state.
5) SWEET's default emissions are based on US equipment cannot be used in developing country scenario.
6) The role of informal sector in the tool is not included.

Hence, there is a critical need to develop a better MSW management decision support tool for developing countries with locally appropriate default values, more accurate GHG emissions estimates, and cost estimates.

2.2.5 *Conclusions and Recommendations*

Above analyzed tables summarizes factors influencing SWM while the result reveal that municipalities and entities having likelihood of usage of LCA tools should collect the data and analyzed based on the individual criterias explained and evaluated in this paper. Most of the tools indicated the significance level of variables and they are presented with tick signs as "significant impact" in the tool and dash ('-') signs as "insignificant" or does not impact the tool in the table.

A tool should be developed that enables local governments in developing countries to select solid waste management systems which reduce greenhouse gas emissions, and to estimate system costs, considering their specific local conditions.

A particularly important component of future work is to consider the integration of type of equipment selection for collection and transportation of waste, route optimization, executing vermi-composting process in controlled conditions, regional landfilling approach, , biomining of the open dumps and converting to landfills, this will enable a better understanding about how it will affect the Solid waste management decision making. Additionally, further investigation is required to evaluate the performance of Python and deep learning algorithms for development of LCA decision making models. Over the past few years, several statistical and uncertainty prediction models like Monte Carlo Simulation were developed to support decision making process for efficient solid waste management. However, there is still a high demand to develop more advanced LCA models which can have defaults of the developing countries with more critical input variables.

Acknowledgement: I am thankful to the University of Texas at Arlington for the immense support and help in the research work.

References

[1] Tacoli, C., 2012. Urbanization, Gender and Urban Poverty: Paid Work and Unpaid Carework in the City. International Institute for Environment and Development: United Nations Population Fund, London, UK.

[2] UNPD, 2012a. World Urbanization Prospects, the 2011 Revision. Department of Economic and Social Affairs, Population Division, United Nations, New York. http://esa.un.org/unup/pdf/WUP2011_Highlights.pdf (accessed 6/2/20).

[3] World Bank, 2012. What a Waste: A Global Review of Solid Waste Management. Urban Development Series Knowledge Papers. http://documents.worldbank.org/curated/en/2012/03/16537275/waste-global-review-solid-waste-management (accessed 6/12/20).

[3, a] UN-HABITAT, 2010. Collection of Municipal Solid Waste in Developing Countries. United Nations Human Settlements Programme (UN-HABITAT), Nairobi.

[4] Guerrero, L., Maas, G., Hogland, W., 2013. Solid waste management challenges for cities in developing countries. Waste Management 33(1), 220–232.

[5] WHO, 2007. Population health and waste management — Scientific data and policy options. Report of WHO workshop, Rome, Italy, 29-30 March 2007. http://www.euro.who.int/_data/assets/pdf_file/0012/91101/E91021.pdf (accessed 6/12/20).

[6] UNEP, 2011. Towards a Green Economy: Pathways to Sustainable Development and Poverty Eradication, United Nations Environment Programme (UNEP).

[6, a] Shah, K. V., Sattler M., (2020). Solid Waste Management Challenges and Solutions in India. Environment Magazine, The Magazine for Environmental Managers, A&WMA, pp. 13-17.

[7] Jha, A., Singh, G., Gupta P., 2011. Sustainable municipal solid waste management in low income group of cities: a review. Tropical Ecology 52, 123-131.

[8] Wilson, D., Rodic, L., Scheinberg, A., Velis, C. and Alabaster, G., 2012. Comparative analysis of solid waste management in 20 cities. Waste Management & Research 30, 237-254.

[9] Levis, J.W., Barlaz, M.A., DeCarolis, J.F., Ranjithan, S.R., 2013. A generalized multistage optimization modeling framework for life cycle assessment based integrated

[9, a] Levis, J.W.; Barlaz, M. Composting Process Model Documentation. Available online: http://www4.ncsu. edu/~jwlevis/Composting.pdf (accessed on June 25th, 2020).

[9, b] Levis, J.W.; Barlaz, M.A.; Decarolis, J.F.; Ranjithan, S.R. Systematic exploration of efficient strategies to manage solid waste in U.S municipalities: Perspectives from the solid waste optimization life-cycle framework (SWOLF). Environ. Sci. Technol. 2014, 48, 3625–3631.

[10] Laurent, A., Bakas, I., Clavreul, J., Bernstad, A., Niero, M., Gentil, E., Hauschild, M.Z., Christensen, T.H., 2014. Review of LCA studies of solid waste management systems - Part I: lessons learned and perspectives. Waste Manag. 34, 573e588. https://doi.org/10.1016/j.wasman.2013.10.045.

[11] Boldrin, A.; Neidel, T.L.; Damgaard, A.; Bhander, G.S.; Møller, J.; Christensen, T.H. Modelling of environmental impacts from biological treatment of organic municipal waste in EASEWASTE. Waste Manag. 2011, 31, 619–630. [CrossRef] [PubMed]

[12] M Banar, Z Cokaygil, A. Ozkan Life cycle assessment of solid waste management options for Eskisehir, Turkey, Waste Management, 29 (2009), pp. 54-62

[13] Winkler, J. (2004). Comparative evaluation of life cycle assessment models for solid waste management. Pirna: Forum für Abfallwirtschaft und Altlasten e.V.

[14] Shah K.V., Shah D.D., (2018), An Approach Towards Sustainable Municipal Solid Waste Management in India, Waste Management and Resource Efficiency'nin İçinde (Ghosh S.K., Ed.), Springer, ss.1067-1076.

[15] Shekdar, A.V., 2009. Sustainable solid waste management: an integrated approach for Asian countries. Waste Manag. 29, 1438e1448. https://doi.org/10.1016/j. wasman.2008.08.025.

[16] United States Environmental Protection Agency. Municipal Solid Waste Generation, Recycling, and Disposal in the United States: Facts and Figures for 2009. United States Environmental Protection Agency.

[17] US EPA. Documentation for Greenhouse Gas Emission and Energy Factors Used in the Waste Reduction Model (WARM)—Organic Materials Chapters. Available online: https://www.epa.gov/sites/production/files/2016-03/documents/warm_v14_ management_practices.pdf (accessed on 15 October 2019).

[18] US EPA. Documentation for Greenhouse Gas Emission and Energy Factors Used in the Waste Reduction Model (WARM)—Background Chapters https://www.epa.gov/sites/production /files/2016-03/documents/warm_v14_management_ practices.pdf (accessed on 15 October 2019)..

[19] Thorneloe, S.A.; Weitz, K.; Jambeck, J. Application of the US decision support tool for materials and waste management. Waste Manag. 2007, 27, 1006–1020.

[20] Thyselius, L. ORWARE—A simulation model for organic waste handling systems. Part 1: Model description. Resour. Conserv. Recycl. 1997, 21, 17–37.

[21] Tolaymat T, US-EPA Sustainable Materials Management Options for Construction and Demolition Debris. (November 13, 2018). See from https://www.epa.gov/smm/sustainable-management-construction-and-demolition-materials.

BIOGRAPHY:

Ketan V Shah *works as a Graduate Research Assistant at University of Texas at Arlington, Civil and Environmental Engineering Department. Ketan received his BE in Chemical Engineering in 2009 from Maharaja Sayaji Rao University, Vadodara, India, and his MBA in Technology Management in 2012 from Centre for Environmental Planning and Technology, India.*

Ketan has experience leading 20-member teams for different projects with combined worth of $ 80 million. Proficient in field sampling, environmental monitoring & management, research & statistical data analysis, Life cycle assessment with 6+ years of experience. TxSWANA Scholar 2020. He is pursuing PhD on Solid Waste Management Decision making tools at The University of Texas at Arlington in Civil and Environmental Engineering Department. Ketan is a member of ASCE, SWANA, AEESP, Institute for Sustainability, AIChE, A&WMA, Tech-Cast Global. He may be contacted at ketan.shah@mavs.uta.edu

Dr. Melanie Sattler *works as a Professor of Civil Engineering Department at University of Texas at Arlington,*

APPENDIX

Table A: *Criteria for selection of SWM LCA tools*

Step 1. 26 tools identified	Step 2. 8/26 tools screened	Step 3: 6/8 tools shortlisted

Based on ✓ Waste composition variability, ✓ Waste collection, ✓ Transportation, ✓ Treatment, ✓ Final disposal, ✓ Emissions in the environment, ✓ Environmental performance, ✓ Cost.	Scrutinized on ✓ The year of development, ✓ Material and documentation available, ✓ Modified relevant environmental emission changes, ✓ Technical support available, ✓ Executing capacity based on geographical location	Tools having process with ✓ Utmost flexibility, ✓ Application variability (e.g., collection, processing, material recovery, recycling etc.).

Table B: *Analysis of Life Cycle Assessment-Decision Support Tool for SWM*

(Stepwise scrutiny of tools; Yellow highlighted, shortlisted for the study, brown highlighted- tool considers cost)

Sr. No.	Name	Aim and Concept	Tool Type	Environmental Impact Indicator	Input fractions	Last Update	Documentation?
1	EASETECH	Comprehensive waste LCA and LCC tool developed at the Technical University of Denmark	Stand alone on Windows	The user can choose various LCIA methods e.g., ILCD	80	2018	Yes
2	SWOLF	Optimizable dynamic life-cycle assessment framework considering future changes in policy requirements, waste composition and energy system.	Stand alone	Landfill diversion, and GHG emissions	41	2014	Yes
3	WARM_v14	Created by the U.S. EPA to help solid waste planners and organizations estimate GHG emission from different waste management practices.	Stand alone	GHG emissions and energy reduction	54	2016	Yes
4	MSW-DST	Aims to calculate life-cycle environmental trade-offs and full costs of different waste management or materials recovery programs.	Stand alone	Energy consumption, and emissions for 32 pollutants	26	2012	Yes
5	WRATE	Environmental assessment of waste management systems	Stand alone	LCIA method CML 2001	52	2017	Yes
6	SWEET_v3	The Solid Waste Emissions Estimation Tool (SWEET) is an excel-based tool that quantifies emissions of	Stand alone	GHG emissions reduction estimated and	86	2020	Yes

		methane, black carbon, and other pollutants from sources in the municipal solid waste sector.		tracking progress over time			
7	Balkwaste	Aims to support the decision maker throughout the various steps of municipal solid waste management planning	Stand alone	Greenhouse gas effect, emission to air, conventional fuel savings,	13	2011	Yes
8	Wasptool	Examines and evaluates the effectiveness of possible waste prevention strategies	Web-tool	Generated waste, waste reduction, diversion from landfill	11	2017	No
9	HOLIWAST	Comparisons of up to five stakeholder views or waste management policies. Data only available for “best available technologies” and user cannot modify the technology data.	Web-tool	NA	NA	2007	No
10	SIMUR	Models different waste collection and treatment options and evaluated the environmental impacts of different scenarios	Stand alone	LCIA impact methodology—CML Version 3,	16	2011	Yes
11	IWM (EPIC/CSR)	Aims to evaluate the environmental and economic performance of the various elements of their existing or proposed waste management systems	Stand alone	Global warming, acidification, urban smog, health risks water	8	2004	Yes
12	IWM - PL	LCA and cost analysis quantifying potential environmental impacts and economic aspects of municipal waste management systems	NA	Ecopoints	6	2011	Yes
13	IWM2, UK	User-friendly model for waste managers	Stand alone	Fuels, Final Solid Waste, Air	9	2001	Yes

14	LCA-IWM	Supports the decision making in the planning of urban waste management systems by allowing comparison of different scenarios.	Stand alone	LCIA method CML 2001	10	2006	Yes
15	ORWARE	Designed for strategic long-term planning of recycling and waste management. It is originally developed for environmental assessment of biodegradable liquid and organic waste, but can also handle treatment of mixed waste.	Stand alone	GWP, acidification, eutrophication, photo oxidants, primary energy carriers,	12	Applied in a study in 2017	Yes
16	SSWMSS, Japan	NA	NA	NA	NA	2009	No
17	WISARD	Aims to assist decision makes when evaluating alternative waste management scenarios	Stand alone	NA	NA	2000, still in use till 2009	No
18	CO2ZW	Calculates GHG emissions emanating from the waste operations of European municipalities.	Stand alone	Carbon footprint	6	2013	Yes
19	LCA-LAND	Landfill model	Stand alone	No impact indicators—only emissions	NA	1999	Yes
20	MSWI	NA	NA	NA	NA	2003	No
21	ARES	NA	NA	NA	NA	2004	No
22	WAMPS	Screening LCA model applied to find optimal waste management solutions and alternative waste treatment technologies. The focus is on environmental aspects but also evaluates economic consequences	Stand alone	GWP, eutrophication,	11	2009	Yes

23	IFEU project	German model for environmental assessment of waste systems based on the software UMBERETO. The detailed biological treatment module is not included in the official library of the UMBRETO software, but in the IFEU project specifically.	Stand alone	NA	NA	2002	No
24	WASTED	Provides a comprehensive view of the environmental impacts of municipal solid waste management systems.	Stand alone	No impact indicators—only emissions	6	2005	Yes
25	VMR	Aims to be an open-source waste management tool with a detailed representation of the effects of operating conditions and input stream characteristics.	NA	Not implemented at this stage	19	Under development	Yes
26	KISS	Calculates carbon footprint of different sorting and treatment systems in waste management. The model has been developed as part of the TOPWASTE project (The Optimal Treatment of Waste)	Stand alone	Carbon footprint, resource recovery and loss	13	2015	Yes

Inclusive circular waste management in the refugee camps – strategy for Rohingya Refugee Camp

Waste-to-Resources
International Symposium MBT & MRF

What is the EEB?

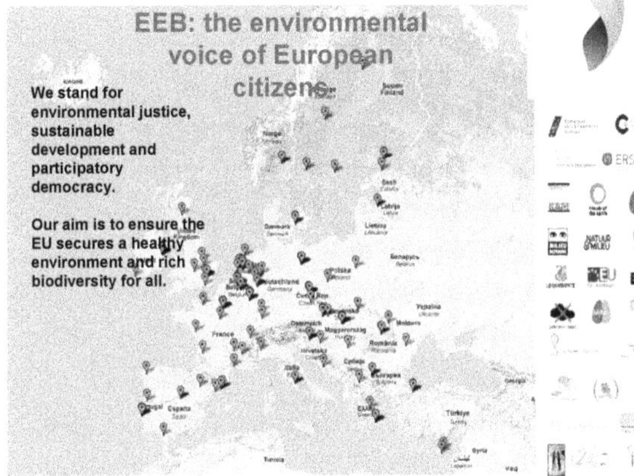

EEB
European
Environmental
Bureau

www.eeb.org

Cox's Bazar refugee camp

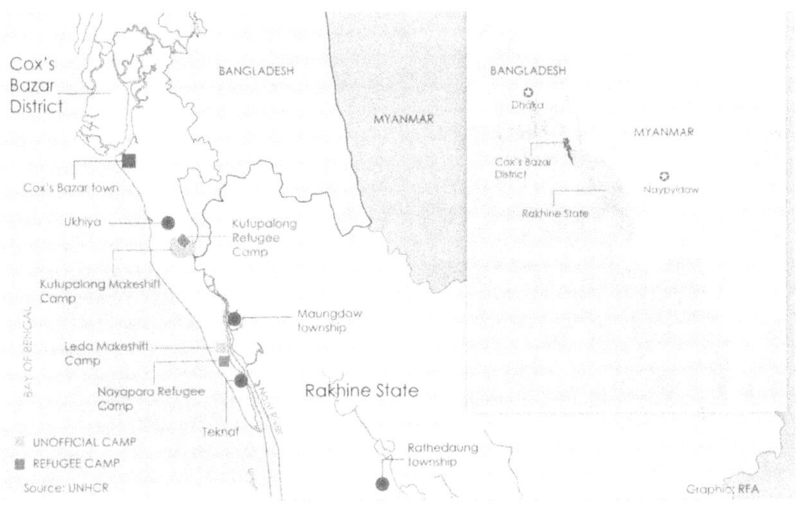
Cox's Bazar District
BANGLADESH
MYANMAR
Cox's Bazar town
Ukhiya
Kutupalong Refugee Camp
Kutupalong Makeshift Camp
Maungdaw township
BAY OF BENGAL
Leda Makeshift Camp
Nayapara Refugee Camp
Rakhine State
Teknaf
UNOFFICIAL CAMP
REFUGEE CAMP
Source: UNHCR
Rathedaung township
BANGLADESH
Dhaka
MYANMAR
Cox's Bazar District
Naypyidaw
Rakhine State
Graphic: RFA

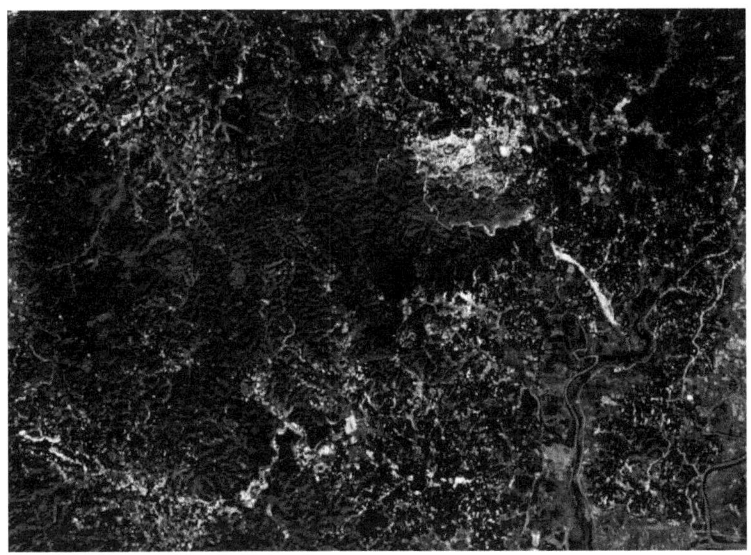

PROBLEM
identification : ...

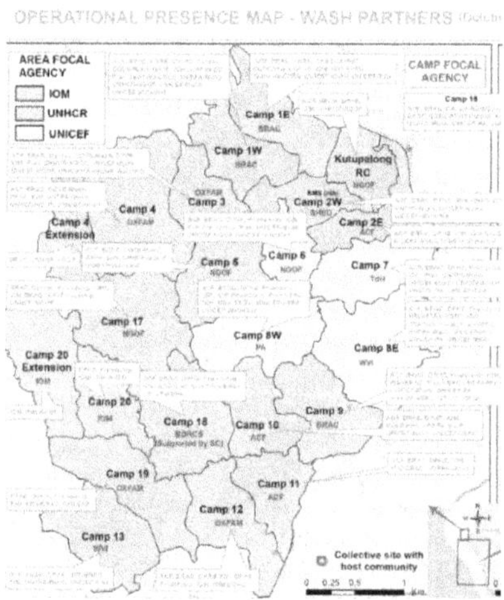
OPERATIONAL PRESENCE MAP - WASH PARTNERS
AREA FOCAL AGENCY
IOM
UNHCR
UNICEF
CAMP FOCAL AGENCY
Camp 1E
Camp 1W
Kutupalong RC
Camp 3
Camp 2W
Camp 4
Camp 2E
Camp 4 Extension
Camp 5
Camp 6
Camp 7
Camp 17
Camp 8W
Camp 8E
Camp 20 Extension
Camp 20
Camp 18
Camp 10
Camp 9
Camp 19
Camp 11
Camp 12
Camp 13
Collective site with host community

ORGANIC WASTES
INORGANIC WASTES

BURNING INORGANIC WASTES IN SAFE SITES

DUMPING IN THE DUMPING SITES

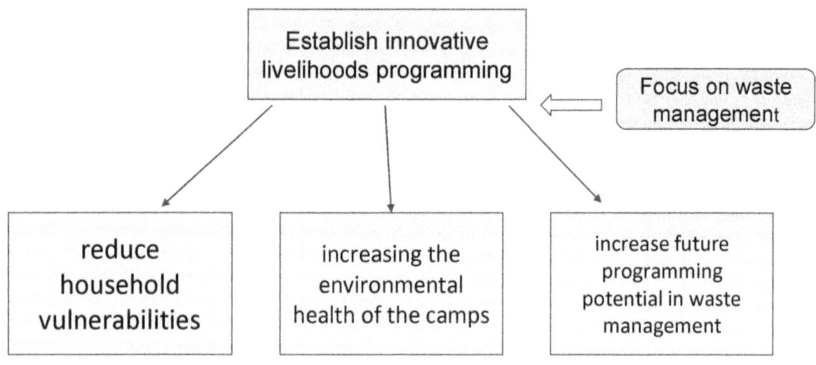

Strategy: 3 pillars

- **WASH** (beyond SPHERE approach as minimum but more ambitious ZW & CE approach)
- **Livelihoods** – waste pickers as people-centred CE, and start-ups to be supported financially
- **Social cohesion** – material and skills flows between the two communities

There is potential for a change

- **biowaste** stream – composting (possibly enhanced) with support for kitchen gardens = result: nutrients recovered for food production

- **plastics** (especially low grade) – downcycling with support for extrusion = result: cleaner environment and production of materials

- Upcycling and community **behavioral change**:
  Zero Waste Camp centres: reuse, repair, create

Increased Livelihoods

- Ideas provide opportunities beyond typical WASH approach and waste picking:
  - Collectors (waste pickers) - formalisation and internalisation, people-centred approach
  - Repairers (e.g. electronics, textiles, home equipment)
  - Retailers and dismantlers (buy items and resell, e.g. plastics, compost, larvaes)
  - Upcyclers (remake into new items and sell, e.g. tyres to shoes, decorations)
  - Transformers (recyclers) - "precious plastic" or composting
- Role of boundary spanning NGOs and social enterprises is critical in generating access to new markets and capturing greater value.
- Significance of the trade in second hand items and repair, upcycling and waste picking as an income model for those without any specialized skills or knowledge beyond that of community resources - NGOS provide trainings

Opportunities for Livelihoods

- Creates job opportunities and can reduce vulnerabilities of households as well as promoting gender equality

- Increase social cohesion: zero waste activities could offer a place for contact interaction

- Behavioral change through increased awareness of the consumption's impact on the environmental health

UNHCR

UNHCR

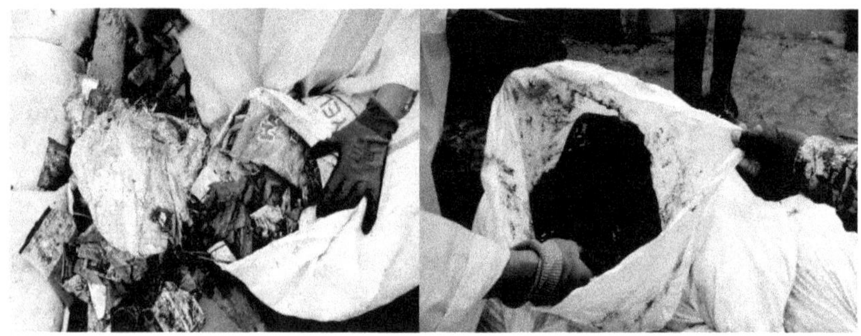

Plastics: we need to start with:

- Identifying in detail types and amounts of plastics not collected for recycling
- Value chain analysis - demand driven pull measure,
- Improve separate collection by engaging people

- tokens, vouchers,

- impact on clean-up and livelihoods

- Safe and efficient temporary storage
- Supporting recycling start-ups in host communities

p.31

RECYCLING OPTIONS (LOW TECH OPTIONS)

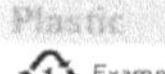

1 PET Examples: bottles for soft drinks, ketchup, salad dressing, water, peanut butter and jam jars, some waterproof packaging, fleece

Industrial (professional) scale solutions (needs machines)	Human scale solutions (no machines, DIY)
• First press the bottles/jars, then using shredder you get flakes, which can be melted and made into granules which can be sold. Also flakes can be sold in some places. 	• If there are no other options, you can use bottles as ecobricks in building. The safety of such use is sometimes questioned (plastics include some substances of concern as softeners/ plasticisers), but it is still a better solution, if the only alternative for plastics is to be burned at dumpsites, which is extremely harmful.

3 PVC Examples: isolation for cables, canalisation and drainpipes, window frames, containers for household chemicals, vinyl records, blister packaging for non-food items

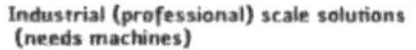

Industrial (professional) scale solutions (needs machines)	Human scale solutions (no machines, DIY)
• Can be melted and made into pipes, mats, traffic cones, marine products, garden furniture.	• Don't heat it up (it contains chlorine, which is toxic and will be released during burning). • It is better to collect PVC separately and hand it to companies dealing with hazardous waste. • Can be used to make new pipes.

https://www.letsdoitworld.org/keepitclean/

PLASTIC EXTRUSION

Brick out of plastics

Housing from recycled plastics

Ecopost: Recycling plastic waste into environmentally friendly fencing posts

http://conceptosplasticos.com/
https://ubuntoo.com/solutions/plastiqube

p.33

Issues with plastics stream

- System of tokest is questionable by UNHCR and OXFAM
- Complying with safety and emissions regulations
- Specifications of products made from this material
- Acceptance by the community
- Bins become redundant
- Tokens speculations
- Costs
- Collection of littered items - need a lot of cleaning

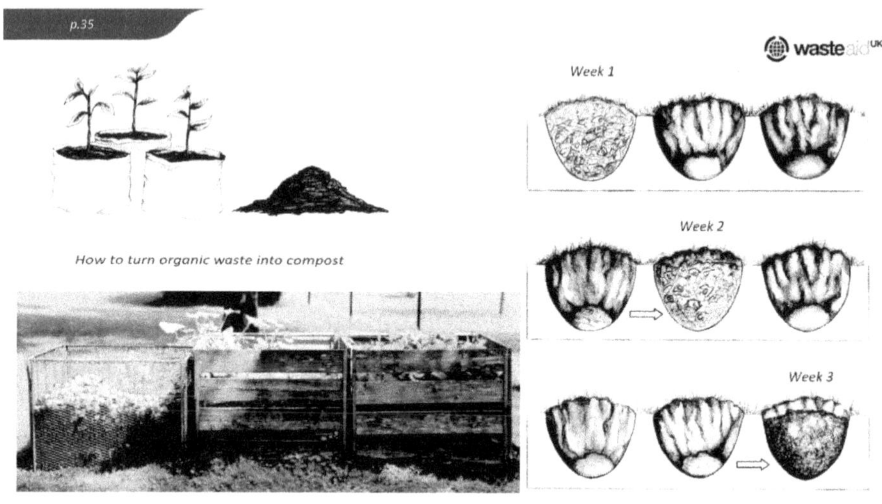

Kitchen gardens

1. **Household**
   - Self-employment
   - Savings
   - Independence

1. **Urban/camp level**
   - Climate change
   - Eating disorders
   - Poverty
   - Employment
   - Biodiversity angle
   - Land management

1. **Macro-Level** Food System
   - Resilience
   - Adaptability

p.38

BLACK SOLDIER FLY IN COMPOSTING

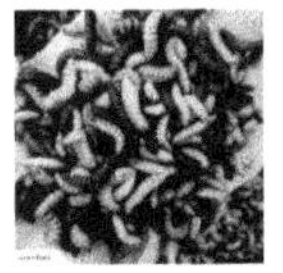

Sources:
https://www.insectsforall.nl/en/projects/insects-as-feed-for-fish-in-bangladesh/
http://www.bluegoldbd.org/innovationfund/project-examples/ento-feed/

p.39

p.41

COMMUNITY REPAIR, REUSE, RECYCLE

REDUCE, REUSE, REPAIR, **RECYCLE**

Using a network of community centres: trainings, "repair cafes", meetings.
Repair of: mobile phones, textiles, umbrelas, home equipment, carpentry, mechanics,

Limitations and Barriers

- Uncertainty of data
- Government policy preventing refugees from engaging informal livelihood opportunities
- Limitation of 16 days per person for cash for work
- Engaging female volunteers with movement restrictions within the camps
- Lack of access to basic rights including education, livelihoods and justice for the Rohingya people in Bangladesh due to lack of legal status, government restrictions and systemic challenges
- Increased violence?
- Congestion, limited space in the camp and absence of additional land
- Absence of a community governance system and lack of accountability in the existing majhi system

10 years of National Solid Waste Policy in Brazil: The Legacy of Technological and Market Approach

Christiane Dias Pereira

Technische Universität Braunschweig, Braunschweig, Germany

WhiteLabel-Tandem Projects UG, Münster, Germany

Abstract

The implementation of the Legal rules of National Solid Waste Policy in Brazil brought a new systematic approach for waste management in the country, particularly regarding the introduction of technology for the waste recovery, articulating an environmental agenda with social inclusion, economic development, citizenship, construction of urban infrastructure, mobilizing efforts under a cross-sectional view, strongly related to sustainable development.

Although the good news, after 10 years, Brazil still has 40 % of final waste disposal being done at wild landfills and less than 3 % of waste recovery. To change traditional practices, we need to open a multidisciplinary discussion integrating multiple market segments to enable the design of tools to implement a sustainable management of municipal solid waste.

Keywords

Waste, Environment, Recovery, Technology, Law, Management, Valorization, Treatment, Landfill.

1 Introduction

The Brazilian National Solid Waste Policy (PNRS) was based on principles found in Europe, as well as the hierarchy of procedures concerning sustainable solutions or the problems associated with the waste management, focusing the resources preservation and climate protection.

The importance of the PNRS is explained by the pioneering legislation in several social and technical aspects. The Policy was outlined 10 years ago with a multifaceted approach allowed by the participation of various sectors and agents of society.

The innovation also stands out in relation to technical issues. No policy so far had treated solid waste in such a broad and in-depth manner. All types of materials that have the possibility of being reused were covered by the regulation, such as domestic, industrial, and electronic waste.

The potential for reuse of waste was seen and ratified for the first time in an institutionalized and regulated way. The law provided a particularly good base for all sectors of society to know how to deal with the waste produced.

Therefore, the National Solid Waste Policy creates a positive agenda emboldening the adaptation of landfills as electric power plants, closing and remediating the wild dumps and promoting waste valorization and social inclusion. Other practices are promoted by the National Policy as, for example, the composting for organic waste from MSW in smaller scales and the decomposition in anaerobic digesters of larger scales, RDF production, resulting in the creation of humus and energy.

Ten years later, the growth of selective collection, the consolidation of Reverse Logistics systems, the engagement of companies and those at the forefront of the national industry with the PNRS commitments, as well as greater awareness of the population concerning the waste generated in their homes, has been verified.

The implementation of the provisions of Waste policy brought a new systematic approach for waste management in the country, particularly relating to the introduction of technology for the waste recovery, imposing a series of new activities, that should be implemented in the short and medium term for adequacy of current practices in alignment according to the Law.

The activities should articulate an environmental agenda with social inclusion, economic development, citizenship, construction of urban infrastructure and the introduction of technology, mobilizing efforts under a cross-sectional view, strongly related to sustainable development. The interaction among the federative entities, external control and regulation agencies, the private sector, and society will allow for greater efficiency in public contracting as well as guarantee the legal security of the contracts signed and the continuity of the operations contracted.

2 Brazilian Waste Management

2.1 General Information

According to ABRELPE 2020, in the period between 2010 and 2019, total MSW generation increased by about 19% in the country, with a 9% growth in the per capita generation rate, reaching about 79 million tons annually. The quantity of solid waste collected in Brazil was 349 kg per person per year in 2019. Of more than 5,570 cities in Brazil, almost 73% have initiatives of selective collection, but mostly either implemented it partially or do not have enough sorting capacity, resulting in collected recyclables being destinated to landfills. Although the selective collection started in the 80s, no more than 3% of the waste is recovered. Forty percent of the generated waste ends up in inadequate places, such as wide dumps. In Brazil sanitary landfill dominates the destination by waste management.

Brazil represents an important market, expecting for 2021 around 5 billion euros turn over in waste management where 80 % is done by private sector, generating 332.000 direct jobs and more than 1 million people involved in informal jobs collecting, sorting and selling recyclable waste.

The extremely high proportion of the organic fraction is typical for emerging market economies, and the consequence is extremely low calorific values. Although the import role played by material recycling considering the need to integrate informal sector, the biotechnologies play an important role as a solution to minimize environmental impact combined with the use of alternative fuels at coprocessing by cement market.

The demand is concentrated in the search for infrastructure, technologies, including technical-operational aspects, as well as the consolidation of the appropriate technologies for the implementation and monitoring of future treatment plants.

The opportunity for new businesses in Brazil is focused on forming partnerships among national operators and foreign technology suppliers, but also increasing the interest of competitive segments of the economy, such as the cement industry, cellulose - and energy industries, that will divide up the market with the traditional operators in the public cleaning sector.

Considering the new legislation, the Brazilian market is not prepared to fulfill its needs, resulting on high dependency in terms of import of machines and technologies. This creates a big chance to integrate the countries through their suppliers, where the multiplier effect can be extremely high, providing services within construction, operation, and monitoring of waste management technologies.

The market development for secondary resources is a consequence of the implementation of mechanisms of the Law, demanding infrastructure, technology, and effective waste management systems.

Regarding the economical sustainability, most of the cities do not have an adequate waste fee, as presented by SNIS 2019, the fragility of financial sustainability remains in the sector, since only 44.8% of the municipalities charge for the services, and the amount collected covers only 57.2% of the costs.

2.2 Private contracts

The National Policy for Solid Waste - PNRS, published in 2010, and the other contracting frameworks for the concession of solid waste management services to the private sector for long periods, between 20 and 35 years, were enough to raise discussions and interest in the sector for the introduction of technological routes for waste valorization.

In this sense, regulatory advances have allowed private investments to lead the waste sector in the path of innovation through technological contribution, favoring the valorization of solid waste and the protection of the environment, with greater preservation of natural resources and generation of employment and income.

Even considering the great potential of the Brazilian market, an analysis of the market leads to fiscal and budgetary limitations, insufficient specialized technical staff, underestimated cost recovery systems, limited operational scale and deficient infrastructure to meet the waste reduction targets in landfills.

Only 40 of the 292 projects propositions, published during 2002 to 2020, were signed and the others were either cancelled/paralyzed or are under development and/or contracting procedure. These contracts totalize 265 million euro per year to manage almost 10 % of the total generated waste in Brazil. The contracts had average gate fee value of 51,00 euro/ton. Gate fee was fixed between 6,00 to 146,00 euro/ton depending on the offered service.

Although the conquests aimed at providing a system of sustainable management of solid residues are not so expressive, they point to a maturing of the market regarding the change of the collection-disposal paradigm.

Another relevant aspect is the observed insufficient technical capacity of municipalities when they are faced with contracts that involve waste valorization practices. This fragility has been mitigated with the adoption of public lines, since 2019, that provide not only financing but also technical knowledge, aimed at the planning phase of the tender.

The interaction among the federative entities, control institutions, the private sector, and society will allow for greater efficiency in public contracting as well as guarantee the legal security of the contracts signed and the continuity of the operations contracted.

2.3 Technologies' tendencies

The contracts awarded cover the full range of recycling and waste treatment technologies in use today, including landfills with high safety standards:

- Waste sorting plants with low and high levels of automatic sorting systems.
- Fermentation and composting plants.
- Mechanical-biological and mechanical waste treatment plants to produce alternative fuels such as Refuse Derived Fuel (RDF) for use in cement plants.
- Incinerator.
- Gasification.

The main objectives of the signed contracts follow the traditional waste management in the form of collection, transport, and landfill, rarely with a landfill gas collection. Guidelines and targets for avoiding and reducing large-scale landfill and treatment technologies, such as fermentation, incineration, and composting, are presented. An important element of flexibility in the contracts studied is additional income from various sources, such as the sale of energy collected from landfills, however there is no precise information regarding the amount of ancillary revenue and little definition about its destination, whether it will be shared with the public partner or not. Furthermore, the contracts do not contain targets for pretreatment adequacy but are already introducing milestones for reduction of waste in landfills.

2.4 Secondary Resources

The recycling of materials and energy recovery saves resources, reduces pollution, restricts the occupation of land for final disposal, creates jobs, contributes to sustainable development, a better environment, and mitigates climate change.

In this context, the amount of waste generated alone would already justify the use of technologies for its recovery. However, the decision making should be based not only on qualitative and quantitative knowledge of the raw material, but, above all, on the potential of the consumer market for the secondary resources because, when there is no consumption, these residues, although stabilized, would be sent to the landfill.

Thus, it is evident that mapping the consumer market of secondary resources becomes indispensable for the technological management not to suffer an interruption.

Because the absence of a feasibility study for integration in the economic chain of the secondary resources generated, may compromise the amount of consideration related to the services provided, first by not generating the expected accessory revenue and when the secondary resources are not sold, representing an increase in the operational costs of the landfill.

Table 1 Market for Secondary Resources

Secondary Resources	Average Price situation	General consideration
Recyclables	167,00 euro/t	There are cooperatives (organized informal sector) in 75 % of the Brazilian cities. Some are responsible for the collection and sorting of recyclables.
Compost	15,00 euro/t	There is no legal limitation to use compost from mixed domiciliary waste. There is a high demand promoted through public policy.
Biogas	50,00 euro/t	The anaerobic digestion is promoted through public policy but considering that there is not selective collection for organic fractions, the mechanical effort to segregate the organic fractions will be high and to limit the implementation of plant with continuously biodigestion technologies.
RDF	21,00 euro/t	The cement market is active to promote coprocessing and has given a large contribution to develop public policies that foment energy recovery.

3 Market overview

During the first 2 years of publication of the legal standard, a series of contracts in concession modalities were signed contemplating technologies extremely high-tech as incineration and fermentation plants but also simple techniques such as manual sorting of recyclable materials and extensive systems for composting. However, through low operational capacity or even low prices for secondary resources, it was identified a change in the profile of acquisition of the market where priorities were given for more simplistic technologies such as the recycling of materials.

Currently, we have a new market orientation where it is evident that the implementation of mechanical-biological treatment plants having as emphasis production of RDF to supply the new demands of the cement industry and composting practices.

Considering the alternative fuels for cement market, we have a sector based on quest for operational excellence dictating the new rules for an industry accustomed to practices of low complexity such as collection and landfill. This partnership represents a change of paradigm, not only to introduce a new feature to the secondary market of waste but basically by ensuring the professionalism of a market accustomed to high profitability in the short term but which are no longer resonance when there are legal requirements for waste recovery.

Even trespassing these discussions, we have the participation of peripheral actors such as public prosecutors, financing banks, environmental agencies and the community, pressing the market to offer its services in accordance with the assumptions of the law.

With regard to the composting practices there is no legal limitation to use compost generated by mixed waste, however, looking back to the past 80s and 90s, Brazilian waste market lived the composting boom that resulted in stabilized mixed waste, full of impurities such as glasses and plastic pieces. Furthermore, to keep producing compost from mixed waste, it will be necessary to conquest the consumers through providing high quality of compost or even to motivate decentralized treatment behavior where the citizens take the responsibility for their own composting method in micro scale focusing on its own use.

In general, it is observed that the law published in 2010 has already led to a series of changes in the sector that created a positive agenda for encouraging waste sector to implement waste valorization and social inclusion and for that will be necessary to build capacity and to develop the secondary resource market as a new business.

4 Conclusion

The opportunity for new businesses in Brazil is focused on forming partnerships among national operators and foreign technology suppliers, but also increasing the interest of competitive segments of the economy, such as the cement industry, cellulose - and energy industries, that will divide up the market with the traditional operators in the public cleaning sector.

Looking at the 10-year trajectory we can conclude that we need strength and firmness from the public and private sectors. The environmental issue must be treated as a matter of public health and as a relevant aspect for the economy, since when there is non-compliance with the Legislation, it is the population's money that is used to repair the problem and the future generations experience the repercussion of environmental impacts.

5 Literature

FRICKE, Klaus; PEREIRA, Christiane; LEITE, Aguinaldo; BAGNATI, Marius (Coords.).	2015	Gestão Sustentável de Resíduos Sólidos Urbanos: transferência de experiência entre a Alemanha e o Brasil. Braunschweig: Technische Universität Braunschweig, 2015, ISBN 978-3-924618-45-2.
ABRELPE - ABRELPE – Associação Brasileira de Empresas de Limpeza Pública e Resíduos Especiais.	2020	Panorama dos Resíduos Sólidos no Brasil 2020. São Paulo: Abrelpe, 2020.
SNIS – Sistema Nacional de Informações sobre Saneamento.	2019	Diagnóstico do Manejo de Resíduos Sólidos Urbanos do SNIS, ano de referência 2018. Brasília: SNIS, 2019.

Author's addresses:

Lawyer and Civil Engineer Christiane Dias Pereira

Technische Universität Braunschweig
Beethovenstrasse 51 a
D-38106 Braunschweig
Telefon +49 175 3426247
E-Mail chrdiasp@tu-bs.de

www.tu-bs.de

WhiteLabel-Tandem Projects UG-
Sentruper Str. 165
D-48149 Münster
Telefon +49 2504 933197
E-Mail christiane@wltp.eu

www.wltp.eu

Green Deal, Tox-Free-Environment and SCIP Database

Beate Kummer

Kummer:Umweltkommunikation GmbH, Bonn/Rheinbreitbach

Abstract: The Green Deal and the Tox-Free-Environment are new approaches of the EU commission which will have great effects for the circular economy. The interface between chemicals and waste legislation is a major problem for the intended circular economy. To reuse or recycle more waste, information on the composition of the waste is needed. The European Waste Framework Directive obliges producers to document the presence of substances of very high concern in a new database (SCIP), which is on line since January 2021. Some products from different industries are of varying complexity and pollutant problems. It is indicated that the new database is of limited use for recycling companies. Further requirements focusing on the desired recycling of used products are discussed.

Keywords: chemicals, waste, substances of concern, REACH, SVHC, recycling, plastics, WEEE, end-of-life vehicles.

1 Introduction Green Deal

Everyone was eagerly awaiting the "Green Deal" of our new Commission President. The European Commission has thus committed itself to adopting a new action plan for the circular economy. Already today, some 4 million people are employed in the environmental services sector, which is an increase of 6% compared to 2012. After all, these activities already generate a value added of almost EUR 147 billion annually.

It is urgent that we change „our way of life" and the consumption of resources; a change in raw materials is more than ever needed. The worldwide consumption of materials has tripled in the last decades and is expected to double in the next 40 years without intervention. Recycled materials cover only 12% of the EU material demand! High recycling rates for paper and glass - published by the statistical offices - conceal the still poor "reuse rates" across all material sectors. However, for reasons of security of supply, a genuine recycling system for all raw materials (metallic, mineral but also bio-based) must be introduced urgently. Especially the so-called "future technologies" are groaning under supply bottlenecks and fluctuating prices. Important raw materials, which are used for technologies such as energy efficiency or digitalisation, are not available in sufficient quantities given the current growth potential. The majority of products marketed in the EU today are mainly based on unsustainable and sub-optimal use of resources. Many products are short-lived, non-reusable and non-repairable. This still leads to a high volume of waste in the EU. Annual increases of 5% have been recorded since 2010. A large proportion of waste is not recycled, a considerable proportion is still landfilled (23 %) and is thus lost to the economic cycle.

But now there is a discrepancy: Can waste be recycled if it contains harmful substances? Can a "Tox-Free-Environment" still succeed in the face of hundreds of different hazardous substances in waste (from former products)? One of the biggest challenges, for example, regarding recycling of plastics from old electrical appliances, is to remove barriers to reuse in EU product, consumer and chemical regulations. Strict regulations for new products cannot be applied 1:1 to recycling. It is hoped that the upcoming discussions will drive forward a policy for "sustainable products" to promote the design, production and marketing of sustainable products. It is right that minimum requirements are needed to prevent environmentally harmful products from being placed on the EU market, but they must not take the air out of the secondary raw materials market.

2 The Tox Free Environment

2.1 SCIP Database Approach

SCIP stands for "substances of concern in products" and is a new database that is set up by the European Chemicals Agency (ECHA) and is accessible in all Member States. It is fed with a number of information on "substances of concern" in products. This information will be made available to waste recyclers, which will then know which waste streams contain substances of concern. The main target of the database is to finally remove such hazardous substances from the economic cycle and thus come a step closer to the EU Commission's goal of a "Tox-Free Environment". It started in January 2021.

However, all manufacturing companies are well advised to get to grips with it as quickly as possible, because data acquisition will be the real challenge. For example, when a car or electronic device is placed on the market in the EU, there are highly complex supply chains during the production process. It is not unusual to find hundreds of suppliers in Asian countries or on other continents. However, the database can only be "fed" if all the necessary information from all suppliers is available. Manufacturers of less complex products such as packaging, toys or textiles are also affected.

SCIP aims to collect information on substances of concern or of very high concern in articles or complex products (such as lead stabilisers or plasticisers in plastic components, brominated flame retardants in electronic equipment, polyaromatic hydrocarbons in tyre granules, etc.). The database has its EU legal basis in the Waste Framework Directive (WFD). SCIP is based on Article 9(i) in conjunction with the recently amended WFD. This is a renewed attempt to more closely link chemicals and waste legislation. Member States are now required to adopt targets for waste prevention measures. The wording in the WFD as legal basis for SCIP is the following

"*(i)* promote the reduction of the content of hazardous substances in materials and products, without prejudice to harmonised legal requirements concerning those materials and products laid down at Union level, and ensure that any supplier of an article as defined in point 33 of Article 3 of Regulation (EC) No 1907/2006 of the European Parliament and of the Council[1] provides information pursuant to Article 33(1) of that Regulation to the European Chemicals Agency as from 5 January 2021.... „

The current WFD is therefore the basis for waste management in the EU; it contains extensive requirements for the proper disposal of waste. It must be implemented into the national law of the EU member states by July 2020. In Germany, the draft of the „Kreislaufwirtschaftsgesetz" takes up this issue with the new Section 62a. According to this, ECHA has until 5 January 2021 to set up the database.

For suppliers of an article containing substances on the Candidate List, Article 9 (1) (i) of the WFD subsequently provides for corresponding notification obligations to ECHA. The so-called Candidate List (Annex XIV of the REACh Regulation) currently contains about 200 substances which must comply with the criteria of Article 57 of the REACh Regulation:

- Substances that are classified as carcinogenic or mutagenic or toxic to reproduction of category 1A or 1B according to Regulation (EC) No. 1272/2008 (CMR substances),

- Substances which are persistent and bioaccumulative and toxic according to the criteria of Annex XIII of REACH (PBT substances),

- Substances which are very persistent and very bioaccumulative according to the criteria of Annex XIII of REACH (vPvB substances),

- Substances which, according to scientific evidence, are likely to have serious effects on human health or the environment but which cannot be classified in the above-mentioned groups - e.g. endocrine disrupters.

The request of the EU Commission to provide recycling and waste management companies with more information on substances of concern is correct. In most cases, disposal companies today do not know which substances of concern are contained in the waste to be disposed of (e.g. electronic scrap, batteries or packaging) when it arrives at the recycling plant. The EU Commission now wants to counter this with SCIP. The aim is to create more transparency regarding the composition or presence of hazardous substances in complex products, and this is fed by the manufacturers who can still use hazardous substances today - in some cases with legally regulated exceptions (e.g. lead in various compounds). This means that companies that supply articles containing substances of very high concern (svhc - substances of very high concern) in a concentration above 0.1% weight by weight (w/w) to the EU market will have to submit information on these articles to ECHA from 5 January 2021. An "article" is an object that contains a specific shape, surface or design during production (e.g. cable or screw on a printed circuit board). This definition can be found in Art. 3 of the REACh Regulation, and "complex objects" will soon also be affected by the reporting obligation. These in turn are products consisting of more than one product (e.g. printed circuit board).

The SCIP database is only applicable if a high number of affected manufacturers participate. SCIP provides, for example, that information exceeding the legal requirements must be entered. Thus, notonly information on substances of very high concern ("svhc") but also on "substances of concern" must be entered, i.e. substances that are restricted in products under regulations other than REACh (e.g. POP Regulation -Regulation (EU) 2019/1021 of 20 June 2019 on persistent organic pollutants). The information (consisting of product names, concentration ranges, localisation of the svhc substance, information on safe use of the substance) in the database will then be made available to waste disposers and consumers. Whether this ambitious project will becrowned with success is more than questionable, however. Car manufacturers and manufacturers of electrical appliances have already announced that it will not be possible to provide billions of individual pieces of information per manufacturer. The database will cause high costs and a very complex information gathering process. But recyclers have also already communicated that SCIP will not be helpful.

2.2 Which examples illustrate the problems of SCIP?

Example data supplier automotive industry: Since the first regulations on end-of-life vehicle recycling came into force, the automotive industry has relied on IMDS - International Material Data System as a database (https://www.mdsystem.com/imdsnt/startpage/index.jsp) for materials and components. According to its own information, it has cost the European industry around 10 billion euros to establish the database. Anyone who asks the operator can get access. In the IMDS material data the material and chemical compositions of components, semi-finished products and materials are declared. The main use of the IMDS material data is to secure and prove legal substance regulations for complete vehicles and their spare parts. Furthermore, IMDS material data are used as a data source for the calculation of the recycling rate according to ISO 22628 in the system approval of the EU type approval. Currently IMDS is accessed by about 130,000 active users and about 14,000 substances are reported. Besides aircraft, passenger cars are certainly the most complex products affected by SCIP. The establishment of SCIP now means that for thousands of individual affected items per vehicle all necessary information on svhc has to be obtained from suppliers in the EU and outside the EU. In the figure below, an electronic component from a vehicle is used to illustrate that even the designation of an "article" or product is not easy because there are often no generally valid proper names. The European automotive industry expects several billion Euros for this complex data collection for substances of very high concern, because the current IMDS structure is not usable.

Example Data supplier to the electrical and electronics industry: Electronic assemblies - e.g. printed circuit boards fitted with electronic components - typically have several hundred assembly positions with a large number of electronic components; these are "articles" according to REACh. The majority of the electronic components used are typically in a weight range of 2 mg to 5 g each (see example in Pict.1). Multiple sourcing (several suppliers per assembly item), a common practice in the electronics industry, further increases the amount of information to be provided to ECHA. In the case of assembled PCBs (printed circuit boards), it is not possible to identify the electronic component to which the information entered in SCIP belongs, because such a component usually does not have any identification features to distinguish it from others. A member of the ZVEI (Zentralverband Elektronindustrie, Frankfurt) has recently made calculations on the effort required. 2651 end products were identified that contain brass components with low lead content (lead is an svhc substance on the REACh candidate list). As part uses of these components in end products there are 426,370 combinations that would have to be created in the database. For all parts and substances concerned, this company would have to employ a total of 23 people for one year in order to map each product variant once in the SCIP database.

Example data users in the recycling industry: The nature of today's recycling technologies (e.g. dismantling facilities for end-of-life vehicles and shredder plants) will hardly allow for the determination of individual material information for each component of a vehicle regarding questionable materials in day-to-day business. Hundreds of different vehicle models of the most varied composition are usually disposed of in one plant. What counts here is essentially a high throughput in tonnage in order to ensure economic recovery in view of the raw material price situation. If questionable substances have to be removed from individual "mining products" (such as screws, solder contacts) in the future, economic recovery is practically impossible, because simply checking for the presence of questionable substances will take far too long. This applies not only to vehicles but also to many other complex products. Furthermore, the question must be answered, is this even necessary to achieve the EU's "Tox-Free-Environment" goal? For example, the svhc material lead as an additive in steel alloys changes into the gaseous aggregate state during the recycling process and does not remain in the melt. A separation of individual alloying elements is thus carried out automatically. Electronic scrap is treated separately by specialised recyclers in accordance with the WEEE Directive, and manufacturers already provide the information required by Article 15 of the WEEE Directive (e.g. via the I4R platform). In the recycling process, only very specific components (such as batteries) are removed manually before the remaining parts go through the sorting and treatment processes. In complex electronic devices, the substance of very high concern is usually present in very small quantities in small sub-particles of the product (see example of assembled PCB in Pict. 1). Detailed information about these tiny sub-particles (product category, material category) is not helpful for recyclers, because knowing the presence of svhc substances would usually not change the final, often metallurgical treatment process. Some svhc materials (e.g. organic hydrocarbon compounds) do not survive the recycling process at all, they are destroyed by the high process temperatures.

In January 2021, ECHA published the database. It can be found under the following link: https://echa.europa.eu/de/scip. Manufacturers or suppliers of articles containing svhc are obliged to submit information to the SCIP database from January 5th 2021. This requires the software IUCLID 6, which already had to be used for the registration of substances and can be obtained free of charge from the ECHA homepage (www.echa.europa.eu). ECHA has announced that webinars on the use of the SCIP database are organised regularly for affected companies. The EU Member States must now establish concrete aspects for the legal enforcement of the SCIP database in national law by summer 2021.

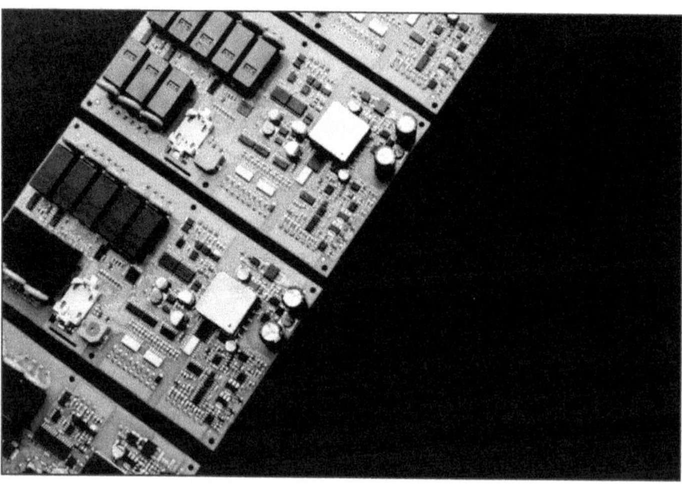

Pict. 1: Printed circuit board (source: www.freepik.com) which contain hundreds of single articles.

3 What are the main points of criticism?

The application of the new database will cause enormous difficulties. First of all, it is of paramount importance to remind all the companies concerned of their obligations. From today's point of view, it is more than questionable whether this will succeed in the foreseeable future in view of the current economic crisis and experience with REACh. In the coming months, business associations, chambers of commerce, ECHA and training institutes would have to offer numerous events to raise awareness of SCIP and demonstrate the use of the prototype. It would be helpful at this point to first of all familiarise oneself with the experiences with REACh and the implementation of the already existing information obligation for products.

In a market surveillance programme, for example, the obligation to apply REACH Art. 33, which has been in force since 2007 (!!) and already includes the obligation to provide information (notification) on substances of very high concern in the supply chain, was reviewed. In 15 countries it was examined whether product manufacturers comply with their obligation to provide information on substances of very high concern (svhc). The result is alarming: there is a high infringement rate of over 80%. This means that more than three quarters of the companies investigated do not receive the necessary information on svhc from suppliers or do not pass it on to customers. Therefore, at this point it must be doubted whether data for SCIP is reported in sufficient detail. Furthermore, the handling will be much more complex and opaque than the current information obligation according to Art. 33 within the supply chain.

The use of the SCIP database by recyclers is also more than questionable. Even if the idea of bringing more transparency into material flows is correct, SCIP will not, from today's point of view, be able to contribute to "shedding more light on the pollutants in waste". Recycling companies work on a tonnage scale, the input into the processing plants is fed by many different models (see for example above: thousands of models of vehicles and electrical appliances). Even with a somewhat simpler material flow such as packaging, the recyclers do not have it much easier. In addition, article descriptions are not communicated to the disposal company, and a sorting system for packaging, for example, is not designed to sort according to article numbers, but according to colour, type of plastic, etc. The recycler will usually not have time to take care of further information if he has to generate sales in a short time.

Before the adoption of SCIP, a broad discussion with all market participants in the EU would therefore have been more than helpful to ensure that a well-intentioned request could be successful. The overlap between waste and chemicals legislation is extremely complex, so a high degree of judgement is required if the well-intentioned goal of a "tox-free environment" in particular is to succeed in the long term. However, an impact assessment is urgently needed for SCIP, and the involvement of data holders and recyclers would be necessary in such a project in order to assess the effort, costs and benefits.

4 Is there an alternative to TOX-Free-Environment Approach and SCIP?

In order to meet the goals of the EU Commission in the area of climate protection and resource conservation, there is no alternative to more recycling and a higher use of recycled raw materials. In order to solve the "pollution problem" in waste, more trust in the economy is necessary. Today, there are a large number of restrictions on pollutants, which are already regulated in chemicals, substances and waste legislation. Here are just a few examples, all of which must be applied when placing new products on the market: Toys Directive, End-of-Life Vehicles Directive, Electronic Scrap Directive and ROHS (EU Directive 2011/65/EU serves to restrict the use of certain hazardous substances in electrical and electronic equipment), Packaging Directive, substance restrictions in Annex XVII of the REACh Regulation, POP Regulation. The application of all of the above-mentioned regulations already existing today will gradually - in connection with REACh - lead to the fact that substances of less and less concern will also arrive in waste management. Even today, there are hardly any cases in which recycled waste has led to a pollution problem in production. Today, the much greater challenge is to bring more waste into high-quality recycling in order to save more, especially non-renewable resources. In the worst case, SCIP will lead to even more pollutant removal in recycling, when it will then lead to even more disposal of waste streams. As a result, large amounts of valuable resources will be lost, which are also necessary for successful climate protection.

The EU must therefore move away from having a completely pollution-free recycling economy and continue to pursue a "Tox-Free Environment". Already today we have ubiquitously occurring pollutants in all environmental media, which represent the so-called background pollution. These loads are caused by water pollution, emissions from industry, traffic, agriculture, etc. However, we have already achieved a great deal through strict environmental and product legislation. Particularly today, in times of foreseeable massive economic and raw material crises and the threat of climate catastrophe, it is not appropriate to spend billions of euros on a new database that will probably not work. Rather, it will be necessary in the future to weigh up with a great deal of judgement how climate gas reduction and resource conservation can be regulated within the EU. Additional burdens for the economy, which cause high costs and bring little benefit, and are also not enforceable, are not needed at present. The chances of changing the SCIP database can currently be classified as extremely low, so manufacturing companies are well advised to look into the matter (https://echa.europa.eu/de/scip-database) and, if necessary, to consult external scientific advice.

5 Contact:

Dr. Dipl.Chem. Beate Kummer, Fachtoxikologin, Umweltauditorin, Dozentin

Kummer:Umweltkommunikation GmbH

Gebr. grimmstr. 17

53619 Rheinbreitbach

Tel: +49 2224 9011480

E-Mail: buero@beate-kummer.de

25. Weniger Abfall. Weniger Deponierung. Mehr Recycling.

Peter Hoffmeyer

Nehlsen AG, Bremen, Deutschland

Less waste. Less landfill. More recycling

Abstract

Increasing population growth means an increasement of waste amount and automatically an increasement of the whole carbon oxygen consumption, worldwide.

In order to keep the air clean and to protect world's resources, innovative and effective solutions for waste prevention and waste management are necessary. Nehlsen set its vision along the solution of these problems.

The key to take responsibility for one's own consumption is to become aware of its impacts and to be ready to change. With powerful PPPs between, on the one hand proved concept-providers and on the other hand willing people to make a change, the necessary steps can be taken to go into a green future.

Inhaltsangabe

Steigendes Bevölkerungswachstum bedeutet einen Anstieg der Abfallmenge und damit automatisch einen Anstieg des gesamten Kohlenstoffverbrauchs weltweit.

Um die Luft sauber zu halten und die Ressourcen der Welt zu schützen, sind innovative und effektive Lösungen für die Abfallvermeidung und das Abfallmanagement notwendig. Nehlsen hat sich die Lösung dieser Probleme als Ziel gesetzt.

Der Schlüssel, um Verantwortung für den eigenen Konsum zu übernehmen, ist, sich seiner Auswirkungen bewusst zu werden und bereit zu sein, sich zu verändern. Mit leistungsfähigen PPPs zwischen, einerseits bewährten Konzeptanbietern und andererseits veränderungswilligen Menschen, können die notwendigen Schritte in eine grüne Zukunft gegangen werden.

Keywords

Recycling, Kreislaufwirtschaft, Deponie- und Abfallmanagement, Energieproduktion, öffentlich-private Partnerschaft, Umweltverantwortung, Aufklärung, Bewusstsein

Recycling, circular economy, waste and landfill management, energy production, Public Private Partnership, environmental responsibility, education, awareness

1 Introduction

1.1 Peter Hoffmeyer

Born 13th March 1959 in Bremen. Married, five children. Graduated Engineer in environmental Technology, University of Berlin.

Since 2020 Chairman of the Advisory Board of Nehlsen AG, Bremen. Former CEO of Nehlsen AG (2001-2019). Among others, former president of the Federal Association of German Waste Management Industry (BDE).

1.2 Nehlsen AG, Bremen

Nehlsen – a family company with a rich history and a bright future. Now in the fourth generation, Nehlsen work to ensure an intact environment by offering pioneering services in the fields of recycling, disposal, industrial services, and logistics in our core sectors. Nehlsen specialises in developing and realising innovative recycling solutions for all types of waste and are committed to driving forward the changeover from waste management to resource management. Nehlsen AG is Co-Founder of Zentek Services GmbH & Co. KG, a community of leading medium-sized disposal companies. A specialized service provider for superregional disposal systems and waste management. Establisher of "Dual system Zentek", a licence system for Recycling and collection of packaging material.

Nehlsen AG was founded in 1923 has 2.750 employees worldwide and in 2020 a turnover of 420m€.

2 What to achieve?

2.1 Targets of the European Commission

As Europe we set us high goals to achieve. Until 2020 we wanted to increase the preparing for re-use and the recycling of waste materials (such as paper, metal, plastic and glass) from households to a minimum of overall 50% by weight.
Furthermore, we wanted to reach an increasement regarding the preparing for re-use, recycling and other material recovery, including backfilling operations using waste to substitute other materials, of non-hazardous construction and demolition waste to a minimum of 70% by weight.

In addition, we now have the chance to reach the last goal set until 2025 subsequently, preparing for re-use and the recycling of municipal waste shall be increased to a minimum of 55%, 60% and 65% by weight by 2025, 2030 and 2035 respectively.

2.2 Waste Hierarchy

The European waste hierarchy was set in 2013, and the way to proceed was quite clear: "Preventing waste is the preferred option and sending waste to landfill should be the last resort."

Figure 1: Waste Hierarchy, EU Commission (www.ec.europa.eu/environment)

3 What currently happens

3.1 Population Growth

The world's population is increasing. Whereas we had around 2.5 billion in 1950, we now had 7.5 billion in 2015 and will have 10 billion in 2050.

It will be quite logical that more people will produce more waste and will consume more carbon oxygen, and this will be a fact, if we will not react and change our mind, now.

3.2 Plastics worldwide and in the EU

The worldwide plastic production has grown exponentially in just a few decades - globally from 1.5 million tons in 1950 to 322 million tons in 2015. This has also been accompanied by an increase in the amount of plastic waste generated. Means, 2.5 billion people produced 1.5 million tons in 1950; 7.5 billion people produced 322 million tons, in 2015.

Less than a third of plastic waste in Europe is recycled, the other 60% end up in the incinerator or in the landfill. Most of the plastics are produced for packaging, but also consumer goods and construction branch play a major role in plastic production. It may be assumed that especially the development of plastics for consumer goods will still play a greater role in future.

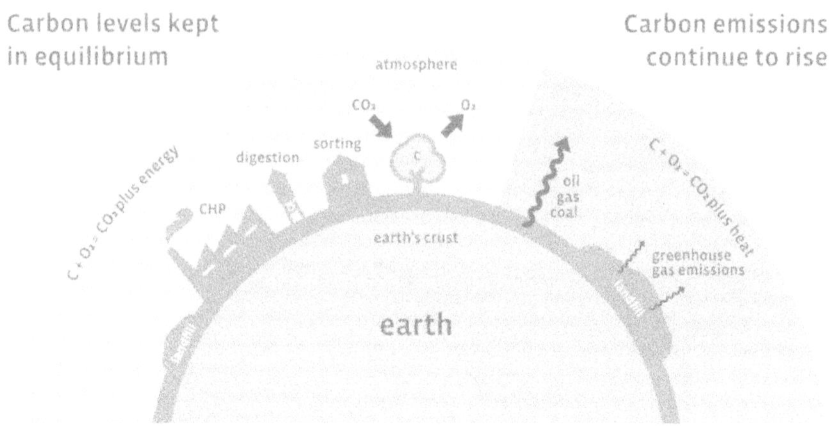

Figure 2: Carbon level (own illustration)

3.3 Development in Waste Management in Germany

Germany has made good progress in waste management in recent years. In 1970 Germany only separated one kind of waste: Household waste on a Landfill. Since 1986 when 'The Law on the Prevention and Disposal of Waste (AbfG – Abfallgesetz) comes into effect, there was a big change and waste was separated in: Household waste, paper, packaging, glass and partly bio waste. The development of new technologies showed alternative ways for waste treatment and for energy production.

The percentage breakdown of the typical content of household waste in Europe shows that a differentiated treatment of waste is necessary and productive. 30% organic waste does not must be incinerated but needs to be exploited via digestion, i.e. with biogas plants. Within the years until now, Germany started to use its waste for different processes: exploitation, digestion, recycling, incineration for energy production.

Rough estimated figures, based on years of experience working in waste industry, show that the development in Germany and also in Europe tends to an increasement of organic waste and paper. I little increasement also within the packaging waste and quite similar amount of glass. This positive impact is due the different collection of bio-waste. An Achievement of high efficiency of waste incineration is the result of different collection.

4 Possible solutions

4.1 Nehlsen waste and landfill management

In order to be up to date, to meet specifications, to be innovative and to be kind of role model, Nehlsen developed and implemented its own waste and landfill management (figure 2) where waste is sorted, occasionally more than three times, where waste is exploited, where waste is recycled, where waste is used to produce energy. All to follow one prior aim – taking responsibility for the environment and making the world a little better.

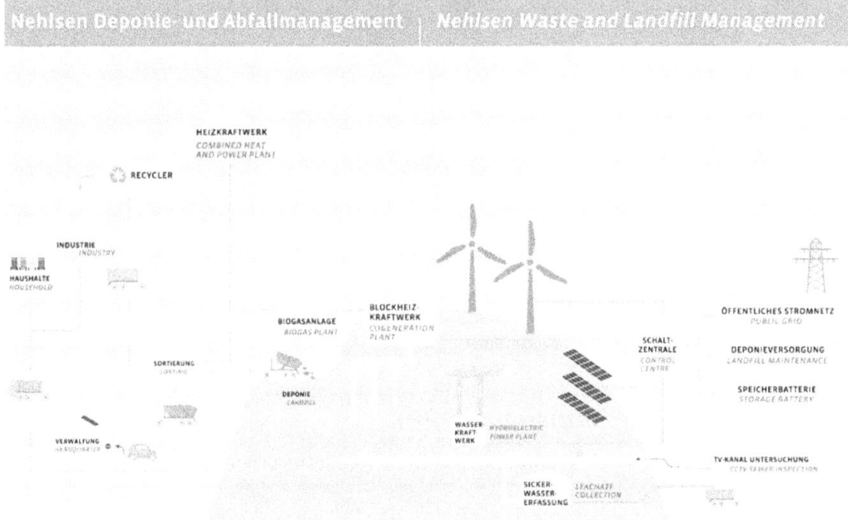

Figure 3: Waste and Landfill Management, Nehlsen AG (own illustration)

4.2 From a recycling to a circular economy

The next step for Germany is to implement a circular economy. The achievement of this aim is prognosed for 2030. Circular economy means to preserve and enhance natural capital by controlling finite stocks and balancing renewable resource flows. It is also a model to optimise resource yields by circulating products, components, and materials in use at the highest utility at all times in both technical and biological cycles. The effectiveness of this system is fostered by revealing and designing out negative externalities. As figure 3 shows, it is a complexify and new system whose implementation requires time, conviction, willingness and perseverance.

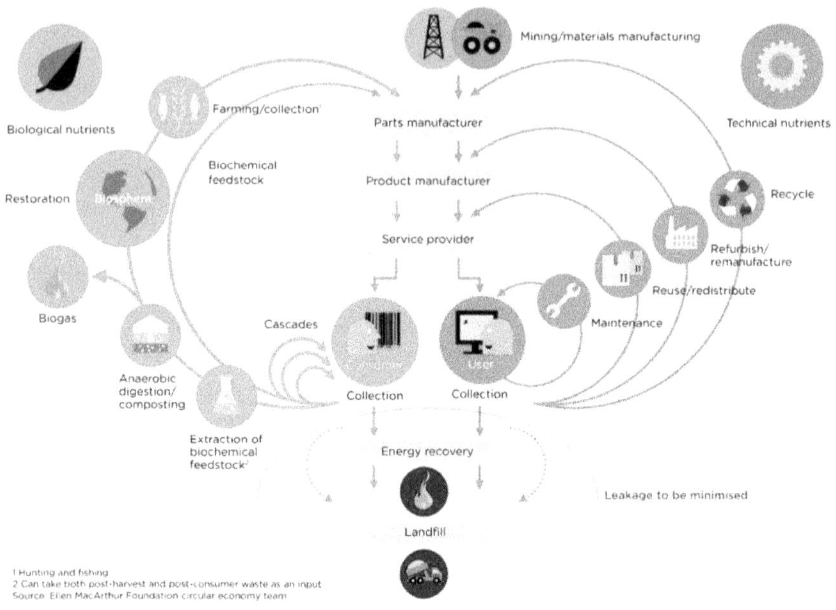

Figure 4: Circular economy (own illustration)

4.3 Public Private Partnership (PPP)

In order to rise the readiness to change, first the cornerstones must be set (governmental, environmental and mental).

Creating awareness, educating people, offering implementable solutions while acting along the three pillars of sustainability is a difficult task, but one that can be mastered. Through a strict and logical law construct (i.e. fee-systems, restrictions in packaging), through uniformly school education and through public campaigns.

All these solution approaches do not need to be newly invented, they worked already in other regions and countries, so why not helping other regions in being good in waste and landfill management? Entering Public Private Partnerships is a fair and good solution in order to spread the know-how and experiences from successful waste management systems. Figure 4 presents the four basic points of working within a PPP, especially the benefits of having fee stability, adopting fundamental know-how and risk sharing are essential points.

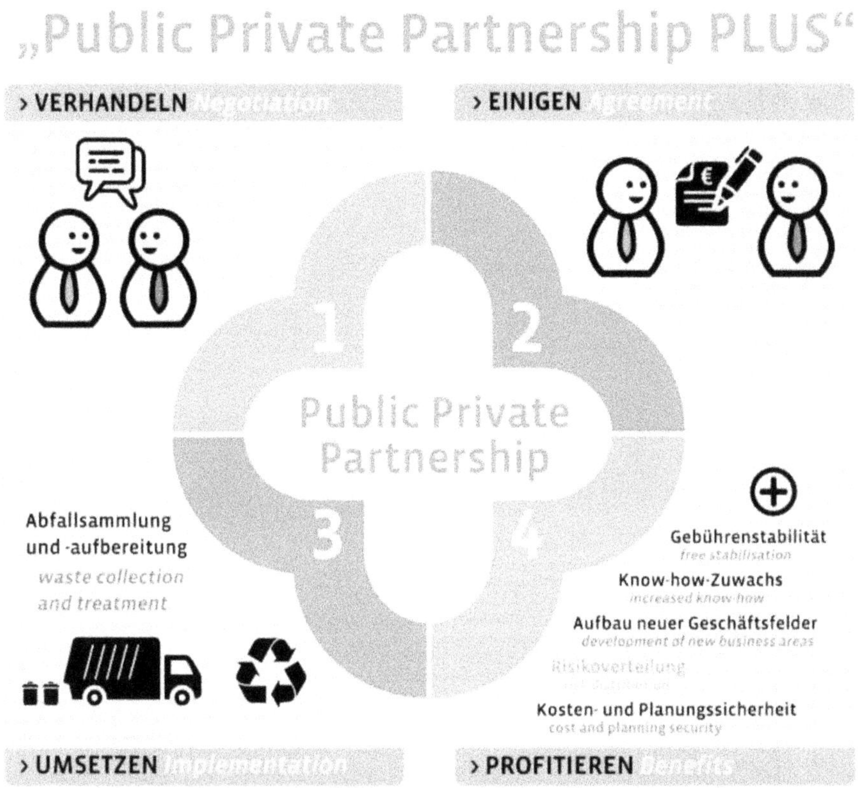

Figure 5: Public Private Partnership (own illustration)

Anschrift des Verfassers:

Peter Hoffmeyer
Nehlsen AG
Konsul-Smidt-Straße 50-52
D-28217 Bremen
Telefon +49 421 898 20 400
E-Mail Peter.Hoffmeyer@nehlsen.com

"More Recycling and More Economical Recycling by Optimising Collection "

Clemens Pues

Sales Director PreZero Recycling

PreZero Deutschland KG

Abstract

PreZero is an international company in waste and recycling management. The company provides waste disposal, sorting, processing and recycling services, combining all the expertise along the value chain under one roof. PreZero therefore positions itself as an innovation driver in the industry with the goal of creating a world in which resources are no longer wasted thanks to closed loops. In addition to a company presentation, the article describes how recycling can be sustainably improved by optimizing collection.

Inhaltsangabe

PreZero ist ein international tätiger Umweltdienstleister. Das Unternehmen bündelt mit der Entsorgung und Sortierung von Abfällen, der Aufbereitung sowie dem Recycling alle Kompetenzen entlang der Wertschöpfungskette. Das Unternehmen sieht sich Innovationstreiber der Branche mit dem Ziel, eine Welt zu schaffen, in der dank geschlossener Kreisläufe keine Ressourcen mehr vergeudet werden. Neben einer Unternehmensvorstellung beschreibt der Beitrag wie durch eine Optimierung der Sammlung das Recycling nachhaltig verbessert werden kann.

Keywords

Recovery, packaging, sorting, recycling, circular economy

Wertstoffe, Verpackungen, Sortierung, Recycling, Kreislaufwirtschaft

1 PreZero

1.1 Company presentation

PreZero is an international environmental service provider engaged in recovery and recycling management. Founded in 2009 under the name GreenCycle, the company now functions as the environmental division of the Schwarz Group based in Neckarsulm, which also includes the retail companies Kaufland and Lidl as well as the Schwarz Produktion.

With over 458,000 employees, the Schwarz Group is represented worldwide. In Europe and North America in particular, the company operates around 12,500 stores and markets. PreZero also operates internationally and currently employs over 4,800 people at around 140 locations in Germany, the Netherlands, Poland, Sweden, Austria, Italy and the USA.

Together with its customers PreZero connects economy with ecology and supports their successful and sustainable development with the key goal to close loops. With the division into its three business units Dual, Recovery and Recycling, PreZero itself already covers key stages along the value chain. With its licensing activities PreZero Dual is the strategic link between the manufacturers and the waste disposal companies. In the business unit Recovery PreZero ensures the separate collection, sorting and professional processing of recyclables from various fractions, which are processed into new products in our busines unit Recycling. In total, around four million tons of recoverable materials were collected by PreZero in 2019 and further processed by recycling. In addition, 175,000 tons of recycled materials were put back into circulation in Europe and the USA. PreZero does not only seek to be a classic waste disposal company but rather a modern recyclable materials manager for the future.

Operationally, PreZero serves more than 9.2 million citizens in Europe as a municipal waste disposal company and is a waste disposal partner for a wide range of commercial customers. Moreover, the company handles recoverable materials management within the Schwarz Group with its subsidiary GreenCycle.

This is supplemented by various departments and brands. The energy and environmental innovation unit within GreenCycle thus designs practical ideas for the energy transition as well as carefully handling our resources. With the PreTurn brand, intelligent multi-use load carriers and pooling services are also being developed. These are reshaping the entire supply chain efficiently and transparently and are thus optimizing it quite decisively. Last but not least, OutNature is also part of PreZero. OutNature is developing new and sustainable fiber and paper products for packaging solutions in retail and industry. It is doing so by using a new material: with the fibers of the cup plant, OutNature is focusing on a raw material that can offer an economically and ecologically viable alternative to virgin wood fibers used in paper production.

1.2 What drives PreZero forward

PreZero is committed to a clean future where an efficient and fully closed recycling loop protects our environment and creates sustainable value. The aspiration is to preserve resources and reduce the amount of non-reusable waste to zero – PreZero. As the environmental division of the Schwarz Group, PreZero pursues the idea of closed loops: from fully recyclable products through trade and disposal to sustainable recycling and reprocessing into new products.

With a clear vision and mission, PreZero has placed its focus on the fundamental challenges of the future and positioned itself as an innovative solution provider.

> The vision: New thinking for a cleaner tomorrow.
>
> The mission: With our innovative environmental services we close loops and thereby preserve resources.

Because the future belongs to those who build it – PreZero thinks and acts innovatively, efficiently, and responsibly. Simultaneously, PreZero is always open to new ideas and utilizes the potential and diversity of all employees. PreZero also engages in fertile cooperation with universities, promotes research and encourages a thirst for knowledge in students through collaborations. By thinking ahead, the company considers which concepts can solve the problems of tomorrow's customers. Therefore, PreZero also consistently invests in modern technologies and is constantly striving to innovate.

This innovative spirit has led to the development of numerous sustainable solutions. Recycled products such as paper from old cardboard within the Schwarz Group, household goods sold at Lidl made from recycled plastic by PreZero or biogas from food waste are sustainable outcomes resulting from increased efficiency. PreZero doesn't let anything go to waste: instead, where others see trash, PreZero sees valuable material. To achieve its waste management goals, PreZero is breaking new ground.

2 New Thinking for a cleaner tomorrow

2.1 The History of PreZero

PreZero is young but combines various experiences. As a digital sales organization of GreenCycle, PreZero.com was created in 2018 as an online platform on which customers could request a real-time quote for the disposal of recyclables and waste of various fractions. What couldn't be foreseen back then was, that just one year later, the PreZero brand would bundle together all of the GreenCycle services (founded in 2009) in the field of waste and recycling management ranging from consulting to disposal and further recycling.

Today, PreZero can already look back on more than ten years of experience in recyclables management. The company was initially founded to take care of managing recyclables from Lidl, Kaufland and the Schwarz Produktion within the Schwarz Group. Owing to increasing know-how and the success of waste management within the Schwarz Group, the company set out to make its range of services available to third parties.

2.2 Entry into the Operational Waste Management Business

Parallel to developing the PreZero brand, initial talks on the acquisition of the Tönsmeier Group took place. The waste disposal and recycling company from Porta Westfalica, a family-owned business at that time being run by its third generation, was founded in 1927 as a so-called "Bahnamtliche Spedition", meaning a forwarding agent entitled by railway authorities. In 1958 it was commissioned for the first time to carry out dust-free waste collection. At the beginning of the 1990s, the first locations were opened in the new federal states of Germany and the company pioneered the newly created system for the Extended Producer Responsibility (ERP) in Germany, the dual system. The company's success was raised to the European level when Tönsmeier opened its first branch in

Poland in 1996. In the 2000s, the company entered the plastics and wood recycling business. This was followed by participation in GRN Glasrecycling in 2005, before the two-flexible packaging (LVP) sorting plants in Porta Westfalica and Oppin were put into operation in 2006 and 2007. Following extensive renovations, the annual sorting capacity of the two plants now amounts to around 240,000 metric tons. 2010 saw the opening of the combined heat and power plant "Energie Anlage Bernburg" (today known as PreZero Energy).

The international business in the field of waste and recycling management was also expanded under the PreZero brand. Shortly afterwards, Sky Plastic, another leading recycling company, was integrated. With two operational sites in Haimburg (Austria) and Fonte (Italy), today's PreZero Polymers AG processes plastic waste and produces PP, PE and PS recycled material from it.

The international expansion of the waste management business continued in 2019 with the groundbreaking ceremony for the construction of a new sorting plant in Zwolle, Netherlands. Since January 2020, PreZero Netherlands has now been processing around 80,000 tons of light packaging per year in the eastern part of the Dutch provincial capital. With PreZero's participation in Kunststoff Recycling Grünstadt, PreZero took another important step towards a closed recycling loop in 2020, not long afterward, PreZero US opened the first LDPE recycling plant in California. In that same year, PreZero was pleased to announce growth in Poland through the purchase of Komart, a locally based disposal company there.

3 The Business Areas of PreZero

3.1 Dual

Europe-wide, product manufacturers are held responsible not only for their products but also for their packaging in terms of avoidance, reusability and recycling. With PreZero Dual, PreZero has successfully established its own dual system on the German market and is a reliable partner for customers for the licensing of sales packaging. By making effective use of existing waste management structures and expanding them with innovative approaches, PreZero wants to actively help shape and advance the EU's requirements for the further development of the circular economy.

3.2 Recovery

With PreZero Recovery PreZero aims to get the maximum potential value out of waste. From separate collection to sorting and processing, the management of recyclable materials forms the basis for the subsequent recycling of waste. A modern fleet of vehicles with the latest safety technology and a wide range of different types of containers are used for collection. Therefore, the employees are supported by the latest digital technologies. Once the collected waste has arrived at one of PreZero's modern sorting facilities, PreZero aims to separate the material streams as effectively as possible for further recycling.

Flexible packaging is currently processed by PreZero at various locations in Europe. Here, innovative, highly efficient separation processes are used to sort plastics, tinplate, Tetra Paks, paper, cardboard, carton and aluminum. However, PreZero is also an expert in sorting and processing recyclables from industry and trade, including paper, wood, glass, scrap and metal, as well as mixed commercial waste, and recycling these waste streams as high-quality raw materials for the market. Waste that can no longer be recycled is used by PreZero as an alternative fuel. From this, PreZero Energy generates climate-friendly energy and thus contributes to a reliable energy supply.

3.3 Recycling

PreZero Recycling ensures that the processed recyclables are turned into new raw materials for industry. In this way, PreZero helps to close loops. In Europe, PreZero is a leader in the recycling of post-consumer plastics with PreZero Polymers. PreZero's recyclable materials are the source materials for new applications in industrial production and are used in the gardening sector, furniture manufacture, household appliances, construction, home and office furniture, and the automotive industry, among others. At PreZero organic waste is processed into biogas and compost in its own composting and fermentation plants. For all other material streams, PreZero works with a partner network of metal and glass works and paper factories in Europe. In the U.S., PreZero has state-of-the-art recycling facilities in California and South Carolina. Plastic waste is converted there into high-quality recycling materials and organic waste into protein-rich animal feed products. These technologies make PreZero a pioneer in loop economy solutions in the US.

4 Collection optimization

4.1 Requirements for the circular economy

The main challenges facing the waste management industry can be summarized in a few key words: the achievement of the national and European recycling quotas and hence the provision of high-quality recyclates for the economic cycle. Whereas established systems are based on primary raw materials, switching to recyclate-based raw material use entails more than just the expected bottleneck of sufficient availability of input material, which can be remedied by investments at a single point in the supply chain.

In fact: At first, waste seems to be sufficiently available. When converting to the use of recycled materials, all players in the supply chain should switch over at the same time and in the most coordinated manner possible. However, since changes have to be made at all points in the supply chain at the same time, a "chicken and egg" problem arises very quickly when making investment decisions: Nobody wants to start or is able to do so, because they do not know whether the assumptions for the investment decision are valid and if they match the decisions of the other actors. The changeover to a different system of "economic activity" must thus involve all areas - and this starts with the collection of waste as new primary raw materials.

4.2 Collection as the basis for the design of downstream circular activities

With the collection of packaging waste, the basic types of design for collection systems raise the question of whether to use a pick-up or delivery system, and in the case of delivery systems, the question of how far households are required to bring their packaging. When deciding on design, in addition to economic considerations - delivery systems are initially cheaper - quantity and quality aspects are especially decisive. Experience has shown that higher quantities are collected in household collection than with delivery systems. Bag systems achieve better quality than container-based systems. When creating a system, deciding to scale down to just one aspect, would, however, be insufficient. When setting up a completely new system, the sorting, processing and reuse structures downstream of the collection are also being set up. With a staggered expansion of the collection system quantities can be increased in a targeted manner over time, for instance, and the structure of sorting and recycling can follow suit.

Figure 1 Examples of collection systems

However, higher collection quantities first lead to a higher input of materials which, as "misdirected waste", do not correspond to the actual purpose of the collection. These substances then increasingly lead to more complex requirements for sorting and recycling. At the same time, however, from a certain point, by expanding the collection system - e.g. in the case of lightweight packaging, by taking in "non-packaging of the same material" when introducing a recycling bin - further secondary raw material quantities can be tapped.

With respect to this, the following figure shows the composition of packaging waste depending on the type of collection, based on an exploratory analysis carried out at the beginning of January 2021. While the proportion of undesired articles was below 20% in the "yellow bag" and "yellow bin" systems, the proportion was higher for material collected with a so-called "recycling bin". The slight difference between the "yellow bag" and the "yellow bin" in this case was also due to the fact that the area with the bin collection had only recently been converted.

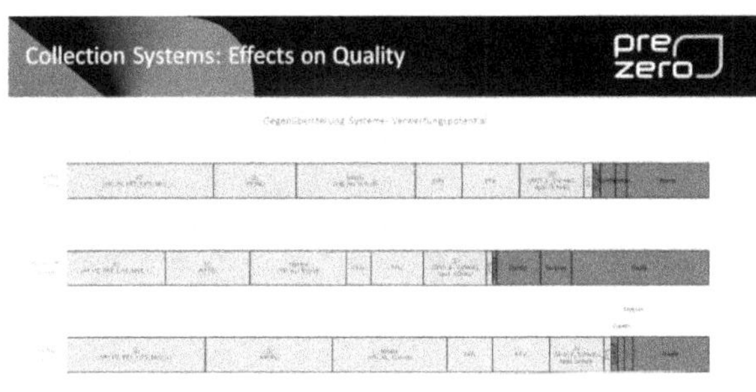

Figure 2 Collection systems and their impact on quality

Without accompanying measures, the quality of the collection usually decreases over time. The development of quantity and quality over time is in itself not an uncertainty that should hinder investment decisions. However, uncoordinated system changes in different collection areas combined with short contract periods for sorting and recycling mean that the composition of collected material becomes increasingly difficult to plan for a mixture. Hence, the sorting and processing plants are being increasingly required to adapt to materials with diverging properties. And this applies in particular if there are strongly fluctuating input qualities in the processing plants between the sorting and processing plants, due to short contract periods or contractual lack of options for establishing a direct supply relationship.

4.3 Collection and sorting

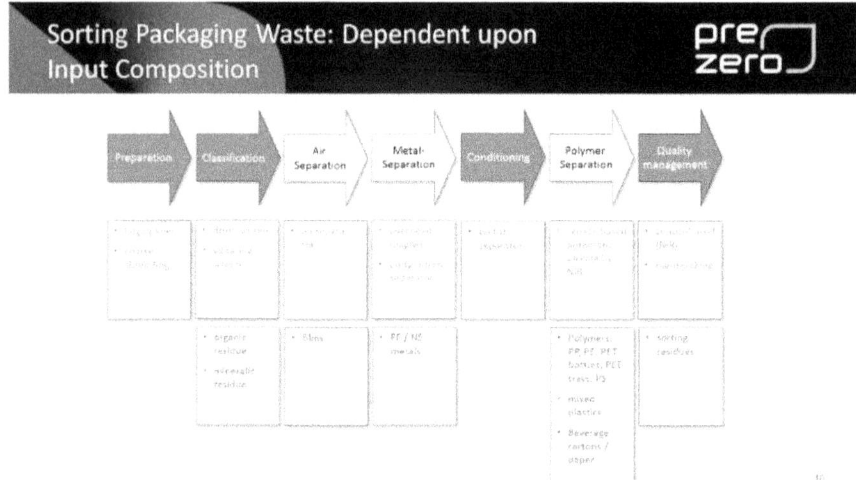

Figure 3 Sorting of separately collected packaging waste

The quality and cost-effectiveness of sorting depends essentially on the composition of the material with respect to specific valuable material components. The percentage distribution between the fractions is less important. The more variable the composition, the more difficult the long-term control of the plant with respect to yield and quality. Therefore, control mechanisms that show respective management changes in real time are all the more important to be able to act accordingly. Of course, the most desirable thing here is to have as constant an input as possible, which is most likely to enable consistent output qualities.

4.4 Collection and processing

Besides the effects of collection on sorting and thus also on processing, which have already been described by way of example, there are other direct effects. This is illustrated by the example of control of collection qualities, which is not done in Germany but has been practiced successfully for years in Norway and the Netherlands, for example.

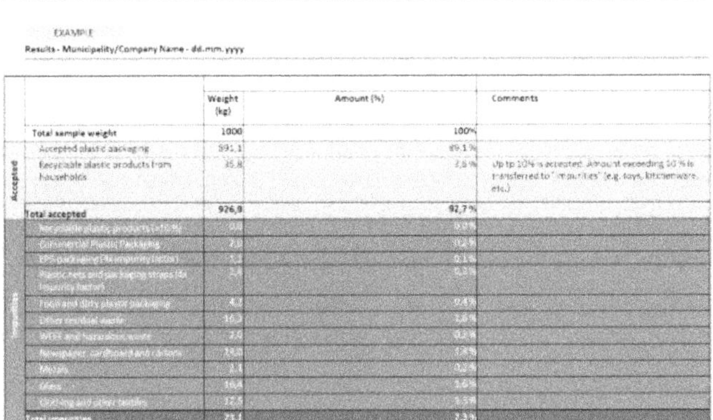

		Weight (kg)	Amount (%)	Comments
	Total sample weight	1000	100%	
Accepted	Accepted plastic packaging	891,1	89,1 %	
	Recyclable plastic products from households	35,8	3,6 %	Up to 10% is accepted. Amount exceeding 10 % is transferred to "impurities" (e.g. toys, kitchenware, etc.)
	Total accepted	926,9	92,7 %	
Impurities	[illegible]	[illegible]	[illegible]	
	Commercial Plastic Packaging	[illegible]	[illegible]	
	EPS packaging	[illegible]	[illegible]	
	Plastic nets and packaging straps	[illegible]	[illegible]	
	Food and dirty plastic packaging	[illegible]	[illegible]	
	Other residual waste	[illegible]	[illegible]	
	WEEE and hazardous waste	[illegible]	[illegible]	
	Newspaper, cardboard and cartons	[illegible]	[illegible]	
	Metals	[illegible]	[illegible]	
	Glass	[illegible]	[illegible]	
	Clothing and other textiles	[illegible]	[illegible]	
	Total impurities	73,1	7,3 %	

Figure 4 Assessment of collection material in Norway

In the Norwegian EPR system, as in Germany, all packaging used by end consumers is subject to licensing. This results - almost automatically -in the right to collect, sort and recycle this packaging in the "dual system". However, the effects of certain types of packaging on downstream sorting and recycling steps, such as nets, or the effects of increasing pollution are known. In the Norwegian system, collected packaging waste is tested on all wrapping at regular intervals according to a uniform scheme and transferred to an evaluation grid. Depending on the degree of deviation, the companies commissioned with the collection are given different requirements to correct negative deviations. A direct comparison of the sorting results in sorting systems with materials from other areas of origin clearly shows that the quality of the waste collected in Norway was rated as best and is therefore "more popular" with processors.

Contact data:

Dr. Clemens Pues

Sales Director
PreZero Recycling Deutschland KG
An der Pforte 2
32457 Porta Westfalica
Telefon + 49 571/9744-312
E-Mail: Clemens.Pues@prezero.com

Material Flow Analysis of Materials for Reuse and Recycling in the Greater Dresden Area

André Rückert, Christina Dornack

Institute of Wastemanagement and Circular Economy,

Technische Universität Dresden, Germany

Abstract

In a transformation experiment carried out in Dresden, a team of organised civil society is trying to make a contribution to resource conservation and climate protection through the mediation of secondary materials from different areas of urban society. According to the current state of research, despite a pandemic and closure for several months, 1,541 kg of residual materials have been brokered so far, resulting in a saving of around 2,426 kg of CO2 equivalents. With the help of such initiatives, an important step can be taken towards a circular economy and, at the same time, an increased ecological and economic added value can be realised in a region.

Keywords

Regional Material Cycles, Resource Conservation, Material Flow Analysis, Reuse, Recycling

1 Introduction

In a report on the use of natural resources in Germany, the UMWELTBUNDESAMT (2018) found that in 2016 around 1,040 million Mg of renewable and non-renewable raw materials were mined or harvested and processed in the economic system. At the same time, only about 13 % of the raw material demand in German industry is covered by secondary raw materials. According to the WUPPERTAL INSTITUT (2017), one reason for this may be the insufficient information on where and when waste is produced that could be used as a secondary raw material. Furthermore, the value of waste depends to a large extent on what is known about the composition of this waste and which potential users there are in an urban society for residual materials. This is because many production residues from small/medium enterprises in private or public ownership and also residues from e. g. trade fair buildings or exhibitions in public museums can be the raw materials for various stakeholders in an urban society. The transition to a circular economy makes it imperative to coordinate material flows and information flows more closely. Information on quantities and especially qualities of products and the raw materials they contain must be collected and maintained. In a transformation experiment in which the city of Dresden serves as a real laboratory, the mediation of residual materials between different stakeholders from urban society (Figure 1) and a mapping of the available residual materials for reuse and further use is being promoted.

Figure 1: Stakeholders in Dresden's urban society. (Source: Own research)

The exchange of used materials takes place in a warehouse located near the centre of Dresden, where stakeholders can drop off and pick up used materials. In the future, a digital representation of the materials in the warehouse via an open-source web platform will enable the exchange of leftover materials not only via a local warehouse, but also directly between the participating actors in the urban society. The main criteria of the material, such as size, external condition and number of pieces, are recorded and displayed online.

2 The Approach

Methodologically, the first task was to identify the materials in Dresden that have a high potential for reuse and recycling and which stakeholder groups are involved in these material transfers. For this purpose, 11 material categories with a total of 55 sub-categories were initially developed, which are based on the guidelines for standardised waste analysis from the SÄCHSISCHES LANDESAMT FÜR UMWELT, LANDWIRTSCHAFT UND GEOLOGIE (2016). In addition, seven possible stakeholders in the urban society were identified. At the beginning of the experiment, the entire inventory of the material storage was weighed and the origin of the materials was noted. In the further course of the experiment, newly added materials and materials leaving the warehouse were also weighed and their deliverers or collectors were noted. The data was collected centrally and processed in a material flow model.

3 First Results

From October 2019 to January 2021, approximately 3,476 kg of residual materials were collected or handed in by individual stakeholders at the warehouse. 1,679 kg of residual material was already in the warehouse before the start of the project. The total mass of residual material collected so far in the period of observation amounts to 5,155 kg. On the other hand, 1,541 kg of materials have been transferred, which leads to a surplus of 3,611 kg that still remains in the warehouse.

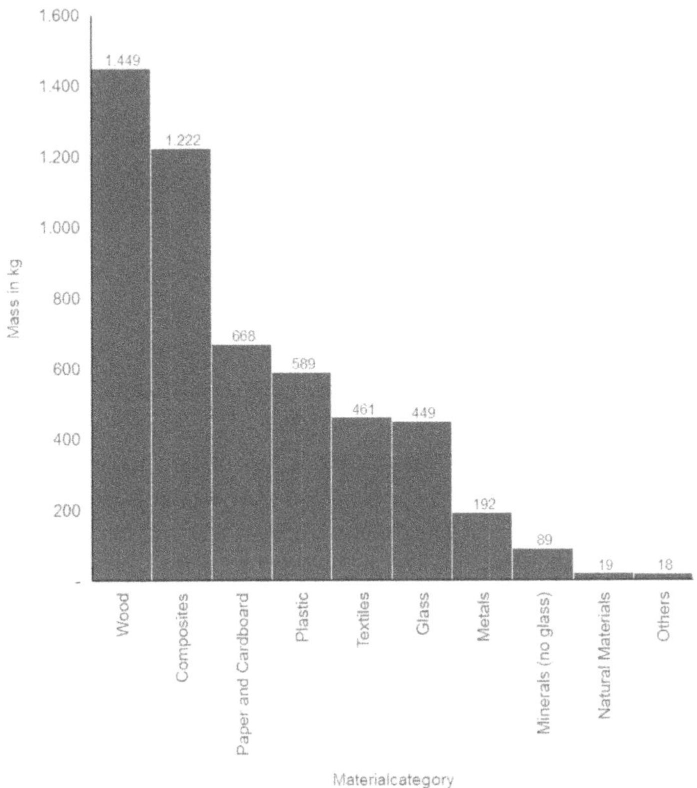

Figure 2: Composition of materials in the warehouse in kg. (Source: Own research)

The material that was delivered the most in terms of volume was wood with 1,449 kg, followed by 1,222 kg of composite materials and 668 kg of paper.

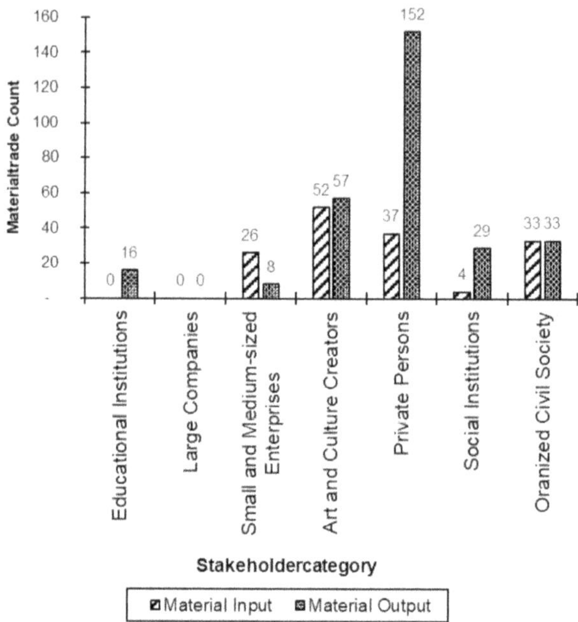

Figure 3: Number of individual material trades and participating stakeholders. (Source: Own research)

In addition to the material quantities, Figure 3 shows the number of deliveries and collections of material in the period under review. A total of 152 individual bookings of incoming material to the warehouse have been registered so far in the project period, compared to a total of 294 individual bookings of outgoing material. In combination with the material flow model shown in Figure 4, a clear redistribution of material between the individual stakeholders in the urban society can be seen. The delivered batches are thus larger in terms of quantity than the individual material outputs.

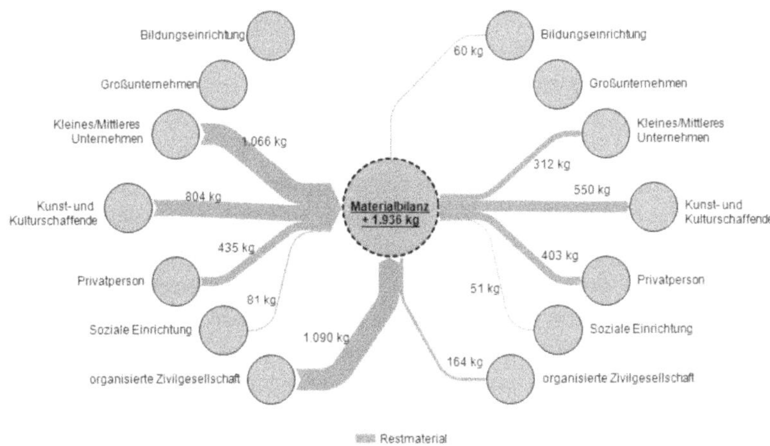

Figure 4: Material flows between stakeholders since the start of the project. (Source: Own research)

Figure 4 shows the material flows in the urban community determined so far. The material balance in the warehouse differs from the 3,611 kg in the previously mentioned balance surplus. This is due to the quantities already in the warehouse before the balance period, the origin of which can no longer be fully traced.

One finding is that individual small and medium-sized enterprises often hand over residual materials from production for reuse and recycling, but they themselves rarely obtain residual materials for this purpose. The main reason is the heterogeneity of the residual materials, which increases the workload for the companies to use them. This is also shown by initial findings from the EU Project FORCE (2016-2020) in which PRANDI et. al. (2017) dealt with waste minimisation through cascade use of waste wood.

Consequently, a reuse of the residual materials has so far mainly been conceivable by smaller start-ups and creatively working art and culture professionals as well as private individuals. Educational institutions and social institutions have so far only participated to a very small extent in the transfer of materials. Large companies have not participated at all in material procurement in the project period to date.

The Covid 19 pandemic had a restrictive effect on the operation of the material agency. This led to the storage facility being closed for several months during the project period and also made it difficult to further disseminate the idea in the urban society through workshops.

4 Risks and Chances

When residual materials are passed on, the associated risks and opportunities must be carefully evaluated. Risks exist, for example, in the danger of pollutants contained in the material from the start and pollutants that could be transferred through contamination during the primary use of the material. Furthermore, there is a risk of pollutants being released when material is repurposed and processed. For example, materials that are intended for indoor use could be used in an outdoor area through creative use, where pollutants could be released into the environment through exposure to environmental factors. At the same time, the use of materials that were originally intended for outdoor use could lead to the evaporation of volatile organic pollutants when used indoors. But also the modification of the residual materials can lead to a release of pollutants. HAUFFE (2020) identified various risks of the release of pollutants from residual materials and identified above all flame retardants, plasticisers and adhesives as possible pollutants that could be problematic when materials are repurposed. Thermal treatment by laser or remelting of plastics was identified as particularly critical. In addition to the ingredients, the size of the material also poses a risk, for example, when it is used in creative workshops with children, as some materials pose a risk of being swallowed. HAUFFE (2020) developed the following recommendations on how to deal with these materials:

- Use the materials in a setting that is as close as possible to their former use.
- The use of materials in handicrafts should be determined on the basis of the possible ingredients of the material. For example, materials containing adhesives should not be passed on to kindergartens, as the adhesive as well as the carrier material with plasticisers could very quickly come into contact with the mouth.
- Avoid passing on small materials to very young "handicraft groups" in order to avoid the danger of swallowing and choking.
- When passing on material whose composition is not completely clear, it should be pointed out that thermal processing can be an increased source of danger and should only take place under professional conditions.
- Material that may have been treated with flame retardants should be handled with particular care. This applies to storage and handling during the material transfer itself as well as to further processing. Thermal treatment is strictly discouraged

The residual materials that are reused are analysed at random for inorganic pollutants.

The risks mentioned are counterbalanced by many opportunities. First of all, the highest goal of the waste hierarchy - waste prevention - can be fulfilled in a region through the regional procurement of secondary materials. This promotes resource conservation by saving primary materials and reduces climate impacts by saving greenhouse gases released during the production phase of the materials. In the course of the project, 2,426 kg of CO2 emissions have already been saved according to defensive calculations. In the long term, an accounting of the avoided CO2 emissions could create new opportunities for actors in the circular economy to participate in CO2 certificate trading, for example.

Besides the positive ecological effects, there are also economic incentives to promote the regional mediation of secondary materials. First of all, the trade of Second-Hand goods and materials can generate proceeds that enable the initiatives to finance themselves. An example of a financially sustainable brokerage of Second-Hand goods and materials are the three "Stilbruch" department stores of Stadtreinigung Hamburg, which achieved a turnover of about 4 million € in 2020 and are financially self-supporting (cf. SIECHAU, 2021). Another example is the "Nochmall" Department Shop of Stadtreinigung Berlin, which also offers Second-Hand goods for reuse. In combination with other instruments of resource conservation, such as repair cafés and affiliated ReUse hubs for the careful collection of secondary materials, these enterprises can contribute to an increase in regional value creation (cf. THÜRMER, 2021). Furthermore, companies that put their residual materials to a second use also benefit financially by reducing waste disposal costs.

In addition to the financial added value, jobs in the circular economy can also be created by passing on leftover material to artists, designers and start-ups in the region.

The fact that the topic of recycling and reuse of residual materials is relevant is also shown by the web-based recycling exchange of the Chamber of Industry and Commerce, which is so far only accessible to companies, as well as the growth of the German ReUse Community. In addition to Dresden's "Zündstoffe" project for material mediation, material mediation initiatives are also emerging in Berlin, Leipzig and many other cities in Germany that want to make their contribution to increased regional value creation and resource conservation.

5 Conclusion

The research results so far show that the mediation of residual materials offers opportunities for increased value creation in an urban society, which brings both ecological and economic benefits. If certain precautionary measures are taken during the mediation and subsequent processing of residual materials, it is possible, according to current knowledge, to further process the materials without health concerns. Furthermore, it must be examined whether the municipalities can make a contribution to resource conservation by supporting or operating the material mediation services. In addition, it must be investigated to what extent residual materials that would be suitable as secondary raw materials but are handed in at recycling centres can be brought into a second value chain before they are considered waste.

6 Acknowledgement

We would like to thank the Federal Ministry of Education and Research for the three-year funding and the City of Dresden for the implementation of the Future City Process, in which the project "Zündstoffe - Materialvermittlung Dresden" has been realised since October 2019.

7 Literature

Hauffe, M.	2020	Recherche zu Schadstoffen in Restmaterialien für das Projekt „Materialvermittlung Dresden". TU-Dresden.
Industrie und Handelskammer	2021	IHK Recyclingbörse, https://www.ihk-recyclingboerse.de/ (zuletzt zugegriffen 19.03.2021)
Prandi M., Righeschi S., Ferrando G., Piersantelli N., Cepolina S., Capannelli G., Incerti E., Caruso M., Pizzorno A. C., Marzoli I.	2017	Four Collection Schemes, EU-FORCE Projekt, https://drive.google.com/file/d/180FBqY71kha-BksnDFM6spR4ZmG-Ed6F/view (zuletzt zugegriffen 19.03.2021)
Siechau R.	2021	Beitrag: Zero Waste: ReUse bei Stilbruch Hamburg Erfahrungen und Ausblick. 17. Kreislaufwirtschaftstage Münster 2021.
Thürmer A.	2021	Beitrag: Zero Waste Stadt Berlin?! 17. Kreislaufwirtschaftstage Münster 2021.
Umweltbundesamt (Hrsg.)	2018	Die Nutzung natürlicher Ressourcen - Bericht für Deutschland 2016.
Wuppertal Institut für Klima, Umwelt, Energie gGmbH (Hrsg.)	2017	Digitale Kreislaufwirtschaft. Die Digitale Transformation als Wegbereiter ressourcenschonender Stoffkreisläufe.

Author's address:

M. Sc. André Rückert
TU Dresden Institut für Abfall- und Kreislaufwirtschaft
Pratzschwitzer Str. 15
D-01796 Pirna
Telefon +49 351 463 441 41
E-Mail andre.rueckert@tu-dresden.de

Zero Waste Cardboard Technology to Support Product Stewardship

Aharon Arakel, Bithi Roy, Rubick Arakel, Warwick Madden

Pact Renewables Pty Ltd, Sydney, Australia

Abstract

The principles of sustainability, as a core measure for reducing the adverse impacts of climate change, are now acknowledged globally. To reduce waste and reliance on landfilling and incineration, the concept of "take-back" (also expressed as "Reverse Logistics", "Extended Product/Producer Responsibility" and "Product Stewardship") is now increasingly being accepted by manufacturing industries and implemented in many jurisdictions, on a voluntary basis or through enactment.

This paper discusses the lifecycle issues associated with application of current product stewardship approaches to sustainable management of paper and cardboard packaging waste. We also provide an example of how technology-based innovation can simplify the application of reverse logistics to achieve tangible paper/cardboard product stewardship. The technology example presented herein involves direct use of waste paper/cardboard in manufacturing of a variety of industrial products and consumer goods that at the end of their useful life can be placed in soil to degrade and provide conditioning effects. By doing so, no waste is generated and reliance on landfilling or incineration of such waste paper/cardboard is avoided. However, as the challenge of waste paper/cardboard is complex and growing other zero waste technologies are also needed for uptake of the best fit options by industry and governments, and to promote product stewardship, design thinking and active engagement of the consumers.

Keywords

Waste paper and cardboard, Product Stewardship, Extended Producer Responsibility, Reverse Logistics, CtP technology

1 Introduction

The adverse impacts of climate change have accelerated the global search for sustainable use of resources and activities to reduce the impacts and preserve limited resources. This search for ethical sustainable governance (ESG) is reflected by numerous recent consumer surveys, that show over one third of consumers are willing to purchase environmentally friendly products. Sustainability is not just another buzzword as in the case of manufacturing of industrial products and consumer goods it requires implementation of reverse logistics for all manufacturers due to the growing environmental concerns, legislation, corporate social responsibility and sustainable competitiveness. Reverse logistics refers to the sequence of activities required to collect the product used by a consumer for the purpose of reuse, repair, re-manufacture, recycle or disposal. Reverse logistics has been expressed in different ways in different jurisdictions including the "take-back" concept, "Extended Product Responsibility" (EPR in U.S.A.) or "Product Stewardship" in Australia, ranging in enforcement from a voluntary basis to enacted laws. A careful review of relevant literature shows that although reverse logistics is still in an evolutionary phase, in fact the product take-back concept, initially developed in Germany and Japan in the early 1990's, has now evolved globally into a broader initiative known as "Extended Producer Responsibility" (EPR). EPR is a policy concept in which a producer's physical and/or financial responsibility for a product is extended to the post-consumer phase of the product's life-cycle (e.g., EU Directive, 2018). EPR policies include take-back mandates, as in Germany, but other types of protocols and instruments may also fall under the EPR umbrella, such as "product stewardship", a voluntary based EPR concept in Australia.

Accordingly, despite their contentious nature, "take-back" or EPR laws and initiatives were first enacted in Germany in 1991 under the so-called product "take-back" law and known as Packaging Ordinance. This required packaging manufacturers and distributors to take back packaging from consumers and ensure that a specified percentage of it is recycled. The law was facilitated by manufacturers and distributors meeting their obligations by joining a "producer responsibility organization" which handles collection and arranges for recycling (PALMER AND WELLS, 2002).

Faced with packaging waste making up between 20 to 30 percent of the total weight of the municipal solid waste stream (TANAKA, 1998; CLEAN JAPAN CENTRE, 2001), Japan was another early adopter of EPR by enacting *The Container and Packaging Recycling Law* in 1995. This law enforces manufacturers to be responsible for meeting phased-in recycling rate targets for glass and PET bottles, followed by targets for paper and plastic containers and packaging. The Japanese refer to the targets as "voluntary" and as in the Netherlands, local government in Japan maintains responsibility for collection of packaging waste and return to industry responsible for paying for recycling. Since then, many European countries have mandated EPR programs under the EU Directives (2018) or their own directives for many consumer goods.

In the United States, there has been resistance to wholesale adoption of the EPR approach as the focus on producer responsibility has been contested by U.S. business interests who consider the idea of replacing the decentralized solid waste collection and recycling system, that already exists in the U.S. as a duplication. This system is a centralised mandatory system in which producers are responsible for collection and recycling their products at end-of-useful life. Despite this, the U.S. Environmental Protection Agency, some state governments and industry groups have commenced supporting some of the ideas behind the EPR concept. There have been some shifts in the U.S. packaging industry groups who have traditionally opposed to mandated producer funding of recycling, towards supporting the EPR concept by means of engaging on the legislative front and support of recycling funding provided that the programs are set up in specific ways and meet certain requirements at the state level. These shifts are partly driven in response to recycling markets being recently thrown into turmoil in the wake of China's *National Sword policy*, which has driven a renewed focus by policymakers, at both the state and federal levels, on trying to stabilize the economics of materials recovery. For example, in a distinct shift from its previous position, in October 2020, Ameripen, a group representing several large U.S companies across the packaging-material spectrum, adopted a new internal policy on how the group would engage on the legislative front when it comes to recycling funding (AMERIPEN, 2021).

Product stewardship is relatively new in Australia, but in the wake of National Sword policy, it is becoming a major concern for state and federal governments as well as communities. Although sporadic, the efforts with product stewardship in Australia has so far reduced the environmental and human health impacts of selected waste streams generated by automotive, electronic, beverage and agricultural industries. In response to the slow progress with industry uptake, the establishment of a national centre of excellence for product stewardship was announced recently (PRODUCT STEWARDSHIP CENTRE OF EXCELLENCE, AUSTRALIA; 2020) to promote collaboration between industry and community groups.

As indicated earlier, the progress with reverse logistics is still is an evolving phase because of the issues related to adoption, implementation, forecasting product performance, outsourcing, networking from a secondary market perspective, and disposition decisions are yet to be thoroughly examined. One drawback hindering a broad-based adoption and implementation of reverse logistics relates to poor application of the reverse logistics concept, as defined presently, to certain waste categories due to the lack of technical solutions for product wastes having short recycling loops which leads to their eventual disposal in landfills or incinerators at the end of their useful life. For example, in contrast to wastes made from glass and plastic with long recycling loops (28-30 rounds), recycling of waste paper/cardboard is problematic because of short loops (2 recycling loops for high quality packaging boxes and up to 8 recycling loops for low quality egg cartons and beverage trays). A compounding problem with paper/cardboard waste is the high lifecycle cost associated with disposing cardboard waste at the end of its useful life, due to its low value and bulky nature.

In the context of reverse logistics, the challenge of waste paper/cardboard has been exacerbated with the exponential increase in the amount of cardboard packaging waste generated through increased online purchasing of consumer goods and prepared food due to the global Covid-19, that ultimately all need sustainable disposal solution. To move forward, innovation is needed to drive product stewardship for waste paper/cardboard in tandem with other waste categories in order to achieve global progress with impactful outcomes.

In this paper, we discuss lifecycle issues associated with current approaches to reverse logistics and the challenges in applying these approaches to addressing the product stewardship challenge for the ever-mounting paper/cardboard-based packaging waste. We also provide an overview of how technology-based innovation can drive reverse logistics to achieve waste paper/cardboard product stewardship. The technology presented herein involves the direct use of waste paper/cardboard for the manufacture of a variety of industrial products and consumer goods that, at the end of their useful life, allow degradation once placed in soil and provide conditioning effects. By doing so, landfilling or incineration of such waste is avoided.

In the following sections, the terms EPR, product stewardship and reverse logistics are used interchangeably, as their concepts are compatible with each other.

2 The challenge

Figure 1 is a schematic example of the conventional approach to reverse logistics proposed for various waste categories (SRIVASTAVA, 2008; ZIELINSKA, 2020). As shown, raw material from recycling of waste is used for production and distribution of new industrial products and consumer goods. In this scenario, the packaging waste from delivered industrial products and consumer goods is recycled through options of (a) repair/reuse, renovation and distribution, or (b) disassembly, service and distribution, or (c) regeneration and production, or (d) recycling of the waste paper/cardboard to produce raw material for manufacturing new packaging products.

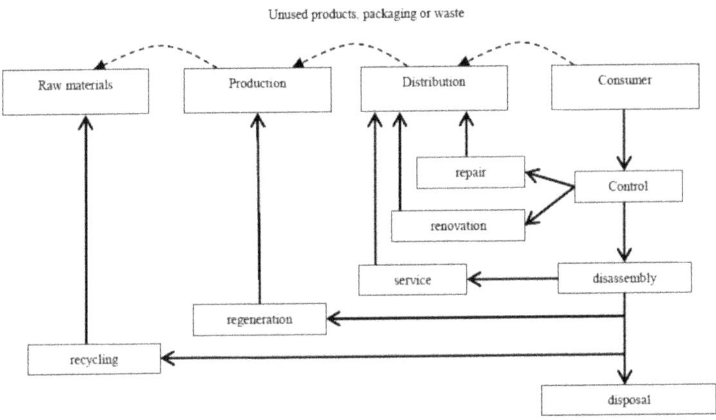

Figure 1 An example reverse logistics scheme used in this paper for discussing lifecycle aspects of paper/cardboard as related to product stewardship (Source: Zielinska, 2020; adopted from Srivastava S.K 2008).

As indicated earlier, these options although relevant to most other waste categories, cannot be adopted satisfactorily for the waste paper/cardboard category as:

- option (a) is not applicable because the quality of cardboard packaging material by simply repair/reuse/renovation will be the lower than a new cardboard packaging material;

- option (b) is also not feasible because of the damage caused to cardboard material during disassembly/dismantling;

- option (c) is also not applicable because inability to regenerate the packaging product after disassembly;

- option (d) although possible is limited to a few recycling loops and will incur substantial life cycle costs.

It should be immediately noted, that whereas option (d) might offer some scope, in practice, the end product of this option is reached just after a few recycling loops, and is commonly in the form of low quality products (e.g. egg cartons and beverage trays) and ultimately end up in landfills.

3 Proposed approach for addressing the challenge

Considering the above constraints, new approaches are needed to address the key issues related to waste paper/cardboard management (i.e., to avoid landfilling and incineration). The new approaches should offer solutions for two fundamental challenges, namely:

- how to achieve zero waste discharge?
- how to make the solution attractive to Industry by producing products from waste paper/cardboard with lower life cycle costs?

With reference to Figure 1, one approach is using the dissembled waste paper/cardboard from any recycling loop and use it as part of a feedstock material for manufacture of industrial products and consumer goods that become degraded once placed in soil. This approach has been used by Pact Renewables Pty Ltd from the outset to develop a zero waste paper/cardboard technology which overcomes the constraints discussed earlier. The technology, known as CtP (cardboard-to-product), diagrammatically shown in Figure 2 and further described below, offers a low lifecycle cost option for applying reverse logistics for sustainable management of waste paper/cardboard and achieve impactful outcomes.

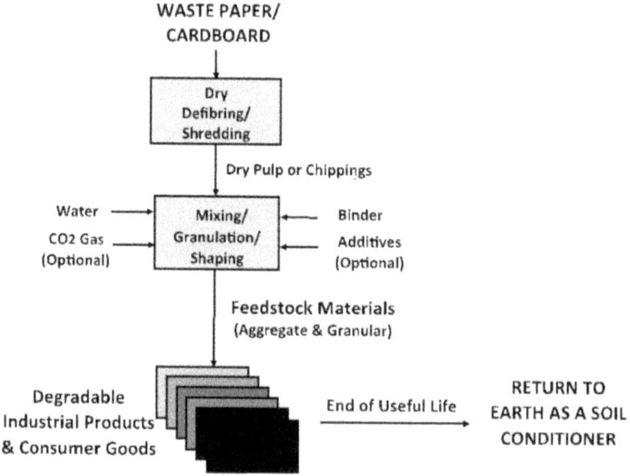

Figure 2 Schematic presentation of CtP zero waste technology, showing the key process steps leading to beneficial use of waste paper/cardboard to produce degradable products.

The technology behind the proposed approach is essentially comprised of two process components, namely, dry defibring of waste paper/cardboard and then mixing the defibred pulp with a proprietary mineral-based binder, water and selected additives in a vessel to produce degradable feedstock materials. The feedstock materials, from which the products and goods are produced are comprised of pulp/chippings, mineral-based binders (made of sulphates, carbonates and hydroxides of Ca, Mg and K), and optionally one or more degradable additives. The feedstock materials, in the form of either aggregate or granules, are used beneficially for producing industrial products and consumer goods using conventional equipment such as moulding machinery for the manufacture of agricultural containers. For some product streams, a conventional or a purpose-built agglomeration unit can be used to produce pebble and granular products of desired sizes, shapes, textural features and functionalities, according to end user requirements.

The amount of cardboard in the feedstock materials depends on the intended application area but generally can be up to 60% by weight of total feedstock material. Additives may include organic and inorganic fillers, pestcides, mineral nutrients, colourants and coating agents. A distinct advantage of the feedstock material of this technology is the ability to absorb and react with CO_2 gas in order to refine the minerology of the binders for producing end products with additional functionalities, such as improved water holding capacity and enhanced compressive strength.

Since the conception of the technology, numerous laboratory and field trials have been undertaken to assess the technology application to various industries and assessment of the performance of the products for target application areas. Table 1 provides a summary listing of currently identified application areas of the technology reflecting its attractiveness to sustainably address the paper/cardboard waste challenge. Furthermore, as the technology assessments progress, new application areas are being identified systematically assessed for product quality, functionalities and degradability.

Table 1 Summary list of identified application areas of feedstock materials produced by CtP technology using waste paper/cardboard

Application Area	Product Types
Agriculture and food production	• Containers for nursery/seedling, horticulture forestry, landscaping and mine site tailings vegetation, etc. horticultural, agricultural containers • Granular and sheet mulch • Granular soil conditioners for revegetating land divisions, decommissioned landfills, mine sites and brownfields, landscaping and commercial orchards, and garden watering and nutrient supply • Grow media for greenhouse/glasshouse farming

Odour control media	• Aggregate for odour reduction in poultries, piggeries, cattle farms, etc. • Aggregate for reduction of malodour from food consumption (household, restaurants, military, cafes, grocery stores, schools, hotels, cruise and cargo ships, hospitals, etc) and food waste management (including composting, garbage collection, waste management centres)
Goods packaging fillers	• Aggregate and granular fillers for goods packaging • Aggregate fillers for padded envelopes
Garden decoration and landscaping	• Colourful and nutritious granules and pebbles for home garden decoration and landscaping
Compost amendments from food waste	Granular compost amendments incorporating fruit /vegetable/fish/meat/dairy product waste generated by household and commercial food production operations

Figure 3 shows images of selected products containing waste paper/cardboard. Long-term field monitoring and laboratory assessments, using a combination of visual and microscopic observations, leachate analysis and mineralogical determination by X-Ray Diffraction methods have been performed to assess the degradability of various product streams. The results point to the effectiveness of the combined action of physical, geochemical and biological processes (e.g., plant root growth through the walls of the agricultural containers) in degrading and integrating the products in the receiving soils over a 6-12 months of placement in garden or forestry soil.

Figure 3 – Examples of products containing waste paper/cardboard. Viewing from left to right – a partially degraded agricultural container; packaging fillers for a consumer good package; and field trial of soil conditioners.

4 Benefits offered by the proposed technology solution

Aside from the economic and environmental benefits arising from using waste paper/cardboard as part of feedstock material formulations, as described earlier, CtP is potentially and enabling technology for the successful application of the principles of reverse logistics for sustainable management of waste paper/cardboard. Accordingly, this zero waste technology avoids landfilling and incineration of low value voluminous waste paper/cardboard by using it to manufacture products which degrade in soil at the end of their useful life.

Further, the products made from CtP technology incur low lifecycle costs and have functionalities similar or even better than the similar products available in markets thus making products using waste paper/cardboard commercially attractive for investments. Additionally, the technology is scalable and adaptable as an end of pipeline solution; hence, production operations using the technology can be located near industries/population centre where waste paper/cardboard is available in sufficient quantities to achieve the desired economies of scale.

Finally, the sustainability of the technology is further enhanced through use of widely available mineral resources in formulating the binders, utilising conventional mineral and food processing equipment.

5 Conclusions

Review of industry news and views and scientific literature point to a significant progress with Community understanding and industry recognition of the need for product stewardship, as related to plastic and glass waste streams. To some extent, this has been at the expense of a slower progress with addressing the challenge of waste paper/cardboard as indicated by the absence of any sustainable and cost effective solution and as a result the bulk of this major waste stream is still ending up in landfills or incinerated.

Paper/cardboard waste is voluminous, low value and being fibrous has a short recycling loop, therefore managing it in a way to avoid landfilling/incineration requires technology-based innovation that enable zero waste discharge. It is further highlighted herein that other technologies, such as the one exemplified in this paper, need to be developed and market tested before their uptake by industry and governments. A whole community engagement is also needed to support the best fit product stewardship schemes, promote product design thinking and drive the process of active consumer engagement for addressing the challenge of this difficult and largely forgotten waste stream.

6 References

Ameripen;	2021	https://resource-recycling.com/recycling/2021/01/19/ameripen-starts-to-shift-on-recycling-policy/
Clean Japan Centre;	2001	Recycling-Oriented Society: Towards Sustainable Development, Tokyo
EU Directive, 2018/852;	2018	Directive of the European Parliament and of the Council of 30 May 2018 amending Directive 94/62/EC on packaging and packaging waste (Official Journal of the European Union, L 150/141, 14.6.2018)
Palmer, K. and Walls. M.;	2002	Economic Analysis of the Extended Producer Responsibility Movement: Understanding Costs, Effectiveness, and the Role for Policy
Product Stewardship Centre of Excellence, Australia;	2020	https://stewardshipexcellence.com.au/product-stewardship/
Srivastava, S.K.;	2008	Network design for reverse logistics, *Omega. The International Journal of Management Science*, 36(4), 535-548
Tanaka, M.;	1998	Waste Management and EPR Practices in Japan. OECD Workshop on Extended and Shared Responsibility for Products: Economic Efficiency/ Environmental Effectiveness, 1-3 December, 1998, Washington, D.C.
Zielinska, A.;	2020	A comparative analysis of reverse logistics implementation for waste management in Poland and other European Union countries

Author's address(es) *like this example:*

Dr Aharon Arakel
Pact Renewables Pty Ltd
P.O. Box 92
Thornleigh, NSW, 2120, Australia
Phone +61 29484 4274
E-Mail info@pactrenewables.com

Vertreiberrücknahme von Elektro(nik)altgeräten – ein Erfolgsmodell?

Dr. Ralf Brüning, Julia Wolf

Dr. Brüning Engineering UG, Brake

Dr. Stephan Löhle, Ute Schmiedel

cyclos GmbH, Osnabrück

Dr. Ines Oehme, Kerim Zaidi

Umweltbundesamt, Dessau-Roßlau

Take back of waste electrical and electronic equipment by distributors – A success story?

Abstract

The directive 2012/19 EU concerning waste electrical and electronic equipment (WEEE) established new obligations for distributors concerning the take back of WEEE. In Germany, the ElektroG2 – that contained those obligations - came into force on the 24th of October 2015. The aim of the project „Assessment of the efficiency of the distributor obligations established in the ElektroG" for the German Federal Agency of the Environment is a systematic evaluation of the new distributor obligations concerning the take back of WEEE, registration and reporting. Based on the results of data evaluations, on-site inspections and stakeholder interviews, policy recommendations for future adjustments of distributor obligations are given.

Inhaltsangabe

Die Richtlinie 2012/19/EU über waste electrical and electronic equipment (WEEE Richtlinie) legt Rücknahmeverpflichtungen für Vertreiber fest. In Deutschland wurden diese durch das ElektroG vom Oktober 2015 umgesetzt. Das im Auftrag des deutschen Umweltweltbundesamts durchgeführte Projekt „Effizienzbestimmung der Vertreiberpflichten nach ElektroG" hatte die Zielsetzung eine systematische Bewertung der neu eingeführten Vertreiberpflichten (Rücknahme-, Anzeige- und Mitteilungspflichten) durchzuführen. Als Ergebnis wurden aus Erkenntnissen einer Datenauswertung, Vor-Ort-Besuchen und Stakeholderbefragungen konkrete Handlungsempfehlungen für eine zukünftig bessere Gestaltung der Vertreiberpflichten abgeleitet.

Keywords

WEEE-Richtlinie, Elektro(nik)altgeräte, EAG, ElektroG, Vertreiberpflichten

WEEE directive, electrical and electronic equipment, WEEE, distributor obligations

1 Einleitung

In der Richtlinie 2012/19/EU über waste electrical and electronic equipment (WEEE Richtlinie) wurde erstmals festgelegt, dass die EU-Mitgliedsstaaten Rücknahmeverpflichtungen für Vertreiber in die nationale Gesetzgebung bzgl. Elektro(nik)geräten (EEG) aufnehmen müssen. In Deutschland wurde die Richtlinie 2012/19/EU durch das ElektroG vom Oktober 2015 umgesetzt.

Nach dessen Inkrafttreten am 24. Oktober 2015 ist die Bundesregierung unter anderem zur Evaluierung der Vertreiberpflichten nach § 17 ElektroG verpflichtet. Weiterhin wurden im ElektroG neue Rücknahme-, Anzeige- und Mitteilungspflichten festgelegt. So sind seit dem 24. Juli 2016 gemäß § 17 Abs. 1 ElektroG erstmals Vertreiber mit einer Verkaufsfläche von mindestens 400 m² für EEG zur unentgeltlichen Rücknahme von Elektronikaltgeräten (EAG) verpflichtet. Dies gilt auch für den Fernabsatzhandel, sofern die Lager- und Versandflächen für EEG mindestens 400 m² erreichen. Vertreiber können zurückgenommene EAG u. a. an Hersteller und/oder öffentlich-rechtliche Entsorgungsträger (örE) übergeben, sofern sie zurückgenommene EAG nicht selbst behandeln. Dem Fernabsatz sowie dem stationären Handel obliegt in diesem Zusammenhang die Möglichkeit sich Dritter, z.B. einem Rücknahmesystem für EAG zu bedienen.

Die Vertreiberrücknahme bezieht sich zum einen auf eine 1:1 Rücknahme eines ähnlichen Altgerätes bei Neukauf am Ort der Abgabe oder in unmittelbarer Nähe hierzu (auch privater Haushalt). Entscheidend ist hierbei, dass das EAG der gleichen Geräteart entspricht und im Wesentlichen die gleichen Funktionen wie das neue Gerät erfüllt. Zum anderen bezieht sich die verpflichtende Rücknahme auf eine 0:1 Rücknahme im Einzelhandelsgeschäft oder in unmittelbarer Nähe hierzu. Die 0:1 Rücknahme gilt für Kleingeräte, die eine äußere Abmessung von 25 cm nicht überschreiten.

Sofern Vertreiber EAG auf freiwilliger Basis oder verpflichtend zurücknehmen, sind sie gemäß § 29 ElektroG ebenfalls verpflichtet, erfasste und ggf. verwertete Mengen an EAG an die stiftung ear zu melden. Die stiftung ear registriert Hersteller von EEG in Deutschland und koordiniert die Abholung von EAG bei den örE. Darüber hinaus sind gemäß § 29 Abs. 4 ElektroG auch zurückgenommene Mengen, die an Dritte (örE, Hersteller) übergeben werden, der stiftung ear zu melden. Die Rücknahmetätigkeit ist gemäß § 25 ElektroG anzeigepflichtig, unabhängig davon, ob die Rücknahme freiwillig gemäß § 17 Abs. 3 ElektroG oder verpflichtend gemäß § 17 Abs. 1 und 2 ist.

In dem Projekt „Effizienzbestimmung der Vertreiberpflichten nach ElektroG“ im Auftrag des deutschen Umweltweltbundesamts wurde erstmals eine systematische Bewertung der neu eingeführten Pflichten bezüglich der Umsetzung der Vertreiberpflichten gemäß ElektroG durchgeführt.

Im Rahmen des Projekts wurden schwerpunktmäßig Auswertungen der an die stiftung ear gemeldeten Vertreiberrücknahmemengen, Untersuchungen zur praktischen Umsetzung der Vertreiberrücknahme in Form von Vor-Ort Besuchen und Befragungen sowie eine Evaluierung der entsprechenden Kosten und Aufwänden durchgeführt. Abschließend wurden Handlungsempfehlungen für eine zukünftig bessere Gestaltung der Vertreiberpflichten formuliert.

2 Effizienzbestimmung der Vertreiberpflichten

2.1 Vertreiberrücknahmemengen

Die Analyse der Vertreiberrücknahmemengen hat ergeben, dass seit der systematischen Mengenerhebung via stiftung ear relevante EAG-Mengen über Vertreiber zurückgenommen werden (in 2017 101.148 Tonnen). Diese Mengen sind zum weitaus überwiegenden Anteil der Gerätekategorie 1 (Haushaltsgroßgeräte) zuzuordnen (82,6 %), die überwiegend bei privaten Haushalten im Tausch gegen ein neues Haushaltsgroßgerät erfasst werden.

Den Mengendaten der stiftung ear folgend, resultiert mit ca. 88,3 % (89.846 Tonnen) der weitaus überwiegende Anteil aus der verpflichtenden Vertreiberrücknahme gemäß § 17 (1) ElektroG. Im Detail zeigt sich, dass mit ca. 82,9 % der Gesamtmenge (83.869 Tonnen), die Rücknahme fast ausschließlich zur Rücknahme verpflichtete Elektrofachmärkte erfolgt. Aus Sicht der Mengenrelevanz spielt, ausgehend von den gemeldeten Daten, die freiwillige Vertreiberrücknahme gemäß § 17 Abs. 2 ElektroG mit 1,8 % (1.811 Tonnen) eine deutlich untergeordnete Rolle.

EAG-Mengen, die über reine Fernabsatzvertreiber zurückgenommen wurden, sind in den Mengendaten zur Vertreiberrücknahme nahezu nicht zu finden (lediglich ca. 37 Tonnen). Da besonders zur Rücknahme von EAG verpflichtete Vertreiber sowohl im Fernabsatz als auch stationär EEG vertreiben, kann im Detail nicht bestimmt werden, welche EAG bei Rücknahme welcher ursprünglichen Vertriebsart zuzuordnen sind.

Nach der Auswertung des Verzeichnisses der stiftung ear über Sammel- und Rücknahmestellen ist anzunehmen, dass die Mehrzahl praktizierender Rücknahmestellen diese Rücknahme nicht gemäß § 25 Abs. 3 ElektroG bei der stiftung ear angezeigt hat.

Es ist ebenfalls anzunehmen, dass nur ein sehr geringer Anteil der gelisteten Rücknahmestellen, an denen freiwillig EAG zurückgenommen werden, letztlich Mengendaten an die stiftung ear melden. Diese Einschätzung bestätigten auch die Erkenntnisse aus den Vor-Ort Besuchen.

2.2 Praktischen Umsetzung der Vertreiberrücknahme

Bei der Auswertung zur praktischen Umsetzung der Vertreiberrücknahme wurden zunächst Internet-Recherchen (stationärer Handel und Fernabsatzhandel) sowie anschließend Vor-Ort Besuche bei praktizierenden Vertreibern (stationärer Handel) durchgeführt.

Nach den Vor-Ort Untersuchungen kann in Bezug auf den stationären Handel positiv festgehalten werden, dass die Vertreiber i. d. R. ihren Rücknahmeverpflichtungen nachkommen. Bei allen Teilnehmern der Vor-Ort Untersuchungen werden EAG angenommen, insbesondere auch bei allen nicht verpflichteten Vertreibern mit Verkaufsflächen unter 400 m². Hinsichtlich der Erfüllung der im ElektroG formulierten Pflichten wird deutlich, dass mehr Akteure 0:1 Rücknahmen anbieten, als dazu verpflichtet wären. Dies gilt für den regionalen Einzelhandel sowie für deutschlandweit tätige Unternehmen. Deutlich wurde, dass in vielen Unternehmen die Rücknahme von EAG schon vor der Novellierung des ElektroG in die Geschäftsprozesse integriert wurde. Hintergrund (und Motivation) hierfür ist die Rücknahme im Rahmen des Kundenservice und die damit verbundene Kundenbindung. Die Vor-Ort Untersuchung hat zudem bestätigt, dass die Haushaltsgroßgeräte den wesentlichen Mengenanteil an zurückgenommenen EAG ausmachen.

In Bezug auf Entsorgungswege wurde deutlich, dass eine Mehrheit der befragten Vertreiber die EAG zu kommunalen Wertstoffhöfen bringt und diese dort dem örE übergibt. Dies gilt insbesondere für regionale Einzelhändler.

Insgesamt ist festzustellen, dass in vielen Fällen die Mitteilungs- und Anzeigepflichten an die stiftung ear durch Vertreiber nicht oder noch nicht ausreichend umgesetzt wurden. Die Datenqualität und -quantität sind verbesserungswürdig. Die Informationspflichten werden weitestgehend von überregional agierenden Vertreibern umgesetzt. Vertreiber, die EAG auf freiwilliger Basis zurücknehmen, sind von der Informationsverpflichtung ausgenommen und stellen den Verbraucher*innen in den überwiegenden Fällen auch keine Informationen bzw. kein Informationsangebot bereit.

Für den Fernabsatzhandel wurden ebenfalls Defizite bei der Erfüllung der Vertreiberpflichten deutlich. Die Zusammenarbeit mit sogenannten „Rücknahmesystemen", welche die EAG-Übernahme als Dienstleister organisieren, ist ausbaufähig. Aus der Analyse geht hervor, dass Fernabsatzhändler zum Teil keine Rückgabemöglichkeiten anbieten. Dies gilt vor allem bei solchen, die kein deutschlandweites Filialnetz haben. Hier wird verstärkter Handlungsbedarf gesehen.

2.3 Kosten und Aufwände

Im Rahmen des Projekts wurden die grundsätzlichen Kosten und Aufwände dargestellt, welche weitestgehend auf Ergebnisse der vor Ort Befragungen basieren. Ferner wurde, um die durch die Novelle des ElektroG 2012 zusätzlichen Aufwände für die erweiterten Mitteilungs- und Informationspflichten zu ermitteln, eine Ex-Post-Messung durchgeführt. Die jährlichen Kosten durch die zusätzlichen Aufgaben der Vertreiber im Zuge der Novellierung ElektroG 2015 belaufen sich demnach auf ca. 5,24 Mio. €.

3 Handlungsempfehlungen

Aus den Erkenntnissen der Datenanalysen, der Internetrecherchen und Erfahrungen der Vor-Ort Besuche wurden Handlungsempfehlungen abgeleitet, um die derzeitige Situation der Vertreiberrücknahme zu optimieren und fortzuentwickeln.

Hinsichtlich der Inkonsistenzen und Abweichungen bei der Meldesystematik wird empfohlen, die Meldemaske der stiftung ear anzupassen sowie das Verzeichnis der Rücknahmestellen neu zu strukturieren.

Da Mengenmitteilungen gemäß § 43 ElektroG von unterschiedlichen Stellen durchgeführt werden können, jedoch Vertreibern oft nicht klar ist, ob und wenn ja, durch wen eine Mitteilung zu erfolgen hat, sollte festgelegt werden, dass Entsorgungsdienstleister, die EAG von Vertreibern übernehmen, diese nur annehmen dürfen, wenn der Nachweis dieser Mengen als Vertreiberrücknahmemenge nachgewiesen werden kann. Ferner wird empfohlen, dass eine summarische Meldung von unterschiedlichen Rücknahmestandorten durch die Unternehmenszentrale durchgeführt werden kann.

Als ein Paradigmenwechsel zur bestehenden Meldesystematik wird eine Zentralisierung von Mengenmitteilungen durch die behandelnden Erstbehandlungsanlagen empfohlen, da letztlich die separat mitgeteilten Mengen dort anlanden. Auch dort können die unterschiedlichen Herkünfte registriert, dokumentiert und an die stiftung ear gemeldet werden. Hinsichtlich EAG-Mengen, die in das Ausland exportiert werden, um beispielsweise dort erstbehandelt zu werden, müsste das Meldesystem auf solche Exporte angepasst werden, um den Melderahmen dokumentarisch zu schließen.

Hinsichtlich der Ausweitung der Kriterien für die verpflichtende Rücknahme wird die Beibehaltung des aktuellen 400 m² Flächenkriteriums empfohlen. Zusätzlich wird empfohlen, relevante Einzelhandelsbetriebe zur Rücknahme zu verpflichten, sofern sie selber EEG vertreiben. Hierfür muss eine prüfbare Größenabgrenzung geschaffen werden, die sich auf die Gesamtverkaufsfläche bezieht. Ferner sollte ein Kriterium hinsichtlich der Mengenschwelle von vertriebenen EEG geschaffen werden, die in Kombination mit dem Flächenkriterium greift.

Für den Fernabsatzhandel wird eine Verpflichtung zur Teilnahme an einem bestehenden Rücknahmesystem empfohlen. Ferner sollten unzureichende oder fehlende Informationspflichten auf den Internetseiten sanktioniert werden können. Ebenso sollte geprüft werden, inwieweit Fernabsatzvertreiber bei den aktuell mangelnden Rücknahmemengen hinsichtlich einer Mitfinanzierung der Erfassung im stationären Handel verpflichtet werden können.

In Hinblick auf die 0:1 Rücknahme ist eine Anpassung des Größenkriteriums an die Kategorie 5 (Haushaltskleingeräte) mit < 50 cm empfohlen. Ferner sollten Vertreiber dazu verpflichtet werden, beim Kauf eines Neugeräts beim Kunden explizit nachzufragen, ob ein Altgerät zurückgenommen werden soll. Für den Fernabsatz sollte dies in den elektronischen Kaufvorgang eingebunden werden. Ohne das aktive Handeln des Kunden sollte der Kauf nicht abgeschlossen werden können. Im Fall eines Posteinsands sollte geprüft werden, ob die automatisierte Generierung eines Versandetiketts vorgeschrieben oder gefördert werden kann.

Flankiert werden sollten die beschriebenen Maßnahmen mit einer verstärkten Aufklärung von Verbraucher*innen bzgl. der EAG Rückgabe sowie von Vertreibern bzgl. der Auswahl geeigneter Entsorgungsdienstleister, um einen Mengen-abfluss in den informellen Sektor zu vermeiden. Darüber hinaus sollten Fernabsatzhändler zusätzlich verstärkt über ihre Rücknahme-, Informations- und Mitteilungspflichten informiert werden.

4 Literatur

WEEE-Richtlinie	2012	Richtlinie 2012/19/EU des Europäischen Parlaments und des Rates vom 4. Juli 2012 über Elektro- und Elektronik-Altgeräte (ABl. L 197 vom 24.7.2012, S. 38), in der Fassung vom 4. Juni 2018.
ElektroG	2015	Gesetz über das Inverkehrbringen, die Rücknahme und die umweltverträgliche Entsorgung von Elektro- und Elektronikgeräten (Elektro- und Elektronikgerätegesetz - ElektroG) vom 20. Oktober 2015 (BGBl. I S. 1739), das zuletzt durch Artikel 16 des Gesetzes vom 27. Juni 2017 (BGBl. I S. 1966) geändert worden ist.

Anschrift der Verfasser(innen)

Dr.-Ing. Ralf Brüning
Dr. Brüning Engineering UG
Kirchenstraße 26
D-26919 Brake
Telefon +49 4401 7049760
E-Mail info@dr-bruening.de

Beitrag biologischer Rest- und Abfallstoffe zur Bioökonomie Vorstellung eines neuen Ansatzes zur Behandlung biologischer Rest- und Abfallstoffe zur Erzeugung biobasierter Stoffe

Kyra Atessa Vogt, Iris Steinberg

Hochschule Darmstadt

Contribution of Biological Residual and Waste Materials to the Bioeconomy

Abstract

A new method for the treatment of biological residues and waste materials, especially kitchen and canteen waste, for the production of bio-based materials is described. Furthermore, first investigations on the production of bio-based materials from kitchen and canteen waste are presented. The intention is to determine the influence of such a treatment before the fermentation of kitchen and canteen waste on the biogas production potential.

Inhaltsangabe

Ein neuer Ansatz zur Behandlung biologischer Rest- und Abfallstoffe, speziell Küchen- und Kantinenabfälle, zur Erzeugung biobasierter Stoffe wird erläutert und erste Ergebnisse von Untersuchungen zur Erzeugung biobasierter Stoffe aus Küchen- und Speiseabfällen werden vorgestellt. Es wird untersucht, welchen Einfluss eine solche Behandlung vor der Vergärung der Küchen- und Kantinenabfällen auf das Biogasbildungspotential hat.

Keywords

Biologischer Abfall, Carbonsäuren, biobasierte Produkte, Bioökonomie, Speiseabfälle
biological waste, carboxylic acids, bio-based products, bioeconomy, food waste

1 Einleitung

Im Jahr 2016 betrug der Rohstoffkonsum in Deutschland rund 1,2 Milliarden Tonnen; ungefähr ein Drittel davon basierte auf fossilen Energieträgern (Statistisches Bundesamt, 2020). Die Reserven fossiler Energieträger sind jedoch endlich (VGB PowerTech, 2018), so dass die Verwendung nicht-fossiler - sog. biobasierter - Rohstoffe, die auf regenerativen Stoffen basieren, immer bedeutender wird. So kann einerseits der Verbrauch an fossilen Energieträgern reduziert und gleichzeitig ein Beitrag zur Bioökonomie, einer biobasierten Wirtschaft, geleistet werden. Hier setzt die nationale Bioökonomiestrategie der Bundesregierung Deutschland an, die eine nachhaltige, klimaneutrale Entwicklung verbunden mit dem Einsatz biogener Rohstoffe zu einer nachhaltigen, kreislauforientierten Wirtschaft postuliert. (Bundesministerium für Bildung und Forschung, 2020).

Derzeit werden zur Produktion biobasierter Stoffe hauptsächlich stärke- und zellulosehaltige Pflanzen, wie Mais, Ölsaaten, Zuckerrohr und Holz, eingesetzt, die vor allem in Konkurrenz zur Nahrungsmittelproduktion stehen (Umweltbundesamt, 2020). Daher ist es bedeutsam die Verwendung bereits vorhandener Biomasse, wie biologischer Rest- und Abfallstoffe, zu betrachten.

2 Zielstellung

Jährlich werden in Deutschland ungefähr 11 Millionen Tonnen biologisch abbaubare Rest- und Abfallstoffe entsorgt, zu denen getrennt erfasste organische Abfälle über die Biotonne, Garten- und Parkabfälle sowie Küchen- und Kantinenabfälle zählen (Statistisches Bundesamt, 2019). Diese werden hauptsächlich über etablierte biologische Behandlungsverfahren der Kompostierung oder Vergärung entweder stofflich als Bodenverbesserer (Kompost) oder zusätzlich energetisch als Biogas verwertet. Jedoch wird bei diesen Verfahren der überwiegende Anteil des in den biologischen Rest- und Abfallstoffen gebundenen Kohlenstoffs als Kohlenstoffdioxid (CO_2) emittiert. Daher gilt es, diese Verfahren hinsichtlich der vollständigen Verwertung des Kohlenstoffs zu optimieren.

In den etablierten biologischen Behandlungsverfahren werden die organischen Abfälle mithilfe von Mikroorganismen ab- bzw. umgebaut. Zu diesen Zwischenprodukten zählen organische Verbindungen mit einer oder mehreren Carboxygruppen (-COOH) (Federle, et al., 2017), die je nach Anzahl der Kohlenstoffatome in kurzkettige (1 - 3 Kohlenstoffatome), mittelkettige (4 - 10 Kohlenstoffatome) und langkettige (>10 Kohlenstoffatome) Carbonsäuren unterteilt werden (Hopp, 2018). In Kompostierungs- bzw. Vergärungsprozessen liegen dabei überwiegend kurzkettige Carbonsäuren vor. Diese werden in dem hier vorgestellten biotechnologischen Behandlungsverfahren mithilfe von spezialisierten Mikroorgansimen und Ethanol als Additive zu mittelkettigen Carbonsäuren aufgebaut. Die Abtrennung der Carbonsäuren erfolgt innerhalb des Behandlungsverfahrens als in-situ Extraktion mittels eines unpolaren Extraktionsmittels. Diese Carbonsäuren können zur Herstellung biobasierter Grundchemikalien verwendet werden und so zu einer biobasierten Wirtschaft (Bioökonomie) beitragen. Diese Art der Bewirtschaftung von Abfällen entspricht der stofflichen Verwertung (Recycling) gemäß § 6 Abs. 1 Kreislaufwirtschaftsgesetz.

Als mögliche Ausgangssubstrate zur Produktion mittelkettiger Carbonsäuren wurden bereits sekundäre Reststoffe wie Sickerwässer aus Kompostierungsanlagen sowie flüssige Gärreste aus Biogasanlagen untersucht (Kannengiesser, et al., 2019). Diese Substrate erwiesen sich als grundsätzlich geeignet; der Ertrag an mittelkettigen Carbonsäuren ist dabei aus den Gärresten im Vergleich zu den unbehandelten Sickerwässern geringer, da bereits während des durchlaufenen Vergärungsprozesses Carbonsäuren in Biogas umgewandelt wurden.

Ebenfalls geeignete Substrate mit einem hohen Potential zur Produktion mittelkettiger Carbonsäuren sind Küchen und Kantinenabfälle (Kannengiesser, et al., 2019). Werden diese Abfälle zur Produktion von Carbonsäuren eingesetzt, reduziert sich einerseits insgesamt der Anteil an für die Biogasbildung zur Verfügung stehenden Kohlenstoffverbindungen, andererseits könnte der vorgeschaltete Prozess auch den Anteil verfügbarer Zwischenprodukte erhöhen und sie für die Vergärung leichter verfügbar machen.

Im Rahmen dieser Arbeit soll daher untersucht werden, wie sich das Biogaspotential der Substrate nach der Abtrennung der mittelkettigen Carbonsäuren – und somit der Beitrag zur energetischen Verwertung - verändert, einhergehend mit der Ermittlung der gebildeten Carbonsäuren als Beitrag zum stofflichen Recycling.

3 Material und Methoden

3.1 Substrat

Es werden flüssige Küchen- und Kantinenabfälle untersucht, die bereits pasteurisiert an einer Vergärungsanlage für Lebensmittel, Speisereste aus Küchen und Kantinen sowie gewerbliche Flotaten und Sudrückständen angeliefert werden.

Für die Bestimmung des Biogasertrags wurden zum Animpfen als Inokulum Gärsubstrat derselben Vergärungsanlage und als Referenzsubstrat reine Zellulose verwendet.

3.2 Versuchsaufbau

Für die Untersuchungen werden ein Versuchsaufbau zur Erzeugung biobasierter Stoffe und ein zweiter zur Bestimmung des Biogasbildungspotentials verwendet.

3.2.1 Erzeugung biobasierter Stoffe

Der Versuchsaufbau zur Erzeugung biobasierter Stoffe besteht aus einem luftdicht verschließbaren Gefäß mit maximal 26 Liter Fassungsvolumen. Das entstehende Gas wird durch einen Gasauslass im Deckel in einen Gasbeutel geleitet. Zur Entnahme von Substratproben ist der Versuchsreaktor mit einer Vorrichtung zur Probeentnahme versehen. Das unpolare Extraktionsmittel setzt sich auf der polaren Substratschicht ab. Die Temperierung des Versuchsaufbaus findet in einem temperierten Raum statt. Die nachfolgende Abbildung veranschaulicht den beschriebenen Versuchsaufbau und die technische Umsetzung.

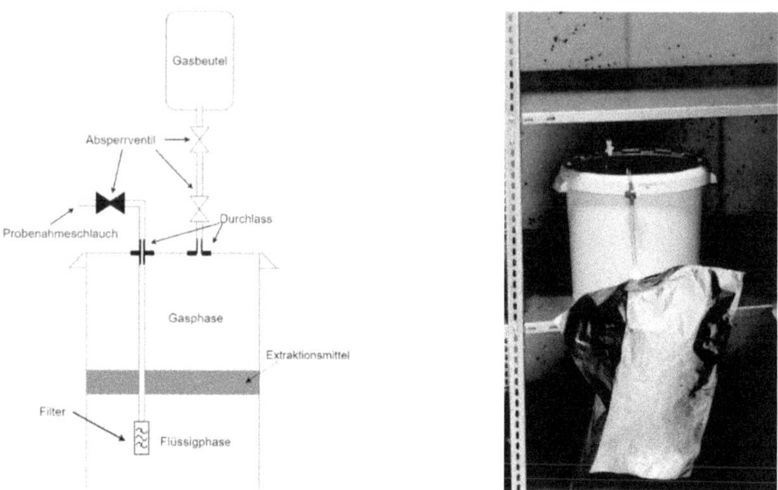

Abbildung 1: Schematische Zeichnung des Versuchsaufbaus (links) (verändert nach Noll, 2020) und technische Umsetzung (rechts)

3.2.2 Biogasbildungspotential

Der Versuchsaufbau besteht aus einer Standflasche mit einem Fassungsvermögen von 2 Liter (max. Füllvolumen 1,8 Liter) mit Rührwerk. Das entstehende Gas wird in einem Gasbeutel zur Analyse aufgefangen. Das Substrat wird mittels eines Wasserbads auf die gewünschte Betriebstemperatur temperiert. Die nachfolgende Abbildung veranschaulicht den Versuchsaufbau (links) und die technische Umsetzung (rechts).

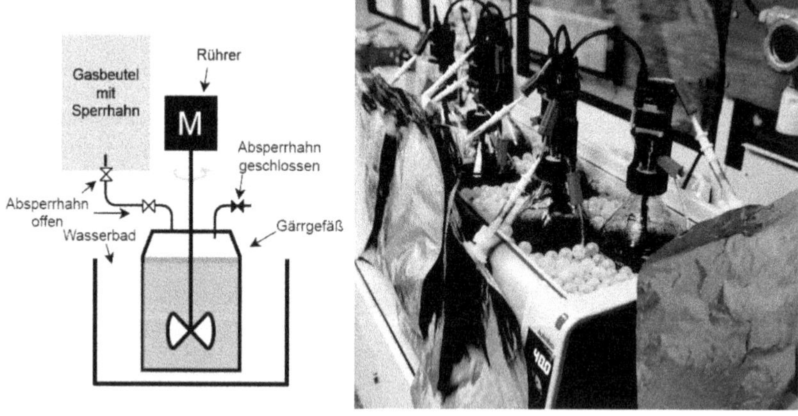

Abbildung 2: Schematische Zeichnung des Versuchsaufbaus (links) und technische Umsetzung (rechts)

3.3 Analytische Methoden

Die Vor-Ort-Analytik umfasst alle Parameter, die umgehend nach der Probenahme bestimmt werden. Dazu zählen der pH-Wert und das Redox-Potential.

Der pH-Wert ist ein wichtiger Parameter für die Erzeugung biobasierter Grundchemikalien aus organischen Abfällen, da er Rückschlüsse auf die Lebensbedingungen für die Mikroorganismen im Substrat zulässt. Daher wird dieser nach jeder Probenahme direkt im Probenahmegefäß genommen.

Das Redox-Potential gibt eine Aussage darüber, ob der Prozess im aeroben oder anaeroben Milieu abläuft. Liegt das Redox-Potential unterhalb -330 mV, so befindet sich das Substrat im anaeroben Milieu (König & Wiese, 2007), welches für die Aktivität der spezialisierten Mikroorganismen notwendig ist

Die Bestimmung der Carbonsäuren erfolgt mittels Gaschromatographie und Flammenionisationsdetektion (GC-FID).

Die Bestimmung des Biogasbildungspotentials wurde nach VDI 4630 „*Vergärung organischer Stoffe - Substratcharakterisierung, Probenahme, Stoffdatenerhebung, Gärversuche.*“ in einem gerührten Batch-Gärgefäß (siehe Abb. 2, Kap. 3.2.2) angesetzt und vorgabegemäß durchgeführt (Verein Deutscher Ingenieure, 2016): Während der Versuchsdurchführung wurden die gebildete Gasmenge ermittelt und die Gaszusammensetzung analysiert.

3.4 Versuchsdurchführung

Die Versuchsdurchführung kann in die folgenden drei Phasen unterteilt werden:

1. Bestimmung des Biogaspotentials des unbehandelten Substrates
2. Behandlung des Substrates zur Erzeugung biobasierter Grundchemikalien
3. Bestimmung des Biogaspotentials des behandelten Substrates

Die nachfolgende Tabelle stellt die Betriebsparameter der einzelnen Versuchsphasen dar.

Tabelle 1: Übersicht der Betriebsparameter

Parameter	**Phase 1**	**Phase 2**	**Phase 3**
Dauer	31 d	31 d	33 d
Volumen	Min. 1,5 l/Max. 1,8 l	10 l	Min. 1,5 l/Max. 1,8 l
Temperatur	40°C	38°C	40°C
Einstellung pH-Wert	-	6 - 6,5	-
Probenahme	-	2x pro Woche	-
Gasanalyse	Nach Bedarf	Nach Bedarf	Nach Bedarf
Homogenisierung	20 Umdrehungen/Min	-	20 Umdrehungen/Min
Zusatzstoffe	-	Ethanol*: 11 ml/l Extraktions-mittel*: 100 ml/l	-
Reaktoren	Inokulum 1 Inokulum 2 Substrat 1 (unbeh.) Substrat 2 (unbeh.) Zellulose 1 Zellulose 2	Blind** Substrat A Substrat B	Inokulum 1 Inokulum 2 Substrat 1 (beh.) Substrat 2 (beh.) Zellulose 1 Zellulose 2

*nicht bei Reaktoren „Blind", **dient als Referenzreaktor für Behandlungsverfahren

Der Gäransatz zur Bestimmung der Biogaspotentiale unterliegt bestimmter Voraussetzungen (siehe Kap. 3.3). Die nachfolgende Tabelle veranschaulicht diese. Es ist zu erkennen, dass alle Voraussetzungen eingehalten wurden.

Tabelle 2: Voraussetzungen an Gäransatz der VDI 4630

Parameter	**Soll**	**Phase 1**	**Phase 3**
Glühverlust Inokulum	>50%	50,55%	50,24%
Org. Massenanteil aus Inokulum	1,5% - 2%	1,89%	2,07%
Trockenrückstand Gäransatz	≤ 10%	Substrat = 5,5% Zellulose = 4,23%	Substrat = 5,57% Zellulose = 4,57%
$\frac{oTM_{Substrat}}{oTM_{Inokulum}}$	≤ 0,5	Substrat = 0,21 Zellulose = 0,21	Substrat = 0,15 Zellulose = 0,20

oTM: organische Trockenmasse (Verein Deutscher Ingenieure, 2016)

Die nachfolgende Tabelle zeigt die Zusammensetzung der Gäransätze der beiden untersuchten Phasen.

Tabelle 3: Zusammensetzung der Gäransätze in Phase 1 und 3

Reaktor	Masse [g]			
	Phase 1		Phase 3	
	Inokulum	Substrat	Inokulum	Substrat
Inokulum 1	1.493,90	-	1.500,00	-
Inokulum 2	1.494,00	-	1.000,00	-
Substrat 1	1.494,50	309,32	1.500,00	300,00
Substrat 2	1.494,50	311,27	1.500,00	300,00
Zellulose 1	1.494,50	6,60	1.500,00	6,60
Zellulose 2	1.494,00	6,60	1.500,00	6,60

4 Ergebnisse und Diskussion

4.1 Erzeugung biobasierter Stoffe

4.1.1 Vor-Ort-Analytik

Der originäre pH-Wert des untersuchten Substrats ist 3,6. Dieser verändert sich während der Versuchsdauer nur minimal und pendelt zwischen 3,6 und 3,7. Der pH-Wert des behandelten Substrates wurde zunächst auf 5,95 eingestellt. Im Verlauf des Versuchs steigt dieser an und pendelt zwischen 6,0 und maximal 6,4. Dies bedeutet, dass optimale Bedingungen für die Mikroorganismen im Substrat vorlagen.

Das Redox-Potential des unbehandelten Substrates (Blind) verläuft während der Versuchsdauer zwischen 17 mV und -75 mV. Dieser Wertebereich weist auf ein aerobes Milieu im Substrat hin. Dahingegen befindet sich das behandelte Substrat während der gesamten Versuchsdauer um den Bereich des anaeroben Milieus mit Werten zwischen -257 mV und -379 mV, was sich positiv auf den Stoffumsatz der Mikroorganismen auswirkt.

Die Zusammensetzung des gebildeten Gases während der Dauer der Untersuchungen deutet auf einen erfolgreichen Behandlungsverlauf hin. Es wurden durchschnittlich rund 5 Vol.-% CH_4, 24 Vol.-% CO_2, 13 Vol.-% O_2, 57 Vol.-% N_2 und 57 ppm H_2S gemessen. Die Messwerte von CO_2 und H_2S weisen deutlich höhere Schwankungen auf im Vergleich zu den anderen Gasen. Die für einen anaeroben Prozess hohen Sauerstoff- und Stickstoff-Messwerte sind auf das Öffnen der Reaktoren zur Probenahme des Extraktionsmittel zurück zu führen. Jedoch haben diese keine Auswirkung auf das anaerobe Milieu in den Substraten, wie die Messwerte des Redox-Potentials belegt. Dies wird durch den luftdichten Abschluss des Substrates mittels des unpolaren Extraktionsmittels erzielt.

4.1.2 Bildung von Carbonsäuren

Die nachfolgende Abbildung 3 veranschaulicht den Verlauf der Bildung von Carbonsäuren und gibt diese in Gramm pro Liter an. Es ist deutlich zu erkennen, dass das unbehandelte Substrat (Blind) hauptsächlich Essigsäure (C2) beinhaltet. Dahingegen ist in dem behandelten Substrat eine deutlich höhere Konzentration von Essigsäure (C2) sowie Propion- (C3), Butter- (C4), Valerian- (C5) und Capronsäure (C6) festzustellen. Es ist deutlich ein stetiger Anstieg von Carbonsäuren mit zunehmender Versuchsdauer zu erkennen. In dem unbehandelten Substrat sind noch geringe Konzentrationen höherwertiger Carbonsäuren (C6 - C12) zu erkennen, wohingegen in dem behandelten Substrat diese weniger nachzuweisen sind. Dies ist vermutlich auf den Übergang dieser Carbonsäuren in das Extraktionsmittel zurück zu führen, welches bei dem unbehandelten Substrat nicht zugefügt wurde. Um dies zu bestätigen, veranschaulicht die nachfolgende Abbildung 4 den Verlauf der extrahierten Carbonsäuren, die im Extraktionsmittel nachgewiesen wurden. Diese zeigt die Konzentration an Carbonsäuren im Extraktionsmittel in Gramm pro Liter. Mit zunehmender Versuchsdauer haben sich Buttersäure (C4) und Capronsäure (C6) aus dem behandelten Substrat in das Extraktionsmittel gelöst. Des Weiteren ist ein geringer Anstieg an Capryl- (C8), Perlagon- (C9) und Caprinsäure (C10) zu erkennen. Dies bestätigt, warum keine höherwertigen Carbonsäuren in dem behandelten Substrat nachzuweisen sind.

In dem unbehandelten Substrat sind nach 31 Tagen Versuchsdauer rund 17 g Essigsäure (C2) pro Liter Substrat nachzuweisen. Die Konzentrationen der weiteren Carbonsäuren (C3 - C12) sind mit unter 1 g/l sehr gering.

Dahingegen haben sich durch die Behandlung des Substrats innerhalb einer Versuchsdauer von 31 Tagen rund 25 g/l Essigsäure (C2), 15 g/l Propionsäure (C3) und 15 g/l Buttersäure (C4) gebildet. Zusätzlich konnten im Verlauf der Versuchsdauer rund 1,6 g/l Buttersäure (C4), 0,8 g/l Valeriansäure (C5), 2,1 g/l Capronsäure (C6) und 0,6 g/l Caprylsäure (C8) extrahiert werden.

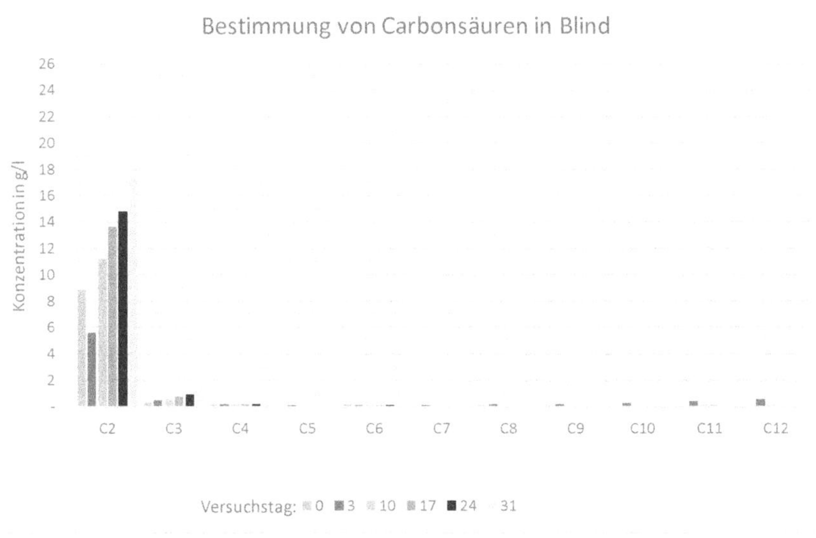

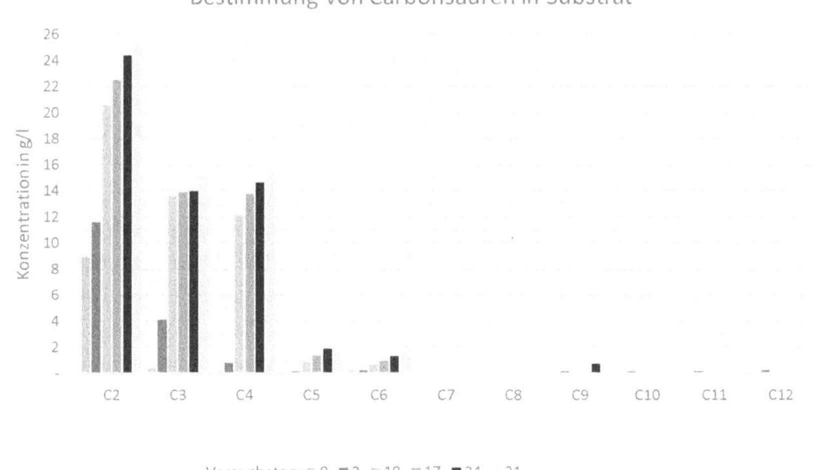

Abbildung 3: Vergleich der Konzentrationen an Carbonsäuren zwischen Blind-Reaktor und behandelten Substrat nach Phase 2

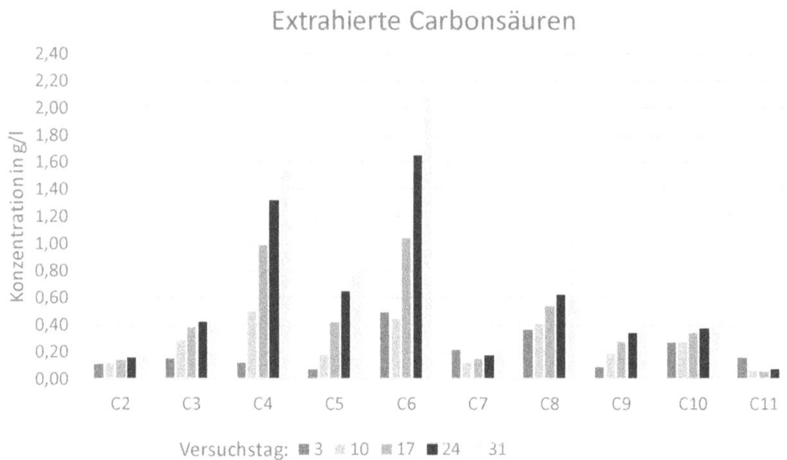

Abbildung 4: Extrahierte Carbonsäuren in Extraktionsmittel

4.2 Biogasbildungspotential

4.2.1 Biogasbildungspotential unbehandeltes Substrat

Zur Beurteilung des Biogasbildungspotentials wird die spezifische Biogasproduktion nach VDI 4630 in Normlitern pro Kilogramm organische Trockenmasse angegeben. Vor der Behandlung des Substrates beträgt die spezifische Biogasproduktion des unbehandelten Substrates 985,11 l_N/kg oTM. Das spezifische Biogasbildungspotential der Zellulose, welche als Referenzsubstrat dient, beträgt 575,54 l_N/kg oTM. Der Kontrollwert für Zellulose liegt bei 745 l_N/kg oTM. Dieser Wert darf um 10% unter- bzw. überschritten werden (Verein Deutscher Ingenieure, 2016). Da das spezifische Biogasbildungspotential des Referenzsubstrats den Kontrollwert um rund 23% unterschreitet, sind die Ergebnisse nicht ausreichend belastbar zur Bewertung des Biogaspotentials.

4.2.2 Biogasbildungspotential behandeltes Substrat

Zur Beurteilung des Biogasbildungspotentials des behandelten Substrates wird ebenfalls die spezifische Biogasproduktion nach VDI 4630 berechnet. Nach der Behandlung des Substrats beträgt die spezifische Biogasproduktion des behandelten Substrates 1.177,75 l_N/kg oTM. Die spezifische Biogasproduktion des Referenzsubstrats, Zellulose, konnte nicht bestimmt werden, da das Netto-Gas-Normvolumen einen negativen Wert aufwies. Somit kann der Kontrollwert nicht eingehalten werden und die Ergebnisse sind als nicht ausreichend belastbar zu bewerten.

5 Schlussfolgerung

Die Untersuchungen zur Erzeugung biobasierter Grundchemikalien erzielte positive Ergebnisse. Die Ausbeuten der hier behandelten Küchen- und Kantinenabfälle betrugen rund 25 g/l Essigsäure (C4) und rund 16 g/l Buttersäure (C4). Damit sind diese vergleichbar mit den Ergebnissen vorangegangener Untersuchungen zu Sickerwässer der Kompostierung, bei denen rund 16 g/l Essigsäure (C2) und rund 18 g/l Buttersäure (C4) produziert wurden (Kannengiesser, et al.,2019). Die Betriebsbedingungen waren über den gesamten Untersuchungszeitraum stabil. Die höchste Konzentration im Substrat wies Essigsäure (C2) mit rund 25 g/l auf. Es wurden rund 8 g/l Essigsäure mehr gebildet im Vergleich zu dem unbehandelten Substrat. Des Weiteren konnten maximal 2,1 g/l Capronsäure (C6) im Extraktionsmittel nach der Behandlung nachgewiesen werden. Die extrahierte Masse könnte noch vergrößert werden, wenn der Versuchsaufbau derart optimiert wird, dass die Kontaktfläche zwischen Extraktionsmittel und Substrat vergrößert wird, um den Übergang der mittelkettigen Carbonsäuren in das Extraktionsmittel zu begünstigen.

Die ermittelten Biogasbildungspotentiale des unbehandelten und des behandelten Substrates sind mit rund 985 bzw. 1.178 l_N/kg oTM über Literaturangaben für vergleichbare Substrate, wie z.B. für fettarme Speisereste rund 643 l_N/kg oTM bzw. 762 l_N/kg oTM für fettreiche Speisereste (Landesanstalt für Landwirtschaft Bayern, 2020). Unter Berücksichtigung der Kontrollwerte nach VDI 4630 sind sie als nicht ausreichend belastbar einzustufen, so dass auf Basis dieser Versuchsreihe keine valide Aussage über den Einfluss der Erzeugung biobasierter Grundchemikalien auf das Biogasbildungspotential flüssiger Küchen- und Kantinenabfälle getroffen werden kann. Daher sind weitere Untersuchungen erforderlich, um bewerten zu können, ob und welche Auswirkungen eine vorgeschaltete Behandlung zur Bildung und Abtrennung mittelkettiger Carbonsäuren auf das Gasbildungspotenzial hat.

6 Literatur

Bundesministerium für Bildung und Forschung — 2020 — Nationale Bioökonomiestrategie - Zusammenfassung, Berlin

Federle, S., Hergesell, S. & Schubert, S. — 2017 — *Die Stoffklassen der organischen Chemie.* Berlin: Springer Verlag, 2017

Hopp, V. — 2018 — *Chemische Kreisläufe in der Natur.* Berlin: Springer Verlag, 2018. Bd. 2.

Kannengiesser, J., Campitelli, A., Vogt, A., Steinberg, I., Jager, J., & Schebek, L. — 2019 — Combination of different treatment methods for biological waste and agricultural residues to generate bio-based products. Proceedings SARDINIA2019 -

		17th International Symposium on Waste Management and Sustainable Landfilling. Forte Village/Santa Margherita di Pula (CA): CISA Publisher.
König, R., Wiese, J.	2007	Prozessbegleitende Fermenterüberwachung auf Biogasanlagen.
Landesanstalt für Landwirtschaft Bayern	2020	www.lfl.bayern.de, [Online] Online verfügbar unter: https://www.lfl.bayern.de/iba/energie/049711/?sel_list=49%2Cb&anker0=substratanker#substratanker, [Zugriff am 22.09.2020]
Noll, P.	2020	*Potential organischer Reststoffe zur Produktion von Ethanol zur Erzeugung von Carbonsäuren (unveröffentlicht).* Hochschule Darmstadt, Darmstadt
Statistisches Bundesamt	2019	Fachserie 19 Umwelt, Reihe 1 Abfallentsorgung
Statistisches Bundesamt	2020	Umweltökonomische Gesamtrechnungen - Aufkommen und Verwendung in Rohstoffäquivalenten - Berichtszeitraum 2000 bis 2016
Umweltbundesamt	2020	www.umweltbundesamt.de, [Online] Online verfügbar unter: https://www.umweltbundesamt.de/biobasierte-biologisch-abbaubare-kunststoffe#22-sind-biobasierte-kunststoffe-nachhaltiger-als-konventionelle-kunststoffe [Zugriff am 05.09.2020].
Verein Deutscher Ingenieure	2016	*VDI 4630 - Vergärung organischer Stoffe - Substratcharakterisierung, Probenahme, Stoffdatenerhebung, Gärversuche.* Berlin: Beuth Verlag GmbH.
VGB PowerTech	2018	www.de.statista.com, [Online] Online verfügbar unter: https://de.statista.com/statistik/daten/studie/152334/umfrage/statische-reichweite-von-ressourcen/ [Zugriff am 05.09.2020].

Anschrift der Verfasser(innen)

Kyra Atessa Vogt, M.Eng., Prof. Dr.-Ing. Iris Steinberg
Hochschule Darmstadt
Fachbereich Bau- und Umweltingenieurwesen
Haardtring 100
D-64295 Darmstadt
Telefon +49 6151 16 38 169
E-Mail: kyra-atessa.vogt@h-da.de, iris.steinberg@h-da.de

Contribution of Biological Residual and Waste Materials to the Bioeconomy
Presentation of a New Approach for the Treatment of Biological Residual and Waste Materials to Produce Bio-Based Materials

Kyra Atessa Vogt, Iris Steinberg

University of Applied Sciences, Darmstadt, Germany

Abstract

A new method for the treatment of biological residues and waste materials, especially kitchen and canteen waste, for the production of bio-based materials is described. Furthermore, first investigations on the production of bio-based materials from kitchen and canteen waste are presented. The intention is to determine the influence of such a treatment before the fermentation of kitchen and canteen waste on the biogas production potential.

Keywords

biological waste, carboxylic acids, bio-based products, bioeconomy, food waste

1 Introduction

Around 1.2 billion tons of raw materials were consumed in Germany in 2016; approximately one third of this was based on fossil fuels (Statistisches Bundesamt, 2020). Though, the fossil resources are limited (VGB PowerTech, 2018). Therefore, the use of non-fossil - so-called bio-based - raw materials based on renewable materials is becoming more and more important. As a result, the consumption of fossil resources can be reduced and at the same time a contribution towards the bioeconomy, a bio-based economy, can be made. This is where the national bioeconomy strategy of the Federal Government of Germany comes in, which postulates sustainable, climate-neutral development combined with the use of biogenic raw materials to create a sustainable, circular economy. (Bundesministerium für Bildung und Forschung, 2020).

Currently, mainly starch- and cellulose-containing plants, such as maize, oilseeds, sugar cane and wood, are used for the production of bio-based materials, which compete with food production (Umweltbundesamt, 2020). It is therefore important to investigate the use of already existing biomass, such as biological residues and waste materials, instead.

2 Objective

About 11 million tonnes of biodegradable residual and waste materials are disposed in Germany every year, including separately collected organic waste via the organic waste bin, garden and park waste, and kitchen and canteen waste (Statistisches Bundesamt, 2019). These are mainly treated by established biological processes of composting or digestion, either materially utilized as soil improvers (compost) or additionally energetically as biogas. But in these processes, the majority of the carbon bound in the biological residues and waste materials is emitted as carbon dioxide (CO_2). Therefore, it is necessary to optimise these processes with regard to the complete use of carbon.

The established biological treatment processes degrade or convert the organic waste with the help of microorganisms. Such intermediate products include organic compounds with one or more carboxy groups (-COOH) (Federle, et al., 2017), which are classified depending on the number of carbon atoms into short-chain (1 - 3 carbon atoms), medium-chain (4 - 10 carbon atoms) and long-chain (>10 carbon atoms) carboxylic acids (Hopp, 2018). In composting as well as in the digestion processes, mainly short-chain carboxylic acids are available. These are converted to medium-chain carboxylic acids with the help of specialised microorganisms and ethanol as an additive in the biotechnological treatment process which is presented here. The separation of the carboxylic acids takes place during the treatment process as an in-situ extraction using a non-polar extraction solvent. These carboxylic acids can be used for the production of bio-based materials and in this way contribute to a bio-based economy (bioeconomy). This type of waste management corresponds to material recycling according to § 6 section 1 of the German Circulation Economy Law (Kreislaufwirtschaftsgesetz).

As a potential feedstock for the production of medium-chain carboxylic acids, secondary residues, leachates from composting facilities and liquid residues from digestion facilities have already been investigated (Kannengiesser, et al., 2019). These feedstocks are suitable in general; the yield of medium-chain carboxylic acids from the digestion residues is lower than from the untreated leachates, as carboxylic acids have already been transformed into biogas during the digestion process.

Kitchen and canteen waste is also a suitable substrate with a high potential for the production of medium-chain carboxylic acids (Kannengiesser, et al., 2019). By using these waste materials for the production of carboxylic acids, the overall amount of carbon compounds which are available for the biogas production is reduced, while on the other hand, this process could also increase the number of available intermediates and make them more available for the fermentation process. Therefore, we investigated how the biogas potential of the substrate changes after the separation of the medium-chain carboxylic acids - and therefore the contribution to energy recovery - together with the determination of the carboxylic acids as a contribution to material recycling.

3 Material and Methods

3.1 Substrate

We investigated liquid kitchen and canteen waste which was delivered already pasteurised to a fermentation plant for food, food waste from kitchens and canteens as well as commercial flotates and brew residues.

For the determination of the biogas yield, we used fermentation substrate from the same digestion plant as inoculum and pure cellulose as reference substrate.

3.2 Experimental setup

For the investigations, we use an experimental set-up for the production of bio-based substances and a second one for the determination of the biogas production potential.

3.2.1 Production of bio-based materials

The experimental set-up for the production of bio-based substances consists of an airtight vessel with a maximum capacity of 26 litres. The produced gas is led through a gas outlet in to a gas bag. The experimental reactor is equipped with a sampling device for taking substrate samples. The non-polar extraction agent settles on the polar substrate layer. The temperature control of the experimental set-up takes place in a temperature-controlled room. The following figure illustrates the described experimental setup and the technical implementation.

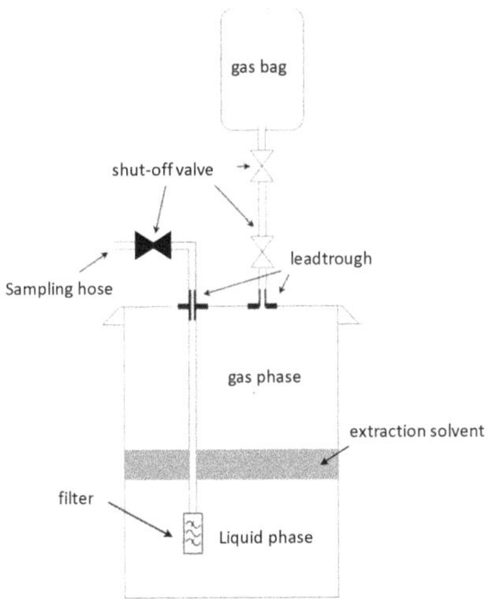

Figure 1: Schematic drawing of the experimental setup (left) (modified according to Noll, 2020) and technical implementation (right).

3.2.2 Biogas production potential

The experimental set-up consists out of a standing bottle with a capacity of 2 litres (max. filling volume 1.8 litres) with stirrer. The produced gas is collected in a gas bag for analysis. The substrate is temperature-controlled to the target operating temperature by a water bath. The following figure shows the experimental set-up (left) and the technical implementation (right).

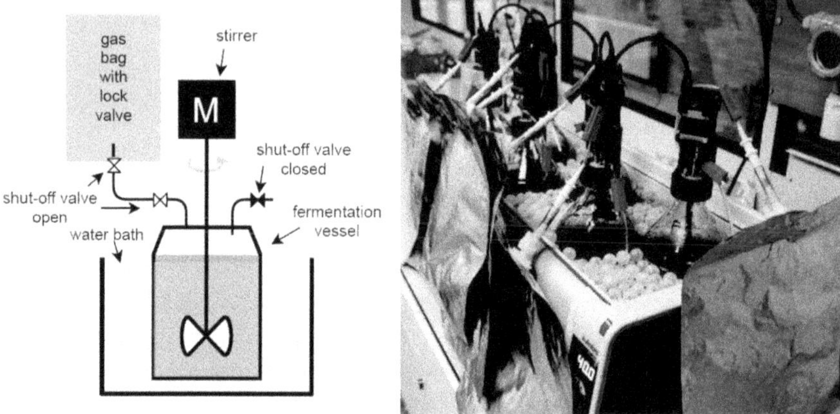

Figure 2: Schematic drawing of the experimental setup (left) and technical implementation (right)

3.3 Analytical methods

The on-site analysis includes all parameters that are analysed immediately after sampling. These include the pH value and the redox potential.

The pH value is an important parameter for the production of bio-based materials from organic waste. It allows conclusions regarding the ideal conditions for the microorganisms in the substrate. For this reason, we take this value immediately after each sampling in the sampling vessel.

The redox potential indicates whether the process takes place in an aerobic or anaerobic milieu. If the redox potential is below -330 mV, the substrate is in an anaerobic milieu (König & Wiese, 2007), which is necessary for the activity of the specified microorganisms.

The carboxylic acids are analysed by gas chromatography and flame ionisation detection (GC-FID). The determination of the biogas production potential was carried out according to VDI 4630 "Fermentation of organic matter - substrate characterisation, sampling, material data collection, fermentation tests." in a stirred batch fermentation vessel (see

Fig. 2, Chap. 3.2.2) and performed as specified (Verein Deutscher Ingenieure, 2016): During the experimental procedure, the amount of produced gas was determined and the gas composition was analysed.

3.4 Experimental procedure

The experimental procedure is divided into the following three phases:

1. determination of the biogas potential of the untreated substrate
2. treatment of the substrate for the production of bio-based materials
3. determination of the biogas potential of the treated substrate

The following table presents the operating parameters of the individual experimental phases.

Table 1: Overview of the operating parameters

Parameter	Phase 1	Phase 2	Phase 3
Duration	31 d	31 d	33 d
Volume	Min. 1.5 l/Max. 1.8 l	10 l	Min. 1.5 l/Max. 1.8 l
Temperature	40°C	38°C	40°C
Adjustment pH-value	-	6 - 6.5	-
Sampling	-	2x per week	-
Gas analysis	As needed	As needed	As needed
Homogenisation	20 rpm	-	20 rpm
Additives	-	Ethanol*: 11 ml/l Extraction solvent*: 100 ml/l	-
Reactors	Inoculum 1 Inoculum 2 Substrat 1 (untreated) Substrat 2 (untreated) Cellulose 1 Cellulose 2	Blind** Substrat A Substrat B	Inoculum 1 Inoculum 2 Substrat 1 (treated) Substrat 2 (treated) Cellulose 1 Cellulose 2

*not in reactors "blind", **used as reference reactor for treatment processes

The fermentation batch for determining the biogas potential underlies certain preconditions (see chapter 3.3). The following table summarises these. All preconditions were fulfilled, as shown in the table below.

Table 2: Assumption on fermentation batch of VDI 4630

Parameter	Target	Phase 1	Phase 3
Ignition loss inoculum	>50%	50.55%	50.24%
Organic mass from the inoculum	1.5% - 2%	1.89%	2.07%
Dry residue fermentation batch	≤ 10%	Substrat = 5.5% Cellulose = 4.23%	Substrat = 5.57% Cellulose = 4.57%
$\frac{oTM_{Substrat}}{oTM_{Inoculum}}$	≤ 0.5	Substrat = 0.21 Cellulose = 0.21	Substrat = 0.15 Cellulose = 0.20

oTM: organic dry-mass (Verein Deutscher Ingenieure, 2016)

The following table shows the composition of the fermentation mixtures of the two investigated phases.

Table 3: Composition of fermentation batch in phase 1 and 3

Reactor	Mass [g]			
	Phase 1		Phase 3	
	Inoculum	Substrat	Inoculum	Substrat
Inoculum 1	1,493.90	-	1,500.00	-
Inoculum 2	1,494.00	-	1,000.00	-
Substrat 1	1,494.50	309.32	1,500.00	300.00
Substrat 2	1,494.50	311.27	1,500.00	300.00
Cellulose 1	1,494.50	6.60	1,500.00	6.60
Cellulose 2	1,494.00	6.60	1,500.00	6.60

4 Results and Discussion

4.1 Production of bio-based materials

4.1.1 On-site analytics

The original pH value of the tested substrate is 3.6. This varies only minimally during the duration of the investigation and stabilises between 3.6 and 3.7. In the beginning, we adjusted the pH value of the treated substrate to 5.95. In the course of the experiment, it increases and fluctuates between 6.0 and a maximum of 6.4. That means that the substrate had optimal conditions for the microorganisms.

The redox potential of the untreated substrate (blind) runs between 17 mV and -75 mV during the investigation duration. This range of values indicates an aerobic milieu in the substrate. In contrast, the treated substrate is in the anaerobic range with values between -257 mV and -379 mV for the entire duration of the investigation, which has a positive effect on the metabolic rate of the microorganisms.

The gas composition during the investigations indicates that the treatment process was successful. On average, about 5 vol.-% CH_4, 24 vol.-% CO_2, 13 vol.-% O_2, 57 vol.-% N_2 and 57 ppm H_2S were analysed. The measured values of CO_2 and H_2S show significantly higher fluctuations compared to the other gases. The high oxygen and nitrogen measurement values for an anaerobic process are due to the opening of the reactors to sample the extraction solvent. However, these have no effect on the anaerobic environment in the substrates, as evidenced by the redox potential measurement. This is achieved by the airtight sealing of the substrate by the non-polar extraction solvent.

4.1.2 Production of carboxylic acids

The following figure 3 shows the progression of the formation of carboxylic acids and indicates them in grams per litre. It can be clearly seen that the untreated substrate (blind) mainly contains acetic acid (C2). In contrast, the treated substrate contains a significantly higher concentration of acetic acid (C2) as well as propionic (C3), butyric (C4), valeric (C5) and caproic acid (C6). A steady increase of carboxylic acids with increasing test duration can be recognised. In the untreated substrate, small concentrations of higher-value carboxylic acids (C6 - C12) can still be detected, whereas these are less detectable in the treated substrate. This is probably due to the transfer of these carbon acids into the extraction solvent, which was not added to the untreated substrate. To verify this, the following figure 4 illustrates the development of the extracted carboxylic acids that were detected in the extraction solvent. This shows the concentration of carboxylic acids in the extractant in grams per litre. As the duration of the experiment increased, butyric acid (C4) and caproic acid (C6) dissolved from the treated substrate into the extraction solvent. Furthermore, a small increase in caprylic (C8), perlagonic (C9) and capric (C10) acid can be seen. This explains why no higher-value carboxylic acids can be detected in the treated substrate.

After 31 days of treatment, about 17 g of acetic acid (C2) per litre of substrate can be detected in the untreated substrate. The concentrations of the other carboxylic acids (C3 - C12) are very low with less than 1 g/l.

On the other hand, about 25 g/l acetic acid (C2), 15 g/l propionic acid (C3) and 15 g/l butyric acid (C4) were produced by the treatment of the substrate within a treatment duration of 31 days. In addition, about 1.6 g/l butyric acid (C4), 0.8 g/l valeric acid (C5), 2.1 g/l caproic acid (C6) and 0.6 g/l caprylic acid (C8) were extracted in the developement of the investigation.

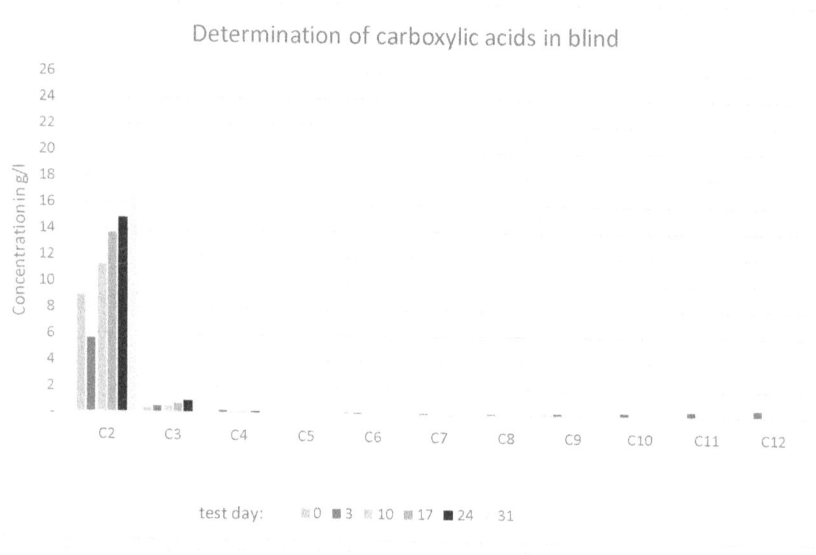

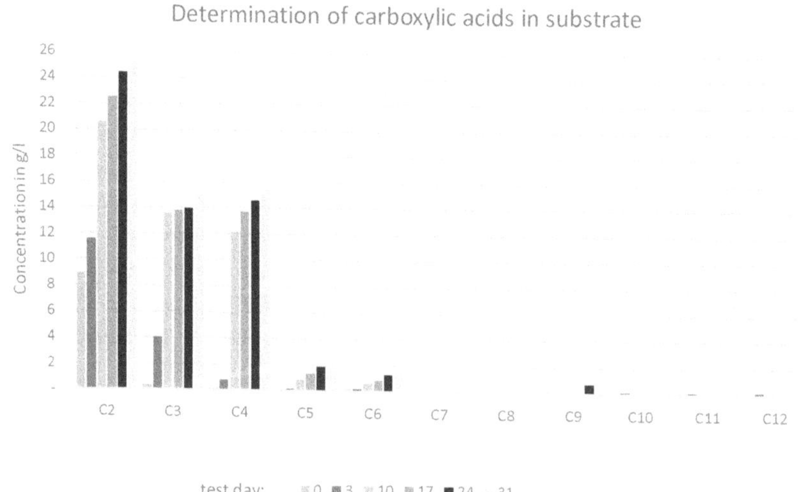

Figure 3: Comparison of the concentrations of carboxylic acids between blind reactor and treated substrate after phase 2

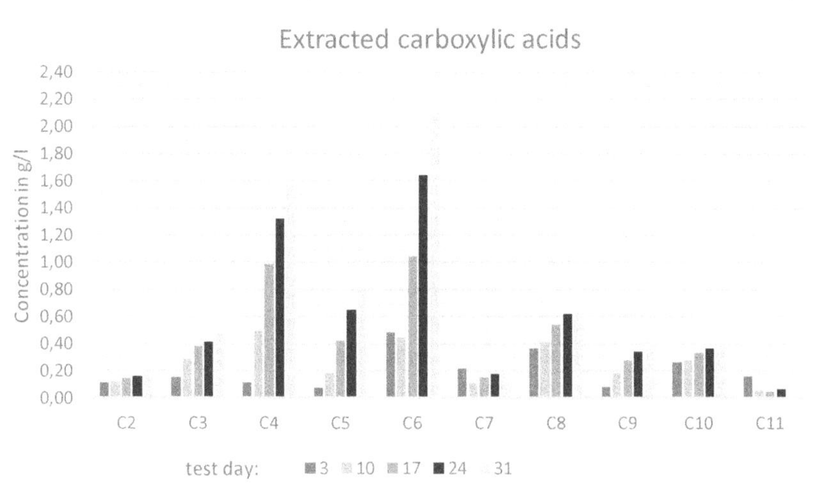

Figure 4: Extracted carboxylic acids in extraction solvent

4.2 Biogas production potential

4.2.1 Biogas production potential of untreated substrate

To evaluate the biogas production potential, the specific biogas production according to VDI 4630 is calculated in standard litres per kilogram of organic dry matter. Before the treatment of the substrate, the specific biogas production of the untreated substrate is 985.11 l_N/kg oTM. The specific biogas production potential of the cellulose, which was used as reference substrate, is 575.54 l_N/kg oTM. The control value for cellulose is 745 l_N/kg oTM. This value can be exceeded or undercut by 10% (Verein Deutscher Ingenieure, 2016). Since the specific biogas production potential of the reference substrate is around 23% lower than the control value, the results are not sufficiently reliable for evaluating the biogas potential.

4.2.2 Biogas production potential of treated substrate

To evaluate the biogas production potential of the treated substrate, the specific biogas production is also determined according to VDI 4630. After the treatment of the substrate, the specific biogas production of the treated substrate is 1,177.75 l_N/kg oTM. The specific biogas production of the reference substrate, cellulose, could not be determined because the net gas standard volume had a negative value. Thus, the control value cannot be fulfilled and the results are not sufficiently reliable.

5 Conclusion

The investigations for the production of bio-based materials generated positive results. The yields of the kitchen and canteen waste that was treated here were about 25 g/l acetic acid (C4) and about 16 g/l butyric acid (C4). These are comparable to the results of previous studies on leachates from composting facilities, which produced around 16 g/l acetic acid (C2) and around 18 g/l butyric acid (C4) (Kannengiesser, et al.,2019). The operating conditions were stable throughout the investigation period. The highest concentration in the substrate was acetic acid (C2) with around 25 g/l. About 8 g/l more acetic acid was produced in comparison to the untreated substrate. Furthermore, a maximum of 2.1 g/l caproic acid (C6) could be detected in the extraction solvent after treatment. The extracted mass could be increased if the experimental set-up will be optimised in such a way that the contact area between the extraction agent and the substrate will be increased in order to favour the transition of the medium-chain carboxylic acids into the extraction agent.

The determined biogas production potentials of the untreated and the treated substrate are, with about 985 and 1,178 l_N/kg oTM, respectively, are above literature data for comparable substrates, e.g. for low-fat food waste about 643 l_N/kg oTM and 762 l_N/kg oTM for high-fat food waste (Landesanstalt für Landwirtschaft Bayern, 2020). Considering the control values according to VDI 4630, they are to be classified as not sufficiently reliable, so that on the basis of this test series no valid statement can be made about the influence of the production of bio-based materials on the biogas production potential of liquid kitchen and canteen waste. For this reason, further investigations are required in order to be able to evaluate whether and what effects a pre-treatment for the production and separation of medium-chain carboxylic acids has on the gas production potential.

6 Literatur

Bundesministerium für Bildung und Forschung	2020	Nationale Bioökonomiestrategie - Zusammenfassung, Berlin
Federle, S., Hergesell, S. und Schubert, S.	2017	*Die Stoffklassen der organischen Chemie.* Berlin : Springer Verlag, 2017
Hopp, V.	2018	*Chemische Kreisläufe in der Natur.* Berlin: Springer Verlag, 2018. Bd. 2.
Kannengiesser, J., Campitelli, A., Vogt, A., Steinberg, I., Jager, J., & Schebek, L.	2019	Combination of different treatment methods for biological waste and agricultural residues to generate bio-based products. Proceedings SARDINIA2019 - 17th International Symposium on Waste Management and Sustainable Landfilling. Forte Village/Santa Margherita di Pula (CA): CISA Publisher.

König, R., Wiese, J.	2007	Prozessbegleitende Fermenterüberwachung auf Biogasanlagen.
Landesanstalt für Landwirtschaft Bayern	2020	www.lfl.bayern.de, Available at: https://www.lfl.bayern.de/iba/energie/049711/?sel_list=49%2Cb&anker0=substratanker#substratanker, [Access on 22.09.2020].
Noll, P.	2020	*Potential organischer Reststoffe zur Produktion von Ethanol zur Erzeugung von Carbonsäuren (unpublished).* Hochschule Darmstadt, Darmstadt
Statistisches Bundesamt	2019	Fachserie 19 Umwelt, Reihe 1 Abfallentsorgung
Statistisches Bundesamt	2020	Umweltökonomische Gesamtrechnungen - Aufkommen und Verwendung in Rohstoffäquivalenten - Berichtszeitraum 2000 bis 2016
Umweltbundesamt	2020	www.umweltbundesamt.de. [Online] Available at: https://www.umweltbundesamt.de/biobasierte-biologisch-abbaubare-kunststoffe#22-sind-biobasierte-kunststoffe-nachhaltiger-als-konventionelle-kunststoffe [Access on 05.09.2020].
Verein Deutscher Ingenieure	2016	*VDI 4630 - Vergärung organischer Stoffe - Substratcharakterisierung, Probenahme, Stoffdatenerhebung, Gärversuche.* Berlin: Beuth Verlag GmbH.
VGB PowerTech	2018	www.de.statista.com. [Online] Available at: https://de.statista.com/statistik/daten/studie/152334/umfrage/statische-reichweite-von-ressourcen/ [Access on 05.09.2020].

Author's addresses

Kyra Atessa Vogt, M.Eng., Prof. Dr.-Ing. Iris Steinberg
University of Applied Sciences, Darmstadt
Department of Civil- and Environmental Engineering
Haardtring 100
D-64295 Darmstadt
Phone: +49 6151 16 38 169
E-Mail: kyra-atessa.vogt@h-da-de, iris.steinberg@h-da.de

How to fail successfully in designing an alternative fuel production

Hubert Baier
WhiteLabel-TandemProjects UG (ltd.), Münster/ Germany

Abstract:

In addition to the waste management approach, the vocabulary must first be standardised, because waste conditioners should need to understand the requirements of co-processing, and kiln operators should know how the pre-treatment process is working and financed, and finally hardware suppliers need to have a vital understanding of both sides.

Unfortunately, there are many MBA investments that are not properly designed due to a wrong approach and financing concept, playing into the hands of the opponents or stressing the patience of the banks.

The cement manufactoring process, quite popular in the waste management, is a very cost and energy intensive chemical conversion process in which clinker and fuel ash are producing the precursor of the standardized cement. All fuels must be compatible to that process, the product and finally to the local air pollution control.

In the following, it is shown why the thermal potential of the waste as well as the technical assessment of the cement plant results in the definition of the MBT's need of a sufficient equipment and derives the financial need and economic viability, and prevent failure.

Finally, an economically outlook will be given with regard to the development on current energy market.

Keywords:
Wording, AFs, waste assessment, technical assessment, specification, design, pre-processing, co-processing, famous fails, viability, green paradox

1 The initial situation

Especially in developing countries, the cement industry is seen as the third pillar of sustainable waste management and is to be supplied with so-called "RDF".
We have already read a lot about this - apart from the fact that very few authors have ever operated a rotary kiln or a MBT plant by themselves - the opinion has become established that this is a cheap energy source that the cement industry will snatch up with the palm of its hand.
As you may guess, many things are here assumed that are unfortunately not quite true.
Why?

2 Babylonian confusion

It starts with an imprecise choice of terms: "RDF" is a term wrongly used to describe a waste-derived fuel (WDF) that has been produced without any reference to its origin or manufacturing, or even to its purpose and quality.
When it was realised that something was missing, so-called quality classes were hastily invented to heal that gap and to let the consumer know what properties and pollutant loads he was supposed to get. Parts of the southern European waste industry wanted to declare falsely their combustible fraction as a "fuel" or even as a "product" bypassing their

customs. Ergo, you still don't have a usable specification that fits to the combustion in a rotary kiln or fossil-fired power plant boiler.
It slowly dawned that it would not work without any quality assurance, after all.
When we founded the BGS e.V. at that time, the first AF to be quality-monitored was of course the fine-grained high-grade SRF for the main burner and not the coarse-grained low-grade RDF for the calciner.
The production of solid AFs from materials of recycling facilities (MRF) or commingled municipal solid waste (MSW), however, takes place in several treating stages, whereby in mechanical-biological (MBT) or biological-mechanical treatment (BMT) organic fractions, impurities and inertia and valuables such as plastic foils and metals are separated from the combustible high calorific fraction (HCF). This HCF can be used directly in the FBC or grate fired waste-to-energy power plants, in a pre-combustion chamber at the calciner of the clinker kiln or as a feedstock for the production of low-grade fuel (RDF) or downstream to high-grade SRF.

3 Technical assessment of the recycling process

In addition to the option on having AFs used in fossil-fired power plants or in Waste-to-Energy or so-called industrial power plant, the cement industry with its almost 4,000 locations have established itself worldwide as the best-known co-processer of alternative fuels; which also leads to the fact that most mistakes are made here:

4 The cement production

Now, we have reached the point to differentiate between the AF types. In many of my articles, I already shown the special features of the clinker burning process several times, but only a few suppliers and operators of the intended waste processing plants pay attention to the pitfalls. This is because the opinion still prevails worldwide that the treatment plants must not cost anything, that the clinker process is benign and that everything that is combustible is already "AF", which the cement manufacturers will accept with enthusiasm due to the expected gate fee... Far from it!

The company, from where the author comes from, still inspects a lot of kilns and pre-treating facilities for optimization and reconstruction all over many countries and has, for example, to rebuild a large MBT plant under time pressure (two years before the German ban of landfill came into force) to produce customized AFs for the local industries of power plants and cement works. The MBT in question was not able to produce AFs of sufficient quality to meet the specification for local customers.

Fig. 1: Converting waste into customised AFs as well as managing its properties requires knowledge from the waste and the recovery process. Only then can and must the processing plant be properly designed (Baier 2020).

There was a problem to solve!

The already commissioned MBT threatened to become a bad investment. Because among the six kilns in the nearest region, there was not a single one with a calciner or even with a chlorine bypass - so all the fuel could only be fed through the main burner – it has to be customized to SRF!

So, what do we learn from this?

1. Despite many mistranslations, cement is not burned in a "cement kiln", but clinker - a precursor product that is ground with gypsum and additives to produce cement that conforms to standards!
2. The burning process in the rotary kiln is an energy-intensive chemical production process and not a disposal process, because the ash also becomes part of the product.
3. Rotary kilns equipped with preheater or operating as a wet process cannot use "RDF" with a grain size of 80 - 100 mm, but only SRF with <25 mm, which shall be fed by the main burner.
4. Cement manufactures do not always know exactly how tolerant their rotary kiln actually is and what happened in their tube, e.g. with regard to water vapour, energy demand, oxygen content at the point of application, etc.
5. If the conditions at the feeding points are specified, the pre-processing plant can now be designed properly and can produce customized fuels,
6. this requires preliminary waste assessment to identify the composition, the amount of impurities and the knowledge of a sensible demand-oriented technology.
7. The chance not to be rejected from the supplier list will increase if this preliminary work has been done properly, and
8. will save the investors' money.
9. This will also include the determination of the trajectories of SRF particles when leaving the burners tip to burn in the power plant boiler, clinker or lime kiln!
10. The better the technical assessment of the customers plant AND the waste assessment are executed, the more the waste processing plant, including investment requirements, can be designed.

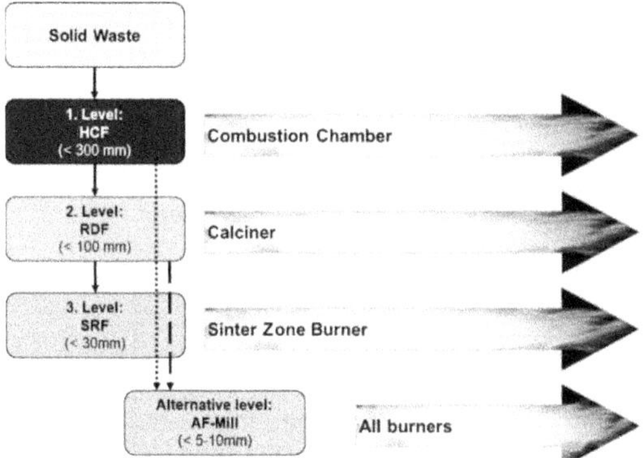

Fig. 2: The production of customized alternative fuels, its designations in order to fuel the right point.

Thus, different names of AF types are reasonable - please do not confuse or even forget them. Waste-derived fuels (WDF) are to be designated:

- ✓ For the low-grade fuel **RDF,** it is for the **calciner** with its long burn-out time due to the size and its lower calorific value of H_i 16-19 MJ/kg
- ✓ **Add-on**s such as HotDisc or StepCombustor provide additional feeding points at the **calciner** to feed coarse **HCF** with a grain size of ~300 – 800 mm and a H_i 14-16 MJ/kg
- ✓ kiln inlet, between the vertical tower and the horizontally rotating kiln tube, in which end-of-life-tyres (H_i 24-28 MJ/kg) can be charged as an option, and
- ✓ at the end of the kiln the sinter zone or **main burner**, which must ensure its approx. 2,000°C hot flame for the cooking of clinker with high-grade **SRF** at <25 mm and H_i >20 MJ/kg.

In the meantime, there is an establishing AF-mill that skips one comminution stage and thereby provides the pyroprocess with the reaction surface of AFs that the combustion kinetics actually require.

The above-mentioned assessment at the identified customer thus provides the basis and tolerances for the fuel specification and the design of the pre-treatment plant.

5 Assessment of the waste composition

An important starting point for co-processing is to determine the thermal potential, waste quantity and composition in the access area from which the waste will originate. The waste streams in question should be analysed by sorting with regard to their shares of usable high calorific fraction and their unusable components such as organics, metals, glass and other interfering materials. Taking into account the regional characteristics (rural, urban, touristic), the season and weather or cultural peculiarities, these have to be reliably collected, too.

Thanks to a global market, packaging is produced worldwide from comparable materials. The main components of the high-calorific fraction are therefore polymers, paper, composites, textiles, etc., whose share in the mixture varies.

Unfortunately, in the mixture, the organic matter carries high water contents and soaks paper, cardboard, textiles, etc., which will have to be dried again and will cause additional costs. It becomes difficult if the identified cement plant in the region is operating a wet kiln process, because at this point at the latest you should talk to the waste owner (usually the municipality) about separate paper collection because of the later SRF specification.

Furthermore, the inputs of product-relevant and emission-relevant (volatile) heavy metals must also be determined. The paths of alkalis, chlorine and sulphur inputs as well as the grain sizes of individual fractions must also be determined.

These data will be mirrored with the results from the technical assessment from the cement plant.

Incidentally, these data will also become the basis for the required permits for both the pre-processor and the customer. It is therefore mandatory to ensure that no dubious values, falsely quoted from literature or determined from one individual analysis, hinder or even prevent the use of alternative fuels.

It had also become very popular to set limited values without specifying the underlying measurement method or its standard... Enjoy your discussions with the monitoring authorities or the billing with your suppliers!

6 Purpose and design of a conditioning plant

Solid waste is usually collected in some consistency, usually mixed, and is tipped at the pre-treatment plant. There is no possibility of qualitative intervention and the plant operator has to accept what comes in.

A MBT respectively BMT plant is ultimately a "splitting plant" to increase the recycling potential according to the results of the waste assessment and technical assessment, and shall separate pollutants and interfering substances, recover residual materials if necessary, and make compost out of organic matter and to produce demand-oriented AFs out of the combustible HCF. It is obvious that the processing depth of each step consequently varies depending on the goal and purposes!

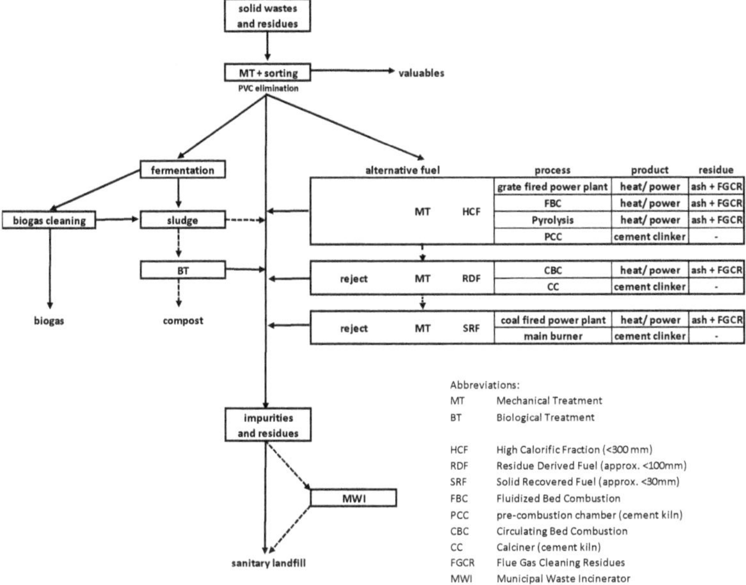

Fig. 3: The strategy and target(s) must be defined before so that the solution as well as its required equipment consequently the financial demand can be derived in a bankability study.

One focus is becoming increasingly important - worldwide. It is the reaction to the increasing plastic pollution of the seas and soil, and the will to reduce CO_{2eq}.

Thus, MBTs are becoming "servants of several masters" and shall serve an important purpose in the interplay of governmental waste and climate policy. However, if the conditioning does not fulfil its purposes, the reputation is quickly gone – also with the banks behind them.

One example is the reaction of the European Environmental Bureau[1] which explicitly addressed in an Eastern European country the unreliability of MBT technology and favoured

[1] The European Environmental Bureau (EEB) is an umbrella NGO that represents over 160 environmental organisations from 35 European countries. The EEB is involved in environmental decision-making processes in Brussels and prepares position papers on environmental policy issues that should receive greater attention on EU level. It represents its members at the European Commission, the European Parliament

the robust and established MWI technology. In that case, a lot of money was invested in the construction of local MBTs, but they produced completely useless RDF(!) without any adjustment regarding specification and demand. Consequently, the local cement industry insisted on continuing to import its alternative fuels from the UK.

In the meantime, however, the local cement industry had built dryers in order to at least covering its own substitution rates by locally processed AFs...

That's scarce...!

In fact, the treatment of solid waste in MBTs has state-wide $CO_{2eq.}$-reduction potentials, which are the sum of several "avoidance potentials":

- ✓ Avoidance of wild dumping
- ✓ No diffuse emission of methane
- ✓ Controlled guidance by banning landfilling untreated waste
- ✓ Increase of disposal fees leads to human escape on saving money and consequently to reduce the amount of waste (voila!)
- ✓ separate fermentation of organics and the recovery of recyclable materials such as biomass for composting, paper, plastics, metals, from which their upstream chains are shortened or even eliminated
- ✓ use of alternative fuels with "emission-neutral", renewable portions such as paper, cardboard, textile, natural rubber etc.
- ✓ by saving primary fuels

If the organic content and thus the main input for moisture is known, the sequence of the process steps (biological-mechanical treatment BMT- before mechanical-biological MBT) can certainly be changed and thus bio-drying can take place before a generous sieving.

With the help of a mathematical image of a RAL-certified MBT, the mass flows and properties can be simulated in advance. And last but not least, there will be a non-negligible quantity of impurities that still need to be disposed of in a sanitary landfill or to be burnt.

7 The most popular failures

A sustainable production of viable AFs is characterised by the following steps and avoidance of popular mistakes:

1) Bag opener:
   to enable an efficient segregation in the sense of cleaning, all waste (MSW as well as MRF) must be broken down.
   In some cases, "bag openers", which are developed for manual sorting of recycables (from the dual collection-system "yellow bag") are misused for comminution the waste for cleaning, which consequently does not work. A high-performance shredder is recommended for shredding the solid waste including its impurities.

2) Segregation:
   In the meantime, it has been shown worldwide that popular drum or trommel sieves

and the Council of the European Union, and also supports and accompanies member organisations' activities on European environmental issues on a national level.

are best for composting due to its design for the separation of bimodally distributed heaps. Trommel sieves are not suitable for efficient separation of organic material, inertia (stones, ceramics, soil, etc.) and other impurities from MSW with an unclear particle size distribution. "Snakes" or "pigtails" are formed, which reduce the throughput capacity and have to be shred again for a sufficient degree of segregation in order to ensure the required HCF purity downstream. Moreover, they actively utilise only ~20 % of their tubular screening surface.

Fig. 4: Compost sieves does not meet the intended separation effect in MSW (Baier 2002).

Finally, five drums would have to be installed, to benchmark flat screens of its separation results, which ultimately leads to higher space requirements and costs. Flat screens require a higher substructure, but make a better use of their total available screening area for an efficient separation of the oversized HCF from impurities.

3) Air separation/ wind sifting:
   For an efficient separation, a) clearly ground particles are required, which b) can be turbulently separated from each other at appropriate wind velocities and that can sediment according to its grain shape or grain size and its small differences in density in a c) sufficiently large settling chamber.

Fig. 5: The screened combustible oversize will be sufficiently cleaned by sifters that can separated and sediment according to its small differences in density in a sufficiently large settling chamber (Baier 2006).

You already guessed it - these compact and popular adaptations from demolition waste sifters unfortunately cannot fulfil all these physical requirements due to their design and will end as a costly thing only for conveying. When sifting, a previously determined degree of freedom must necessarily be fixed, i.e. the range of grain size should be kept as narrow as possible by efficient screening in order to be able to separate light (HCF) from heavy impurities more effectively.

In the meantime, the market offers many interesting innovations, but they all have one thing in common: They obey the laws of nature!

So in general, if you want to design, tender or buy equipment, first have a look unto the physical principles behind and let it seriously explain. Avoid equipment with combined effects or with additional auxiliary insides (separating drum, nozzles, baffles, etc.).

4) metal separation:
   Magnets are well known and are often planned in a large number. What always surprises me is that they are often installed at a 90° angle to the direction of the discharge belt. Magnets extract ferro-magnetic metals normally more efficiently at the point of their weightlessness and then at the end of the belt by the magnetic force.
   Eddy current separators shall always be mounted after magnets, because iron is disturbing the effect.

As an excuse, we often hear that the plant had to be assembled in the already approved cubature and therefore no space could be found... Question: For whom are you building the plant - for the authorities?

5) NIR-sorting:
   The separation of PVC serves exclusively to reduce organically bound chlorine. Inorganic chlorine compounds such as table and pickling salt, cannot be detected and removed by means of NIR during pre-processing. Therefore, it must be clarified with the customer before planning what their tolerance threshold is or, at the cement plant whether a chlorine bypass is ideally installed, already.

If not, these steps beforehand lead to low-grade RDF, which can only be used in a (hopefully existing) calciner or a W-t-E power plant, and its use will be limited by the quality.

High-grade SRF can only be obtained by further processing of RDF, which requires further steps:

6) Fine shredder:
   If any paper, cardboard, textile etc. saturated by moisture has survived the upstream process, they are now shredded down to a particle size of <30 mm or smaller.
   A deeper look at the combustion kinetics explains that all fuel particles have to pass through the same sequence, but with different time requirements: Drying, pyrolysis, ignition and burnout until one of the reactants is used up will define the shape and length of the flame.

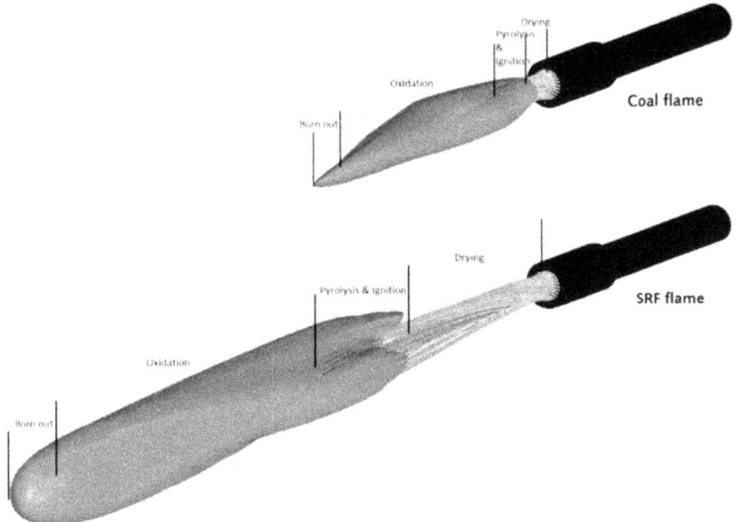

Fig. 6: In a CFD analysis, the sequence of drying, pyrolysis, ignition and bumout can be well understood for both coal and SRF. The strong prolongation of the flame results from the time required for water evaporation as well as the conversion of the coarse fuel particles (Process optimisation/ ZKG 02-2012).

Since these steps influence the temperature profile and the clinker formation process, large combustion surfaces are mandatory for a proper conversion of SRF.

7) If the moisture content in SRF is still unacceptably high, the alternative fuel must be dried at the conditioning plant or at the kiln side with potentially available waste heat (has to be clarified during the technical assessment). In both cases, special attention must be paid to the management of the vapours.

8) 3D-separation:
This separation is mandatory for suitable SRF, and can only be ensured by
a) using a so-called "Rocket Mill" as a robust fuel mill with a drying effect, or
b) by air classifying of fluffy particles that will offer a large reaction surface
(note: Size reduction and screening is even counterproductive here, as the remaining 3D particles still do not get enough time to burn out due to their unfavourable surface/size ratio and end up unburnt in the clinker bed, leading to the well-known reductive conditions).

9) Stockpiling is often regarded as "dead capital". But, if you want to increase the TSR by feeding more SRF, we observe more often that the kiln loses its tolerance and performance to quality fluctuations, massively.

I.e. homogenisation and quality control lead to a safe kiln operation, which is why the docking stations with moving floor trailers originally installed for "experimental purposes" are becoming insufficient and would have to be replaced by proper storage facilities with "police filter" and homogenisation capacities.

8 Speaking of money:

A popular statement that we hear again and again is that the client assumes the plant can be re-financed by the sale of the recovered recyclable materials and alternative fuels.

And, hopefully this price is not linked to quality criteria such as ash, water or chlorine and its energy content... If this works out, I can really congratulate to such a contract!

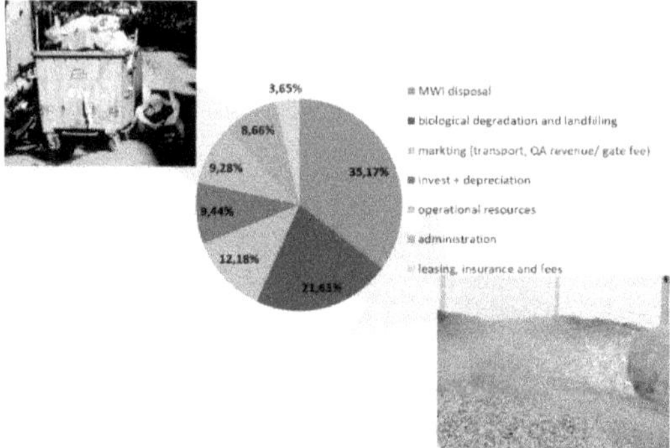

Fig. 7: Cost structure of a MBT plant for the production of SRF from mixed solid waste. The highest costs result from the disposal fees of impurities that can only be deposited of in a Municipal Waste Incinerator (MWI), and followed by the biological processing for sanitary landfilling (Baier 2010).

For the energy and world market prices, the renowned economist Prof. Hans-Werner Sinn 2008 formulated, based on his researches, that we will end up in a so-called "green paradox". Countries, that reduce the demand for fossil fuels (e.g. by phasing out) thus drive the suppliers to offer their output at lower prices elsewhere (see Germany selling its lignite to the Czech Republic or Norway, Saudi Arabia, the USA or Russia outbidding each other in oil and gas production). Simply said: The more is produced, the less those who do not reduce their primary energy consumption have to pay. It's the basic law of economy of supply and demand!

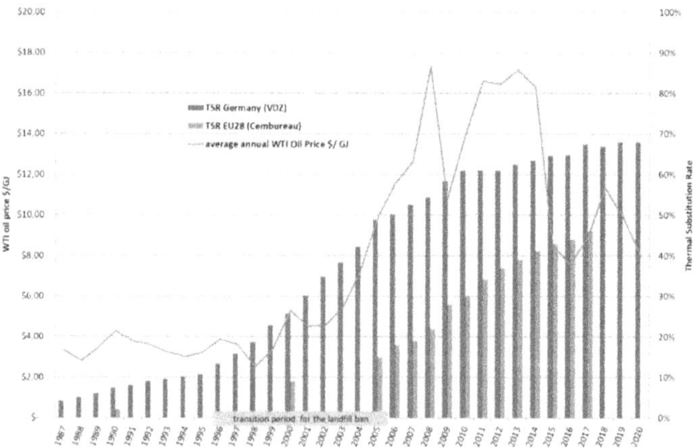

Fig. 8: While the WTI-based world energy costs fluctuate strongly, the German efforts for increasing the thermal substitution rate (TSR) are not viable any more. It has taken its biggest increase in preparation before the enforcement of the landfill ban in 2005, while today energy costs are decreasing, and the supply on currently suitable AF-qualities is steadily decreasing (based on VDZ, CEMBUREAU and statista).

We must also apply this to the relationship between the cement and waste industries: In principle, the cement industry, as a large and global energy consumer, always buys at the cheapest energy price (e.g. for coal), while the waste management industry is a local competitor who wants to get rid of its combustible AFs as an "energy source".
The MBT operators will be flooded with the cheapest plastic items and have to refinance the operation by a minimal of disposal fees. They try to generate additional revenue from sorting recyclables, which must be pushed into an oversaturated market and can only accommodate their (unattractive) AFs by increasing the gate fees (additional payment).

This only works in countries where the polluter-pays principle is valid and the polluter will finance the entire chain from "bin to fire" where the gate fee is already included in the disposal fee. Indeed, a political mammoth task!

But in the meantime, we are confronted with the problems ahead:

1) Raw materials become cheaper for the plastics industry (at the same amount of supply) the more plastics we recycle or ban, and
2) the more recycling and processing facilities we need to cope with the growing flood of plastics, and
3) in a mid-term, co-processing is only attractive if the disposal fee will cover the gate fee to the work as well.

Otherwise, the waste will remain floating in the seas, wildly dumped and burnt.

Unless, taxes are introduced on using primary hydrocarbons or a withholding tax on the capital gains earned by resource owners on financial investments - worldwide!

9 Literature

Hubert Baier, Hans-Otto Gardeik	1999	Utilisation of secondary materials in Cement Plants – Wishful Thinking and Reality, in Barrage, A.; Edelmann, X. (Hrsg.), in Congress Proceedings R´99, EMPA, Vol. II, S. II.40 – II.45, Genf
Hubert Baier	2004	Sekundärbrennstoffaufbereitung in der Praxis – Die SBS-Anlage Ennigerloh, in Wiemer, K., Kern, M. (Hrsg.), Witzenhausen-Institut, Neues aus Forschung und Praxis, Bio- und Restabfallbehandlung VIII, biologisch-mechanisch-thermisch, Fachbuchreihe Abfall-Wirtschaft des Witzenhausen-Instituts für Abfall, Umwelt und Energie, S. 118 – 124, Witzenhausen
Hubert Baier	2005	Erzeugung von Ersatzbrennstoff für den Einsatz in Zement- und Kraftwerken, in Thomé-Kozmiensky, K.J., Beckmann, M. (Hrsg.), Ersatzbrennstoffe 5 - Herstellung und Verwertung, TK Verlag, S. 321 – 336, Neuruppin
Hubert Baier	2006	Produktion und Vertrieb von Ersatzbrennstoffen (EBS) - Beispiel EBS-Anlage der Ecowest in Ennigerloh in Internationale 6. ASA-Abfalltage, Mechanisch-Biologische Restabfallbehandlung, MBA in der Bewährung, 1. – 3. Februar 2006 Congress Centrum Hannover, ASA Arbeitsgemeinschaft Stoffspezifischer Abfallbehandlung e.V. (Hrsg.), S. 287 – 304, Hannover
Hubert Baier	2007	Abfallverwertung durch anwendungsorientierte Herstellung von Ersatzbrennstoffen – Herstellung und Vertrieb von EBS aus

		Siedlungs- und Gewerbeabfall am Beispiel der Anlage Enniger-loh, in Schriftenreihe des IFAAS - Institut für angewandte Abfall-wirtschaft und Stoffstrommanagement Suderburg e.V., Suder-burger Abfall Seminar 8./9. März 2007 an der Universität Lüne-burg – Campus Suderburg, Kap. 6, Suderburg
Hubert Baier	2007	Zwischenlager für heizwertreiche Fraktionen, in Flamme, S., Gal-lenkemper, B., Bidlingmaier, W., Doedens, H., Kranert, M., Steg-mann, R. (Hrsg.), Münsteraner Schriften zur Abfallwirtschaft, Band 11, 10. Münsteraner Abfallwirtschaftstage, LASU der Fach-hochschule Münster, S. 311 – 317, Münster
Hubert Baier	2010	Disruptive substances and the burning behaviour of solid alterna-tive fuels, in Zement-Kalk-Gips International, Volume 63 (2010), Nr. 6, S. 58-67
Hubert Baier, Karl Menzel	2011	Potentials of AFR Co-processing, in Proceedings of 53rd IEEE-IAS/ PCA Cement Industry technical Conference, 22-26 May; At-lanta (GA)
Hubert Baier	2012	Utilization of alternative fuels in the cement clinker process, 52-59, CEMENT INTERNATIONAL 4, 1/2012, 4 VOL. 10
Hans-Wer-ner Sinn	2012	The Green Paradox, (fundamentally revised translation), MIT Press, Cambridge, Massachusetts (US)
Hubert Baier	2014	Moderne Zementwerke und strategische Ansätze zur Aufberei-tung von Ersatzbrennstoffen - Eine aktuelle Bestandsaufnahme, in Thomé-Kozmiensky, K.J., Beckmann, M. (Hrsg.), Energie aus Abfall, Band 11, TK Verlag, S. 859 - 869, Neuruppin
Hubert Baier	2015	Big Bang Theory and Practice, ZKG international, Gütersloh/ Germany; 68, 1/2; 20-23
Hubert Baier	2018	Unlocking Morocco's co-processing potential Part I, International Cement Review, March 2018, p. 34ff. Tradeship Publications Ltd. Dorking, Surry (UK)
Hubert Baier	2018	Unlocking Morocco's co-processing potential Part II, International Cement Review, April 2018, p. 53ff., Tradeship Publications Ltd. Dorking, Surry (UK)
Hubert Baier	2018	From waste to co-processing, International Cement Review, Sep-tember 2018, p. 71ff, Tradeship Publications Ltd. Dorking, Surry (UK)

Author's address:
WhiteLabel-TandemProjects UG (ltd.)
Dr. Hubert Baier
Sentruper Str. 165
D-48165 Münster/ NRW
Phone: +49 2504 9331 – 97
eMail: hubert.baier@wltp.eu
homepage: www.wltp.eu

Organic Waste for the Production of Hard Carbon for Batteries – A Life Cycle Assessment Perspective

Huiting Liu[1], Claudia R. Tomasini[2], Linghan Lan[4], Jun Li[4], Xiang Zhang[5], Niklas von der Aßen[3], Marcel Weil[1,2]

[1] Institute for Technology Assessment and Systems Analysis (ITAS), Karlsruhe Institute of Technology (KIT), Karlsruhe, Germany

[2] Helmholtz Institute Ulm (HIU), KIT, Ulm, Germany

[3] Institute of Technical Thermodynamics, RWTH Aachen University, Aachen, Germany

[4]Chongqing University, Chongqing, China

[5]Central South University of Forestry and Technology, Changsha, China

Abstract

As conventional graphite anode is facing challenges of resource scarcity and environmental contamination, synthetic carbon emerges to be one of the promising anode materials for lithium-ion batteries and especially post-lithium batteries. Hard carbons derived from organic wastes possess comparable electrochemical performance, advantages of abundant and green raw materials, and potential of low cost. This work using life cycle assessment evaluates and describes the environmental impacts of different hard carbon materials. It is concluded that hard carbon gained from coconut shells presents favorable environmental performance, owning to the well-designed production processes. In contrast, hard carbon obtained from bamboo waste will lead to significant environmental burden, unless the production processes are optimized. It highlights the need for further research on raw material screening and process configuration at early phases of material design.

Keywords

Hard Carbon, Battery, Organic Waste, Life Cycle Assessment, Sustainability, Pyrolysis, Activation, Bamboo Waste, Coconut Shell, Apple Pomace

1 Introduction

With the development of energy transition, the demand for batteries is increasing rapidly owning to its widely applications such as in electric vehicles and stationary storage. However, problems in terms of performance, resource, cost and sustainability put existing battery system under pressure considering mid- and long-term development. For example, raw materials such as cobalt, lithium, and natural graphite are identified as critical by EU and U.S., which implies potential scarcity and consequential price increase. [*Blengini et al., 2020; Fortier et al., 2018*] Nickel, as another important component of battery, is also considered as critical given the scenario of high demand in the next 30 years. [*Weil et al., 2018*] Lithium-ion batteries (LIBs) have been commercialized for over 30 years. However, upstream supply chain and production of LIB have exerted significant negative influence on environment. [Weil et al., 2020] The critical raw materials are usually imported from politically unstable territories, for example, cobalt is found in large quantities in Congo, which makes manufacturers run the risk of unsecured supply chain. Furthermore, LIBs supplying for European countries is still highly dependent on imports mainly from East Asian producer at present. These challenges stimulate research on new battery systems based on more abundant metals, such as sodium-ion and potassium-ion batteries, and new battery materials with improved performance and sustainability, such as hard carbon anode materials.

A key component to pave the way is the material for negative electrode, also known as anode. Carbon-derived anode materials have been substantiated by stable electrochemical performance and reliable safety over the past 30 years. At present, commonly used graphite presents theoretical gravimetric capacity of 372 mA h g^{-1} in existing lithium-ion batteries. [*Chen et al., 2020*] However, graphite is still hanging in doubt due to limited electrochemical performance, potential scarcity, and environmental problems. Especially the performance of graphite in sodium-ion batteries is not satisfactory. Sodium cannot be efficiently inserted into graphite, which results in significant lower capacity at 35 mA h g^{-1} in sodium-ion batteries (SIBs) than in LIBs. [*Yu et al., 2020*] As for lithium- and potassium-ion battery, graphite is also suffering limited specific capacity, poor rate capability and short cycling life, and it becomes worse in fast-charging application, which hinders the development of high-power batteries. [*Chen et al., 2021*]

On the other hand, the mining and purification of natural graphite has been reported causing serious amount of dust, sever landscape damage and critical contaminants to surrounding. [*Dolega et al., 2020*] Comparatively, production of synthetic graphite requires considerable energy, as the graphitization process takes up to weeks at an elevated temperature over 2500°C, and the precursors of synthetic graphite are side products from petroleum and coal industry. [*Dunn et al., 2015*] Potential high greenhouse gas (GHG) and hazard emissions are generated along with upstream material and electric energy production. [*Dolega et al., 2020*] The pollutants originating from the graphite industry can harm human health, and it is reported that human toxicity potential is the category with the highest impact in natural graphite anode production. [*Zhang et al., 2017*] The negative impact on environment also imperils surrounding plant and animals through habitat destruction, loss of biodiversity and pollution of water resource. [*CES, 2017*]

Hard carbon, also known as non-graphitizable carbon, is a promising alternative to graphite for battery anode. In contrast to graphite-based materials, hard carbon cannot be graphitized even at a temperature over 3000°C. [*Dou et al., 2019]* This material presents disordered structure with random arrangement of graphene layers and thus can offer high capacity for ions. [*Dou et al., 2019]* Moreover, with large space for improving energy density, hard carbon is a suitable option for high power batteries and fast-charging application. Nevertheless, the mechanisms of the ion-intercalation into hard carbon are still controversial, due to the complexity of structure. [*Chen et al., 2021; Dou et al., 2019*] Hard carbon can be derived from biomass and carbonaceous waste, such as glucose, cellulose, lignin and phenolic resin. In order to obtain sufficient electrochemical performance, precursors are usually pyrolyzed at 600-2000°C with proper pre- and post-treatments, which means large potential of energy saving compared to graphite synthesis. [*Xie et al., 2020*] Although hard carbon is still facing hurdles such as high initial irreversible capacity and low market maturity, the abundant and low-cost precursors, low synthesis temperature and competitive electrochemical properties makes this material a promising candidate, especially in the context of sustainable development of energy system. [*Peters et al., 2020*]

In order to investigate potential negative impacts and hotspots involved in the hard carbon production and supply chain, this paper presents a comparative life cycle assessment on manufacture of hard carbon derived from different organic waste precursors. Previous studies on the biowaste hard carbon are firstly reviewed and summarized. Based on that, life cycle assessment (LCA) is introduced. Within the LCA framework, manufacture system is modelled, and system scope of the assessment is identified. Through inventory modelling and impact assessment, environmental profile of different hard carbon is presented, and contribution of specific process stages is analyzed by taking the ratio of impacts. Finally, conclusions are drawn to contribute advises on raw material selection and system design. Organic wastes that release less negative impact are screened. The identified hotspots can imply optimization potential for material and battery design. As a result, this work provide support in the early design phases to build a more sustainable battery system.

2 Literature review

Although production of other carbonaceous materials such as activated carbon and graphite have drawn academic attention because of large demand and significant carbon footprint, environmental analysis with particular regard to hard carbon manufacturing is still very limited. Peters et al. conducted the first environmental assessment on the hard carbon anode materials. Three precursors i.e., waste apple pomace, waste tires and synthetic resin respectively representing organic waste, non-organic waste, and non-waste materials were compared with an LCA framework. [*Peters et al., 2019*] Results shows that the hard carbon made from organic waste material leads to less negative impact than others. One reason is that waste materials could create added value and save environmental costs from upstream supply chain. Moreover, processing organic materials are relatively less complicated than non-organic materials, which require higher energy demand and more chemical inputs for pre- and post-treatment. Besides apple pomace, there are abundant biowaste materials that has been identified suitability for hard carbon synthesis. [*Peters et al., 2019; Yu et al. 2020*] However, an in-depth research that exclusively compares environmental performance of different organic waste precursors has not yet been established. Therefore, it is still not clear if organic raw materials and corresponding processes have impact on the life cycle environmental impact of hard carbon products.

From an electrochemical point of view, there has been various research that focused on the biowaste hard carbons and revealed corresponding performance in the full batterie cells. Two studies investigated hard carbon derived from apple waste via a facile synthesis route, without pre-treatment such as acid washing. [*Wu et al., 2016*; *Dou et al., 2018*] The researchers even pointed out, that extra acid treatment on pectin-free apple pomace had led to drastically reduction on hard carbon yield. Hard carbon anode materials were prepared from coconut shells involving activation by acid washing and physical treatment with CO_2. [*Jayaraman et al., 2017*; *Nita et al., 2020*]. So far, coconut shell is the only organic waste applied to produce hard carbon material in industrial scale. As informed from experts, the Japanese company Kuraray™ and Sri Lankan company HayCarb™ are employing coconut shells to yield hard carbon products. Bamboo waste was also converted into microtubular carbon fibers as anode for lithium and sodium ion batteries and gained convincing performance in sodium-ion battery. [*Zhang et al., 2019*] Though the material is not specifically defined as so-called “hard carbon”, as confirmed by the author, it refers to the consistent structure of non-graphitizable carbon. Since the definition and terminology of hard carbon is still under debate [*Dou et al., 2019*], the term hard carbon is not universally accepted despite referring identical material structure. Additionally, there are numerous other organic waste materials that have been utilized as anode materials, such as peanut shell [*Dou et al., 2018*], orange peels [*Saha et al., 2020*], green tea [*Pei et al., 2020*] etc. Review articles [*Górka et al., 2016*; *Dou et al., 2019*; *Yu et al. 2020*] summarized and described relevant studies in a systematic way.

Through literature review and the abovementioned stakeholder interviews, this work identifies the apple pomace, coconut shell and bamboo waste as three research objectives to be investigated. Comparative life cycle assessment is conducted on the hard carbon production from the three target precursors, to evaluate the attributional environmental impact and identify the environmental hotspots.

3 Life Cycle Assessment

3.1 Framework

In this study, life cycle assessment (LCA) that takes the whole industrial systems involved in the products´ life cycle is performed according to the ISO standard 14040/44. This method can be used to assess environmental aspects and potential impacts associated with a product such as hard carbon materials. [*Baumann et al., 2004*] With help of the software openLCA, the assessment is performed. Detailed procedure of system and inventory modelling and impact analysis are presented in following sections. Finally, the results of different scenarios are interpreted into graphs and analysed based on process design.

3.2 System specification

3.2.1 Process modelling

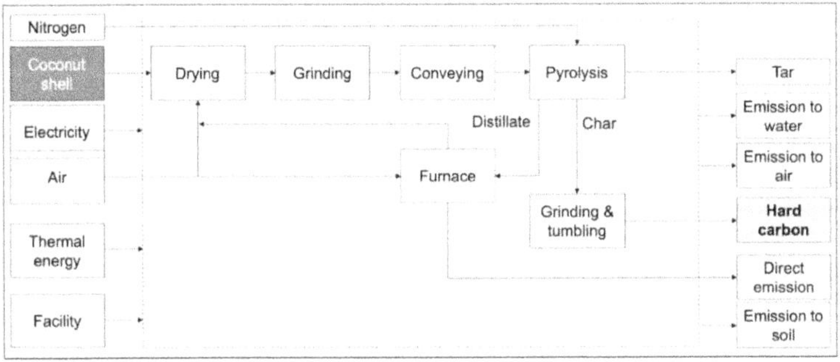

Figure 1 *System boundary of hard carbon production from coconut shell*

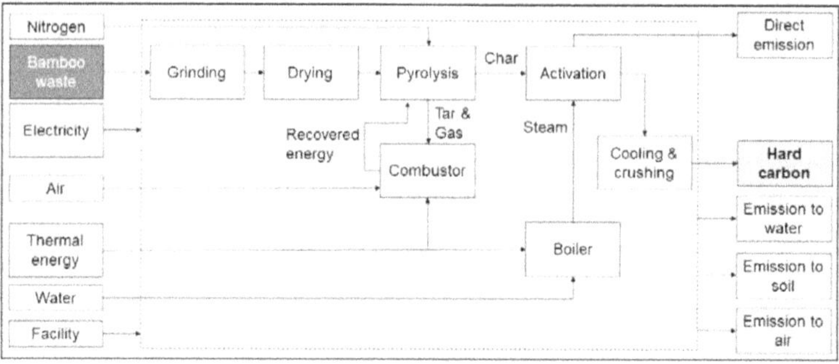

Figure 2 System boundary of hard carbon production from bamboo waste

Three different biowaste precursors and their production processes are considered: coconut shell, apple pomace and bamboo waste. For a comparison, non-organic waste tire precursor is included within the research scope. Synthesis of coconut shells and bamboo wastes are specifically modelled in this work. Corresponding flowsheets are depicted in Figure 1 and 2, and their system boundaries are hereby defined. The apple pomace and waste tire are modelled by adapting established procedures from [*Peters et al., 2020*]. The corresponding flowsheets could be found in [*Peters et al., 2020*]. The same functional unit, system boundary and analysis method are applied consistently to four precursors, to ensure comparability of results.

Details about original concepts of coconut shell and bamboo waste are described as follows. Production processes are modelled from literature review and laboratory test. The energy and material flows are simulated in industrial scale of a hypothetical plant. The synthesis route of coconut shells is more concise than others. Coconut shells are firstly pre-treated, including drying and grinding, and then conveyed to pyrolysis furnace. The temperature of furnace is given at 500°C, which is much lower than other cases. After that, the distillate of volatile gas and tar are transported to power the drier. In this way, energy of the distillate is recovered.

Analogously, in the case of bamboo waste, pyrolysis tar and gas from furnace are also combusted. The different is that recovered heat in this case is used to heat pyrolysis furnace. As a result, energy consumption of pyrolysis is reduced. The pyrolysis temperature for bamboo waste is at 800°C, which is between that of coconut shells and apple pomace. After pyrolysis, the char-based material from bamboo waste is further activated through steam, which is generated by a boiler and powered by nature gas.

3.2.2 Goal and scope definition

For the purpose of comparison, functional unit is defined as 1kg of hard carbon produced. Four hard carbon materials produced from different precursors are investigated as described in section 3.2.1. To identify environmental hotspots along with production chain, this study specifies the scope from raw material until factory gate, namely cradle-to-gate analysis. [*ISO 14040, 2006; ISO 14044, 2006*] Use and end-of-life phases are out of research scope, due to absence of experimental establishment in regard to electrochemical performance of batteries. Cut-off approach is used for process model, which allows waste feedstocks free of environmental burden for further use. [*Ruiz et al., 2015*] Ecoinvent database in version 3.5 are utilised to model life cycle inventories. [*Wernet et al., 2016*] Impact assessment method ReCiPe midpoint with the hierarchise perspective [*Huijbregts et al., 2016*] is used to assign and assess environmental impact.

3.3 Life cycle inventories

Table 1 Process inventory for hard carbon from coconut shell

Item	Amount	Unit
Inputs		
Fresh coconut shell	6.7000	Kg
Chemical factory, organics	4E-10	Items(s)
Electricity, medium voltage	2.1600	MJ
Heat, district or industrial, natural gas	1.3300	MJ
Outputs		
Hard carbon from coconut shell	1.0000	kg
Carbon dioxide, biogenic	3.7800	kg
Carbon monoxide	0.0015	kg
Naphtalene	0.7830	kg
Nitrogen	9.9000	kg
Nitrogen oxides	0.0012	kg
Oxygen	0.6900	kg
Water	2.3600	kg

Table 2 Process inventory for hard carbon from bamboo waste

Item	Amount	Unit
Inputs		
Fresh waste bamboo	6.7800	kg
Chemical factory, organics	4E-10	Items(s)
Nitrogen, liquid	1.0691	kg
Heat, district or industrial, natural gas	63.8843	MJ
Tap water	3.2453	kg
Outputs		
Hard carbon from waste bamboo	1.0000	kg
Carbon dioxide, biogenic	6.2552	kg
Carbon dioxide, fossil	3.2103	kg
Methane, biogenic	0.02635	kg
Methane, fossil	6E-5	kg
Nitrogen	1.0691	kg
Nitrogen oxides	6E-6	kg

On the basis of primary lab data and literatures, life cycle inventories (LCI) of production processes are modelled bottom-up in an industrial scale. As mentioned before, the LCI of apple pomace could be found in [*Peters et al., 2020*]. Mass and energy balance of coconut shell hard carbon are calculated based on reference process, i.e., activated carbon production from coconut shells in Indonesia from [*Arena et al., 2016*] and general chemical engineering guidelines. The inventories for bamboo hard carbon are obtained from both laboratory determination and reference process of activated carbon production from bamboo from [*Liao et al., 2019*]. Background system for upstream processes was modelled by taking available inventories from database Ecoinvent. [*Wernet et al., 2016*] Detailed life cycle inventory data of hard carbon derived from coconut shells and bamboo waste are respectively provided in Table 1 and 2.

4 Results

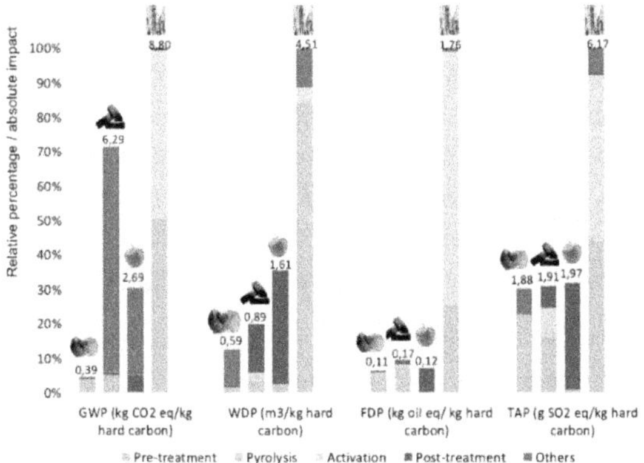

Figure 3 Relative environmental impact of hard carbon (HC) anode materials from different organic waste precursors. HC-waste tire as reference from [Peters et al., 2020]

To provide a baseline for comparison, the waste tire precursor is taken as reference for results shown in Figure 3. Preliminary results present relative environmental impacts of four hard carbon derived from respectively: coconut shell, waste tyres, apple pomace, and bamboo waste. The precursors are represented by small icons on the top of corresponding results. Four impact categories are investigated: global warming potential (GWP), water depletion potential (WDP), fossil depletion potential (FDP) and terrestrial acidification potential (TAP). For a detailed contribution analysis, the impacts are parcelled out into several compartments as a function of synthesis route: pre-treatment, pyrolysis, activation, post-treatment and others. Specifically speaking, pre-treatment consists of cutting, drying, shredding, grinding, etc., while post-treatment refers to off-gas scrubbing, cooling, etc., and others refers to facilities, direct emissions from pyrolysis, etc. To facilitate comparison, the impacts are normalized as ratio of absolute value to maximum value in the same category, so the impacts of bamboo waste have been set to 100% in all sections. However, corresponding absolute value is also listed on the top of each bar.

As illustrated in Figure 3, manufacturing hard carbon from bamboo results in highest negative impacts in all sections. The main driver is direct emission from activation process, while pyrolysis also contributes to drastically increasing impacts. The reason is that natural gas burned for pyrolysis and steam boiler leads to large amount of emission and energy consumption, which gives rises to significant GWP, FDP and TAP associate with pyrolysis and activation process. The noticeable WDP of pyrolysis process is originated from supply chain of nitrogen, which is applied to provide an inert environment for pyrolysis.

As to other two precursors, only coconut shells outperformed non-organic waste tire precursor and presents lowest impacts. The strategy of low temperature pyrolysis and combustion of pyrolysis distillate have saved large amount of energy. On the other hand, the synthetic route of coconut shells is simpler than other precursors, without additional post-treatment or activation, why the environmental burden is relative lower. In this case, electricity consumption for shredding the coconut shell is the primary contributor in categories GWP, FDP and TAP, while facilities for chemical plants being the main driver for WDP.

Although apple pomace delivered solid results in other three categories, the water depletion that is higher than waster tire cannot be easily neglected. Here, post-treatment contributed majorly to the categories WDP, FDP and TAP, owning to sodium hydroxide used to scrub residual gas from production. Meanwhile, direct emission from pyrolysis furnace contributes to the major part of GWP.

5 Conclusion

Three hard carbon materials derived from different biowastes have been investigated in this work and presented distinct environmental profiles. The results reveal that organic waste precursors of hard carbon do not continues to outperform non-organic wastes in terms of environmental impacts. Among 3 biowaste precursors, only coconut shells exhibit favorable performance in all categories. The result of coconut shells implies the importance of proper process design, which is typified by a rather simple production route. In this case no further treatment or activation is required, and the raw material is carbonized at a lower temperature, which provides potential for energy saving. Additionally, energy is recovered through combustion of pyrolysis by-products. Such plant configurations and production route show lower environmental burdens.

Comparatively, hard carbon obtained from bamboo waste presents highest negative impact, with activation process as a predominant hotspot. Activation could be energy and resource-consuming and thus causes environmental burden. In addition, it might exert adverse impact on battery performance if not designed in a proper way. Therefore, the effect of activation and associative emission need to be understood in order to screen a sustainable and well-performance anode material.

The impact of apple pomace hard carbon lies in between. Results of apple pomace indicate that post-treatment should be considered carefully. It can eliminate hazardous emission such as sulfur dioxide. Whereas in the meantime, consumption of chemical products results in significant strain on environment, which is the reason for 93% excessive water depletion. In this work, it also be noted that models are up scaled based on laboratory-scale data and such scale effects are not taken into consideration. Given the case of large-scale production, post-treatment of residual gas might become valuable and necessary. Therefore, this trade-off should be further investigated taking account of scale effects.

Within this work, a preliminary comparison of organic wastes for hard carbon production has been conducted in prospective way. However, upstream supply chain of waste and biomass products are modeled as free of burden and the associated environmental impacts are avoided. Consequential effects coming from allocation of wastes have not been taken into account. To close this gap, the research scope should be further extended.

It should be emphasized that the presented life cycle assessment is mainly built on literature and limited laboratory data. Use phase of the anode materials and thus electrochemical performance have not been considered yet. Therefore, further investigation that integrates the battery electrochemical testing and chemical process simulation is needed, in order to provide solid environmental assessment. In this way sustainability aspects can be considered within the development phase of emerging battery systems and support potential scale-up of production processes.

Acknowledgement

This work contributes to the research performed at CELEST (Center for Electrochemical Energy Storage Ulm-Karlsruhe) and was funded by the German Research Foundation (DFG) under Project ID 390874152 (POLiS Cluster of Excellence), and was financially supported by the Initiative and Networking Fund of the Helmholtz Association within the Network of Excellence on post-Lithium batteries (ExNet-0035)

Author's address(es)

M. Sc. Huiting Liu
Institute for Technology Assessment and Systems Analysis (ITAS),
Karlsruhe Institute of Technology
Karlstraße 11
D-76139 Karlsruhe
Telefon +49 721 608-23977
E-Mail huiting.liu@kit.edu

Dr. Marcel Weil
Institute for Technology Assessment and Systems Analysis (ITAS),

Helmholtz Institute Ulm (HIU),
Karlsruhe Institute of Technology
Karlstraße 11
D-76139 Karlsruhe
Telefon +49 721 608-26718
E-Mail marcel.weil@kit.edu

Reference

Arena, N., Lee, J., Clift, R.,	2016	Life cycle assessment for activated carbon production from coconut shells. Journal of Cleaner Production, 2016, 125, 68-77
Baumann, H., Tillmann, A.	2004	The Hitch Hiker´s Guide to LCA – An orientation in life cycle assessment methodology and application. ISBN 91-44-02364-2
Blengini, G. A., Latunussa, C. EL et al.	2020	European Commission, Study on the EU's list of Critical Raw Materials – Final Report (2020)
Chen, K.-H., Goel, V., Namkoong, M. J., Wied, M., Müller, S., Wood, V., Sakamoto, J., Thornton, K., Dasgupta, N. P.	2021	Enabling 6C Fast Charging of Li-Ion Batteries with Graphite/Hard Carbon Hybrid An-odes. Adv. Energy Mater. 2021, 11, 2003336
Chen, Z., Bresser, D. et al.	2020	The success story of graphite as a lithium-ion anode material – fundamentals, remaining challenges, and recent developments including silicon (oxide) composites. (Review Article) Sustainable Energy Fuels, 2020, 4, 5387-5416
Coastal & Environmental Service Limited Mozambique Lda (CES)	2017	Environmental impact assessment of ancuabe graphite mine. Non-technical summary-
Dolega, P., Buchert, M., Betz, J.	2020	Environmental and socio-economic challenges in battery supply chains: graphite and lithium, project report, 2020, Oeko-Institut, Germany
Dou, X., Geng, C., Buchholz, D., Passerini, S.	2018	Research Update: Hard carbon with closed pores from pectin-free apple pomace waste for Na-ion batteries featured. APL Materials 6, 047501 (2018)

Dou, X., Hasa, I., Saurel, D., Jauregui, M., Buchholz, D., Rojo, T., Passerini, S.	2018	Impact of the Acid Treatment on Lignocellulosic Biomass Hard Carbon for Sodium-Ion Battery Anodes. ChemSusChem 2018, 11, 3276
Dou, X., Hasa, I., Saurel, D., Vaalma, C., Wu, L., Buchholz, D., Bresser, D., Komaba, S., Passerini, S.	2019	Hard carbons for sodium-ion batteries: Structure, analysis, sustainability, and electrochemistry. Materials Today, Volume 23, 2019, Pages 87-104. ISSN 1369-7021
Dunn, J. B., James, C., Gaines, L., Gallagher, K., Dai, Q., Kelly, J.C.	2015	Material and energy flows in the production of cathode and anode materials for lithium-ion batteries
Fortier, S.M., Nassar, N.T., Lederer, G.W., Brainard, Jamie, Gambogi, Joseph, and McCullough, E.A.	2018	Draft critical mineral list—Summary of methodology and background information—U.S. Geological Survey technical input document in response to Secretarial Order No. 3359: U.S. Geological Survey Open-File Report 2018–1021, 15 p
Górka, J., Vix-Guterl, C., Matei Ghimbeu, C.	2016	Recent Progress in Design of Biomass-Derived Hard Carbons for Sodium Ion Batteries. C 2016, 2, 24
Huijbregts, M.A.J. et al.	2016	ReCiPe 2016 v1.1. A harmonized life cycle impact assessment method at midpoint and endpoint level. Report I: Characterization
ISO	2006	ISO 14040 – Environmental management – Life cycle assessment – Principles and framework, International Organization for Standardization, Geneva, Switzerland
ISO	2006	ISO 14044 – management – Life cycle assessment – Requirements and guidelines, International Organization for Standardization, Geneva, Switzerland
Jayaraman, S., Jain, A., Ulaganathan, M., Edison, E., Srinivasan, M.P., Balasubramanian, R., Aravindan, V., Madhavi, S.	2017	Li-ion vs. Na-ion capacitors: A performance evaluation with coconut shell derived mesoporous carbon and natural plant based hard carbon. Chemical Engineering Journal, Volume 316, 2017, Pages 506-513, ISSN 1385-8947
Liao, M., Kelley, S., Yao, Y.	2019	Generating Energy and Greenhouse Gas Inventory Data of Activated Carbon Production Using Machine Learning and Kinetic Based Process Simulation. ACS Sustainable Chemistry & Engineering, 2019, 8, 1252-1261.
Nita, C., Zhang, B., Dentzer, J., Ghimbeu, C.M.	2021	Hard carbon derived from coconut shells, walnut shells, and corn silk biomass waste exhibiting high

		capacity for Na-ion batteries. Journal of Energy Chemistry, Volume 58, 2021, Pages 207-218, ISSN 2095-4956
Pei, L., Cao, H., Yang, L. et al.	2020	Hard carbon derived from waste tea biomass as high-performance anode material for sodium-ion batteries. Ionics 26, 5535–5542 (2020)
Peters, J. F., Abdelbaky, M., Baumann, M., Weil, M.	2020	A review of hard carbon anode materials for sodium-ion batteries and their environmental assessment. Matériaux & Techniques, 107 5 (2019) 503
Ruiz, E. M., Lévová, T., Bourgault, G., Wernet, G.	2015	Documentation of changes implemented in ecoinvent database 3.2, Ecoinvent Centre, Zürich, Switzerland
Saha, A., Sharabani, T., Evenstein,E., Nessim, G. D., Noked, M., Sharma, R.	2020	Probing Electrochemical Behaviour of Lignocellulosic, Orange Peel Derived Hard Carbon as Anode for Sodium Ion Battery. J. Electrochem. Soc. 167 090505, 2020
Weil, M., Peters, J., Baumann, M.	2020	Stationary battery systems: Future challenges regarding resources, recycling, and sustainability. In The Material Basis of Energy Transitions, Academic Press, Chapter 5, Pages 71-89
Weil, M., Ziemann, S., Peters, J.	2018	The Issue of Metal Resources in Li-Ion Batteries for Electric Vehicles. Behaviour of Lithium-Ion Batteries in Electric Vehicles. Ed.: G. Pistoia, 59–74, Springer Inter-national Publishing.
Wernet, G., Bauer, C., Steubing, B., Reinhard, J., Moreno- Ruiz, E., Weidema, B.,	2016	The ecoinvent database version 3 (part I): Overview and methodology, Int. J. Life Cycle Assess. 21(9), 1218 (2016)
Wu, L., Buchholz, D., Vaalma, C., Giffin, G. A., Passerini, S.	2016	Apple-Biowaste-Derived Hard Carbon as a Powerful Anode Material for Na-Ion Batteries. ChemElectroChem 3, 292 (2016)
Xie, F., Xu, Z., Guo, Z., Titirici, M-M.	2020	Hard carbons for sodium-ion batteries and beyond. Prog. Energy 2 042002
Yu, H.-Y., Liang, H.-J., Gu, Z.-Y., Meng, Y.-F., Yang, M., Yu, M.-X., Wu, X.-L.	2020	Waste-to-wealth: low-cost hard carbon anode derived from unburned charcoal with high capacity and long cycle life for sodium-ion/lithium-ion batteries. Electrochimica Acta, 137041
Yu, P., Tang, W., Wu, FF. et al.	2020	Recent progress in plant-derived hard carbon anode materials for sodium-ion batteries: a review. Rare Met. 39, 1019–1033 (2020)

Zhang, Q., Gong, X., Meng, X.	2017	Environment Impact Analysis of Natural Graphite Anode Material Production. Materials Science Forum. ISSN: 1662-9752, Vol. 913, pp 1011-1017
Zhang, X., Hu, J., Chen, X., Zhang, M., Huang, Q.	2019	Microtubular carbon fibers derived from bamboo and wood as sustainable anodes for lithium and sodium ion batteries. Journal of Porous Materials (2019) 26:1821–1830

Waste management challenges in Romania during the COVID-19 pandemic

Florin-Constantin Mihai

"Alexandru Ioan Cuza" University, Iasi, Romania

Abstract

This work points out the main challenges of municipal and medical waste management sectors in the context of the COVID-19 pandemic in Romania. Key environmental threats are identified based on waste management deficiencies related to illegal dumping practices, waste collection schemes, landfills, and hazardous waste incinerators, waste statistics gaps, and unsound environmental awareness. COVID-19 related waste flow is estimated taking into consideration the potentially infectious waste from healthcare facilities and municipal waste generated in quarantine places and self-isolated persons at national levels. Some best practices are highlighted in management COVID-19 related waste flow and future perspectives are discussed.

Keywords

COVID-19, municipal waste, medical waste, waste incinerators, waste management, environmental pollution, waste statistics,

1 Introduction

Municipal and medical waste management systems were facing several deficiencies in Romania before the COVID-19 pandemic. These are essential public services that could increase the virus spread in the community if not properly managed. The COVID-19 related waste stream is additional pressure to such waste management systems which countries around the world must cope with it (Das et al., 2021) and complex interactions must be revealed (Jiang et al., 2020). The first Covid-19 case was registered on 26 February 2020 in Romania and on 16 March 2020, the emergency state was enforced. People working abroad or tourists are obliged to stay in quarantine places or self-isolation in households in return. On 18 March 2020, the National Public Health institute declared the household waste generated in quarantine places as infectious waste. Therefore, this household waste must be collected and treated as a medical waste stream by specialized economic agents and disposed of by authorized waste incinerators. People in self-isolation must avoid source-separated collection and dispose of their waste in double residual bags and be directly transported to landfill sites.

This paper aims to examine the municipal and medical waste management challenges in Romania during the COVID-19 pandemic in line with waste management infrastructure and current gaps.

2 Municipal waste management

Romania is still a landfill-based country and the recycling rate of municipal waste is around 11 % in 2018 (NEPA, 2020) and the EU target (50 %) set up for 2020 is far to be achievable in the current state (EC, 2020).

Municipal waste management deficiencies are related to:

- Source-separate collection schemes are not widespread in smaller urban areas and rural communities
- Biowaste fraction is the main fraction of municipal solid waste stream in Romania but source-separation facilities are limited even in larger urban areas
- Limited waste collection coverage in some rural regions of Romania
- Regionally integrated waste management systems are facing delays with repercussions to recycling and recovery facilities at the Romanian counties level.
- Plastic pollution, illegal dumping, and open burning practices associated with the lack or poor efficiency of waste management systems
- Environmental pollution associated with informal and illegal recycling/recovering activities (e.g. metal waste) from open burning of e-waste items/end-of-life vehicles and/ or manual dismantling activities.
- Lack of reliable waste statistics database at local administrative units level (LAU2) including urban and rural municipalities

COVID-19 pandemic effects on municipal waste management systems:

- In the first stage, waste management workers lack PPE equipment
- Waste generated by quarantine places are disposed of by waste incineration plants
- recyclable waste generated by self-isolated persons are mixed collected and disposed of through landfills
- increasing amounts of household waste from the population (national lockdown, self-isolation persons, working from home, regional quarantine areas) since

March 2020 compared to similar waste generated by economic agents (closed or restricted activities)

- Illegal dumping sites which pollute the natural environment
- Facemasks littered in urban parks, streets, near freshwater bodies
- Dry recyclables in stock and lower demands for recyclables
- Increasing amounts of waste diverted from material recycling to cement factories as opposed to upper circular economy mechanisms.

Table 1 shows the implications of COVID-19 in Oradea city, where material recycling is declining compared to the previous year (2019) while larger amounts of waste are sent to energy recovery operations such as Holcim cement factory with associated environmental pollution risks. Due to restrictions imposed in 2020, the amounts of street waste and similar waste (economic agents + institutions) are lower compared to 2019. On the other hand, amounts of household waste generated by the population are larger while the construction sector had no restrictions in 2020.

Table 1 Impact of COVID-19 pandemic on urban waste management system from Oradea in 2020

Waste type	2020	2019	Observations
Waste collected from population	17 000 t	15 000 t	Lockdown, working from home
Similar waste collected from economic agents/institutions	30 132 t	35214 t	Closed units or Restricted activities
Street wastes	17.628 t	18 628 t*	Around 1000 t less than 2019
Construction wastes	5992 t	4174 t	Renovation activities, the construction sector was not restricted in 2020
Packaging waste recycled and recovered	553 t	886 t	Material recycling is declining
Waste sent to Holcim cement factory (energy recovery)	8200	3400 t	WtE plants fed with recyclable waste

Data source: https://www.ebihoreanul.ro/stiri/reciclare-cu-deficit-pandemia-de-covid-19-a-puspiedici-si-in-reciclarea-deseurilor-oradenilor-161744.html

This situation is also susceptible to other urban areas of Romania. Furthermore, Romania is still facing the Illegal traffic of non-recyclables items from older EU countries (eg. Germany, Italy) as the destination or transient country. The delays in the development of upper circular mechanisms such as material recycling and composting activities will affect circular economy transition in the medium and long term.

3 Medical waste management

Medical waste management systems are based on waste incineration plants and thermal decontamination facilities at low temperatures (+105°C - +177°C) as treatment and disposal options for hazardous fractions under current legislation. In both cases, the ash from incinerators and treated waste resulted from decontamination facilities (thermal sterilization + mechanical treatment) are further disposed of in landfills. Hospitals can have their thermal decontamination facilities, but in Romania, only 9 public hospitals (56 units questioned) have such operational facilities (Competition Council of Romania, 2021). In the Covid-19 pandemic context, the waste incineration plants must cope with the surplus of COVID-19 related waste flow generated by hospitals compared to previous years plus wastes generated in quarantine places declared as potential infectious wastes since 18 March 2020. There are geographical disparities concerning the access of hospitals and other healthcare facilities to such disposal facilities as shown in table 2.

Table 2 Private medical/hazardous waste management facilities in Romania at regional levels

NUTS-2 Region	Hazardous waste incineration (only)	Thermal decontamination facilities	Incineration + thermal decontamination
North-East	1 (Demeco)	no	1 (Mondeco)
South-East	1 (Eco Fire Sistems)	no	no
South Bucharest -Ilfov	2 Eco-Burn Enviro Eco Business	2 Perfect Courier Pro Air Clean	1 Stericyle
South-West	2 Medline Exim Xtreme Econergy	1 BioHazard	no
West	no	2 Ecolomedgd Stericycle	1 Pro Air Clean Ecologic
North-West	no	no	no
Center		2 Stericycle AKSD	

Data source: Competition Council of Romania (Preliminary Report 2021)

North-West region has no access to waste incineration or thermal decontaminant facilities and the South-East region is served by one incineration plant. Most of the waste incinerators are concentrated in the South-Bucharest-Ilfov region. COVID-19 hospitals are required to contact authorized economic agents to transport COVID-19 waste to be disposed of to waste incinerator plants. Therefore, the increasing demand for such services and long distances are additional burdens for both hospitals and waste disposal facilities that lead sometimes to environmental crimes. Hospitals that have their decontamination facilities can treat part of medical waste flow by thermal sterilization and mechanical processing (shredding, chopping, grinding, compacting) except anatomopathological, chemical, pharmaceutical, cytotoxic, and cytostatic wastes that must be sent to incineration facilities (Competition Council of Romania, 2021).

Medical waste management deficiencies during COVID-19 pandemic refer to:

- Illegal dumping sites on peri-urban areas containing medical waste flow (yellow plastic bags)
- Medical waste overlapping on waste incineration backyards
- Closure or temporary suspended activities of hazardous waste incineration facilities due to environmental crimes investigated by National Environmental Guard
- Additional costs to hospitals budget associated with COVID-19 related waste flow
- Lack of reliable medical waste statistics
- Lack of covid-19 statistics at subnationals levels to properly estimate the COVID-19 related waste flow in Romania

A preliminary report conducted by the Competition Council of Romania (2021) that supervises the medical waste management facilities shows several gaps related to: (i) there is no approved strategy and national medical waste management plan (ii) legislation framework related to medical waste management is unclear and needs to be adjusted (iii) three economic operators managed 70 % of the total medical waste flow in 2019 (iv) waste incinerators account 79 % of the total capacity of disposal facilities compared to thermal decontamination plants (21 %) (v) the use rate of these disposal facilities was under 40 % in 2018. Therefore, these facilities could cover increasing demands associated with the COVID-19 pandemic.

4 COVID-19 related waste flow in Romania

COVID-related waste flow is estimated to be 4312.81 tons at the national level during 26 February - 15 June 2020. Most of this waste stream is associated with self-isolation persons (3305 t) compared to quarantine waste (553.75 t) while COVID 19 patients generated 454 t of medical waste of which 27.21 t in intensive care units (ITU) according to table 3 based on a previous study (Mihai, 2020). This table reveals that around 1008 t of COVID19 waste could have been disposed of in waste incineration plants until 15 June 2020. However, this model needs to be updated to reflect the current situation where more actives cases and ITU patients are registered than in the first half of 2020. In this regard, assessment of Covid-19 related waste flow at subnational levels is a better choice, but this approach strongly depends on the availability and transparency of COVID-related statistics. Also, the estimation method needs to be adjusted because quarantine /self-isolated persons data are provided until 3 July 2020.

Table 3. Covid-19 waste flow at the national level from 26 February -15 June 2020

Period / Tons generated	ICU patients waste	Medical waste COVID-19	Quarantine waste	Self-isolation waste	Total Covid-19 related waste
Prior emergency state (26 Feb -15 March)	0	0.44	5.79	109.72	115.96
Emergency state (16 March -14 May)	17.17	286.04	453.44	1893.56	2633.05
Alert state (15 May-15 June)	10.1	167.47	94.51	1301.81	1563.79
Total period (26 Feb - 15 June)	27.28	453.95	553.75	3305.1	4312.81

Source: Mihai FC (2020). CC-BY

Therefore, the assessment of COVID-19 waste flow in Romania depends on publicly available data. However, both municipal and medical waste sectors have issues with waste statistics data even before the COVID-19 event.

5 Good practices and future perspectives

COVID-19 pandemic draws attention around the world about the importance of municipal and medical waste management sectors as essential public services in reducing public health threats and environmental pollution (Liang et al., 2021). As outlined in the above sections, Romania was already facing several waste management deficiencies. To cope with additional challenges related to the COVID-19 pandemic multi-sectoral efforts (Government, local authorities, NGOs, waste operators, citizens) are required.

Some good practices related to the COVID-19 pandemic and waste management sector

- Waste workers equipped with PPE equipment, public campaigns on how the waste should be collected, and adjusted schedules
- Districts 4 and 6 of Bucharest capital city bought special vehicles to collect household wastes generated in quarantine places and self-isolated persons and transport them in bins at -4C to hazardous waste incinerators
- A special guide to promoting responsible behaviors in multi-apartment buildings with a special section dedicated "How to collect the waste responsibly " https://fiipregatit.ro/ghid/covid19-locatari-la-bloc/
- Separate collection of waste from self-isolated persons or materials with high infectious potential (masks, gloves, dressings, diapers, contaminated wipes, hygiene items, etc.) in special bins.
- Separate collection of facemasks in schools
- Field investigations of the National Environmental Guard regarding the transport and disposal of medical wastes across the country (waste incinerators, economic agents authorized for hazardous wastes transportation)
- Field investigations by inter-institutional collaborations about the illegal dumping, open burning, and illegal traffic of various waste types (including from abroad) during the COVID-19 pandemic on Romanian territory
- Competition Council of Romania performs an overview report regarding the medical waste management facilities in Romania, the medical waste disposal market, and the geographical coverage.
- Several NGO's requests through an open letter more transparency related to COVID-19 statistics and to provide a set of data at subnational levels https://covid19.geo-spatial.org/manifest which could be further examined by GIS tools

Romania must continue the implementation of regional integrated waste management systems (in each county) to provide full coverage of population to source-separated waste collection schemes and to accelerate waste diversion from landfills through circular economy mechanisms with sound facilities (transfer stations, sorting lines for dry recyclables, composting facilities, anaerobic digestion and MBT plants for biowaste, crushing plants for construction and demolition waste stream, urban mining centers, etc) to feed recycling industries with reliable secondary materials. Legislative incentives such as the Compost Law (Law no 181,2020), which entered in force from 20 February, oblige local authorities to provide biowaste collection schemes in urban and peri-urban areas to fed composting facilities or to support home composting activities in rural areas. Also, Environmental NGOs request an approved law for the deposit-refund scheme for packaging waste stream to improve source separation collection, extended producers responsibility (EPR), and material recycling operations. The new " National Recovery and Resilience Plan" recognized the waste management deficiencies in Romania and stipulate that the circular economy is still at an early stage (NREP Romania, 2021). However, environmental NGO's critic this plan to fund from public budget producers and importers of packaging materials against the "polluter pays" principle as part of their EPR policies, therefore, these financial resources should be redirected to improve source-separated collection schemes across urban and rural municipalities (Declic, 2021). Companies must take initiatives to fund deposit-refund schemes, as in the other EU countries, under the EPR framework. Despite some investment directions mentioned in this strategic document none refer to the medical waste management sector. Romania must also upgrade the medical waste management facilities and look for safer and environmentally friendly disposal alternatives compared to some current waste incineration plants that were involved in several environmental crimes in Romania during the COVID-19 pandemic. Environmental authorities intend to expand air drone cameras acquisitions to support environmental crimes investigations (as shown in figure 1) on a national scale

Figure 1 Piles of medical waste on waste incinerator backyard (Brazi, Prahova County) Source: National Environmental Guard

Financial and technical support for hospitals that need thermal decontamination plants to treat some of their medical waste flow, particularly in counties without access to private waste disposal facilities. A study provides some estimations regarding the investments required in hazardous medical waste management facilities between 14.67 million € and 18.00 mil. € (Platon et al., 2020).

There is an urgent need for a strategy and national medical waste management plan based on current data and approved by environmental authorities. Better monitoring systems of environmental factors related to medical and municipal waste management facilities are required in the context of the COVID-19 pandemic. Also, better public access and transparency to waste statistics data are required to properly assess the Romanian progress towards circular economy transition.

Literature

Competition Council of Romania. 2021. Preliminary report regarding the medical waste . A sectoral investigation.

Declic 2021. Stop financing large beverage producers through PNRR. Redirect these funds for separate waste collection infrastructure https://www.declic.ro/pnrr-pentru-infrastructura-de colectare/?fbclid=IwAR1G1v9VR90CGcdfJSg--hDDRDp3BjV_VtT4YVYDIhHNzxYkJwvkdQbUIKY

Das, A. K., Islam, Md. N., Billah, Md. M., & Sarker, A. 2021. COVID-19 and municipal solid waste (MSW) management: a review. Environmental Science and Pollution Research. https://doi.org/10.1007/s11356-021-13914-6

European Commission 2020. Country report of Romania in 2020.

Ebihoreanul 2021 Covid19 pandemic effects on recycling activities in Oradea city https://www.ebihoreanul.ro/stiri/reciclare-cu-deficit-pandemia-de-covid-19-a-pus-piedici-si-in-reciclarea-deseurilor-oradenilor-161744.html

Jiang, P.; Klemeš, J.J.; Fan, Y.V.; Fu, X.; Bee, Y.M. **2021** More Is Not Enough: A Deeper Understanding of the COVID-19 Impacts on Healthcare, Energy and Environment Is Crucial. *Int. J. Environ. Res. Public Health* , *18*, 684. https://doi.org/10.3390/ijerph18020684

Law no. 181 from 18 August 2020 regarding the management of non-hazardous compostable wastes http://legislatie.just.ro/Public/DetaliiDocument/229273

Liang, Y., Song, Q., Wu, N., Li, J., Zhong, Y., & Zeng, W. 2021. Repercussions of COVID-19 pandemic on solid waste generation and management strategies. Frontiers of Environmental Science & Engineering, 15(6). https://doi.org/10.1007/s11783-021-1407-5

Mihai, F.-C. 2020. Assessment of COVID-19 Waste Flows During the Emergency State in Romania and Related Public Health and Environmental Concerns. Int. J. Environ. Res. Public Health, 17, 5439. https://doi.org/10.3390/ijerph17155439

NEPA 2020. Report on environmental status in Romania in 2019. National Environmental Protection Agency

Platon, V., Frone, S., Constantinescu, A., & Jurist, S. 2020. Challenges in Adequate Management of Hazardous Medical Waste to Reduce Impact of the COVID-19 Epidemic in Romania. International Conference Innovative Business Management & Global Entrepreneurship (IBMAGE 2020). International Conference Innovative Business Management & Global Entrepreneurship. https://doi.org/10.18662/lumproc/ibmage2020/16

Author's address:

Dr. Florin-Constantin MIHAI
CERNESIM Research Center
Department of Sciences
Interdisciplinary Research Institute
"Alexandru Ioan Cuza " University
Carol I Blvd no 11
Iasi city, Romania
E-Mail: mihai.florinconstantin@gmail.com

Using computer vision to strengthen the resilience of waste sorting infrastructure during the pandemic

Victor Dewulf

Recycleye, London, United Kingdom

Abstract

An exploration into the current state-of-the-art identification and sorting techniques alongside recent technological developments in machine vision identified the application of convolutional neural networks (the most advanced machine learning algorithms) to waste management as a yet unexplored area of research. To understand how and with what implications convolutional neural networks (CNNs) could transform the waste management industry a proof-of-concept neural network was developed as an affordable system to: identify and classify recyclables, improve sorting efficiency, reduce human drudgery, and drive a waste data revolution. By following the concept of integrated sustainable waste management, potential implementations of the Recycleye algorithm were evaluated in terms of their impact in the case of the COVID-19 pandemic.

Keywords

Computer Vision, Artificial Intelligence, Covid-19, Waste Management, Recycleye

1 Introduction

Manual waste pickers in the industry are no longer equipped handle the 2 trillion tonnes of waste produced each year globally. The repetitive nature of waste sorting has exposed pickers to long-term health conditions, leading to a high labour turnover. Health hazards have been further exacerbated by Covid-19, as the spread of the virus has threatened entire plant closures.

If Material Recovery Facilities (MRFs) are to ensure resilience and agility of their existing sorting infrastructure in the longer term, the industry will need to transition towards automated infrastructure through the deployment of computer vision and robotics. Recent developments in computer vision are enabling the creation of ultra-low-cost sensors. An affordable camera (similar to that found on most smartphones), combined with a powerful AI can now detect items on a material, object and even brand level – replacing the sensing equivalent of a near-infra-red sensor, eddy currents, magnets, and humans for quality control. The deployment of computer vision systems will also streamline reporting into a centralised cloud-based data storage. Computer vision systems are necessary for moving MRFs towards a tech-enabled model where most operational decisions are made by machines, allowing facilities to adapt dynamically and rapidly to changing conditions, such as the pandemic.

2 Literature Review

The research explores the current state-of-the-art identification and sorting techniques alongside recent technological developments in machine vision. Traditional recycling systems were included in the investigation such that the machine vision processes could be benchmarked against them. While the study tries to address techniques that are applicable to all types of waste streams it acknowledges a skew towards municipal solid waste (MSW), due to the larger number of research papers published in that area. However, since MSW is by far the most complex and varied waste stream, any developments that help its sorting will likely be applicable to the other waste streams such as industrial waste, electronic waste, construction waste, demolition waste, and most significantly, medical and PPE (Personal Protective Equipment) waste.

2.1 Existing Sorting Infrastructure

2.1.1 Reliance on Expensive Manual Labour

Categorizing each item on a conveyor into 100 different classes with 94% accuracy under 180ms as performed by Recycleye is not achievable by any human cognition. Machine vision systems offer further advantages over human workers in terms of fatigue, throughput, speed and accuracy (RAHMAN ET AL.,2011). Automating the manual and routine tasks in waste management may put over 60% of jobs in the industry at risk by 2033 (MANYIKA ET AL.,2017). The low-cost technology may even put the 15-20 million informal sector workers at risk in the long run (WILSON ET AL., 2015). On the other hand, as machine vision systems help move the economy towards a circular economy an expected 9 to 25 million new jobs will be created (WILSON ET AL., 2015). The advent of Covid-19 has increased the health risk posed to workers, subsequently the automation of waste sorting lines will eliminate health and safety risks.

2.1.2 Automatic identification and data capture

An alternative method for creating a 'removal chain' that is as efficient as the 'supply chain' is by tagging all trash (GREENGARD, 2010; AIDIC 100, 2012). Barcodes have been used to link recyclables to databases that provide exact disassembly information and help transfer the cost to the manufacturer (IKUO, 2004). While barcodes have become ubiquitous on packaging, they are still rarely placed on the actual appliance or device. They are also prone to line of sight issues. Radio-Frequency Identification (RFID) solves some of those problems, but with a 5 to 10 cents increase in per item costs (FINKENZELLER, 2010). RFID item-level implementation is seen by many as the key to transition towards a circular economy. Its potential for pay as you throw based billing has also been explored (GNONI, LETERRA & ROLLO, 2013). Several studies have cautioned about the environmental impact of imbedding billions of tags in products due to the risk of dissipating toxic and valuable substances (WAGER ET AL., 2005; ABDOLI, 2009).

2.2 Computer vision systems

All machine vision systems contain two components: sensors for image acquisition and algorithms for image processing, as illustrated in Fig. 2.1 For waste sorting the system is often complemented by a third component; a mechanism that removes items the vision system identified and located.

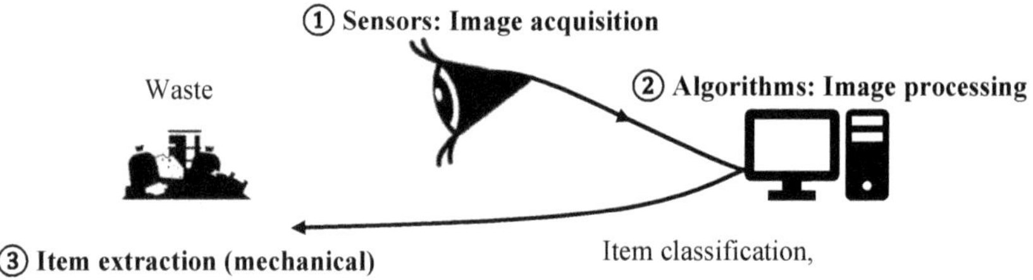

Fig. 2.1 Typical computer vision system architecture (THE NOUN PROJECT, 2018).

2.2.1 Neural networks

Neural networks work by replicating the neurons inside a brain (LI, JOHNSON & YEUNG, 2017). These algorithms are the enablers of product identification (using only RGB images), as opposed to material identification (using XRT or LIBS). Their accuracy is unaffected by surface contaminants such as oils, dust or labels, that are often found in waste streams (KOYANAKA & KOBAYASHI, 2011). In an experiment comparing several algorithms to extract bottles from a conveyor belt, neural networks outperformed all other algorithms by an average of 12.5% due to their capacity in dealing with inexact measurements of the bottles' geometrical features (WAHAB ET AL., 2006) This ability to identify materials within distorted and chaotic waste streams was substantially confirmed in independent studies (MATTONE, CAMPAGIORNI & GALATI, 2000). Other use cases have included detecting wood veneer defect and sorting up to 12 different classes of fish (catch from by-catch) with a 78% accuracy compared to a 42% human accuracy (DOWLATI ET AL., 2012).

Convolutional neural networks (CNNs) are a recent subclass of neural nets that have been responsible for the current explosion in computer vision, following the seminal paper by Krizhevsky, Sutskever & Hinton (2012) that has been cited over 23,700 times. Researching its application in waste management has mainly been done by industry and kept confidential due to the high value of the IP (YANG & THUNG, 2016). to material classes, was able to achieve a higher accuracy of 70% (SINGH ET AL., 2017).

The industry has started to adopt CNNs, but one needs to be cautious in identifying real breakthroughs from unfounded marketing. The below robotic picker (Fig 2.2) extracts wood, stone and metal from construction and demolition waste using high resolution 3D laser sensors, NIR, metal detectors and high resolution RGB cameras (KUJALA, LUKKA & HOLOPAINEN, 2016).

*Fig 2.2 Robotic waste picker (*RECYCLEYE*, 2021).*

Relative to standard RGB cameras, the advanced sensors do not only have a higher CAPEX but their OPEX is also larger, for they require more maintenance and calibration due to their sensitivity to dusty environments. One must also note that commercial recycling is easier to achieve than residential recycling due to the lower level of contamination and variability of waste. Due to the complexity of MSW, current systems are only used in the reject stream for high value item cherry picking or at the end of MRFs for quality control. Other niche application systems include Optisort developed by Refind (REIMERS, 2018) which sorts different batteries. While most of the systems above have tremendously high accuracies they operate on the end stages of MRFs where the waste stream has already attained a certain level of purity. Moreover, they all use expensive sensors as opposed to a single RGB camera. Several projects have developed more affordable 'smart bins' whereby bins will automatically classify garbage and route it to the right container using only one RGB camera (WAGNER, ZSIGMOND & YOSHAZAWA, 2006). Oscar by Intuitive (MURAD, 2018) and Trashbot by CleanRobotics (2017) attempted to classify recyclables from landfill waste, albeit only for common high street items such as bottles and coffee cups. Rad et al. (2017) developed a vision system that can distinguish cigarette buds from leaves for street cleaning.

3 Opportunities and implications of Computer Vision in Waste Management since Covid-19

The advent of Covid-19 has seen a drastic change in the composition of material in addition to reduced output levels. As an MRF is an interdependent system of multiple modules that have hidden relationships, computer vision will be able to illustrate and visualise the relationship between modules. This will enable facilities to use live waste data to view the composition of incoming waste changes and enable immediate parameter changes accordingly that will ensure smooth running of the entire MRF. Computer vision also enables the streamline of reporting through centralised data storage: data is collected 24/7 and available in the cloud to view and share remotely. Especially in times of drastic changes, such as the pandemic, computer vision will move facilities towards a tech-enabled model where most operational decisions are made by machines allowing the company to adapt dynamically and rapidly to changing market conditions and waste compositions, gaining tremendous competitive advantage over traditional businesses.

Fig. 3.1 Recycleye's computer vision solutions trained using deep learning to identify items by material, object, and brand in an industrial setting.

This paper has also found that using computer vision to detect and classify all items on waste streams, as evident in Fig 3.1, can improve the volume and accuracy of back-end quality control checks, removing the need for manual pickers. Computer vision systems, which can identify all items in waste streams – broken down by material, object and even brand exceed human performance. Manual pickers are not only costly, but the onset of the pandemic has resulted in the risk of exposure to Covid-19 to pickers reducing the efficiency as a result of increased bureaucracy that ensures Covid precautions are taken by all employees in facilities. Instead, computer vision systems are able to provide 24/7 item passport logs without the risk of occupational hazards.

Computer vision has the capability to distinguish outgoing waste flows between food-grade HDPE plastic and non-food-grade HDPE plastic. Using artificial intelligence to make this distinction, is enabling facilities to provide higher value bales and subsequently leverage this data to charge their clients based on the individual composition of every ton. Such detection is enabling facilities to capture higher value out of their outgoing waste flows, and providing safer alternatives to manual picking.

4 Conclusion

To meet the European Union's targets for material reuse, the pandemic has only drawn greater attention to the urgent need for a viable solution. European governments have since taken action to tackle the challenges imposed, albeit lacking more long-term solutions.

This paper has discovered an effective and highly employable solution to the current waste industry's most prominent issues. Using AI computer vision to sort various materials, including PPE and other hazardous medial equipment, in a highly cost-effective manner. Moreover, the current fragmented infrastructure of MRFs, in which a different machine is required to sort each type of material, has prevented the waste management sector from responding quickly to changing supply and demands. Robotics and computer vision provide a solution to the challenges and inefficient working practices which arise from manual picking, most notably worker fatigue and exposure to viruses from potentially contaminated waste. As such, vision AI and robotic systems enable for MRFs to achieve consistently high levels of output while minimising occupational hazardous risks. Additionally, the ability for decentralisation means that waste can be sorted at the source of generation, limiting the risk of further transmission of Covid-19 as the need for transportation to centralised facilities will be eradicated.

The current global pandemic has highlighted the need for the waste management industry to minimise worker interaction for both safety and cost purposes. The sector needs to pivot investment towards robotic and vision automation as part of a process to decentralise MRFs in order to maintain the running of removal chains and enable a more circular economy.

5 Literature

CleanRobotics. 2017 *Introducing Trashbot.* Available from: http://www.cleanrobotics.com/ [Accessed Jun 29, 2018].

Dowlati, M., de la Guardia, M., Dowlati, M. & Mohtasebi, S. S. 2012 *Application of machine-vision techniques to fish-quality assessment. TrAC Trends in Analytical Chemistry. 40 168-179.*

		Available from: http://www.sciencedirect.com/science/article/pii/S016599361200221X. Available from: doi: //doi.org/10.1016/j.trac.2012.07.011.
Finkenzeller, K.	2010	*RFID Handbook: Fundamentals and Applications in Contactless Smart Cards, Radio Frequency Identification and near-Field Communication. Third edition edition. Hoboken, John Wiley & Sons. Available from: https://onlinelibrary.wiley.com/doi/book/10.1002/9780470665121 .*
Gnoni, M. G., Lettera, G. & Rollo, A.	2013	*A feasibility study of a RFID traceability system in municipal solid waste management. International Journal of Information Technology and Management. 12 (1-2), 27-38. Available from: https://www.scopus.com/inward/record.uri?eid=2-s2.0-84873869067&doi=10.1504%2fIJITM.2013.051632&partnerID=40&md5=047ed47acca0a40db5f63ee5a7157477.*
Greengard, S.	2010	*Commun.ACM. 53 (3), 19-20. Available from: http://doi.acm.org/10.1145/1666420.1666429. Available from: doi: 10.1145/1666420.1666429.*
Ikuo, H.	2004	*Recovery method of advance waste disposal expense for streetscape. JP2004062837A (Patent).*
Krizhevsky, A., Sutskever, I. & Hinton, G. E..	2012	ImageNet Classification with Deep Convolutional Neural Networks. In: F. Pereira, C. J. C. Burges, L. Bottou & K. Q. Weinberger (eds.). *Advances in Neural Information Processing Systems 25*. Curran Associates, Inc. pp. 1097-1105.
Koyanaka, S. & Kobayashi, K.	2011	Incorporation of neural network analysis into a technique for automatically sorting lightweight metal scrap generated by ELV shredder facilities. Resources, Conservation & Recycling. 55 (5), 515-523. Available from: https://www.sciencedirect.com/science/article/pii/S0921344911000036. Available from: doi: 10.1016/j.resconrec.2011.01.001.

Kujala, J. V., Lukka, T. J. & Holopainen, H.	2016	Classifying and sorting cluttered piles of unknown objects with robots: A learning approach. 2016 IEEE/RSJ International Conference on Intelligent Robots and Systems (IROS). pp.971-978.
Li, F., Johnson, J. & Yeung, S.	2017	Lecture 1 | Introduction to Convolutional Neural Networks for Visual Recognition. Stanford University School of Engineering.
Manyika, J., Chui, M., Miremadi, M., Bughin, J., George, K., Willmott, P. & Dewhurst, M.	2017	*A future that works: Automation, Employment and Productivity*. McKinsey Global Institute.
Mattone, R., Campagiorni, G. & Galati, F.	2000	*Sorting of items on a moving conveyor belt. Part 1: a technique for detecting and classifying objects. Robotics and Computer-Integrated Manufacturing. 16 (2), 73-80.*
Murad, H.	2018	*Interviewed by Dewulf, V (4 July 2018).*
Rad, M. S., von Kaenel, A., Droux, A., Tieche, F., Ouerhani, N., Ekenel, H. K. & Thiran, J.	2017	*A Computer Vision System to Localize and Classify Wastes on the Streets. Available from: http://arxiv.org/abs/1710.11374. Available from: doi: 10.1007/978-3-319-68345-4_18.*
Rahman, M. O., Hussain, A., Scavino, E., Basri, H. & Hannan, M. A.	2011	*Intelligent computer vision system for segregating recyclable waste papers. Expert Systems with Applications. 38 (8), 10398-10407. Available from: http://www.sciencedirect.com/science/article/pii/S0957417411003186. Available from: doi: //doi.org/10.1016/j.eswa.2011.02.112.*
Recycleye	2021	*Recycleye robotic picking system*

Singh, S., Mamatha, K. R., Anusha, N., Kavita, R. D., Shalini, R. & Susmi, Z.	2017	Waste segregation system using artificial neural networks. *HELIX the Scientific Explorer.* 7 Available from: http://helix.dnares.in/wp-content/uploads/2018/03/50_Helix_2053-2058.pdf.
The Noun Project.	2018	Icons for everything. Available from: https://thenounproject.com/ [Accessed Aug 5, 2018].
Wager, P. A., Eugster, M., Hilty, L. M. & Som, C.	2005	Smart labels in municipal solid waste - A case for the Precautionary Principle? Environmental Impact Assessment Review. 25 (5 SPEC. ISS.), 567-586. Available from: https://www.scopus.com/inward/record.uri?eid=2-s2.0-20444443618&doi=10.1016%2fj.eiar.2005.04.009&partnerID=40&md5=627205aefa84c22aba6fb8a2e41ae1c8. Available from: doi: 10.1016/j.eiar.2005.04.009. [Accessed 18 May 2018].
Wagner, F., Zsigmond, F. & Yoshazawa, G. L.	2006	Method and system for disposing of discarded items. 7,086,592 B2 (Patent).
Wahab, D. A., Hussain, A., Scavino, E., Mustafa, M. M. & Basri, H.	2006	Development of a Prototype Automated Sorting System for Plastic Recycling. American Journal of Applied Sciences. 3 (7), 1924-1928. Available from: http://www.thescipub.com. Available from: doi: 10.3844/ajassp.2006.1924.1928.
Wilson, D. C., Rodic, L., Modak, P., Soos, R., Carpintero, A., Velis, K., Iyer, M. & Simonett, O.	2015	Global Waste Management Outlook. UNEP
Yang, M. & Thung, G.	2016	Classification of Trash for Recyclability Status. Stanford University.

Households' willingness to pay for a local recycling program: A case study from Lebanon

Mary Abed Al Ahad[1,4], PhDc; Ali Chalak[2], PhD; Souha Fares[3], PhD; Rima R. Habib[1], PhD

[1]Department of Environment Health, American University of Beirut, Beirut, Lebanon

[2]Department of Agriculture, American University of Beirut, Beirut, Lebanon

[3]Hariri School of Nursing, American University of Beirut, Beirut, Lebanon

[4]School of Geography and Sustainable Development, University of St Andrews, United Kingdom

Abstract

Heavy reliance on unsanitary landfilling practices and a lack of a comprehensive solid waste management strategy in Lebanon have resulted in a trash crisis that erupted in July 2015. Introducing a solid waste recycling program will reduce the reliance on landfills and serve as a sustainable solution for the trash crisis. Our study aimed to assess the public's willingness to participate in and pay for a local solid waste management program that includes a recycling service in Jdeidet Ghazir village in Mount-Lebanon. A cross-sectional survey was administered to 228 local residents. Results showed that most of respondents (n=181, 79.39%) were willing to participate in and pay for the local solid waste recycling program. Willingness to pay was positively associated with their nationality (being Lebanese) and the perceived responsibility to be involved in solid waste management as active citizens.

Keywords

Solid waste management; recycling; composting; landfilling; trash crisis; willingness to pay; Lebanon

1 Introduction

In the last few decades, increased urbanization, rapid population growth, economic development, improved living standards, and changes in consumption patterns and lifestyles have led to an increase in the amount of solid waste generated worldwide (Minghua et al., 2009; Singh, Laurenti, Sinha, & Frostell, 2014). The Increased amounts of solid waste has often gone hand in hand with inadequate solid waste management practices that mainly depend on landfilling with minimal recycling practices; particularly in countries that lack political, financial, and technical support required for proper solid waste management (Chakrabarti & Sarkhel, 2003; McAllister, 2015; Ogwueleka, 2009). Open dumping and open burning of solid waste are the two most common inadequate solid waste management practices in low income countries (Alam & Ahmade, 2013). These practices are associated with various adverse environmental effects including soil pollution, surface and ground water pollution, air pollution, acid rain, failure of agricultural crops, ecosystem deterioration, and global warming (Alam & Ahmade, 2013; Azar & Azar, 2016; Ejaz, Akhtar, Hashmi, & Naeem, 2010). This in turn may potentially lead to acute health outcomes including nausea, vomiting, asthma, malaria, cholera, and chemical poisoning; in addition to chronic health outcomes such as low birth weight, congenital defects, neurological disorders, and cancer among others (Alam & Ahmade, 2013). Dioxins, a highly toxic group of persistent environmental chemical compounds resulting from open burning of solid wastes, have been associated with various adverse health impacts such as chloracne, neurological disorders, cancer, endocrine disruptions, congenital defects, cleft palate, low birth weight and stillbirths (Ejaz et al., 2010; Zhang, Buekens, & Li, 2017).

Low solid waste collection frequency, another manifestation of inappropriate solid waste management practices, is present in many low income countries, which results in the accumulation of solid wastes on the sides of streets and riverbanks leading to unhygienic conditions, foul odors, and aesthetic nuisance to the public (Ejaz et al., 2010). Additionally, inappropriate solid waste management practices that lack recycling programs and mainly depend on open dumping/landfilling have been associated with increased risk of injury and infection among people who collect and sell recyclables (Alam & Ahmade, 2013). Therefore, improved solid waste management systems that include separation of solid wastes at the sources of generation, recycling and composting services, and final disposal in sanitary landfills are highly needed, particularly for developing countries, to protect the environment and the public health.

1.1 Solid Waste Management in Lebanon

Lebanon, a middle eastern country located on the Mediterranean Sea, is divided into eight governorates: Aakkar, Baalbeck-Hermel, Beirut, Beqaa, Mount-Lebanon, Nabatiyeh, North-Lebanon, and South-Lebanon (MOE, EU, & UNDP, 2014). Based on the latest available data from the 2014, Lebanon had an estimated population of 7 million (5.6 million Lebanese and 1.4 million Syrian refugees) and produced around 2,040,000 tons of municipal solid waste per year (MOE et al., 2014; SWEEP-Net, 2014). Most of the municipal solid waste in Lebanon (about 60%) is generated by Beirut and Mount Lebanon governorates (SWEEP-Net, 2014). The composition of the generated municipal solid waste consists mainly of organic biodegradable matter (50%-55%). The rest of the waste composition is distributed between recyclables including: paper and cardboard (15%-17%), plastic (10%-13%), glass (3%-4%), metal (5%-6%) and other miscellaneous material (10%-12%) (SWEEP-Net, 2014).

Since 1994, the solid waste management system in Lebanon has been operated by the central government in cooperation with contracted private companies in Beirut and Mount Lebanon governorates (SWEEP-Net, 2014). This national solid waste management system is characterized by high net costs, up to $130 annual cost per ton of solid waste with a heavy reliance on landfilling activities and minimal treatment and negligible recycling and composting services (SWEEP-Net, 2014). In July 2015, the Naameh landfill, which is the main landfill of the solid waste management system was shut down because it had reached its full capacity (Massoud & Merhebi, 2016; Menhall & Joseph, 2017). Consequently, Lebanon witnessed a solid waste management crisis that led to the accumulation of solid wastes on the sides of the streets and in riverbanks, especially in Beirut and Mount Lebanon, posing serious environmental and public health threats (Massoud & Merhebi, 2016). Additionally, the absence of a governmental plan to manage the crisis prompted several municipalities to implement primitive solutions to deal with the accumulating garbage piles in their territory such as open dumping and open burning practices (Abbas, Chaaban, Al-Rabaa, & Shaar, 2017; Morsi et al., 2017). This situation led to the eruption of public protests asking the government to find solutions for the solid waste crisis (Massoud & Merhebi, 2016). Consequently, in March 2016, the government offered a temporary solution which entailed dumping the collected solid waste from Beirut and Mount Lebanon in the two previously closed coastal landfills of "Costa Brava" and "Bourj Hamoud". Upon reaching the maximum capacity of these landfills, another solid waste crisis will unfold (Khawaja, 2017).

Six years have passed since 2015, and the solid waste management crisis in Lebanon is still ongoing and has been further aggravated due to the economic collapse and the COVID-19 pandemic (Human-Rights-Watch, 2020).

Many factors have led to improper solid waste management in Lebanon and consequently to the ongoing solid waste crisis. These factors include the weak political-economic infrastructures in the country, the personal interests of the political elites, and most importantly, the absence of legislations and policies that deal directly with solid waste management (El Harakeh, Madi, & Bardus, 2017; Menhall & Joseph, 2017; SWEEP-Net,

2014). In Lebanon, there are only two laws that deal with solid waste management indirectly; decree 8735 of 1974 which states that "solid waste management is the responsibility of the municipalities" and decree 9093 of 2002 which emphasizes "granting incentives to municipalities that host a solid waste management facility" (SWEEP-Net, 2014). Nonetheless, these laws are rarely enforced due to political corruption, unclear responsibilities, lack of coordination and enforcement (SWEEP-Net, 2014). In 2018, a partial decentralization policy (administrative decentralization) of solid waste management was discussed at the council of ministries as a solution for the solid waste crisis in Lebanon (MOE, 2018). This policy comprised the decentralization of solid waste reduction, reuse, separation at source, and collection to the municipalities. In addition, the municipalities' responsibility for sorting, treatment, and final disposal of their solid waste is conditional upon obtaining prior approval from the Ministry of Environment (MOE, 2018).

Taking into consideration the current solid waste management situation in Lebanon, introducing a solid waste recycling program that is operated locally by municipalities will not only reduce the reliance on landfills- the initiator of the solid waste crisis- but will also serve as a sustainable solution for the ongoing solid waste crisis. Therefore, this study aimed to assess the public's willingness to participate in and pay for a local solid waste management program, operated by the municipality of "Jdeidet Ghazir" – a village located in the Mount-Lebanon governorate and its associated factors including participants' nationality, perceiving solid waste management as an urgent problem, walking distance from the nearest solid waste bin, participants' preference of the solid waste handling sector, possessing knowledge about the term of composting, perceiving solid waste management as a shared responsibility between the households and the government, perceiving solid waste as a merely governmental responsibility, and perceived connection between the household members' disease history and the ongoing solid waste crisis. The proposed solid waste management program includes separation of the generated solid waste at the source (ie. at the household level) and a recycling/composting service. The village of "Jdeidet Ghazir" serves as a good case-study because its existing solid waste management system is typical of the rest of Mount Lebanon and it is handled by the central government; it includes collection of mixed solid waste and disposal in the operating landfills, with no recycling/composting services.

2 Methods

2.1 Study Design

A cross-sectional survey was developed to assess the public's willingness to participate in and pay for a local solid waste management program, operated by the municipality of "Jdeidet Ghazir". The survey was divided into three sections. The first section included questions about the respondents' socio-demographics and household characteristics. The second section incorporated general questions about respondents' recycling/composting awareness and their opinion concerning the current situation of the solid waste management in Lebanon and in their village, "Jdeidet Ghazir". The third and final section of the survey included a hypothetical scenario of a proposed local solid waste management project that involves separation of solid wastes by each household into two solid waste bags (one bag for recyclables and one bag for organic/food waste) followed by the collection of these bags from each household and treatment through a recycling/composting service; while sending the wastes that cannot be recycled to sanitary landfills. Following the hypothetical scenario, respondents had to answer whether they are willing to participate in and pay for the proposed local solid waste management project in yes/no question format, taking into consideration that this project can be a solution for the ongoing solid waste crisis in Lebanon.

2.2 Study Population and Data Collection

The study population included all the households that are located in Jdeidet Ghazir village. Based on the municipality and the researcher field observation, the total number of households in this village is 334. A cadastral map, which was obtained from the municipality that maps all the streets and sub-streets of the village and the location of these 334 households, was used as a guide to efficiently navigate between the 334 households starting with the households that are located on the main street and followed by the households located on the sub-streets.

Data was collected using the developed cross-sectional questionnaire within a three-week period, from March 23 till April 09, 2018. The questionnaire was administered in the form of face-to-face interviews with any senior household member (the homemaker or the household head) who consent voluntarily to participate in the study. The interviews were conducted by the researcher in the Arabic language (native language of participants). Anyone who refused to participate in the survey was marked as a non-response.

2.3 Study Variables

Respondents' willingness to participate in and pay for a local solid waste management project that involves a recycling/composting service was the outcome in this study and it was assessed as a binary Yes/No variable (0=No; 1=Yes). In addition to the willingness to participate in and pay outcome, this study involved a set of explanatory independent variables that were selected based on the relevant literature (Table 1).

Table 1 Description of the Independent variables used in this study and their expected association with the respondents' willingness to participate in and pay outcome

Variable	Variable description	Variable coding	Expected sign of association (+ positive or – negative) based on relevant literature	
Solid waste management urgency	Whether respondents identify solid waste management problem as an urgent environmental problem in the village that needs immediate attention and action	No=0; Yes=1	+	(Alhassan & Mohammed, 2013; Challcharoenwattana & Pharino, 2016)
Walking distance	The average walking distance time measured in minutes between the household and the nearest solid waste bin	Less than 5 minutes=1; 5 to 10 minutes=2; More than 10 minutes=3	+	(Alhassan & Mohammed, 2013; Ezebilo, 2013)
Handling sector	Respondents preference on which sector should handle the solid waste management services in the village	Public sector=1; Private sector=2; Both in cooperation=3	+	(Ezebilo, 2013)
Composting awareness	Whether respondents have knowledge about the term "composting"	No=0; Yes=1	+	(Afroz & Masud, 2011)
Solid waste management household responsibility	Whether respondents consider the responsibility of proper solid waste management to be a shared responsibility between the households and the government	No=0; Yes=1	+	(Challcharoenwattana & Pharino, 2016; Rahmaddin, Hidayat, & Yanuwiadi, 2015)
Solid waste management government responsibility	Whether respondents consider that solid waste management is merely a governmental responsibility	No=0; Yes=1	-	(Wang, He, Kim, & Kamata, 2014)
Disease history	Whether respondents perceive a connection between the household members' disease history and the ongoing solid waste crisis	No=0; Yes=1	+	(Khattak, Khan, & Ahmad, 2009)
Nationality	Whether the respondent is of Lebanese nationality or not	Other nationalities=0; Lebanese nationality=1	+	A variable added by the researcher to examine whether Lebanese individuals will be more willing to participate in a local recycling program in their village

2.4 Data Analysis

A multivariate binary logistic regression model was carried out to examine the association between the willingness to participate in and pay for the proposed local solid waste program and the list of independent variables described in Table 1. We accounted for the heteroscedasticity of data in the multivariate logistic regression model using the robust function. Results are reported in terms of odds ratios (ORs) and 95% (CIs) confidence intervals. Statistical significance was considered at a P-value of 0.05. All data analysis was carried out using the STATA software (version 14.0).

3 Results

3.1 Response Rate and Socio-Demographics

In total, 228 out of the 334 households agreed to participate in the survey and gave complete responses; therefore the response rate was 68%. Most respondents were females (n=135; 59%), married (n=178; 78%), aged from 31 to 59 (n=125; 55%), and had a university educational qualification (n=101; 44%) (Table 2). Most respondents worked in the private sector (n=74; 32%) or were unemployed (n=99; 43%). Out of those who were unemployed, 84% were female homemakers. As for respondents' nationality, the majority were Lebanese (n=195; 86%) (Table 2).

Table 2 Description of the socio-demographic characteristics of respondents (N=228)

Variable	Category	Count	Percentage
Gender	Female	135	59.21
	Male	93	40.79
Marital status	Single	50	21.93
	Married	178	78.07
Age group	30 years and below	58	25.44
	31-59	125	54.82
	60 and more	45	19.74
Employment	Private sector	74	32.46
	Own business	41	17.98
	Government sector	14	6.14
	Not working	99	43.42
Education	Intermediary school	72	31.58

	Secondary and technical school	55	24.12
	University	101	44.30
Nationality	Lebanese	195	85.53
	Syrian	31	13.59
	Egyptian	1	0.44
	Indian	1	0.44

3.2 Solid Waste Management Knowledge and Practices

Most respondents indicated that the two urgent environmental issues in the village that require immediate attention and action are the solid waste management problem (n=198, 87%) and the wastewater treatment problem (n=186, 82%). The results also showed that 95% (n=217) and 91% (n=207) of respondents reported being knowledgeable about the terms of recycling and composting, respectively. Despite the high proportion of knowledge about the terms of recycling and composting, only 17% (n=38) of respondents indicated that they source-separate their generated solid waste (ie. separating plastics, glass, paper/cardboard, and metals from the organic food remains at the household level) and send the recyclables to organizations and companies that can benefit from them; while the majority (n=190; 83%) did not source-separate their generated solid waste. The reasons cited by the participants for not separating their solid waste are summarized in Table 3. The most common reason for not practicing source-separation of solid waste was reported to be that separation at source is not useful because solid waste is being collected as mingled waste by mixing recyclables with non-recyclables (Table 3).

Table 3 Distribution of respondents' reasons for not separating their generated solid waste at the source (N=190)*

Respondents' Reasons	Count	Percentage
Separation at source is not useful because solid waste is being collected as mingled waste (mixing recyclables with non-recyclables)	178	93.68
The municipality does not provide colored solid waste recycling bags to the households and there is no available recycling solid waste bins in our village	143	75.26
We do not have enough time to source separate our generated solid waste	11	5.79

Source-separation is not a social norm in our society and it is socially acceptable not to source separate our generated refuse	10	5.26
We don't have the required knowledge/awareness for the correct solid waste source-separation	5	2.63
The current solid waste management system is not making use of the source separated solid waste and all solid wastes end up mainly in landfills	5	2.63

N=190* includes only respondents who are not source separating their generated solid waste. Percentages does not add up to 100 because multiple answers are allowed for each respondent.

Most of respondents (n=147; 64%) indicated that the solid waste bins - dedicated for throwing the household generated solid wastes - are not far from their household and they can easily reach them by walking (strictly less than 5 minutes' walk). Nevertheless, the majority of respondents (n=127, 56%) were not satisfied with the current solid waste collection and management services in their village because the solid waste collection frequency is only twice (n=88, 39%) or three times (n=83, 37%) per week, which is insufficient and leads to the accumulation of solid wastes on the sides of the bins, spreading bad odors, attracting pests and wild animals, and causing diseases. In fact, 20% (n=46) of respondents indicated that some members of their household recently suffered from a disease (allergy: n=12; cold/flu: n=26; fly bite related disease: n=4; food poisoning: n=1; heart disease: n=1; and respiratory problems: n=2) and they perceived this disease to be connected to the poor solid waste management practices in the village and Lebanon.

The survey also revealed that 18% (n=42) of respondents prefer the private sector to handle the solid waste management sector in the village and Lebanon because the public sector in Lebanon is corrupt; the public sector has been handling the solid waste management for the previous years and the results were far below proper management; and the private sector is more reliable, has more expertise, better technical and financial abilities, and provides an overall better service. On the contrary, 22% (n=50) of the respondents indicated that the public sector (government and municipalities) should handle the solid waste management because the municipality is already taking a yearly fee for the solid waste services; knows the needs of the village; and can protect its residents from the private sector exploitation. However, most respondents (n=136, 60%) preferred both the public and private sector in cooperation to handle the solid waste management because the private sector has the needed competence, expertise, and reliability, while the public sector can monitor and control the private sector. This will result in a cost-efficient system that is fair to the public.

Results also showed that nearly half of the respondents (n=103, 45%) considered that proper solid waste management is a shared responsibility between the households – being the primary solid waste producers – and the government and the municipality. On the contrary, more than one quarter of respondents (n=62; 27%) considered solid waste management as a merely governmental responsibility.

3.3 Willingness to Participate in and Pay for a Local Solid Waste Recycling Program

The majority of the households (n=181, 79%) were willing to participate in and pay for a local solid waste management program that involves a recycling and composting service. The binary logistic regression analysis showed a significant positive association between respondents' willingness to participate/pay and their nationality (being Lebanese) (OR=5.96; 95%CI=2.10, 16.96), and perceived responsibility that households should be involved in solid waste management as active citizens in cooperation with the government and municipality (OR=3.31; 95%CI=1.41, 7.76). On the contrary, households perceiving solid waste management to be a sole governmental responsibility were less likely to participate/pay for a local recycling solid waste management program (OR=0.26; 95%CI=0.12, 0.57) (Table 4). Additionally, respondents who indicated that they have to walk for more than 10 minutes to reach the nearest solid waste bin showed lower odds (OR=0.25; 95%CI=0.08, 0.76) of participating/paying for a local recycling solid waste management program in comparison to those who have to walk for less than 5 minutes. Similarly, participants who perceived a connection between their household disease history and the ongoing trash crisis, were less likely to participate/pay (OR=0.34; 95%CI=0.14, 0 .82) for a local solid waste management program that involves recycling services (Table 4).

Table 4 The results of the multivariate binary logistic regression model (N=228)

Independent variables		**Distribution of willingness to participate/pay for a local solid waste program**		**ORs**	**P-value**	**95% CI**
		No	Yes			
Solid waste management urgency	No	10 (33.33%)	20 (66.67%)	Ref		
	Yes	37 (18.69%)	161 (81.31%)	2.44	0.098	[0.85, 7.03]
Nationality	Non-Lebanese	11 (33.33%)	22 (66.67%)	Ref		
	Lebanese	36 (18.46%)	159 (81.54%)	5.96	0.001**	[2.10, 16.96]
Walking distance	Less than 5 minutes	31 (21.09%)	116 (78.91%)	Ref		

	5≤ Walking ≤10 minutes	8 (13.56%)	51 (86.44%)	1.81	0.296	[0.59, 5.55]
	More than 10 minutes	8 (36.36%)	14 (63.64%)	0.25	0.014*	[0.08, 0.76]
Handling sector	Public	15 (30.00%)	35 (70.00%)	Ref		
	Private	10 (23.81%)	32 (76.19%)	0.88	0.818	[0.29, 2.67]
	Both in co-operation	22 (16.18%)	114 (83.82%)	2.10	0.086	[0.90, 4.91]
Composting awareness	No	7 (33.33%)	14 (66.67%)	Ref		
	Yes	40 (19.32%)	167 (80.68%)	2.04	0.245	[0.61, 6.75]
Solid waste management household responsibility	No	37 (29.60%)	88 (70.40%)	Ref		
	Yes	10 (9.71%)	93 (90.29%)	3.31	0.006**	[1.41, 7.76]
Solid waste management government responsibility	No	26 (15.66%)	140 (84.34%)	Ref		
	Yes	21 (33.87%)	41 (66.13%)	0.26	0.001**	[0.12, 0.57]
Disease history	No	34 (18.68%)	148 (81.32%)	Ref		
	Yes	13 (28.26%)	33 (71.74%)	0.34	0.016*	[0.14, 0 .82]

**represents significance at 1%; *represents significance at 5%

4 Discussion

Our study attempted to assess peoples' willingness to participate in and pay for a local solid waste management program that involves a recycling and composting service in Jdeidet Ghazir village, in response to the ongoing solid waste management crisis in Lebanon. Results showed that most respondents were willing to participate in and pay for the local solid waste management program. This shows that residents of Jdeidet Ghazir are motivated to improve the solid waste management practices in their village. The factors associated with an increase in peoples' willingness to participate/pay for the proposed program were perceiving solid waste management as a shared responsibility between households and the government/municipality and being Lebanese. Similarly, the literature has shown that people's willingness to participate in and pay for solid waste management improvements increase when they consider proper solid waste management as a shared responsibility between the households that are the primary producers of solid wastes and the government (Rahmaddin et al., 2015). Additionally, it was anticipated that respondents with a Lebanese nationality will be more willing to participate in and pay for a local recycling program because this will improve the solid waste management situation in their village, their permanent home, and will pave the way toward a healthier future.

Our study also showed that perceiving solid waste management as a merely governmental responsibility was negatively affecting the willingness of respondents to participate in and pay for the local recycling program. Similarly, previous research has shown that when people perceive solid waste management as a sole governmental responsibility, their willingness to participate in and pay for solid waste management improvements decrease (Wang et al., 2014). This is mainly attributed to the fact that governments in some countries, like Lebanon, are considered to be unreliable institutions with clashing political interests (Menhall & Joseph, 2017).

Contrary to the literature (Alhassan & Mohammed, 2013; Khattak et al., 2009), our analysis revealed that perceiving the family recent disease history to be connected with the solid waste crisis, and the inconvenient location of solid waste bins were negatively associated with peoples' willingness to participate in and pay for the local recycling program. One explanation could be the negativity bias phenomenon (Vaish, Grossmann, & Woodward, 2008) whereby the negative individuals' experiences with solid waste services might result in a negative attitude/response of unwillingness to contribute toward future solid waste management improvements.

In addition to assessing the peoples' willingness to participate in and pay for a local recycling program as a solution for the ongoing solid waste crisis in Lebanon, our study attempted to provide a descriptive profile on the peoples' solid waste management knowledge and practices. Results showed that even though most people reported having knowledge about the concepts of recycling and composting, only few respondents source-separated their generated solid waste into recyclables which were sent to organizations that can make use of them. The main reason behind the lack of source-separation of solid wastes was the unavailability of a recycling program in the village which makes separation at source a non-useful practice because solid waste is being collected as a mingled waste at the end of the day. This highlights the importance of establishing a local recycling program in the village of Jdeidet Ghazir given that a high percentage of respondents have the needed knowledge and awareness for the successful implementation of the program. Our study also showed that the majority of respondents prefer the public and private sectors in cooperation to handle the solid waste management in the village. Respondents pointed out that for solid waste management services to be cost effective for the public, the public sector (ie. the government and municipality) should monitor and control the private sector, which in turn has the needed competency, expertise, technical resources, financial resources, and reliability. In fact, literature has shown that municipality-private sector partnerships that invest in local recycling and composting services are recommended to enhance the efficiency of solid waste management services in developing countries (Al-Khateeb, Al-Sari, Al-Khatib, & Anayah, 2017; Murad, Raquib, & Siwar, 2007). Nevertheless, successful and sustainable public-private partnerships in the field of solid waste management depend mainly on having appropriate governance structures in place (Kruljac, 2012).

In considering the findings of this study, it is equally important to discuss the encountered limitations. First, this study investigated the peoples' willingness to participate in and pay for a local solid waste management program that involves a recycling service in one Lebanese village, Jdeidet Ghazir village, which might not be generalizable to the whole country. Thus, for future research, it is recommended that similar studies to be carried out in other rural and urban areas of the country with large sample sizes to allow for better generalization. Second, in some of the survey questions such as the family members' disease history, recall bias may have been encountered since respondents might find it difficult to recall such information. Finally, and despite the benefits of face-to-face interviews, social desirability bias could have occurred. Social desirability bias is a common phenomenon in population and environmental health research whereby people tend to give modified answers based on what they perceive to be in line with the researcher hypothesis and point of view (Latkin, Edwards, Davey-Rothwell, & Tobin, 2017). Nevertheless, the researcher who conducted the face-to-face interviews attended many training workshops on how to conduct research interviews, which reduces the social desirability bias.

5 Conclusion

Our study showed that solid waste management is perceived as an environmental priority among residents of Jdeidet Ghazir. The results provided evidence that households are willing to participate in and pay for a local recycling program operated in their village, and this was mainly associated with perceiving solid waste management as a shared responsibility between the households and the government and municipality. This emphasizes the need for redirecting public resources and establishment of a new solid waste program that involves the public through source-separation of the generated solid waste, recycling, and composting services. This program will reduce the dependence on landfills and provide a sustainable solution for the ongoing trash crisis in Lebanon. For future research, it is recommended to assess the public's willingness to participate in and pay for a local solid waste management program using a mixed method approach; through survey questionnaires and also through qualitative focus groups and interviews with the public and community stakeholders.

6 References

Abbas, I. I., Chaaban, J. K., Al-Rabaa, A.-R., & Shaar, A. A. (2017). Solid waste management in Lebanon: challenges and recommendations. *Journal of Environment and Waste Management, 4*(2), 53-63.

Afroz, R., & Masud, M. M. (2011). Using a contingent valuation approach for improved solid waste management facility: evidence from Kuala Lumpur, Malaysia. *Waste Manag, 31*(4), 800-808. doi:10.1016/j.wasman.2010.10.028

Al-Khateeb, A. J., Al-Sari, M. I., Al-Khatib, I. A., & Anayah, F. (2017). Factors affecting the sustainability of solid waste management system-the case of Palestine. *Environmental monitoring and assessment, 189*(2), 93. doi:10.1007/s10661-017-5810-0

Alam, P., & Ahmade, K. (2013). *IMPACT OF SOLID WASTE ON HEALTH AND THE ENVIRONMENT*.

Alhassan, M., & Mohammed, J. (2013). Households' Demand for Better Solid Waste Disposal Services: Case Study of Four Communities in the New Juaben Municipality, Ghana. *Journal of Sustainable Development, 6*(11), 16.

Azar, S. K., & Azar, S. S. (2016). Waste related pollutions and their potential effect on cancer incidences in Lebanon. *Journal of Environmental Protection, 7*(6), 778-783. doi:10.4236/jep.2016.76070

Chakrabarti, S., & Sarkhel, P. (2003). Economics of solid waste management: A survey of existing literature. *Economic Research Unit Indian Statistical Institute, 30*, 1-9.

Challcharoenwattana, A., & Pharino, C. (2016). Wishing to finance a recycling program? Willingness-to-pay study for enhancing municipal solid waste recycling in urban settlements in Thailand. *Habitat International, 51*, 23-30. doi:https://doi.org/10.1016/j.habitatint.2015.10.008

Ejaz, N., Akhtar, N., Hashmi, H., & Naeem, U. A. (2010). Environmental impacts of improper solid waste management in developing countries: A case study of Rawalpindi city. *The sustainable world, 142*, 379-387.

El Harakeh, A., Madi, F., & Bardus, M. (2017). Development of Strategies to Promote Solid Waste Management in Bourj Hammoud, Lebanon. *HastingsG. DomeganC.(Eds.), Social Marketing: Rebels with a Cause (*390-405.

Ezebilo, E. (2013). Willingness to pay for improved residential waste management in a developing country. *International Journal of Environmental Science and Technology, 10*(3), 413-422.

Human-Rights-Watch. (2020). Lebanon: Huge Cost of Inaction in Trash Crisis. Retrieved from https://www.hrw.org/news/2020/06/09/lebanon-huge-cost-inaction-trash-crisis

Khattak, N., Khan, J., & Ahmad, I. (2009). An Analysis of Willingness to Pay for Better Solid Waste Management Services in Urban Areas of District Peshawar. *Sarhad Journal of Agriculture, 25*, 529-535.

Khawaja, B. (2017). Lebanon needs to clean up its act, the country lacks a long-term waste management strategy. *Executive* Retrieved from https://www.executive-magazine.com/opinion/lebanon-needs-to-clean-up-its-act

Kruljac, S. (2012). Public–Private Partnerships in Solid Waste Management: Sustainable Development Strategies for Brazil. *Bulletin of Latin American Research, 31*(2), 222-236. doi:https://doi.org/10.1111/j.1470-9856.2011.00659.x

Latkin, C. A., Edwards, C., Davey-Rothwell, M. A., & Tobin, K. E. (2017). The relationship between social desirability bias and self-reports of health, substance use, and social network factors among urban substance users in Baltimore, Maryland. *Addictive behaviors, 73*, 133-136. doi:10.1016/j.addbeh.2017.05.005

Massoud, M., & Merhebi, F. (2016). Guide to municipal solid waste management. *American University of Beirut (AUB), Publisched by the American University of Beriut-Nature Conservation Center, Lebanon.*

McAllister, J. (2015). Factors influencing solid-waste management in the developing world. Retrieved from https://digitalcommons.usu.edu/gradreports/528/

Menhall, E., & Joseph, N. (2017). *Decentralization as a policy option in Lebanon: the case of the waste management crisis and local level solutions.*

Minghua, Z., Xiumin, F., Rovetta, A., Qichang, H., Vicentini, F., Bingkai, L., . . . Yi, L. (2009). Municipal solid waste management in Pudong New Area, China. *Waste Management, 29*(3), 1227-1233. doi:https://doi.org/10.1016/j.wasman.2008.07.016

MOE. (2018). *Policy Summary on integrated solid waste management in Lebanon Report.* Retrieved from http://www.moe.gov.lb/

MOE, EU, & UNDP. (2014). Lebanon Environmental Assessment of the Syrian Conflict & Priority Interventions. In: Ministry of Environment, Lebanon Beirut.

Morsi, R. Z., Safa, R., Baroud, S. F., Fawaz, C. N., Farha, J. I., El-Jardali, F., & Chaaya, M. (2017). The protracted waste crisis and physical health of workers in Beirut: a comparative cross-sectional study. *Environmental Health, 16*(1), 39. doi:10.1186/s12940-017-0240-6

Murad, M. W., Raquib, M. A., & Siwar, C. (2007). Willingness of the Poor to Pay for Improved Access to Solid Waste Collection and Disposal Services. *The Journal of Environment & Development, 16*(1), 84-101. doi:10.1177/1070496506297006

Ogwueleka, T. (2009). Municipal solid waste characteristics and management in Nigeria. *Journal of Environmental Health Science & Engineering, 6*(3), 173-180.

Rahmaddin, M., Hidayat, T., & Yanuwiadi, B. (2015). Knowledge, attitude, and action of community towards waste management in river bank of Martapura. *International Journal of Applied Psychology, 5*(4), 96-102. doi:10.5923/j.ijap.20150504.03

Singh, J., Laurenti, R., Sinha, R., & Frostell, B. (2014). Progress and challenges to the global waste management system. *Waste Management & Research, 32*(9), 800-812. doi:10.1177/0734242X14537868

SWEEP-Net. (2014). Country report on solid waste management in Lebanon. In.

Vaish, A., Grossmann, T., & Woodward, A. (2008). Not all emotions are created equal: the negativity bias in social-emotional development. *Psychological bulletin, 134*(3), 383-403. doi:10.1037/0033-2909.134.3.383

Wang, H., He, J., Kim, Y., & Kamata, T. (2014). Municipal solid waste management in rural areas and small counties: An economic analysis using contingent valuation to estimate willingness to pay for Yunnan, China. *Waste Management & Research, 32*(8), 695-706. doi:10.1177/0734242X14539720

Zhang, M., Buekens, A., & Li, X. (2017). Open burning as a source of dioxins. *Critical Reviews in Environmental Science and Technology, 47*(8), 543-620. doi:10.1080/10643389.2017.1320154

Sustainability Challenges in Solid Waste Management for Future Prospective in Gaza Strip: Rafah City Case Study

Samir Alnahhal[a], Samir Afifi[b]

Institute of Water and Environment, Al Azhar University

Abstract

Waste management services in the Gaza Strip, Palestine are currently under a great stress due to a complicated political and economic situation. The improper management of solid waste has been identified as one of the key factors that contribute to public health risks and serious environmental problems. This paper presents major challenges towards sustainability of urban solid waste management or services in the Gaza Strip in general and Rafah City in specific. Two case studies are outlined in the paper that can pave the way towards integrated solid waste management. The first case study focuses on source separation of organic waste for 78 households over nine weeks, which indicates about 70% of daily-generated waste is organic. The second study highlights the development of a locally designed and manufactured semi-automated separation, sorting, and processing pilot model. The model was constructed at the temporary transfer station in Rafah city, which receives about 48% of 120 tons generated daily in the city. The model was operated for a period of three months in order to check the performance of the model and to optimize the operation and maintenance phases. During the optimization period, sorting and separation of mixed municipal solid waste into different compositions was carried out by skilled and trained labors. The bulk quantity of Rafah municipal waste was organic with a 58%, which was handled to produce compost inside the model site.

Keywords: Gaza Strip; management; at-source separation; solid waste; sustainability

---

1. Introduction

Since the beginning of human life, waste has been a product of his activity (NRCS, 1992). "Domestic" or "municipal" waste is defined as the unwanted material produced by households, institutional and business buildings (Rushton, 2003). Municipal Solid Waste (MSW) consists of several things, including: food and garden wastes, paper, cardboard, wood, textiles, glass, metals, plastics (Rushton, 2003), leather, nappies, ash, and slug (Tozlu et al., 2016).

Generally, it is well recognized that rapid population growth, rising living standards, urbanization, and changes in advanced technologies have a key role in increasing the daily generated Solid Waste (SW), and in changing its characteristics and constituents (Al-Khatib et al., 2010; Seo et al., 2004; Tozlu et al., 2016; Zhang et al., 2010).

The solid waste management (SWM) includes collection, transport, processing, reuse, and sanitary disposal (Al-Khatib et al., 2007). It is a sophisticated task and its effective management highly depends on the cooperation and coordination between local municipalities, households, entire communities, and others (Afifi and Barhum, 2009)

Emerging countries suffer significantly from currently inadequate SWM and its growing threat in the future. On the community level, poverty and the lack of education are the fundamental barriers for better application of SW practices and regulations, whilst the limited financial resources are the key obstacles for local authorities (Al-Khatib et al., 2007).

In emerging countries, human health and environmental problems that are especially associated with current SW practices are dumping in open and in ad-hoc places; disposing in poorly equipped landfills; and open air burning. This is all in addition to other managerial and institutional problems that could face the local authorities due to a lack of recycling facilities and absence of related national strategies (Tozlu et al., 2016). Frequently, private businesses do not engage in responsible disposal practices due to shortages in verifiable and robust baseline data, and a lack of opportunities for markets to sell processed materials, i.e., recyclables either locally or abroad (Afifi and Barhum, 2006; Afifi and Barhum, 2009),

The study aims to shed light on major challenges towards sustainability of urban solid waste services in the Gaza Strip in general and Rafah City in specific in the hope that it may propose efficient and effective recommendations for Waste Resource Management (WRM). This paper focused on two different case studies implemented in Rafah City at end of year 2016. The first study focused on source separation of domestic organic and nonorganic solid waste and the second study represented the development of a locally designed and manufactured semi-automated separation, sorting, and processing pilot model for domestic solid waste.

2. Solid waste management in Gaza Strip

In the Gaza Strip, according to recent field surveys, the proportion of SW produced is 48% from households, 28% is produced from institutional and businesses, and 24% results from construction and industrial sector (Jouda and Sha'at, 1995). According to previous studies (EQA, 2012; PMOP, 2007; UNDP - PAPP, 2012a), the average of percent weights of different solid waste fractions shows the composition of solid waste in the Gaza Strip as presented in Figure 1.

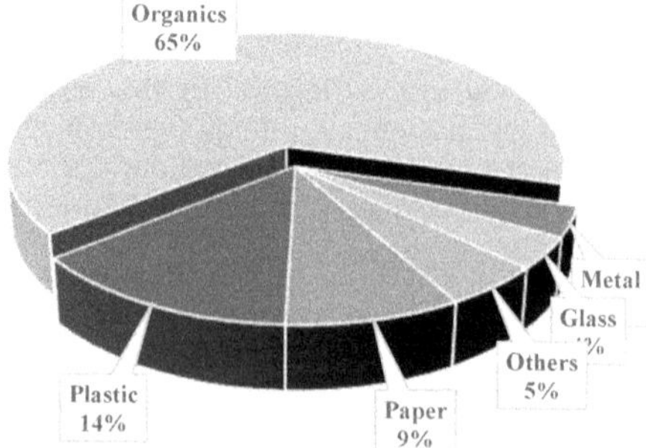

Figure 1: Composition of solid waste in the Gaza Strip, Palestine

Usually, SWM is deemed as a significant responsibility of local municipalities (Afifi and Barhum, 2009). Collection of SW is a major operational component of SWM, which usually presents a great challenge to local municipalities in the Gaza Strip (Al-Khatib et al., 2010). It requires relatively high running costs that should be supported by a budget base all times of the year, i.e., cash-basis. The rate of collected fees from local communities are about 20-50% of required money for effective collection (Afifi and Barhum, 2009). Nevertheless, in Rafah governorate, about 75-95% of MSW is still collected (Alslaibi et al., 2012), (Ouda, 2013).

In the Gaza Strip, daily-generated SW quantities are dumped in three (3) central landfills/disposal sites: "Johr El Deek" in Gaza city, "Deir el Balah" in middle area, and "el Sufa" between Khan Younis and Rafah. These sites are currently overloaded and their maximum capacities were reached several years ago. However, annually, they still receive 1.5 - 2 million tons of all kinds of wastes, including demolition waste, tires, slaughterhouse waste and car scrap metals (UNDP - PAPP, 2012b).

Although SWM in the Gaza Strip is a relatively simple process as it only includes the collection and disposal of material (Ouda, 2013), the collection points are changing, as the typical shared container system is being gradually replaced by house-to-house collection (Afifi and Barhum, 2009).

At present, the municipalities' vehicles transport more than 1000 SW tons for a distance estimated to be about 11-22 km from temporary transfer stations located inside Gaza Strip main cities, i.e., Gaza city, Middle Areas, Khan Younis city, and Rafah city, to central landfills/disposal sites located outside these cities (Alslaibi et al., 2012). The operational efficiency of transporting vehicles is generally poor, which is strongly related to inability of municipalities to perform proper and regular maintenance for these vehicles, which require urgent rehabilitation (EUNIDA, 2009).

3. Sustainability challenges of solid waste in Gaza Strip

Besides the most common problems of SWM in developing countries, Gaza Strip has a complicated situation as it suffers more due to political restrictions and economic constraints, which in turn have a considerable influence on adequate application of (WRM) (Ouda, 2013; UNDP - PAPP, 2012b),. Therefore, the main aim of this paper is to draw attention to the main challenges that retard sustainability of SWM in the Gaza Strip.

3.1. Social dimension.

The social dimension is associated with all issues that affect or have a concern to the public who receive a certain service, i.e. SWM (Padilla-Rivera et al., 2016, p. 102). Human health is seriously threatened by direct exposure to hazardous wastes and indirectly by exposure to pollution generated by the same wastes (NRCS, 1992). In Gaza Strip, there is no proper separation between domestic SW and other hazardous wastes, which could cause human health problems (Afifi and Barhum, 2006; El Baba and Smedt, 2013). Moreover, the three landfill/disposal sites are located in or near the buffering zone with Israel, which often does not allow municipalities to access these sites, particularly during the conflicts and wars time. This then causes accumulation of wastes in local streets for several days, posing severe human health and environmental problems (UNDP - PAPP, 2012b). Other reason for accumulation of wastes is limited financial capacities of local municipalities for following a regular and consistent collection program.

The social dimension is a critical challenge on the sustainability of SWM. Tangible changes are required in public attitudes and their current behaviour. However, the majority of local residents are aware that the current practices of SWM in Gaza Strip causes human health and environmental problems as reflected in the results of a field survey conducted by Applied Research Institute – Jerusalem (ARIJ) (ARIJ, 2005).

Public participation and community involvement is a significant factor for achieving an efficient SW recovery of recyclable material and achieving effective at-source separation approach, which would reduce collection efforts, time, and costs (Hassan et al., 2000). Furthermore, health promotion and environment awareness programs are mostly dependent on external donations and do not cover whole areas within Gaza Strip (Afifi and Barhum, 2006).

3.2. Economic dimension.

Currently, the financial situation of local municipalities is dire and the Palestinian government monetary assistance is modest towards SWM. Local municipalities of Gaza Strip spend about 40-60% of their total budget to provide SW services (UNDP - PAPP, 2012b). Local private sector and investors are not interested in offering financial support for SWM due to the high risks that rise from current political restrictions, shortage of recycling infrastructure, and acute problems in local marketing of recyclables. Therefore, recycling of SW materials is so seldom and/or only low quality of recycling by-products are generated (Afifi and Barhum, 2006).

Most municipalities and the Palestinian government are supported by international agencies, e.g., United Nations, European Commission, and World Bank, which grant external funds to manage the SW sector. For instance, based upon a recent study by United Nation Development Programme (UNDP), Rafah municipality needs about USD$10 million just to extend and/or close and relocate the existing central dumping site “el Sufa”, which have been requested from these international funding agencies (UNDP - PAPP, 2012b). Moreover, the costliest SW collection and transportation processes are significantly dependent on fees collected from the serviced communities. However, poverty and low financial incomes of the majority of Palestinians lead to deep shortages in collecting fees for SW services. Besides that, a lack of municipality resources and their financial capacities make real weaknesses in provision of adequate solid waste services and obstacle the application of WRM (Afifi and Barhum, 2009).

3.3. Environmental considerations:

The environmental dimension is basically linked to a contamination of the environment due to improper management and inadequate disposal approaches of SW. In Gaza Strip, current practices cause severe environmental problems (El Baba and Smedt, 2013), which are associated with improper collection, storage, transportation and disposal systems; seldom use of recycling and reuse options for different types of recyclables; and production of low quality of recycling by-products (Afifi and Barhum, 2006). Therefore, sever environmental problems rises due to the following:

1. Open/random burning practices that causes atmosphere pollution by smoke and harmful gases. It was estimated in 2010, that open burning causes an emission of about 186 thousand tons of CO_2 in Palestinian areas, (West Bank and Gaza Strip) (PCBS, 2014).
2. SW dumping in ad-hoc places leads to soil pollution, landscape nuisance, and groundwater contamination that could occur due to infiltration of toxic waste leachate.

4. Case studies in Rafah City

In literatures, there are many options for each SW process, which can be selected based upon many factors considerations, particularly, the local conditions and circumstances to achieve the IWSM (ARIJ, 2005).

Through this study, the authors would like to shed light on the results of two pilot projects implemented in Rafah city, which have focused on two different SWM options. The two options were investigated through pilot projects conducted in Rafah city. The common aim of both pilot projects was to separate, sort, and process different SW fractions in order to reduce the waste quantities that need transportation and final disposal. By the waste reduction, the cost associated with waste transportation is expected to be lowered; the required area for building disposal sites is foreseen to be reduced; and the life span of current disposal sites (landfills) and transportation means is predicted to be extended. Another goal of both pilot projects is to maximize benefits of recycling and recovery of reusable products, i.e., the organic fraction, which can be processed to produce a soil amendment substrate or what called as "compost".

Through the first pilot project, an at-source separation of solid waste is achieved. While through the second project, a semi-automatically sorting, separation and processing of solid waste is implemented in Rafah city.

4.1. At- source separation for organic and non-organic fractions - case study

Previously, several studies were carried out to estimate percent of daily-generated organic fraction to the total daily quantities of SW in the Gaza Strip as shown in Table 1. A detailed study was carried out to analyse and determine the percentages of different SW fractions in Rafah city in 2007 (Afifi and Barhum, 2006). However, updating such percentages of different solid waste fraction is a key issue for solid waste management, particularly the percent of the Organic Solid Waste (OSW) fraction. Therefore, through this pilot case study, percentage of OSW was updated through implementing at-source separation by households' housekeepers, who separate their daily generated organic and non-organic waste in separated bags. The bags were collected and transported to the processing site based on a specified routing system, where donkey karts were utilized to daily collect the OSW from 78 targeted households. The collection and routing system was coordinated with Rafah Municipality in order to collect only the non-organic fractions from targeted households.

Table 1: SW generation rate and organic fraction percentage in Gaza strip

SW generation rate (*Kg/Person. Day*)	OSW fraction (*%*)	Reference
0.35-1.0	65%	(Nassar and Jaber, 2007)
0.4-1.2	60%-70%	(Al-Hmaidi, 2002)
0.7-0.8	75%	(Jouda and Sha'at, 1995)
0.4	80%	(PNA, 2010)
0.35-1.0	-	(El Baba and Smedt, 2013)

Through this case study, a knowledge rising program was provided to the 78 households about SW, material types of organic and nonorganic, proper ways of separation, included in a training and guidance manual distributed to the targeted and interested households. The knowledge rising program aimed to ensure the effective and efficient at-source separation of organic fraction from other non-organic ones.

As incentives for facilitating the proper at-source separation, coloured plastic bags and buckets of 20 L volume capacity were also distributed to the targeted households. For updating the OSW percent, daily collected bags containing organic waste from each household were separately weighed two times in the processing site. First time, bags were weighed before removing any nonorganic materials. While the second time, bags were weighed after removing nonorganic materials. This two-time weighting was carried out to mainly investigate to what extent households are aware of the "at-source separation" process, and to what level households are committed to provide their daily-generated wastes and to separate organic and nonorganic fractions. The two-time weighting of collected bags would acquire robust and concrete results about the SW quantities.

Results has revealed that the targeted (78) households were highly committed to the OSW quantity as they provided their all quantities with a high percentage of 92% on average as presented in Figure 2. The households were also highly committed to the OSW quality as they separated the SW into organic and no-organic fractions with a high percentage of 95% on average, as presented in Figure 2. Figure 2 shows that the percentages of households commitment to the quantity and quality of OSW during the piloting period of nine weeks. As shown in Figure 2, both curves for households' commitment almost obey the same behaviour along the piloting period. However, both curves were slightly ascending in the first five weeks, while both curves were descending the remaining four weeks. The ascending part could be emphasized as the people were new to the at-source separation approach. While, the descending part was not expected. Nevertheless, it could be explained as people might become unencouraged and tired from separation as no incentives were provided to the households after the fifth week. The incentives include plastic bins, plastic bags, and hygiene materials such as soups, shampoo, and towels.

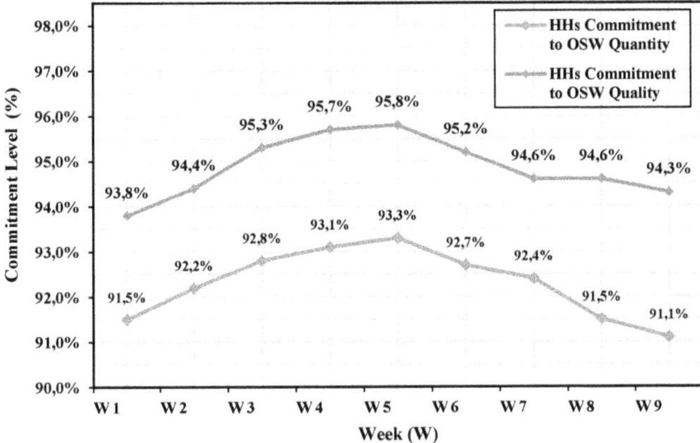

Figure 2: Households' commitment to organic solid waste quantity and quality

Through this case study of SW at-source separation, registration and documentation of the daily-generated quantities of organic and non-organic SW fractions, separated in 78 households in Tal El Sultan area, Rafah City, in order to determine the following:

1. Average quantities of organic waste per household and per person.
2. Average percent of organic fraction to total SW and nonorganic SW.

Figure 3 represents the weekly average quantities of OSW collected daily from all (78) households. Results have revealed that the weekly average at-source separated OSW quantities form all 78 households was about 1077 kg. Based on demographic details of the 78 households, on average, about 4.2 persons live in a household. The daily average production of OSW per household and per person was 1.79 kg and 0.42 kg with a standard deviation of 50 g, respectively. This daily average per person falls within estimated ranges of previous studies and reports shown in Table 1.

Although high commitment of households to OSW sorting quantity and quality as shown in Figure 2, the weekly variation in OSW collected quantities from all households was relatively high with an average of 12%, increasing quantities in the first 5 weeks and decreasing in the last 4 weeks as shown in Figure 3. Considering the weekly average as 1077 kg for all (78) households, this variance is still considered reasonable on the household level as the daily variation did not exceed on average 200 g per household.

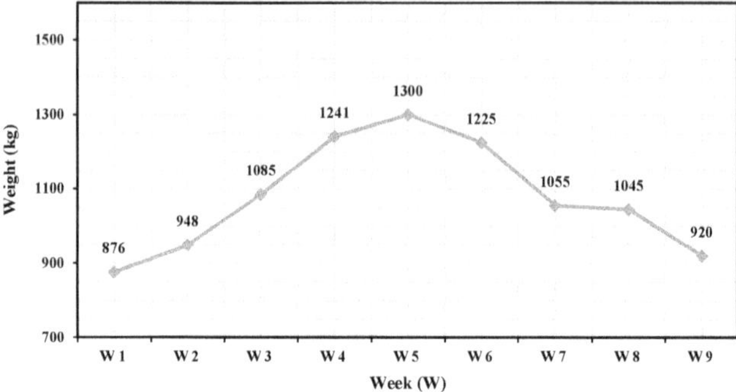

Figure 3: Weakly average at-source separated organic solid waste form 78 households

From another point of view, the average percent of the OSW to the total SW and/or non-organic SW fraction was determined through randomly collecting and weighting both organic and non-organic SW from all (78) households for three out of the nine weeks (whole period of the project) as shown in Table 2.

Table 2: Average percentages of different SW fractions collected from 78 households in three weeks.

SW fractions (%)	Week 2	Week 5	Week 8	Average
Organics	76.4%	71.1%	70.6%	72.70%
Paper	7.8%	7.3%	18.1%	11.07%
Metals	0.7%	0.5%	0.9%	0.70%
Plastic	9.8%	3.6%	3.9%	5.77%
Glass	0.9%	0.5%	0.9%	0.77%
Fabric	1.3%	0.9%	1.4%	1.20%
Sand	0.0%	8.7%	1.4%	3.37%
Others	3.3%	7.3%	2.9%	4.50%

As shown in Table 2, results show that the organic waste has the highest percentage, which reaches 72.70%. While the other remaining non-organic fractions encompass 27.3% in total. Therefore, it would environmentally friend and economically cost effective to handle the OSW for production of soil amendment substrate or compost. This high percent of OSW could pave the way to "at-source separation" and production and use of compost as an abroad strategy throughout the Gaza Strip for agro-farming and reduce the high use of chemical fertilizers. The use of compost would also improve the food production due to its scientifically approved characteristics and enriched element content for enhance the soil properties/conditions, such as elements/minerals content and water holding capacity. This also would result in reduction of the overall SW quantities that require transportation and disposal in final landfills. Hence, associated financial costs and needed efforts would also be minimized.

Figure 4 compares the percentages of different SW fractions resulted by this study with that obtained by a previous study in 2004 which was made by Rafah Municipality (Afifi and Barhum, 2006).

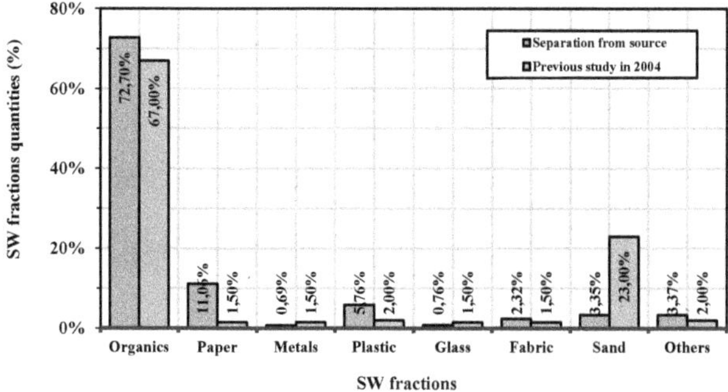

Figure 4: Comparison between different SW fractions resulted by the current study and another study in 2004

As shown in Figure 4, no major differences between the percentages of the various SW fractions in both studies in the OSW fraction, with exception to sand and paper percentages, which were higher in the current study through "at-source separation". This could be emphasized as at-source separation practices reduce the sand content compared to mixed SW collection by different equipment and tools such as wheel loaders and trucks. The increase of paper content collected by "at-source separation" could be due to increase use of paper in the daily life activities of the people comparing to that in 2004. Another study presented a wide variation in the paper percentage for year 2010 and 2011 as 14.7% and 6.3% , respectively (Nassar, 2015).

4.2. Pilot semi-automatic sorting/ separation facility for domestic municipal solid waste case study

Based upon the current population density and estimated daily generation of SW per person, Rafah municipality has estimated that 100-120 tons of SW is being generated daily in the city (Afifi and Barhum, 2009). About 30-40% of this quantity is disposed of at the "Tal El Sultan" temporary transfer station (Afifi, 2017) . The rest of the SW is either disposed of directly to "El Sufa" landfill or randomly dumped and burned in open areas, which is often practiced especially in rural areas.

Although an "at-source separation" of different SW fractions has several benefits in facilitation proper management of Municipal Solid Waste (MSW), it deems a time consuming task as it needs more efforts and time to change people's behaviour. In contrast, integrated management of mixed municipal SW requires undergoing several steps and long processes, including waste generation, onsite storage, house to house collection, transfer and transportation, processing and recovery, and disposal in landfills (Rao et al., 2017). Through an external fund provided from Japan Government through United Nations Development Program, a pilot small to medium-scale semi-automatic separation, sorting, and processing facility was built by a nongovernmental organization, named "Palestinian Environmental Friends Association (PEF)" by the end of 2011 in Rafah city. The facility was built on an area of about 8000 m^2 owned by Rafah Municipality in the transfer station at "Tal El Sultan area". The facility structure and equipment was designed and built up locally, utilizing locally available resources and expertise as shown in Figure 5(a & b) and Figure 6 (a & b).

Figure 5: (a) skilled labours do installation; (b) semi-automated sorting barracks under construction;

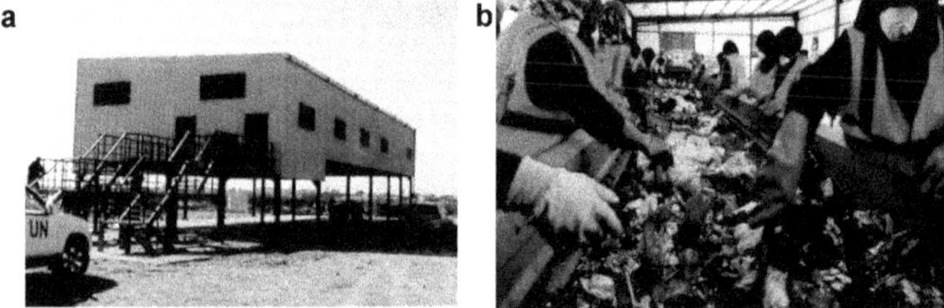

Figure 6: (a) semi-automated sorting barracks ready; (b) trained labours sort and separate different fractions of MSW

The results obtained during a recent operational period of three months have revealed that Rafah Municipality collected and transported about 57ton MSW /day on average to the pilot sorting, separation, and processing facility, as shown in Figure 7.

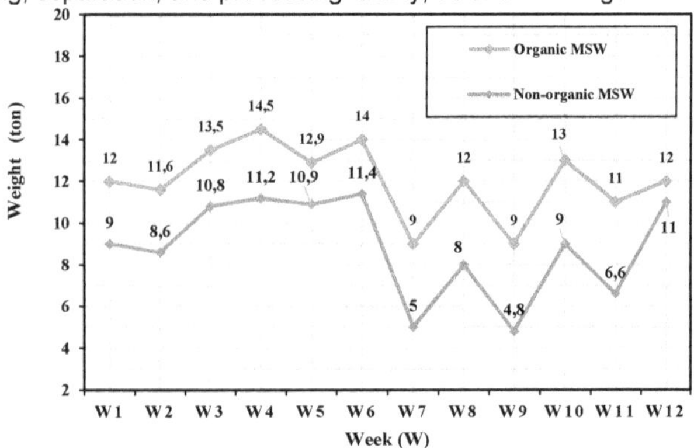

Figure 7: Weights of organic and nonorganic MSW separated at Rafah semi-automatic separation unit

As shown in Figure 7 and , about 48% of 120 ton generated at Rafah City is transported to the transfer station or processing site. However, based on previous study (Afifi and Barhum, 2009), Rafah Municipality collected and transported about 20-30% to the transfer station (Afifi and Barhum, 2009).

Pilot separation unit was designed to process 120 tons of MSW daily. However, as presented in Figure 7, only about 21 tons were daily handled, which is 37% and 18% of daily received MSW and facility capacity, respectively. This also means that, the facility was running for only four hours daily due to Gaza context of electricity shortages, and regular maintenance in the experimental and optimization processes in the first operational period.

Results have revealed that the weight of organic fraction constitute the largest component in the MSW, which was on average about 58% of total MSW. While, the rest 42% is nonorganic waste including both recyclable and non-recyclable materials. The percent of non-recyclable waste, such as, rubber, cloths, and baby diapers, was about 20-30% and was transported to the "El Sufa" landfill as rejected materials.

Other recyclables such as plastic, paper, and metals were compressed for volume minimization before sending for further recycle and recovery processes or exporting. The organic fraction undergoes further treatment, i.e., humification process for soil amendment substrate or compost production at the facility site.

5. Conclusion

Provision of waste services in Gaza strip is a vast challenge for the key stakeholders due to the current complicated political and economic situations.

Waste minimization and management through enhancing at-source separation approach is not an easy task and requires more time and effort to change socio-cultural behaviour of the Gaza people. Moreover, it requires high levels of investment for substantial operation and maintenance costs, and for provision of the necessary equipment and infrastructure that collects, transports, and processes the diversity of recyclables in the waste stream. However, this study showed that at-source separation was successfully achieved in 78 households in Rafah governorates through awareness raising programs.

Waste management through a semi-automatic separation, sorting and processing system on small to medium scale presented a viable solution at present. This acts as a transition period from current practices to a fully integrated and effective solid waste management. The Rafah semi-automatic processing facility is still in its piloting phase and it requires an optimization period of one to three years for preparing adequate feasibility studies and business models for further expansion of the program.

Since the organic waste fraction has the major weight of about 60-70% of the whole SW quantity produced in Gaza, it paves the way for the soil amendment substrate or compost production and industry. This also will minimize the total quantities that need finally dumping sites and landfills. Resulting in increasing the life span of the landfill by at least three times and reducing substantially associated transportation costs. These evidence-based results indicate that replicating and up-scaling such initiatives and pilot projects could be one of the accepted solutions under the context of the Gaza Strip.

Acknowledgements

The authors would like to acknowledge the NGO, Palestinian Environment Friends Association (PEF) and Rafah Municipality (grant number: 2015-16/PEF-RAF-PS/3-02-

01) for their successful implementation of pilot projects and provision of the valuable information.

Conflict of interests

The authors declare that they have no conflict of interest.

Ethical approval:

This study was approved by the Local Human Ethics Committees of the Palestinian Environmental Fiends Association, Rafah, Palestine. All procedures performed in studies involving human participants were in accordance with the ethical standards of the institutional and/or national research committee and with the 1964 Helsinki declaration and its later amendments or comparable ethical standards.

Informed Consent

All of the participants were involved in at-source solid waste separation and at-site manual sorting of solid waste fractions program in Rafah, Palestine. The authors fully informed all participants (households' housekeepers and labours) concerning all of the study features. Participants signed informed consent document that was approved by the Local Human Ethics Committees of the Palestinian Environmental Fiends Association, Rafah, Palestine, in compliance with the Helsinki Declaration.

References

Afifi, S., 2017. Environmental and Social Management Plan For Rafah Solid Waste Transfer Station. Municipal Development and Lending Fund, Gaza. http://www.mdlf.org.ps/Files/solid-waste-gaza/GSWMP_ESMP_Rafah%20Transfer%20Station_June%202017.pdf. Accessed 9 October 2018.

Afifi, S., Barhum, A., 2006. Assessment of technical and environmental performance of municipal solid waste management in Rafah City-Palestine. EGYPTIAN JOURNAL OF SOIL SCIENCE 46, 91–102.

Afifi, S., Barhum, A., 2009. Evaluation of Economical and Social Aspects of Municipal Solid Waste Management in Rafah City-Palestine, in: 2nd International Conference on the Palestinian Environment. 2nd International Conference on the Palestinian Environment, West Bank, Palestine. 06/10/2009, An-Najah Scholars, pp. 165–175.

Al-Hmaidi, M., 2002. The development of a strategic waste management plan for Palestine.: Review of the current situation. handling, transportation and disposal of waste. Negotiations Support Unit, Negotiations Affairs Department, Palestine.

Al-Khatib, I., Arafat, H., Basheer, T., Shawahneh, H., Salahat, A., Eid, J., Ali, W., 2007. Trends and problems of solid waste management in developing countries: A case study in seven Palestinian districts 27, 1910–1919. 10.1016/j.wasman.2006.11.006.

Al-Khatib, I., Monou, M., Abu Zahra, A., Shaheen, H., Kassinos, D., 2010. Solid waste characterization, quantification and management practices in developing countries. A case study: Nablus district – Palestine. Journal of Environmental Management 91, 1131–1138. 10.1016/j.jenvman.2010.01.003.

Alslaibi, T., Abustan, I., Mogheir, Y., Afifi, S., 2012. Quantification of leachate discharged to groundwater using the water balance method and the Hydrologic Evaluation of Landfill Performance (HELP) model. Waste Management & Research 31, 50–59. 10.1177/0734242X12465462.

ARIJ, 2005. Analysis of Waste Management Policies in Palestine: Domestic Solid Waste and Wastewater. ARIJ Applied Research Institute – Jerusalem, Bethlehem, Palestine. http://www.arij.org/. Accessed 4 April 2016.

El Baba, M., Smedt, F. (Eds.), 2013. Solid Waste Management and Practices in Gaza Strip (Palestine). ASCE American Sociaty of Civil Engineers.

EQA, 2012. Overview of solid waste management in the Gaza Strip. EQA - Palestinain Environmental Quality authority, EQA library, Gaza, Palestine.

EUNIDA, 2009. ANNEX B: Programme Fiches by Sector: Final Report - Damage Assessment and NeedsIdentification in the Gaza Strip. EuropeAid. https://ec.europa.eu/. Accessed 4 April 2016.

Hassan, M., Abdul Rahman, R., Lee Chong, T., Zakaria, Z., Awang, M., 2000. Waste recycling in Malaysia: Problems and prospects. Waste Management & Research 18, 320–328. 10.1177/0734242X0001800404.

Jouda, A., Sha'at, A., 1995. Emergency Action Plan for Solid Wastes in the Gaza Strip. Ministry of Planning and International Co-operation, Gaza.

Nassar, A., 2015. Potential of Solid Waste Composting in the Gaza Strip-Palestine. JAERI 4, 18–24. 10.9734/JAERI/2015/15558.

Nassar, M., Jaber, A., 2007. Assessment of Solid Waste Dump Sites in Gaza Strip. Environment Quality Authority (EQA) and Japanese International Cooperation Agency (JICA, Gaza.

NRCS, 1992. Environmental Epidemiology: Vol. 1: Public Health and Hazardous Wastes. National Academies Press, Washington, 1 online resource.

Ouda, O., 2013. Assessment of the Environmental Values of Waste-to-Energy in the Gaza Strip. Curr World Environ 8, 355–364. 10.12944/CWE.8.3.03.

Padilla-Rivera, A., Morgan-Sagastume, J.M., Noyola, A., Güereca, L.P., 2016. Addressing social aspects associated with wastewater treatment facilities. Environmental Impact Assessment Review 57, 101–113. 10.1016/j.eiar.2015.11.007.

PCBS, 2014. World Environment Day, Ramallah, Palestine.

PMOP, 2007. Sectoral planning for solid waste. Gaza Strip. PMOP - Palestinian Ministry of Planning, PMOP library, Gaza Palestine.

PNA, 2010. National Strategy for Solid Waste Management in the Palestinian Territory2010-2014. PNA - Palestinian National Authority, Palestine. http://www.molg.pna.ps/. Accessed 8 April 2016.

Rao, M., Sultana, R., Kota, S., 2017. Solid and hazardous waste management: Science and engineering. Butterworth-Heinemann; distributed by BS Publications, Oxford England, India, xii, 332 pages.

Rushton, L., 2003. Health hazards and waste management. British Medical Bulletin 68, 183–197. 10.1093/bmb/ldg034.

Seo, S., Aramaki, T., Hwang, Y., Hanaki, K., 2004. Environmental impact of solid waste treatment methods in Korea. Journal of Environmental Engineering 130, 81–89.

Tozlu, A., Özahi, E., Abuşoğlu, A., 2016. Waste to energy technologies for municipal solid waste management in Gaziantep. Renewable and Sustainable Energy Reviews 54, 809–815. 10.1016/j.rser.2015.10.097.

UNDP - PAPP, 2012a. Feasibility study and detail design of solid waste management in the Gaza Strip. UNDP - PAPP United Nations Development Programme in Programme of Assistance to the Palestinian People, UNDP - PAPP library - Gaza Office, Gaza, Palestine.

UNDP - PAPP, 2012b. Feasibility Study and Detailed Design for Solid Waste Management in the Gaza Strip: Final. UNDP - PAPP United Nations Development Programme in Programme of Assistance to the Palestinian People, Gaza. http://www.mdlf.org.ps/. Accessed 4 April 2016.

Zhang, D., Tan, S., Gersberg, R., 2010. Municipal solid waste management in China: Status, problems and challenges. Journal of Environmental Management 91, 1623–1633. 10.1016/j.jenvman.2010.03.012.

Author's address(es)

[a]Institute of Water and Environment, Al Azhar University –Gaza, P.O. Box 1277. Jamal Abdel Naser Street, Gaza, Palestine. Tel.: +972-594-390904; fax: +972-8-213-1506. E-mail address: Dr.samir.alnahhal@hotmail.com
[b]Environment and Earth Science, Fuculty of Science, Islamic University of Gaza, P.O. Box 108. Jamal Abdel Naser Street, Gaza, Palestine. Tel.: +972-599-465665;. E-mail address: safifi@iugaza.edu.ps

Waste Prevention and Recycling Strategy for Tbilisi City, Georgia

Virginie Herbst, Dr. Ludwig Streff, Dr. George Tavoularis

ICP Ingenieurgesellschaft Prof. Czurda und Partner mbH, Karlsruhe, Germany and ENVIROPLAN S.A. Consultants & Engineers, Athens, Greece

Abstract

In order to support the Municipality of Tbilisi City (municipal solid waste generation of approx. 416,000 t/y in 2019) to have its own strategy on prevention, separate collection, recycling and recovery activities, the European Bank for Reconstruction and Development has funded the so called Tbilisi Municipal Solid Waste Strategy Project. The project started in January 2020 with a duration of 20 months.

The aim of the project is to develop a waste prevention and recycling strategy, to propose an action plan and to support the City with its implementation.

In this context, a waste analysis has been carried out, scenarios for separate collection (1 to 6 bins), as well as for complete solid waste management system (collection, treatment, landfilling) have been evaluated.

Keywords

Prevention, recycling, separate collection, MBT, MRF, sorting, circular economy, landfill.

1 Introduction

The country of Georgia is located at the intersection of Eastern Europe and Western Asia. It is a part of the Caucasus region, bounded to the west by the Black Sea. The actual population is about 4.01 M. Tbilisi, the capital city, is a self-governing city in Georgia and represents 30% of the country's population (1.17 M. inhabitants in 2019).

In the current decade, Georgia and Tbilisi City in particular have managed to improve the solid waste management sector. However, this remains one of the greatest environmental challenges. In practice, the majority of management of waste all over Georgia still relies mainly on landfilling.

The necessity of developing solid waste management plans and strategies for Georgia's municipalities in general and for its Capital City in particular is derived from the Association Agreement signed by Georgia and the European Union on the 27th of June 2014. By entering this Agreement, the country has invested in a strong partnership aimed at deepening the political and economic relations with its largest trade partner.

In order to support Tbilisi City Municipality, that generates approximately 416,000 t/y of municipal solid waste (MSW) in 2019, to have its own strategy on prevention, separate collection, recycling and recovery activities, the European Bank for Reconstruction and Development has funded the assignment called Tbilisi Municipal Solid Waste Strategy. The project started in January 2020 with a duration of 20 months.

The objective is achieved through four tasks, namely to: (1) prepare a baseline study, (2) develop the waste prevention and recycling strategy, (3) analyse costs and revenues, (4) propose an action plan and support the City with its implementation.

In this context, a waste analysis has been carried out, scenarios for separate collection with 1 to 6 bins, as well as for complete solid waste management system (SWM) system (collection, treatment, landfilling) have been evaluated and will be described in this abstract [ICP-ENVIROPLAN NO.1, 2020].

2 Basic data

During the baseline study [ICP-ENVIROPLAN NO.2, 2020], an updated description of the current waste management system in Tbilisi City (including future projections) has been carried out to identify suitable waste prevention and waste separation / recycling activities for the Municipality.

The baseline study focussed on the current MSW management system (timeline 2019-2031) including all relevant aspects, problems, and gaps for:

- Legal, administrative and institutional issues,
- Organizational issues,
- Technical issues (from collection and transport of different waste streams, till treatment (if existent) and final disposal) and
- Financial issues (including tariff system, system revenues and expenditures).

The main figures of the baseline study are given in the table below.

Table 1 Basic data of Tbilisi City

Category	Basic data
Population in 2019	Total: - 1,171,100 inhabitants (30% of Georgia's population) - Urban: 1,140,700 inhabitants - Rural: 30,400 inhabitants Growth per year: - Urban: + 1.24% - Rural: +0.26%

<table>
<tr>
<td>Waste quantities and streams in 2019</td>
<td>Total MSW:
- Generated: 415,889 t/y
- Collected: 411,730 t/y (99%)
Waste generation index of MSW:
- Urban: 0.99 kg/inh/d
- Rural: 0.40 kg/inh/d
Waste stream shares in total MSW collected – estimates:
- Household waste ≈ 62%
- Commercial waste ≈ 30%
- Street sweeping ≈ 4%
- Green waste ≈ 1%
- Others ≈ 3%
Construction and Demolition Waste (CDW) collected: ≈ 111,560 t/y
Hazardous Waste (HW) estimates: ≈ 45,000 t/y</td>
</tr>
<tr>
<td>Waste composition in 2020 [ICP-ENVIROPLAN NO.1, 2020]</td>
<td>Waste composition carried out in October 2020:
<table>
<tr><td>Biowaste from households and commerce</td><td>45.70%</td></tr>
<tr><td>Garden and park waste</td><td>1.50%</td></tr>
<tr><td>Paper and cardboard</td><td>14.80%</td></tr>
<tr><td>Plastic</td><td>15.30%</td></tr>
<tr><td>Ferrous metals</td><td>1.10%</td></tr>
<tr><td>Non-ferrous metals</td><td>0.30%</td></tr>
<tr><td>Glass</td><td>2.60%</td></tr>
<tr><td>Wood</td><td>1.70%</td></tr>
<tr><td>Textile and leather</td><td>2.20%</td></tr>
<tr><td>WEEE</td><td>1.00%</td></tr>
<tr><td>CDW</td><td>1.00%</td></tr>
<tr><td>Rest</td><td>12.80%</td></tr>
<tr><td>TOTAL</td><td>100.00%</td></tr>
</table>
</td>
</tr>
</table>

After analysing the available data on the current MSW management system of Tbilisi City, the following can be stated:

- Household and commercial waste are collected mixed together;
- Share of commercial waste inside MSW is not known;
- Share of recycling materials represents approx. 38%;
- No separate collection of hazardous and bulky waste;
- Green waste from public parks and garden is collected separately;
- CDW is collected and landfilled separately; no further treatment or recycling;
- Collection frequency is not in line with the available container volume;

- Tariff collection is more than 90% and is collected via the electricity bill;
- Actual fee is approx. 0,63 EUR per person per month, max. 2,5 EUR per household;
- Fee is independent from waste production;
- Waste management services in Tbilisi are well organised. The waste collection rate as well as the fee collection rate are impressively high.

3 Targets and quantification

The national targets set by the Ministry of Environmental Protection and Agriculture of Georgia till 2030 are based on the EU Action Plan for the Circular Economy [EU CIRCULAR ECONOMY, 2015]. Therefore, the targets recommended for Tbilisi City follow the national targets which have been set sometimes even higher than in Europe, especially for recycling (EU reduction target for recycling is set by 65% of municipal waste by 2035).

The table below gives the targets recommended for Tbilisi City, as well as an overview of the waste quantities that represent the waste reduction targets to be achieved. The quantification is based on the data gathered in the baseline study and described in the previous chapter 2.3.

Table 2 Targets and their quantification for Tbilisi City (status March 2021)

Category	Targets	Quantification
Prevention	At least 2 prevention initiatives	-
Collection	Collection of MSW: - 100% by 2025 - 100% by 2030 Collection of hazardous waste (HW): - 75% by 2025 - 100% by 2030	MSW: 490,024 tons by 2030 HW: 55,000 tons by 2030
Landfill / Biodegradable waste	Reduce biodegradable waste going to landfill to a maximum of 35% by 2040	Max. 120,174 tons of biodegradable waste can be disposed by 2040
HW	80% of HW should be treated by 2025 [EMPRESS, 2017]	44,000 tons by 2030
CDW	50% of CDW should be recycled or re-used by 2025 [EMPRESS, 2017]	67,500 tons by 2030

Recycling targets	No general targets	-
Paper and cardboard	80% by 2030	58,000 tons by 2030
Glass	80% by 2030	10,200 tons by 2030
Metals	90% by 2030	6,200 tons by 2030
Plastic	80% by 2030	60,000 tons by 2030
EPR targets	The respective regulation is under development	-

4 Waste prevention

4.1 Priority waste streams

Four priority waste streams have been selected for the prevention strategy of the City.

4.1.1 Food waste

In the EU, almost 20% of the total food produced is wasted. The issue of food waste (FW) is from social and environmental perspective no longer acceptable. Reducing food waste has enormous potential for reducing the resources we use to produce the food we eat. Being more efficient will save food for human consumption, save money and lower the environmental impact of food production and consumption. Especially one of the main objectives is to save nutritious food for redistribution for those in need, helping to eradicate hunger and malnutrition, as some 43 M people in the EU cannot afford a quality meal every second day.

Therefore, food waste is selected for Tbilisi to be a forerunner for whole Georgia. In the waste composition of Tbilisi presented in the chapter 2.3, biodegradable waste from households and commerce represents a share of 47.2% of which almost all is edible or inedible (peelings, fruit core, etc.) food waste. It is reminded that both categories are included under the new definition. If 45% is taken as FW (the rest being other organic material), generation of food waste in Tbilisi is estimated to 187,500 t/y.

4.1.2 Paper waste

Paper recycling for paper fibre production is also beneficial by replacing virgin wood fibre and by savings in water and energy consumption. Paper waste is a material that can be collected and recycled easily and so far very high yields were achieved in Europe. However, if becoming wet in contact with kitchen waste it is immediately downgraded and mostly undesirable by the paper industry.

Paper is a stream that can be targeted by simple measures that are detailed in the chapter 4.2.

Considering the waste composition indicated in the chapter 2.3, generation of paper waste is about 4.8% (cardboard 10%) and represents approximately 20,000 t/y.

4.1.3 Packaging waste

Packaging waste has long been a target stream for the regulatory authorities. Currently, ambitious targets for recycling are set, which are dealt with via Extended Producers Responsibility (EPR) schemes and to a less degree by Deposit Refund Systems (DRS). Even though there are different systems for separate collection applied throughout EU, many municipalities currently provide for separate collection of paper, glass, metal and plastic rather than commingled collection of all recyclables. Source separation is a critical precondition for generating high-quality secondary raw material from waste.

In any case, prevention and reuse of packaging remain a big challenge. Because of its importance, it is selected for Tbilisi's prevention strategy. Packaging waste in Tbilisi represents a share of approximately 30%, means 125,000 t/y.

4.1.4 Electrical and electronic waste

Usable and functioning equipment is frequently found in the waste bin, either because of consumption habits – replacement of mobile phones when a new model appears - or because repairing is difficult or not available.

Waste of electrical and electronic equipment (WEEE) such as computers, TV-sets, fridges and cell phones is one the fastest growing waste streams in the EU. In the European Waste Catalogue, various EEE items are marked with an asterisk indicating hazard. In this respect, an initial depollution step is required during treatment of discarded electrical and electronic equipment. EN Standards are active that address the collection, transport and treatment under the extended scope of EU's WEEE Directive.

In Europe, re-use centers are becoming popular and function as repair and second-hand shops. In case it is unrepairable, some spare parts may be further utilised. In addition, EU legislation is restricting the use of hazardous substances in EEE, requiring heavy metals, to be substituted by safer alternatives.

It is not known whether the import of EEE to Georgia complies with EU or other similar hazardous restriction standards. The share of WEEE in Tbilisi is estimated at approx. 1%, means 4,200 t/y.

4.2 Waste prevention activities

The table below describes the waste prevention activities recommended for Tbilisi City.

Table 3 Waste prevention activities recommended for Tbilisi City

Waste streams / Prevention activities	Responsible	Short-term	Medium-term	Long-term
Food waste				
1) Support local NGOs and charities to systemise food re-distribution	Tbilisi City	✓		
2) Pilot project to establish a “social cuisine” operated with donated food from restaurants	Tbilisi City	✓	✓	
3) Self-monitoring on food losses and reporting	Business sector	✓		
4) Commit to food waste reduction and include in the organisation’s vision statement	Business sector	✓		
5) Affiliate with charities	Business sector	✓	✓	
6) Organise staff training on food hygiene/safety	Business sector	✓		
Paper waste				
1) Transition to electronic governance and promotion of other paper-less applications	Tbilisi City		✓	
2) Collection of schoolbooks and other material (pens, etc.) for reuse	Tbilisi City	✓		
3) Adoption of Environmental Management Systems	Business sector	✓	✓	
4) Promotion of EU-Ecolabel products	Business sector & Tbilisi City	✓	✓	
5) Adoption of best practices in the offices	Business sector	✓	✓	
Packaging waste				
1) Gradual ban of plastic bags, replacement with textile or bio-based bags and introduction of a fixed tax	Supermarkets	✓	✓	
2) Voluntary agreements to replace single-use plastics with other type of material	Food service sector	✓	✓	
3) Promote reusable tertiary containers and euro-pallets for business and industry	Commercial sector	✓		
Electrical and electronic waste				
1) Reuse and/or repair of WEEE and establishment of one or two Green Points	Tbilisi City	✓		
2) Donation of used mobile phones	Tbilisi City	✓		

Horizontal measures				
1) Promotion and adoption of green criteria in municipal tenders	Tbilisi City		✓	
2) Pilot program for the implementation of Pay-As-You-Throw scheme	Tbilisi City		✓	✓
3) Participation on EU platforms and forums	Tbilisi City	✓		
4) Provision of information on waste prevention	Tbilisi City	✓		
5) Elaboration of Public Awareness campaigns	Tbilisi City	✓		

4.3 Potential impacts of waste prevention activities

The impacts of waste prevention activities can be revealed only on a medium – term and on the condition that other activities are also implemented, such as separate collection. In this way, a clear message is conveyed to Tbilisi citizens that waste management are tackled seriously. Therefore, no significant diversion from landfilling can be expected before at least 3 years.

During the first transitional period, the impact will be educational for the need to gradually change behaviour. After this period and provided that all actions are completed successfully, a 2-3% total reduction by weight is estimated. Reduction in volume terms is more pronounced, namely 3-4%.

Positive impacts can be anticipated in 2025, as follows:

- Activities for food waste abatement must be intensified and reach at least 10%, as a contribution to the UN's target. In the long term, FW must be further decreased with additional measures taken by the Government of Georgia, such as legislation modernization, tax deduction or exemption when donating food, use as animal feed, etc. FW reduction can thus reach 10% means 21.000 t by 2025;
- Through economic instruments, behaviour change and reuse activities in households and business sector, reduction in packaging and paper waste can reach 3% in 2025, namely 4.000 t of packaging waste and 2.000 t of paper waste;
- Even though EEE is used extensively in Georgia until almost break down, it can be anticipated that 10% of WEEE can be refurbished in small workshops and rerouted from landfills, namely 460 t by 2025.

While some of the measures described can be very easily implemented, others require a more careful planning and monitoring. However, they have a significant potential and can deliver desired results on a long-term. Such measures will influence behaviour change and therefore have beneficial effects to more than one waste streams.

5 Waste Management System

This chapter describes different suitable scenarios of a fully integrated solid waste management (SWM) system for each main component:

- Collection (1 bin, 2 bins or 3 bins system)
- Treatment (mechanical treatment and/or biological treatment)
- Disposal (of stabilized or non-stabilized waste)

The main scenarios are:

0. Scenario 0 – Current situation (1 bin system without treatment)
1. Scenario 1 – 1 bin system (mixed MSW) with treatment
2. Scenario 2 – 2 bins system (recyclables + residual) with treatment
3. Scenario 3 – 3 bins system (recyclables + residual + organic) with treatment

Sub-scenarios for each main scenario are:

a. Sorting
b. MBT with sorting and stabilization
c. MBT with sorting, bio-drying and RDF production
d. MBT with sorting, biogas production and stabilization

An overview of the various scenarios is given in the table below:

Table 4 Overview of scenarios for collection, treatment and landfilling for Tbilisi City

Scenarios	Collection			Treatment				Landfill	
	1 bin	2 bins	3 bins	Sorting/ MRF	Aerobic	Anaerobic (biogas)	Bio-drying	Non-stabilized	Stabilized
S 0	✓							✓	
S 1a	✓			✓				✓	
S 1b	✓			✓	✓[1]				✓
S 1c	✓			✓			✓[2]		✓
S 1d	✓			✓	✓[1]	✓			✓
S 2a		✓		✓				✓	
S 2b		✓		✓	✓[1]				✓
S 2c		✓		✓			✓[2]		✓
S 2d		✓		✓	✓[1]	✓			✓
S 3a			✓	✓	✓[3]			✓	
S 3b			✓	✓	✓[1][3]				✓
S 3c			✓	✓	✓[3]		✓[2]		✓
S 3d			✓	✓	✓[1][3]	✓			✓

[1] Aerobic biological treatment for stabilization of residual waste before landfilling
[2] Aerobic biological treatment for RDF production through bio-drying system
[3] Aerobic biological treatment for composting of separate collected biowaste

The recycling potential (material and energetic) as well as the waste reduction potential of waste to be landfilled has been assessed for each scenario and compared on basis on the projected total waste amount for the year 2025 in Tbilisi City, namely 460,961 t/y. The results are shown in the table below:

Table 5 Comparison of scenarios for collection, treatment and landfilling in %

Scenarios	Recyclables	RDF	Compost	Landfill
S 0	0%	0%	0%	100%
S 1a	18%	0%	0%	82%
S 1b	20%	0%	0%	47%
S 1c	20%	41%	0%	31%
S 1d	20%	0%	0%	37%
S 2a	28%	0%	0%	72%
S 2b	29%	0%	0%	42%
S 2c	29%	36%	0%	27%
S 2d	29%	0%	0%	34%
S 3a	28%	0%	17%	44%
S 3b	28%	0%	17%	26%
S 3c	31%	21%	17%	17%
S 3d	30%	0%	17%	21%

In addition, Scenario 2 and 3 have been combined. As it is estimated that about 10% of waste produced in Tbilisi is coming from zones with single houses and garden – suitable for separate collection of biodegradable waste in biowaste bin - the share of all waste generated in Tbilisi has been distributed as follows:

- Scenario 2: 90% with 2 bins system in all zones except zones with single houses and garden;
- Scenario 3: 10% with 3 bins system (in addition to 2 bins system an additional bin for separate biowaste collection) in all zones with single houses and garden.

The results are shown in the table and figure below:

Table 6 Comparison of the combination of Scenario 2 (90%) and Scenario 3 (10%)

Scenarios	Recyclables	RDF	Compost	Landfill
S 2a+3a	28%	0%	2%	69%
S 2b+3b	29%	0%	2%	41%
S 2c+3c	30%	35%	2%	26%
S 2d+3d	29%	0%	2%	33%

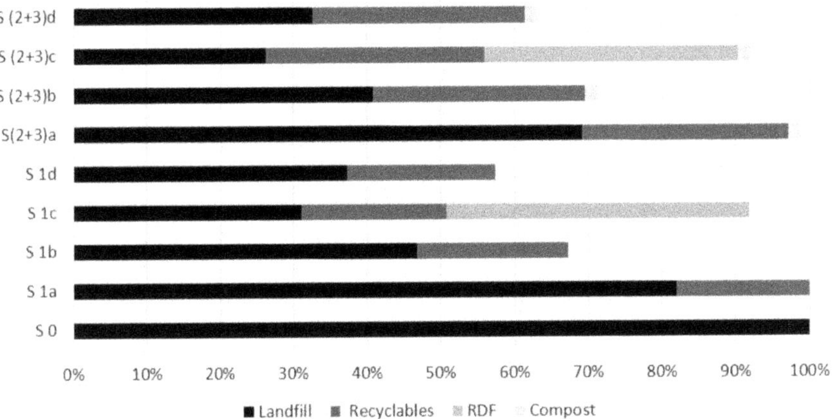

Figure 1 Comparison of the combination of Scenario 2 (90%) and Scenario 3 (10%)

After comparison of all scenarios, it can be concluded that:

- In addition to separate collection schemes the integration of suitable waste treatment components (e.g. sorting plant, MBT, MRF, composting, biogas plant) positively increases the impact on recycling and waste reduction. However, to achieve the challenging recycling targets set in Georgia, it is necessary to implement the 2 bins system in order to achieve approx. 30% of material recovery.
- Although the scenario 3 with 3 bins system (especially sub-scenarios 3c and 3d) shows highest efficiency for material/energetic recycling and waste reduction, it should be stated that a third bin for separate organic waste collection – in the most cases due to not sufficient additional surface area for additional containers – only should be implemented where sufficient surface area is given and where access to the containers is directly connected to single households.
- Due to local given frame conditions (especially limited surface area for additional containers) it is recommended to implement one of the 2 bins scenarios (recyclables and residual waste - scenarios 2a – 2d) in the whole city together with:
  - proposed measures for separate collection:
    - establishment of recycling yards;
    - separate collection of bulky waste;
    - separate collection of garden/park waste;
  - low equipment needs for collection (due to existing overcapacities);
  - a waste treatment facility for dry and wet waste (recyclables and residues).
- It is recommended to implement the 3 bins system (+ 1 bin for organic waste) stepwise in zones with single houses (about 10% of the population) together or after implementation of the 2 bins system chosen;

- If energy production of waste through biogas is not an option for Tbilisi City, the most advantageous scenarios are scenario 2c and the combination of 2c+3c (separate waste collection + MRF + MBT with Bio-drying + landfill + composting) with the highest recycling and waste quantity reduction going to landfill;
- If energy production of waste through biogas is an option for Tbilisi City, the most advantageous scenarios are scenario 2d and the combination of 2d+3d (separate waste collection + MRF + MBT with Biogas + landfill + composting) with a high recycling and waste quantity reduction going to landfill.

6 Summary and conclusion

In the current decade, Georgia and Tbilisi City in particular (1.17 M. inhabitants in 2019) have managed to improve the solid waste management sector. However, this remains one of the greatest environmental challenges. In practice, the majority of management of waste all over Georgia still relies mainly on landfilling.

In order to support the Municipality of Tbilisi City, that generates approximately 416,000 t/y of municipal solid waste (2019), to have its own strategy on prevention, separate collection, recycling and recovery activities, the European Bank for Reconstruction and Development has funded the assignment called Tbilisi Municipal Solid Waste Strategy. The project started in January 2020 with a duration of 20 months.

The objective is achieved through four tasks, namely to: (1) prepare a baseline study, (2) develop the waste prevention and recycling strategy, (3) analyse costs and revenues, (4) propose an action plan and support the City with its implementation.

In this context, a waste analysis has been carried out, scenarios for separate collection with 1 to 6 bins, as well as for complete SWM system (collection, treatment, landfilling) have been evaluated.

The results show that, in addition to separate collection, the integration of suitable waste treatment components positively increases the impact on recycling and waste reduction. Depending on the scenario chosen, 30% of input waste could be recycled, 40% used for energy recovery and the landfilled waste quantities decreased up to 80%. Thus, waste prevention activities could lead to a 3% reduction of the generated waste weight.

Due to local given frame conditions, it is recommended to:

- introduce a 2 bins system (recyclables and residual waste) in the whole city and a 3 bins system (+ 1 bin for organic waste) stepwise in zones with single houses,
- establish recycling yards,
- implement separate collection for bulky and green waste,
- build waste treatment facilities for material recovery and residual waste, and
- implement waste prevention activities.

7 Literature

Empress	2017	Tbilisi Green City Action Plan, September 2017, available at: http://www.tbilisi.gov.ge/page/green-city?lang=en
EU Circular Economy	2015	EU Action Plan for the Circular Economy, 2015, available at https://ec.europa.eu/environment/circu-lar-economy/first_circular_economy_action_plan.html
ICP-Enviroplan no.1	2020	Waste Prevention and Recycling Strategy for Tbilisi City, December 2020, part of the Tbilisi Solid Waste Project – Tbilisi Municipal Solid Waste Strategy
ICP-Enviroplan no.2	2020	Inception Report - Baseline Study and Future Projection, June 2020, part of the Tbilisi Solid Waste Project – Tbilisi Municipal Solid Waste Strategy

Author's address

M. Eng. Virginie Herbst
ICP Ingenieurgesellschaft Prof. Czurda & Partner mbH
Auf der Breit 11
D-76227 Karlsruhe
Telefon +49 721 944 77-33
E-Mail herbst@icp-ing.de

New Orientation of the Waste Management Strategy in Durban, South Africa as Contribution to the City's Ambition to Achieve Carbon Neutrality

Wolfgang Pfaff-Simoneit

WPS Consult UG (LLC)

Logan Moodley

DSW Cleansing and Solid Waste - eThekwini Municipality

Federico di Penta

C40 Cities – Climate Leadership Group

Abstract

The eThekwini Municipality (City of Durban) wants to make use of the great potentials of modern waste management to create additional, meaningful jobs, to contribute to the development of a green economy, to reduce greenhouse gas emissions and to save resources. Ten potential scenarios for the future solid waste management strategy were defined, taking into consideration the City's current financial and operational setting. Large-scale, centralised solutions such as MRF, incineration or RDF generation / stabilisation of waste do not appear appropriate, as they are much more expensive, generate less jobs and do not provide any flexibility for the adaptation of the concept in the future ("lock-in effect" of large waste infrastructure). The smaller-scale approaches of separate collection and recycling are considerably cheaper, have a high potential for creating additional, meaningful jobs and improve the climate balance to a very large extent.

The study was intended as a tool to support informed decision-making rather than as a detailed feasibility study. It is based on existing information; a time-consuming and costly comprehensive data collection was deliberately omitted. Although the results may be subject to a certain degree of uncertainty, this rapid assessment provides a sufficiently accurate basis for deciding on the city's future waste management strategy.

Keywords

Solid waste management strategy, Rapid assessment, Job creation, Greenhouse gas mitigation, Cost, Cost recovery, Source segregation, separate collection and comprehensive utilisation of waste

1 Background and Objectives

The eThekwini Municipality (City of Durban) is presently in the process of developing several overall municipal planning documents, such as a Climate Action Plan to achieve carbon neutrality and a new Integrated Waste Management Plan, which will formalise the City's overall ambitions with regards to waste management, at least in the short and medium term. Multiple options for new waste treatment infrastructure are currently being considered, including composting, anaerobic digestion, materials recovery facilities, incineration etc.

Although the city already represents a best practice of waste management in Africa, operates an efficient collection, transport and disposal service and has assured disposal capacities for several decades to come, it needs to further improve its waste management system. Changing framework conditions and overarching objectives require an increasing diversion of waste from landfill through intensified recovery of waste and recycling. The city wants to make use of the great potentials of modern waste management to create additional, meaningful jobs, to contribute to the development of a green economy, to reduce greenhouse gas emissions and to save resources.

The German consulting firm 'WPS Consult UG' has been assigned to carry out a rapid assessment of the current waste management system and to develop and analyse possible options for the future waste management strategy. The consultancy was sponsored by C40 Cities, with funding from the Climate and Clean Air Coalition and Citi Foundation. C40 connects city practitioners and Mayors to enable stronger collective climate action. C40 supports cities to collaborate effectively, share knowledge and drive meaningful, measurable and sustainable action on climate change.

In order to define a potential "way forward" for waste management in the City, a number of potential scenarios were developed, taking into consideration the City's current financial and operational setting:

- **Status-quo:** The current system was analysed in terms of costs, jobs, GHG emissions as a reference with other possible scenarios
- **Scenario 1: Improvement of cost efficiency – no additional infrastructure**
  - a) Direct transport to landfill with C200
  - b) 2 shift collection
- **Scenario 2: Increased source segregation of dry recyclables**
  - a) Recycling
  - b) Recycling + RDF-production
- **Scenario 3: Material recovery facility**
  - a) Recyclables, RDF and cover material
  - b) Recyclables, RDF, digestion + cover material
- **Scenario 4: Energy-from-waste strategy**
  - a) Incineration
  - b) RDF-production

- **Scenario 5: Source segregation and comprehensive utilisation of waste***
  - a) Recycling + Composting
  - b) Recycling + Digestion

* Scenario 5 is based on a labour-intensive door-to-door collection supported by immediate post-sorting for quality assurance of recovered materials. Besides recyclable materials, organic waste shall be collected separately and recycled through composting and/or digestion. A similar approach was piloted in Vietnam, which is reported in these proceedings by the same author. The theoretical foundations of this approach are described in this paper.

For all scenarios, the financial impact they would have (i.e. they would cost "more" or "less" than the current model), the impact they would have on recycling / composting waste prevention, the number of jobs each scenario would create and the impact on greenhouse gas emissions were calculated. Finally, possibilities and instruments for covering costs besides user fees were considered. The aim of this study was to determine, at a high level and based on available information, the financial, labour and environmental implications of different scenarios, or 'directions', the city could potentially pursue. The study is intended as a tool to support informed decision making, rather than as a detailed feasibility study.

The assessment was based on available data and information provided by the City. Missing or incomplete data were supplemented by estimates and assumptions based on experience and plausibility criteria. Therefore, due to these requirements, the calculations conducted may be subject to variations in excess of 10% as more detailed calculations are performed. They are based on international experience, which has been adapted to the prevailing conditions in South Africa, especially with regard to construction and operating costs.

2 Basic Data

Population: 3,600,000 people / 1,000,000 households,

thereof living in	Urban areas:	25%
	Peri-urban areas:	30%
	Rural areas:	45%

Informal settlements: 12%

Landfilled waste:	Total:	1,341,500 t/a	5,160 t/d
Collected municipal waste:		644,800 t/a	2,480 t/d
DSW waste		569,100 t/a	2,190 t/d
Garden refuse:		54,000 t/a	210 t/d
Light type refuse		1,400 t/a	10 t/d
Recyclables		18,200 t/a	70 t/d
Informal sector		20,800 t/a	80 t/d

Table 1 Municipal waste composition and properties

Fraction	Assumed composition	Waste characteristics	
Fine material	6%	Bio-degradable material	53%
Kitchen + garden waste	52%	Inert material	9%
Nappies	4%	Recyclables	27%
Paper, cardboard	8%	Others	11%
Glass	2%	**Total**	100%
Metals	2%	**Thermal properties**	
Plastics	10%	Calorific value [MJ/kg]	7,3
Compound packaging	4%	Combustibles	48%
Textiles	1%	Ash content	16%
Timber, leather, rubber	2%	Water content	36%
Inert materials*	4%	**Total**	100%
Electric / electronic scrap	3%	**GHG - parameters**	
Other materials	3%	Carbon content [g/kg]	237
Total	**100%**	Fossil carbon [g/kg]	90
		Organic carbon [g/kg]	147

* Stones, sand, gravel, street-sweepings, porcelain, tiles and the like

3 Financial Impact and Cost Recovery

Figures 1 and 2 summarise the financial impact of each scenario. The cost figures take into account full costs, i.e. capital and operating costs (CAPEX + OPEX), the expected average revenues from sales of recyclables, energy and / or RDF (Refuse Derived Fuels) and the resulting cost, meaning costs minus revenues. Revenues from carbon credits are separately presented in vertically shaded bars.

Whereas the cost of scenarios 2, 3 and 5 are more or less in the same range as the present system, scenarios 3 – Material Recovery Facility - and 4 – Energy from waste - are associated with considerably higher cost, which reach up to a doubling of the present costs.

Multiple opportunities to recover costs are discussed, from grant financing over EPR, feed-in tariffs for electricity, carbon credits and other instruments. It turns out that all scenarios require additional funding through user charges, in no case can the cost be covered by revenues from other sources alone. However, all scenarios could be financed in eThekwini, provided a variable increase of fees - which would from an objective point of view still be still affordable - is deemed to be politically feasible.

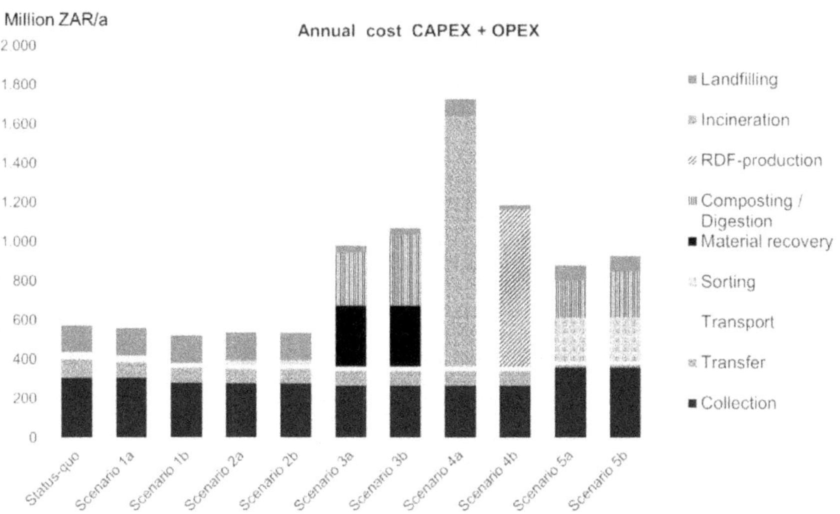

Figure 1 Cost and cost structure of the assessed scenarios

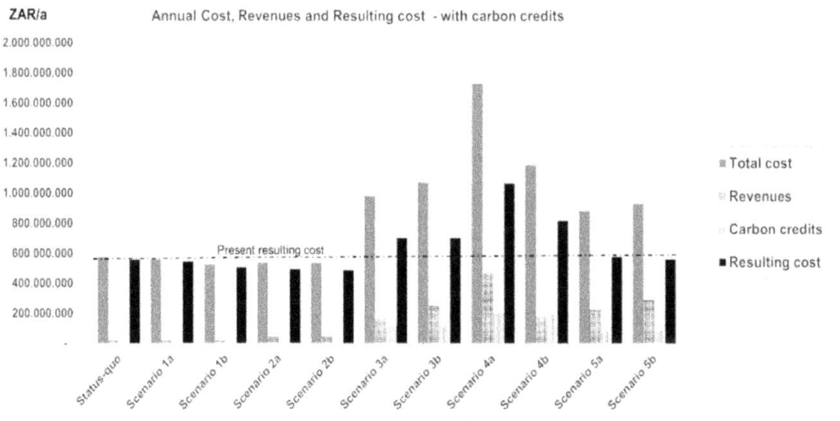

Figure 2 Annual cost, revenues and resulting cost

The financial impact of each scenario is considered within a lower and upper boundary. The lower boundary considers al possible financial instruments to cover additional costs (e.g. including potential revenue from EPR, carbon credits, saved CAPEX through grant financing of investments, provision of disposal site for new facilities free of charge etc.,), while the upper boundary only takes into consideration revenues and proceeds from recycling and feed-in tariffs for electricity generated.

Table 2 shows the impact on user charges in a traffic light system. A light grey dot signals no change of the user charges would be required, medium grey dot means a slight increase of up to 15%, q dark grey dot means a sharp increase of more than 15% would be necessary.

Table 2 Required increase of user charges in the different scenarios

Range of user charges	Scenario									
	1a	**1b**	**2a**	**2b**	**3a**	**3b**	**4a**	**4b**	**5a**	**5b**
Lower end	○	○	○	○	○	○	●	◍	○	○
Upper end	○	○	○	○	●	●	●	●	◍	◍

Key: ○ No increase ◍ slight increase (< 15%) ● sharp increase

Findings

- Scenarios 1 and 2 can be implemented without any impact on the user charges.
- Scenarios 3 – MRF – could only be implemented without impact on user charges if in particular grant investment financing would be available. Without such a cost recovery contribution, these scenarios would require a sharp increase of user charges.
- Scenario 4a – incineration - would require a sharp fee increase in any case.
- For Scenario 4b – RDF production - the same statement applies as for an MRF – fees increase would only be moderate only if grant investment financing could be used.
- Scenarios 5 would require no increase of user charges, if alone carbon finance were available – an option which is becoming very probable in the near future. Without such a cost recovery contribution, Scenario 5 would still only require a slight increase of fees. However, this seems feasible given the enhanced service envisaged in Scenario 5.

A key opportunity to increase fee revenues was identified in the introduction of a minimum fee for all users according to the ‘polluter pays’ principle, with the exception of lowest-income groups. If the internationally recognised affordability limits for waste charges were fully tapped and the current payers' charges were raised accordingly, eThekwini could double its waste fee revenues. However, any increase in fees needs to be gradual and connected to real improvements in the waste management system and needs to be well communicated to the citizens and actually perceived as such by them.

4 Job creation impact

Figure 2 shows the potentials of the scenarios examined for the creation of additional jobs. Scenario 5 offers the greatest opportunities, here the number of people employed in waste management can almost be doubled, compared to current levels. The jobs are created in particular at waste collection and sorting of source separated waste.

Scenario 3 also has considerable employment potential. However, a high proportion of these are jobs in the MRF, which are less pleasant and where employees are exposed to high degrees of odour, dust and noise. In addition, a significant number of jobs are created in the biological part of the process chain.

The employment effects of the other scenarios are rather small. Only scenario 4b, which envisages the production of RDF through waste stabilisation, offers some additional jobs on a significant scale, but predominantly only for skilled workers.

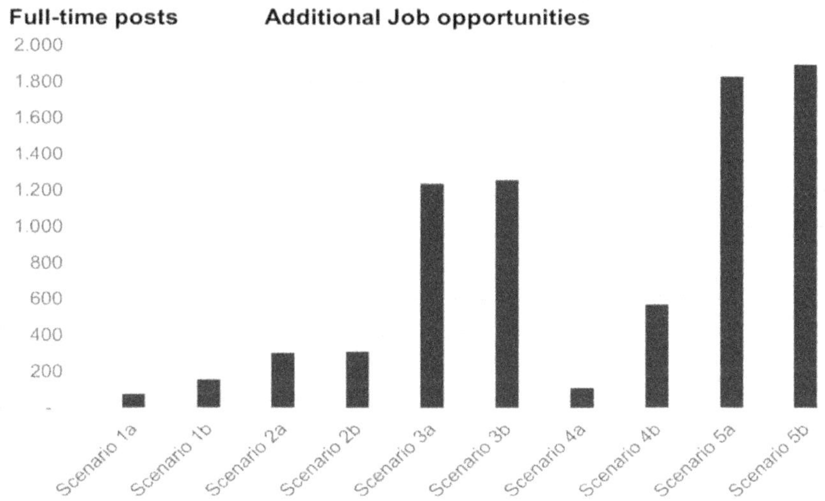

Figure 2 Additional job opportunities of the different scenarios

5 Greenhouse Gas Emissions Impact

The greenhouse gas balances of the scenarios examined show extreme differences. While in the status quo and Scenario 1 net emissions amount to about 500,000 tons of CO2-equ., Scenario 4b would relieve the climate balance by almost 300,000 tons of CO2-equ.

The net greenhouse gas emissions savings in Scenario 5 are the second highest and would achieve an overall reduction of more than 90%, compared to the status quo.

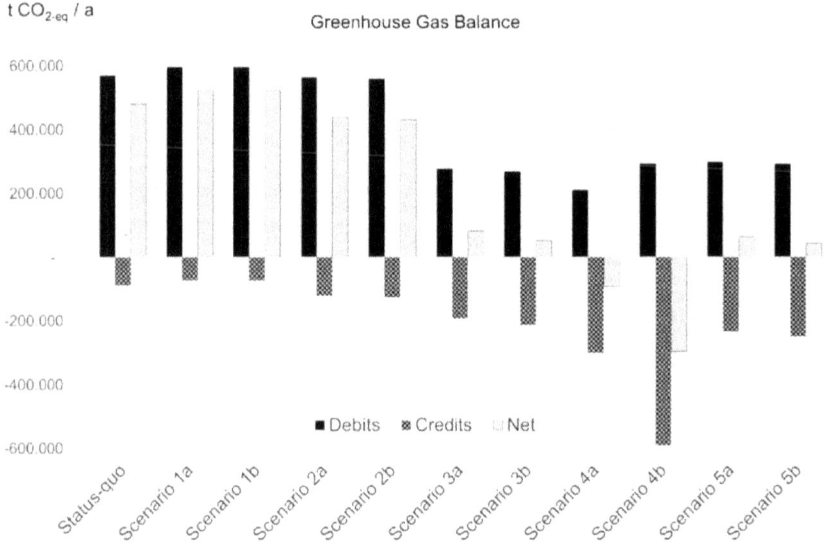

Figure 3 Greenhouse gas balance of the different scenarios

6 Key Findings

6.1 Scenario 1: Improvement of cost efficiency

The examination shows that by directly transporting the waste collected by C200 trucks to the landfill, about 5-10% of the total waste management costs could be saved. However, this requires a more elaborate organisation, as the waste collectors should not go with the collection truck to the landfill.

Another almost 10% of disposal costs could be saved by introducing collection in two shifts. In this way, the waste trucks, whose capital cost account for about 20% of the collection costs, would be used more efficiently.

Due to the somewhat more complex organisation, scenario 1 offers a slightly higher number of jobs.

There is no diversion from landfill since no measures for recycling of waste are assumed in this scenario.

The measures do lead to an increase of 42,000 t CO2-eq greenhouse gas emissions, but this is attributed to the fact that this scenario entails an increase in waste collection in areas which are currently underserviced. Scenarios 1a and 1b do not provide opportunities to generate carbon credits.

6.2 Scenario 2: Increased waste segregation and production of RDF

The examination shows that by increasing the share of separately collected recyclables and re-utilisation, cost savings of around 10% could be achieved.

Provided there is a market for Refuse Derived Fuels (RDF) and cement industry would be ready to pay a fair price for the materials, some slight additional revenues could be realised.

Both variants of Scenario 2 offer job opportunities for around 300 additional jobs for some drivers, waste collectors and in particular for sorting the collected recyclables.

Diversion from landfill is rather low, at around 50,000 tonnes per year, which corresponds to a rate of 6%.

The assumed measures in Scenario 2 would lead to greenhouse gas emission savings of around 40,000 t CO2-ec per year in Variant a) and about 50,000 t CO2-ec per year in Variant b). Nevertheless, Scenarios 2a and 2b do not offer opportunities to generate carbon credits.

6.3 Scenario 3: Material Recovery Facility

The construction and operation of a Material Recovery Facility (MRF) would entail very high costs, which cannot by far be compensated by a higher material yield and increased sale of recyclables. Due to the generally poor material qualities, the revenues are lower than those of single-variety collected recyclables and the sensitivity of the revenues against changing demand is significantly higher.

Provided there is a market for Refuse Derived Fuels (RDF) and cement industry would be ready to pay a fair price for the materials, some slight additional revenues could be realised.

The biogas produced would be sufficient to operate a power plant with 8 MW_{el} capacity.

Both variants of Scenario 3 offer considerable job opportunities, with around 1,200 to 1,300 additional jobs, mostly located in the MRF. However, these are jobs with poor working conditions with potential health and safety problems.

Diversion from landfill is high with about 75%, equivalent to 520,000 tonnes per year.

The construction and operation of a MRF would reduce the greenhouse gas emissions by 83 % corresponding to almost 400,000 t CO2-ec per year in Variant a) and 89 % in Variant b) which equals 410,000 t CO2-ec savings per year.

Scenarios 3a and 3b offer considerable opportunities to generate carbon credits. The additional revenues could cover about 12% of the total cost.

6.4 Scenario 4: Energy from waste

Energy-from-waste strategies generate the by far highest cost both in terms of investment and operational cost. Even if high feed-in tariffs for electricity are assumed, the cost can by no means be compensated.

The resulting costs of RDF production could be even higher than those of incineration, as the cement industry in most countries is not prepared to pay a fair price for the product.

The incineration of waste would allow to operate a power plant with a 50 MW_{el} capacity.

The direct incineration of waste offers limited job creation opportunities, projecting an additional 110 jobs, the RDF production Scenario 4b) instead would generate an additional 570 jobs.

Diversion from landfill is the highest of all Scenarios with 95% in Variant a) and about 85% in Variant b). However, the incineration of waste generates about 130 tons per day of flue gas cleaning residues, which have to be disposed in a hazardous waste landfill, plus 360 tons of bottom ash, which can be disposed of on a sanitary landfill.

Incineration would reduce overall greenhouse gas emissions, as the credits for the substitution of fossil fuels for electricity fed into the grid more than compensates for the direct greenhouse gas emissions of waste incineration.

The greenhouse balance of Variant b) is even better than Variant a). It represents a sink for about 300,000 t CO2-ec per year.

Scenarios 4a and 4b offer considerable opportunities to generate carbon credits. The additional revenues could cover about 12 – 15% of the total cost.

6.5 Scenario 5 – source segregation and comprehensive utilisation of waste

Although the comprehensive recycling of waste generates considerably higher collection costs, these are almost completely offset by higher revenues. The overall cost of this scenario is comparable to current waste management costs, when revenues from carbon credits are also taken into consideration.

The proposed labour-intensive waste collection scheme generates about 3-times higher employment rates than the present system. Almost 1,900 new meaningful jobs could be created, the majority of which would go to more vulnerable and low-income groups.

The already practised CBC-based collection concept in Durban offers favourable opportunities for the implementation of such approaches. People who presently collect recyclables on an informal basis could get a formalised job in waste management, provided the development of adequate waste worker formalisation policies.

It is recommended to carry out pilot projects in order to test the feasibility of such a concept and to obtain experiences and planning data for a later roll-out to other quarters.

Diversion from landfill achieves almost 50% in both Variant a) and Variant b).

The greenhouse gas emissions can be reduced by about 87 % in Variant a) and up to 91 % in Variant b), in particular due to the credits for recycling and the avoidance of uncontrolled methane emissions through composting resp. digestion + composting.

Scenarios 5a and 5b offer considerable opportunities to generate carbon credits. The additional revenues could cover about 10 - 15% of the total cost.

7 Conclusions

Each of the scenarios could realistically be implemented given eThekwini's current financial and operational means. However, the scenarios are differently suited for their application in the city. The rapid assessment of different scenarios clearly indicates that waste management in eThekwini should be further developed towards a circular economy. Large-scale, centralised solutions such as MRF, incineration or RDF generation / stabilisation of waste do not appear to be the most suitable option given the context in eThekwini, as they are much more expensive, generate less jobs and do not provide any flexibility for the adaptation of the concept in the future ("lock-in effect" of large waste infrastructure). The GHG mitigation impact is – with the exception of Scenario 4b (RDF-production) which represents a considerable sink of GHG - only slightly higher.

The smaller-scale approaches of separate collection and recycling, as considered in Scenario 2 and Scenario 5, appear to be much more suitable for the city of eThekwini. They are considerably cheaper, have a high potential for creating additional, meaningful jobs and improve the climate balance to a very large extent. The slightly lower diversion from landfill rate is of less importance in eThekwini, as the city still has considerable landfill air space already secured.

The rapid assessment has proven to be a suitable approach to support informed decision-making. Based on existing information, estimates and assumptions based on experience and plausibility criteria, time-consuming and costly comprehensive data collection can be omitted. Although the findings may be subject to some uncertainty, they provide a sufficiently solid basis for the political deliberations and strategic decisions to be taken. They are sufficiently precise to answer the fundamental question of whether and how eThekwini can make the new waste infrastructure financially viable and what strategic directions the city should take.

8 Literature

Bilitewski, B., Wagner, J., Reichenbach	2009	Best Practice Municipal Waste Management, CD-Rom, Umweltbundesamt (Hrsg.) 2009 https://www.umweltbundesamt.de/themen/abfall-ressourcen/abfallwirtschaft/abfalltechnologietransfer
IFEU	2010	Klimarechner Abfallwirtschaft - Tool for Calculating Greenhouse Gases (GHG) in Solid Waste Management (SWM), https://www.ifeu.de/projekt/klimarechner-abfallwirtschaft/

Pfaff-Simoneit, W	2010	Sectoral Approaches in Solid Waste management to link development and climate change mitigation; ISWA World Congress 2010; http://www.iswa.org
Pfaff-Simoneit, W.	2013	Development of a sectoral approach to establish sustainable SWM systems in developing and emerging countries in light of climate change and increasing shortage of resources' (edited in German language). Doctoral Thesis. ISBN 978-3-86009-203-3
Pfaff-Simoneit, W.	2017	Adapted selective waste collection concepts for developing and emerging countries, Waste-to-Resources 2017
Silpa, K., Yao,L.C., Bhada-Tata, P., van Woerden, F.	2018	World Bank Group (Ed.), What a Waste 2.0: A Global Snapshot of Solid Waste Management to 2050, ISBN 978-1464813290
Wilson, D.C. (Ed.)	2015	Global Waste Management Outlook, United Nations Environment Programme, International Solid Waste Association, ISBN: 978-92-807-3479-9

Author's addresses

Dr.-Ing. Wolfgang Pfaff-Simoneit
WPS Consult UG (LLC)
Tel. +49 (0)6151 9698 185
Mob. +49 (0)160 8369 408
Email: Wolfgang.pfaff-simoneit@t-online.de

MSc. Logan Moodley
DSW Cleansing and Solid Waste - eThekwini Municipality
Deputy Head
Tel: +27 (0)31 322 4575
Cell: +27 (0)83 259 3688
Email: Logan.Moodley2@durban.gov.za

Federico di Penta
C40 Cities – Climate Leadership Group
Programme Manager - Sustainable Waste Systems
Landline: +44 (0)203 525 7535
Mobile: +44 (0)7741 649 142
Email: fdipenta@c40.org

Plastic Wastes as a Resource: Lebanese Experience

Arwa Anouti ElZein

LACECO, Beirut, Lebanon

Abstract

Rapid increase in the plastic wastes generation is observed globally. This is mainly due to economic growth and changing consumption and production patterns as well as rapid urbanization. Thus plastic wastes are being considered a major stream in the solid wastes streams generated worldwide especially that the world's annual consumption of plastic materials has increased 20 times over the last 50 years. Thus more resources are being used to provide for the increased demand.

This paper is trying to provide insight on the plastic wastes being produced from the municipal waste stream generated from Greater Beirut & Mount Lebanon over the time period spanning from 2015-2020. It presents the quantity and the composition of the collected plastic wastes and their recycling potential as an economic opportunity for converting this waste stream to a resource.

Keywords

Beirut & Mount Lebanon, Plastic Waste generation, Circular Economy, Recycling

1 Introduction

Plastic is considered one of the modern innovations with multiple uses such as packaging, electrical and electronic equipment (EEE), construction, agriculture, healthcare and automotive (John N. Hahladakisa, March 2020). Plastic production has increased drastically over the past 50 years and thus plastic wastes became a major stream in solid wastes (IETC, 2009). Efficient management of plastic waste, is crucial for developing sustainable and functional cities (John N. Hahladakisa, March 2020).
The purpose of this paper is to assess the quantity and the composition of the collected plastic wastes and their recycling potential as an economic opportunity for converting this waste stream to a resource during the time frame spanning from 2015 to 2020 within the Beirut & Mount Lebanon Governorates (excluding Jbeil).

1.1 Background Information

During the earlier years, waste generation in Lebanon intensified considerably due to the escalation in the living standards, urbanization, arrival of Syrian refugees as well as population growth (Ismail I. Abbas, 2017).
Lebanon is located Eastern coast of the Mediterranean Sea with a surface area of 10,452Km². Lebanon includes 6 administrative units or Governorates (Mohafazats). These are Beirut, Mount Lebanon, North Lebanon, South Lebanon, Nabatiyeh, and Beqaa. The National Survey of Household Living Conditions conducted by the Central Administration of Statistics, Ministry of Social Affairs and the UNDP (2007) shows that Beirut and Mount Lebanon jointly accomodate around 50% of the population (UN-Habitat, 2011).

The population of Lebanon is estimated to be 4,223,553 (World Bank, 2010). The Governorates of Beirut and Mount Lebanon jointly embrace around 2 million Lebanese in addition to 309,112 Syrian and Palestinian refugees (OCHA, 2018). The Governorates are predominantely urban and peri-urban. However; Mount Lebanon Governorate harbors some rural areas. Beirut is the administrative and economic capital of the country, hosting central Government institutions. Both Governorates extend over an area of 2,031 km2 . Beirut is its own district, while Mount Lebanon is composed of six districts: Jbeil, Kesrwane, el Meten, Baabda, Aley, and Chouf (OCHA, 2018).

Figure 1 Lebanon Governorates(https://creativecommons.org/licenses/by-sa/3.0/deed.en)

According to the Country Report issued in 2014 (Sweep Net, 2014), Beirut & Mount Lebanon produce 52% of the waste stream generated in Lebanon. This is further elaborated in the subsequent representation.

Table 1 - Background Information:Country Profile on Solid waste Management in Lebanon (Sweep Net, 2014)

Population	5.6 million (projected to 2013)
MSW Generation	2.04 million tons (projected for 2013)
Per Capita MSW Generation	1.05 kg/day (weighted average over the country)
Urban Areas	0.95 – 1.2 kg/day
Rural Areas	0.8 kg/day
MSW Generation Growth	1.65%
Medical Waste Gnereation	25,040 tons/year
Industrial Waste	188,850 tons/year

Municipal Solid Waste stream composition in Lebanon shows that the highest waste fraction is the organic component comprising 52% of the MSW stream while recyclables constitute 37% (paper and cardboard 16%, plastics 12% whereas metals constitute 6%, glass 3%) and others 11%.

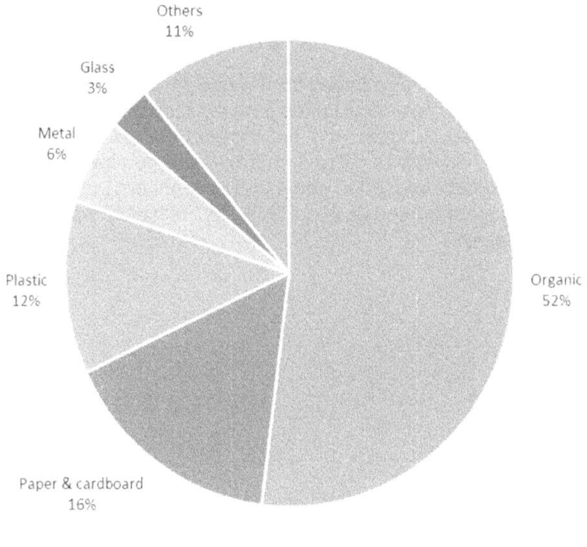

Figure 2 Waste Composition in Lebanon **(Sweep Net, 2014)**

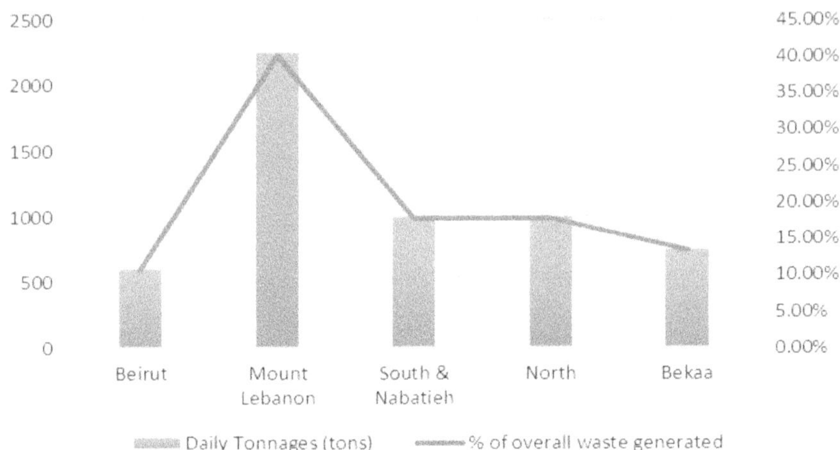

Figure 3 – Waste Distribution by Governorate **(Sweep Net, 2014)**

Scavengers collect all types of recyclables from waste bins or dumpsites and even directly from households. The main recyclables that are scavenged include cardboard, water bottles and other plastics, scrap iron, tin, copper from electric wires, white goods[1] car batteries and old tires as well as Styrofoam trays. The number of working scavengers is estimated between 2,000 and 4,000 scavengers in Lebanon (ILO, 2011). Landfilling is the most common end destination of the municipal wastes streams in Lebanon is represented below.

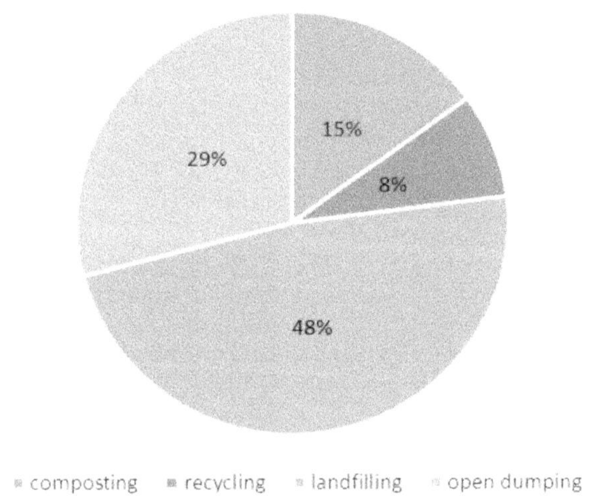

Figure 4 – End destination of municipal wastes at the national level

[1] fridges, washing machines, TV sets, etc

1.2 Current Municipal Solid Waste Management System in Beirut &Mount Lebanon (BML)

In Beirut and Mount Lebanon (BML), commingled municipal wastes are collected from curb side containers and transported to two dirty MRFs for processing. Manual sorting is employed to retrieve recyclables (except for metals that are recovered using magnets and eddy currents) while mechanical separation (using trommel screens) is utilized to retrieve the organic component (Laceco, 2020). Organics are then diverted to the composting facility[2]. Residual waste is then baled and wrapped for landfilling (Ismail I. Abbas, 2017). Bulky items are manually removed and they are picked for recyclables while the discards are landfilled (Laceco, 2020). This is represented in the subsequent figure.

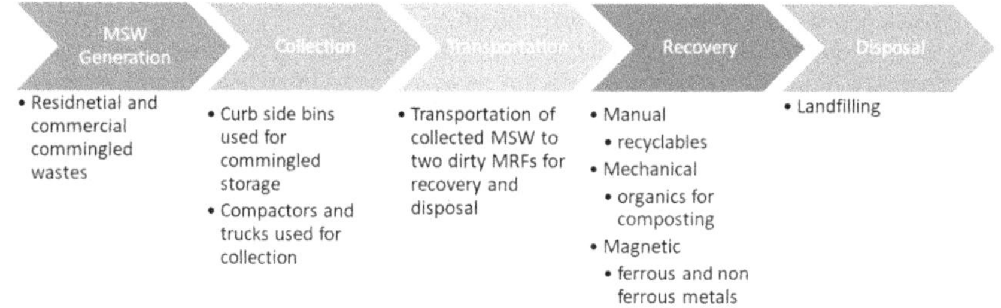

Figure 5 - Schematic representation of waste management & processing in BML

The main processes available in the sorting facilities are:

- Mechanical separation for separation of organic waste
- Manual sorting to recover recyclables
- Magnetic and eddy current separators to extract metals

Informal scavengers in Beirut & Mount Lebanon collect recyclable material from urban centers (ILO, 2011).

2 Materials & Methods

2.1 Approach

The aim of this paper is to analyse the quantity and the composition of the collected plastic wastes and their recycling potential within the Beirut & Mount Lebanon Governorates (excluding Jbeil) during the time frame extending between 2014-2020. Therefore, records of the database for the bulk municipal solid waste (MSW) from the daily unloaded waste collection vehicles received at the two sorting facilities and the quantities of plastic wastes recovered from the commingled waste stream were examined.

[2] One Sorting Facility & the composting facility are damaged due to Beirut Port blast and out of order since 5 August 2020

Also the types and quantities of plastics vended to the recyclables clients were also inspected (IETC, 2009).
Within the framework of this analysis, the entire amount of MSW collected within the BML area over the period of 2014-2020 is considered. This approach can help draw conclusions that relate to the geographic regions under study but not to the national level nor to specific municipalities.
The classification of plastic type considered for this study was according to the Resin Identification Code (RIC) system designed for material recycling and recovery by the Society of the Plastics Industry, namely, (1) Polyethylene terephthalate (PET), (2) High-density polyethylene (HDPE), (3) Polyvinyl chlorine (PVC), (4) Low-density polyethylene (LDPE), (5) Polypropylene (PP), (6) Polystyrene (PS), and (7) OTHER (acrylic, nylon, polycarbonate, poly lactic acid) (Chinnathan Areepraserta, 2016).

2.2 Limitations

The following limitation is associated with the study at hand:
- Scavengers in Beirut & Mount Lebanon collect recyclable material from urban centers (ILO, 2011) thus affecting the composition of the waste stream received at the sorting facility.
- The unavailability of clear data that relates to the informal scavengers sector presents a limitation to this study.

3 Discussion & Results

3.1 Municipal Waste Composition in Beirut & Mount Lebanon (BML)

Data extracted from Ramboll- Laceco study that was conducted in 2012 indicate the following composition of the municipal waste stream entering the sorting facilities as presented below:

Table 2 – Municipal Waste Composition in BML

Waste Category	*(%)*
Food	*53.4*
Glass	*3.4*
Metals	*2*
Papers	*15.6*
Plastics	*13.8*
Textile	*2.8*
Wood	*0.8*
Nappies	*3.6*
Others	*4.6*
Total	*100*

Thus the composition of the waste stream entering the sorting plants within the geographic boundaries of the study is more or less in line with the national composition as presented in figure 2.

3.2 Incoming Waste

The sorting facilities have received incoming quantities of commingled municipal streams as shown in the subsequent tabulation.

Table 3 – Incoming wastes

Date	Waste In (tons)
Year 2014	1,100,747.66
Year 2015	864,777.52
Year 2016	893,745.20
Year 2017	1,062,422.08
Year 2018	1,217,998.52
Year 2019	1,149,928.22
Year 2020	592,696.66

However, the operational years 2015 and 2020 are not indicative as the first witnessed the Lebanese waste crisis and thus witnessed some interruptions. This crisis had extended some of its shadows on the year 2016 in terms of decreased geographic boundaries. Also the year 2020 is not an indicative especially after the Beirut Port Blast that left the waste management plan within the geographic boundaries of the study lacking several of its basic components(one sorting facility and the composting facility extremely damaged beyond the capacity to operate).

3.2.1 Expected Waste Plastic Presence in Incoming Wastes

Based on the BML waste composition as presented in the section (3.1) and the quantities of incoming wastes as presented in the section (3.2), the expected waste plastic in the incoming waste stream is subsequently calculated.

Table 4 – Expected Waste Plastic Quantities in BML

Date	Waste In (tons)	Waste Plastic Capacity (tons)
Year 2014	1,100,747.66	151,903.18
Year 2015	864,777.52	119,339.30
Year 2016	893,745.20	123,336.84
Year 2017	1,062,422.08	146,614.25
Year 2018	1,217,998.52	168,083.80
Year 2019	1,149,928.22	158,690.09
Year 2020	592,696.66	81,792.14

3.3 Plastic Waste Recovery and Recycling

During the years of 2014-2019, the recovered waste plastics recovered from the incoming waste streams are presented in the subsequent tabulations.

Table 5 – Recovered Waste Plastics and in relation to the expected waste plastic quantities.

Date	Recovered waste plastics (tons)	Missed waste plastics recovery opportunity
Year 2014	20,876.94	86.26%
Year 2015	16,576.84	86.11%
Year 2016	9,396.96	92.38%
Year 2017	15,893.24	89.16%
Year 2018	14,879.59	91.15%
Year 2019	16,377.54	89.68%

Table 5 shows that the recovered amounts of plastics are around one tenth the expected waste plastic quantity within the incoming commingled waste stream. This is mainly due to the commingled nature of the incoming waste stream that makes recyclables including plastics in undesirable condition; thus the recycling industries refrain from buying these recyclables (Ismail I. Abbas, 2017). Based on that, the municipal waste

sorting facilities would resort to landfilling the waste plastics in case the recycling industry is not interested to buy acquire them.
However, it is noteworthy to mention that during the years 2014-2020, the quantities of plastic wastes recovered from the total recyclable stream recovered on site range between 20% to 28% as shown in the subsequent tabulation.

Table 6 – Quantities of Recovered Waste Plastics from B&ML

Date	Recovered recyclables (tons)	Recovered waste plastics (tons)	%
Year 2014	86,981.46	20,876.94	24.00%
Year 2015	81,375.56	16,576.84	20.37%[3]
Year 2016	63,098.72	9,396.96	14.89%
Year 2017	68,654.01	15,893.24	23.15%
Year 2018	64,583.14	14,879.59	23.04%
Year 2019	60,101.40	16,377.54	27.25%
Year 2020[4]			

3.4 Plastic Waste Stream Composition

The most commonly recovered plastics in Lebanon are: a) HDPE b) PE and c) PET. This is shown in the subsequent tabulation that is extracted from the Ramboll-laceco study of 2012.

Table 7 – Waste Plastic Composition in BML (laceco, 2012)

Plastic Type	%
PE transparent	16.31%
PE colored	28.40%
HDPE	29.30%
PS	4.50%
PET	21.50%
Total	100%

[3] 2015 is not considered a regular year since it had interruptions due to the waste crisis in the study geographic boundaries

[4] 2020 is also not considered as an indicator year due to Beirut Port Blast that rendered the solid waste management plan with the study area lacking the operation of major components

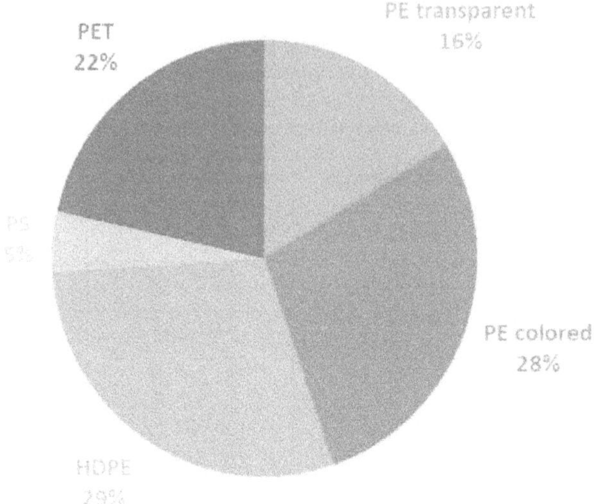

Figure 6 – Waste Plastic Composition in BML (Iaceco, 2012)

The waste plastic composition in BML is in line with findings reported in similar studies where PET, PE, PP and PS are the predominant polymer types found in household waste. (Yanli Zhu, 2020)

3.5 Plastic Waste Stream Market

Site records that are investigated indicate that the main marketed item is PE followed by PET and shredded HDPE. This is subsequently represented.

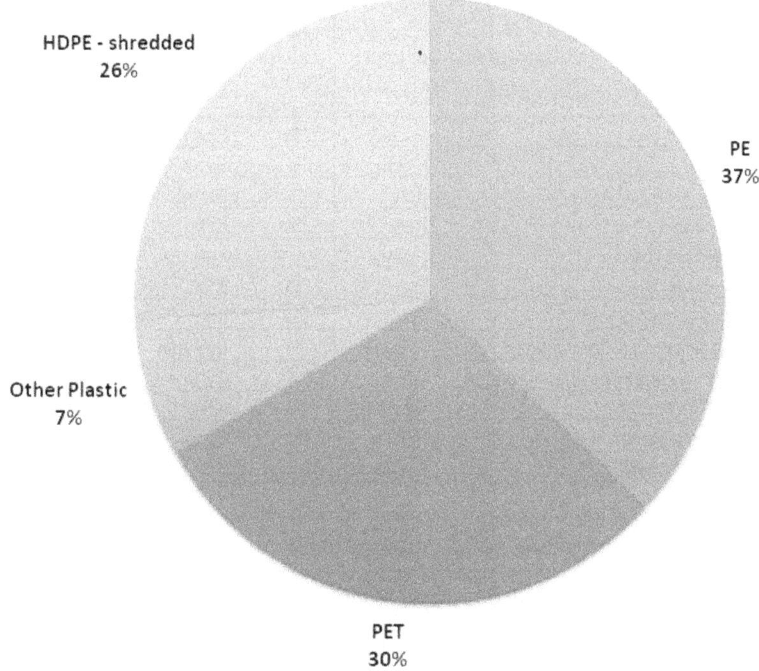

Figure 7 – Plastic wastes from BML acquired by recycling industry

The output plastic waste stream that is exiting the sorting facilities in BML to the recycling industry is similar to the findings of other studies that indicate that PET and PE products are most commonly sorted in plastic sorting facilities and generated as output streams (Martijn Roosen, 2020).

According to Investment Development Authority of Lebanon (IDAL), plastic manufacturing is included within the chemical industry sector. Thirty seven percent (37%) of the companies in the chemical sector manufacture plastic. Tweny six percent (26%) of the Lebanese Chemical Products exported are plastics and articles thereof. While plastics and articles thereof constitute 40% of total imports of chemical products. Manufactured plastic items are the third largest Lebanese exported products (IDAL, 2016). Thus trade deficits exist in the plastic industry sector despite the growth noticed in the sector (Ezzeddine, 2017). This is shown in the tabualtion below.

Table 8 – Lebanon Main Trade Partners for Chemical Industry (IDAL, 2016)

Industry	Main Country of Import	% of total imports from country	Main export destination	% of total exports to this country
Inorganic chemical	Turkey	22%	Spain	37%
Organic chemical	China	27%	Syria	22%
Fertilizers	Italy	16%	Bangladesh	49%
Paints & derivatives	Italy	12%	KSA	17%
Perfumary & derivatives	France		Iraq	25%
Soaps & derivatives	Egypt	22%	KSA	21%
Glue, adhesives & derivatives	Germany	15%	Iraq	13%
Explosives, pyrotechnic & derivatives	France	30%	Ethiopia	84%
Photographic or cinematographic materials	China	28%	Syria	35%
Miscellaneous Chemicals	Germany	16%	KSA	35%
Plastics & articles thereof	KSA	21%	Syria	20%
Rubber & articles thereof	China	31%	KSA	11%

Lebanese chemical production has been limited to the manufacturing of polymers and specialty chemicals as intermediate inputs to several industries. The majority of plants in this sector are mixing plants while manufacturing of basic materials has been minimal given due to lack of raw materials (IDAL, 2016). Thus plastic waste recycling is considered as a potential investment opportunity cosidering the environmental and the economic benefits it presents. This is particularly considered for PET and HDPE (IDAL, 2016).

3.6 SWOT Analysis of Plastic Waste Recovery Market

Strength	Opportunities
▪ Plastic is one of the major recyclable items found in BML waste stream ▪ Plastic is one of the main commodities used and manufactured locally	▪ Increased demand for recycled plastics with increase in import fees on raw material and swelled exchange rates rendering the cost of virgin material extremely high. ▪ Increased demand for construction material for reconstruction of bordering Syria
Weakness	**Threats**
▪ Manual segragation of waste plastics ▪ Lack of specialized plastic retrieval facilities ▪ Commingled waste streams that decrease the waste plastic retrieval ▪ Commingled waste streams that yield a lower grade plastic	▪ Unavailability of a legal framewrok that organizes the recycling sector and thus no control to ensure product safety and compliance with international standards ▪ Unavailability of a legal and institutional framework that encourages investment like incentives for the recycling companies ▪ Competition from countires that have a well developed petrochemical industry

	in the region

4 Conclusion

This paper has focused on the current state for municipal waste plastic recovery in Beirut & Mount Lebanon. It provides some insight on the issue as well as leading the way for further in depth research about the issue of circular economy and waste plastic recovery and recyclability.

The main consclusions that can de derived:

Waste plastic poses a major risk to the Lebanese Environment particularly the marine environment. Waste plastics accumulate in landfills with low recycling rates. However, it is one of the major recyclable streams generated with high recyclability potentials in view of the rise in the cost of virgin resources and rising exchange rates for the USD with respect to the Lebanese Lira. However in order to be able to apply the introduction of circularity into the waste plastic recycling sector, the proper infrastructure in terms of appropriate technology application as well as an encouraging institutional and legal framework is needed to support the system. The regulatory framework needed is multi-dimensional:

- To regulate the use of plastics and thus decreasing the plastic wastes generated
- To provide incentives to the industry to encourage the shift towards manufacturing rather than mixing.
- To issue standards that regulates the quality of the recycled plastics as well as the quality of the input flow. This would ensure the product safety and complaince with international standards

Public awareness to encourage source separation is a milestone to achieve a high quality plastic wastes.

5 Literature

Chinnathan Areepraserta, J. A. 2016 Municipal Plastic Waste Composition Study at Transfer Station of Bangkok and Possibility of its Energy Recovery by Pyrolysis. *3rd International Conference on Energy and Environment Research, ICEER 2016, 7-11 September* (pp. 222 – 226). Barcelona, Spain: Energy Procedia 107 (2017) Elsevier

D-Waste 2014 *Country Report on the Solid Waste Management in Lebanon.* Sweep Net

Ezzeddine, Nancy. 2017 *Policy Brief.* Beirut: The Lebanese Center for Policy Studies.

IDAL 2016 *Chemical Industry Factbook.* Beirut: Investment Development Authority of Lebanon (IDAL).

IETC 2009 *Converting Waste Plastic into a Resource.* Osaka: UNEP.

ILO 2011 *Green Jobs Assessment in Lebanon.* UNDP.

Ismail I. Abbas, J. K.-R.-R. (2017). 2017 Solid Waste Management in Lebanon: Challenges and Recommendations. *Journal of Environment and Waste Management*, 53-63.

John N. Hahladakisa, E. I. 2020 John N. Hahladakisa, E. I. (March 2020). *Plastic waste in a circular economy.* Research gate.

Laceco 2012 *Waste composition study.* Beirut: Unpublished data.

Laceco 2020 Monthly Progress Reports, Beirut: Laceco

Martijn Roosen, N. M. 2020 Detailed Analysis of the Composition of Selected Plastic Packaging Waste Products and Its Implications for

Mechanical and Thermochemical Recycling. *Environmental Science & Technology*, 13282−13293

OCHA, U. O. (2018). 2018 *Situation Report.* OCHA

UN-Habitat 2011 *Lebanon Urban Profile: A desktop Review.* UN Habitat

Yanli Zhu, Y. Z. 2020 A review of municipal solid waste in China: characteristics, compositions, influential factors and treatment technologies. *Environment, Development and Sustainability.*

Author's address

Arwa A. ELZein
laceco
Korom Center, 4th Floor, Dimitri Hayek Street
Sin El Fil, Mount Lebanon, Lebanon
T: +961-1500330 /+961-1500972
F: +961-1500791
E-Mail: arwa.elzein@laceco.me

41 – Ivo Budde – Grate for Riddlings

Ivo Budde

Hitachi Zosen Inova AG

Grate-for-Riddlings – improved fine bottom ash treatment and effects on the CO_2 footprint of waste incineration plant

Abstract

Das Grate-for-Riddlings ist ein System, welches die letzte Zone eines Vorschubrostes in einer Verbrennungsanlage als «Sieb» für die Schlackenfeinfraktion <10 mm umfunktioniert. Dadurch können die Vorteile von nass und trocken ausgetragener Schlacke kombiniert werden. Die Feinfraktion wird separat ausgetragen und ist darum trocken ausgetragen. Ein grosser Vorteil von trocken ausgetragener Schlacke ist, dass die darin enthaltenen Metalle besser zurückgewonnen werden und sich auch die Qualität der mineralischen Fraktionen erhöht.

Die ökonomisch wertvollen Schwermetalle wie Kupfer, Silber und Gold zeigen in der Fraktion < 10 mm eine deutliche Anreicherung. Mit dem HZI Dry Mining Tower können diese gezielt zurückgewonnen werden.

Zudem fällt auch eine leichte Nichteisenmetall-Fraktion an, welche vor allem Aluminium enthält. Diese führt dazu, dass je Tonne Schlacke (Trockenmasse) ca. 100 kg CO_2 Äquivalent eingespart werden können.

Beim Betrachten eines möglichen Business-Case für den HZI Dry Mining Tower fällt auf, dass der wichtigste Faktor die (Schwer)Metallpreise sind, welche am Markt gehandelt werden. Eine untergeordnete Rolle spielen Faktoren wie lokale Deponiekosten oder Betriebsstunden und Durchsatz.

The Grate-for-Riddlings is a system that re-purposes the last zone of a reciprocating grate in an incineration plant as a "screen" for the bottom ash fine fraction <10 mm. This allows to combine the advantages of wet and dry discharged bottom ash. The fine fraction is discharged in a separate way and therefore discharged dry. A major advantage of dry-discharged bottom ash is that the metals it contains are better recovered and the quality of the mineral fractions is also improved.

The economically valuable heavy metals such as copper, silver and gold show a clear enrichment in the fraction < 10 mm. With the HZI Dry Mining Tower, these can be recovered in a targeted manner.

In addition, a light non-ferrous metal fraction also accumulates, which mainly contains aluminum. This leads to a saving of approx. 100 kg CO_2 equivalent per ton of bottom ash (dry matter).

When considering a possible business case for the HZI Dry Mining Tower, it becomes apparent that the most important factor is the metal prices for the heavy fraction of non-ferrous metals that are traded on the stock market. Factors such as local landfill costs or operating hours and throughput play a subordinate role.

Keywords

Schlacke, Feinschlacke, Recycling, Aufbereitung, Grate-for-Riddlings, Dry Mining Tower

Incineration Bottom Ash, Fine Fraction of Incineration Bottom Ash, Recycling, Treatment of Bottom Ash, Grate-for-Riddlings, Dry Mining Tower

1 Grate-for-Riddlings

Nearly every waste incineration plant is equipped with a wet bottom ash extractor. This system is well-known and has a couple of advantages. One major advantage is that there are nearly no dust emissions at the bottom ash outlet and the further treatment, and the material can be easily landfilled. On the other hand – especially in Switzerland – the idea came up to use dry bottom ash extractors instead. The advantages of this technology are that the weight of the bottom ash is reduced due to the lack of water, an increased quality of the bottom ash itself (no clumping of the mineral fraction) as well as the possibility to increase the recovery rate of metals from the bottom ash (especially the fraction < 10 mm).

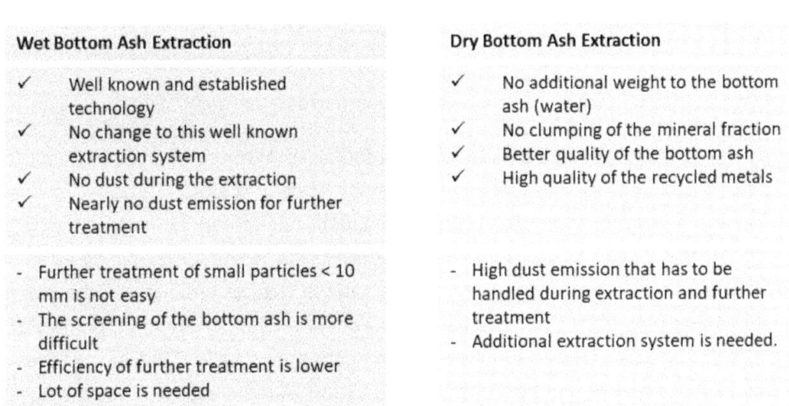

Wet Bottom Ash Extraction	Dry Bottom Ash Extraction
✓ Well known and established technology ✓ No change to this well known extraction system ✓ No dust during the extraction ✓ Nearly no dust emission for further treatment	✓ No additional weight to the bottom ash (water) ✓ No clumping of the mineral fraction ✓ Better quality of the bottom ash ✓ High quality of the recycled metals
- Further treatment of small particles < 10 mm is not easy - The screening of the bottom ash is more difficult - Efficiency of further treatment is lower - Lot of space is needed	- High dust emission that has to be handled during extraction and further treatment - Additional extraction system is needed.

Figure 1: Advantages and disadvantages of wet and dry bottom ash extraction.

The Grate-for-Riddlings is a hybrid solution to combine the advantages of the dry and the wet bottom ash extraction. The solution includes a sieve function in the last zone of the grate. A special grate block has been designed for this purpose. The grate block is designed such that particles which are smaller than 10 mm are sieved out and bigger particles follow the "normal way" towards the wet extractor (see Figure 2).

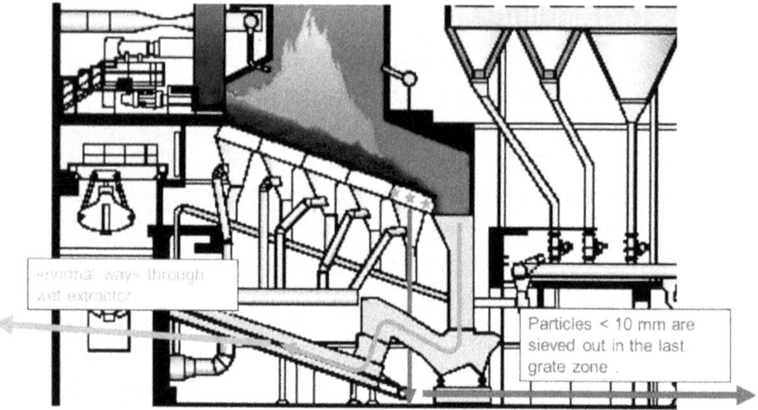

Figure 2: Schematic overview of an installed Grate-for-Riddlings.

Normally, the primary air is supplied through a funnel located underneath the grate and flows into the combustion chamber through the air holes in the grate blocks. To enable screening and the discharge of the screened fraction, the primary air is re-targeted. For this purpose, the design of the primary air supply in the Grate-for-Riddlings zone was slightly adjusted: A plate is welded into the primary air hopper to divide the hopper into two parts. One part is used to discharge the screened fraction. There is no primary air in this part. The other part is used for the targeted primary air supply. That is why there are corresponding outlets and connections on the new longitudinal beams of the grate.

1.1 Further treatment

The main economic advantage in discharging the fine fraction of the bottom ash is that the major part of the heavy non-ferrous (HNF) metals are in the fraction < 10 mm.

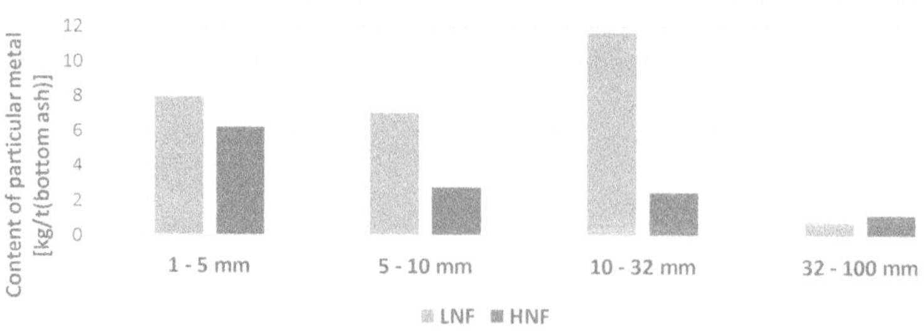

Figure 3: Content of LNF and HNF in the bottom ash for different particle sizes. [intern 2016]

The fine fraction, which is separated by the Grate-for-Riddlings, can be treated in the HZI Dry Mining Tower. This sorting system was developed especially for this kind of fine bottom ash.

Figure 4: Schematic overview of the HZI Dry Mining Tower.

The Dry Mining Tower is designed in a modular way such that it can be implemented or adapted to an existing system. Moreover, the further treatment does not necessarily need to be located next to the incineration plant but can be installed externally to combine the further treatment of multiple plants.

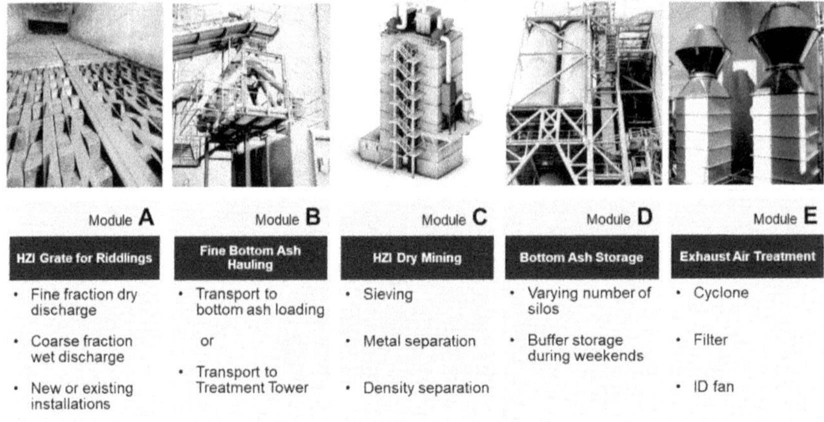

Figure 5: Schematic overview of the whole GfR/Dry Mining Tower System: ***A****: Grate-for-Riddlings, to be installed in the last zone of the grate.* ***B****: possible fine bottom ash loading into container for further external treatment.* ***C****: HZI Dry Mining Tower.* ***D****: Possible bottom ash storage system.* ***E****: Exhaust Air System.*

1.2 Reference Project

The Grate-for-Riddlings can be implemented in a new plant or can be "retro-fitted" to an existing plant. A retrofit was done at Kebag AG in Zuchwil, Switzerland in 2016. Since then the system is in continuous operation. The fine fraction of the separately discharged bottom ash is transported to the ZAV Recycling AG for further treatment. Since 2017 a continuous amount of around 2'500 tons of bottom ash per year is sent for further treatment, resulting in cost savings because no landfilling is required and because the NF can be sold.

In the case of the Kebag AG it was a retrofit project. The existing last zone of the grate was modified with the Grate-for-Riddlings system. In addition, a collecting and transport system was installed under the incinerator. After a steel plate conveyor, which transports the fine fraction out of the building, a loading system with containers was installed.

Figure 6: The left image shows the newly installed grate in April 2016 and the right image shows the same grate after 3.5 years of operation.

A series of measurements performed by UMTEC (Institut für Umweltschutz- und Verfahrenstechnik, Rapperswil) showed an increase of the NE-metals of 8% compared to the total amount of the bottom ash (0.5 – 32 mm) [Bunge 2018].

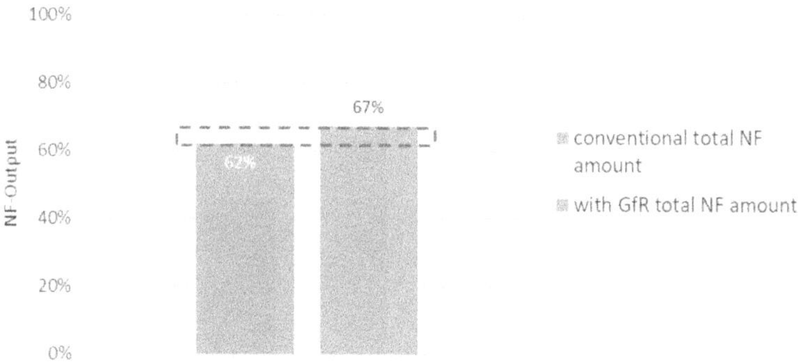

Figure 7: The NF-Output is increased (marked with the red box) with the GfR-System by 8 % compared to the conventional system [Bunge 2018].

One major concern, beside the ejection of the fine fraction from the bottom ash stream, was to ensure that the TOC content of the fine fraction does not increase significantly compared to the "normal way". The measurements which were recorded internally during the test phase after installation showed that the TOC content remained below 1 wt% [Gablinger 2016]. Another external series of measurements confirmed this finding [Bunge 2018].

2 Ecological Benefit

The pressure from society and politics for a company to become more climate friendly increased enormously in recent years. Therefore, some investigation on the improvement of the CO_2-footprint was done. Which amount can be saved if the fine fraction < 10 mm of the bottom ash is separated with a Grate-for-Riddlings system and subsequently treated in a further treatment plant? By far the biggest impact is due to the recycled aluminum. This is since the aluminum is discharged dry and not wet and therefore remains in its elementary state.

From an ecological point of view, it seems sensible to extract metals from the bottom ash in order to replace primary metals. HZI conducted an internal study on this. The aim of this study was to quantify the ecological benefit of the HZI Dry Mining technology (treatment of the fine fraction) compared to conventional wet bottom ash treatment. An incinerator that today discharges its bottom ash wet and retrofits a Grate-for-Riddlings grate thus reduces the "wet mass flow" by 30%. For one ton of bottom ash (TS = 100%), this reduction leads to a reduction of 100 kg CO_2 equivalents. The main contribution to this is, as mentioned, the substitution of primary aluminum production (see Figure 8). This reduction corresponds to about one fifth of the total CO_2 emissions produced on average by the incineration of one ton of municipal waste. The recovery of copper has been considered as a second element in the study. However, it had a lower environmental benefit. Other metals were not considered.

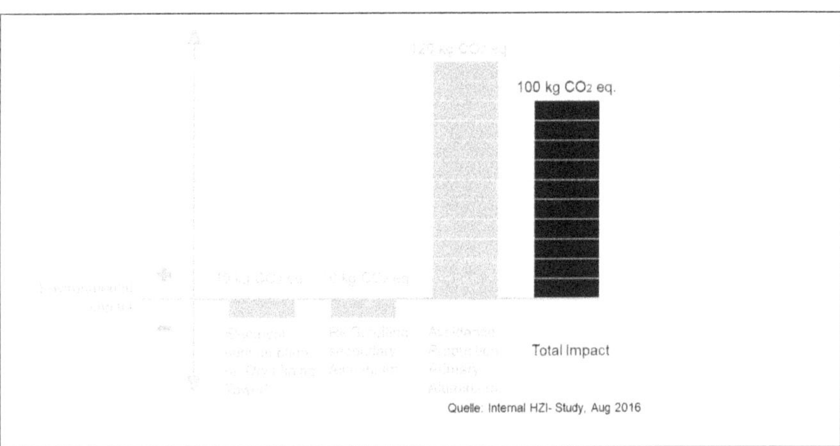

Figure 8: Simplified CO_2 footprint in kg of CO_2 per ton of total bottom ash (dry).

In total, the calculated environmental impact is negative, meaning that the partial dry-discharged amount of the fine bottom ash (Grate-for-Riddlings system), which was subsequently processed using the HZI Dry Mining Tower, is more environmentally friendly than processing a wet-discharged bottom ash.

3 Business Case for an HZI Dry Mining Tower

As with every investment, the question of the business case rises at some point. Looking on the HZI Dry Mining Tower the following factors are the most important ones.:

- Actual metal price
- Composition of HNF in the bottom ash
- Local landfill tax and gate fees
- Throughput and annual operating hours

Looking on the two possible revenues (profit metal price and saved deposition costs) the one of the sold metals is by far the most important. With that the business case is strongly linked with the actual metal price. The dependency is shown on a little case study. Therefore, the following assumptions where made:

- 500'000 t waste throughput per year (operation 8'000 h/y, 20% thereof resulting in total bottom ash)
- 5'255 €/t benefit HNF-mix 0.2 – 10 mm fraction (calculated of average metal price Cu, Ag and Au between 03/20 and 03/21 as well as assumptions for discounts metal smelter and transport)
- The disposal costs for the minerally fraction was set 0 because the fraction must be disposed independent if there is a Dry Mining Tower or not.
- An OPEX (Dry Mining Tower running at 4'000 h/y) and CAPEX where calculated out of experience from other similar plants.

Out of this assumption an IRR over 10 years can be calculated and the HNF-mix price as well as the LNF- and Fe-price can be adapted by +/- 50 % where CAPEX and OPEX remain.

Table 1: Calculated IRR over 10 years for the investment of the Dry Mining Tower with the average metal price between 01/20 – 01/20 and a fictive case of + and – 50% of the metal prices.

	-50%	**average metal-price 03/20-03/21**	+50%
IRR (10y)	-3%	**20%**	38%

Looking back on the prices for metals at the stock market a fluctuation of +/- 50% over 10 years is quite reasonable.

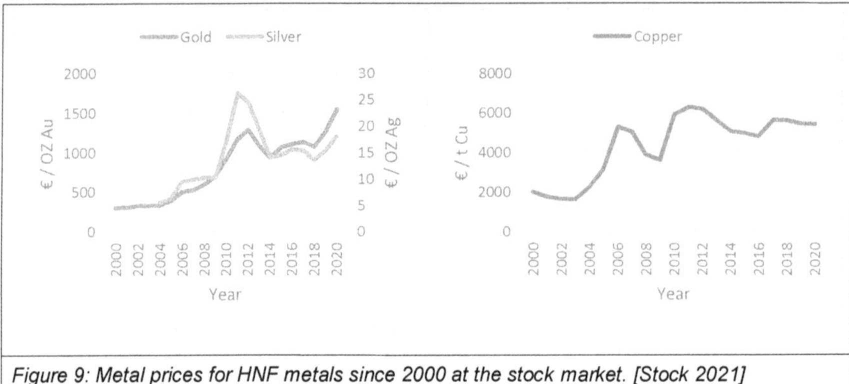

Figure 9: Metal prices for HNF metals since 2000 at the stock market. [Stock 2021]

Another fact that is directly linked with the benefit coming out of the HNF-mix is the content of the single HNF metals in the bottom ash. The challenge is to analyze this content correctly. First of all, the sampling of such a huge and inhomogeneous mass flow is difficult or coupled to some effort respectively [Skutan 2014]. Second, there are several different methods which are used currently to analyzing the HNF-content in the bottom ash. A plant operator is often interested in the maximal concentration of selected elements or components to fulfill landfill legislation. There are several studies about the NF metal content in bottom ash which measured a concentration between 1 up to 5 % [Syc 2020]. In the light of improved recovery of secondary raw materials such analysis should be made and further benefits clarified.

The influence of the local landfill taxes or gate fees is by far of minor importance compared with the metal prices. In addition to that the amount of the fraction, which must be landfilled, that is generated by the GfR system is small compared with the overall amount of generated bottom ash.

The last point mentioned here is concerning the OPEX. The Dry Mining Tower is designed for 10 t/h throughput. Due to the recommended yearly amount which should be treated the operating hours per year can then be estimated.

4 Conclusion

The Grate-for-Riddlings is a system which combines the advantages of wet and dry extracted bottom ash. The major advantage of dry extracted bottom ash is the fact, that the heavy metals can be recycled with with an higher value than the wet extracted bottom ash. For a further treatment of the fine bottom ash fraction < 10 mm the HZI Dry Mining Tower was designed.

With the further treated fine fraction of the bottom ash about 100 kg CO2 equivalents can be saved due to the fact, that a noticeable amount of Alumina is treated in a dry way.

Looking on the business plan it can be said that the business case of the Dry Mining Tower for treating the fine fraction < 10 mm of dry extracted bottom ash (with e.g. Grate-for-Riddling system) is strongly dependent on the metal prices and coupled with that on the metal content in the specific bottom ash. Further points as local land fill taxes play a subordinate role. Circular economy activities should favour reuse of such secondary sources for metals instead of ore mining for primary metals.

5 Literature

Gablinger H., Juchli M.	2016	KEBAG-Trockenschlackeaustrag für Feinschlacke. Vortrag Informationsveranstaltung ZAR
Bunge R.	2018	Untersuchung der aufbereitungsrelevanten Eigenschaften von KVA-Schlacke aus einem Trocken/Nass-Hybridaustrag. In: https://www.igen-ass.ch/uploads/media/Factsheet_IGEN-ASS_TNH.pdf
Stock market	2021	Historische Metallpreise, www.boerse.de
Intern	2016	Internal measuring series of GfR System
Skutan S., Gloor R.	2014	Sampling, sample preparation and analysis of solid residues from thermal waste treatment and its processing products
Syc	2020	Syc et al., Journal of Hazardous Materials, *393*, **2020**, https://doi.org/10.1016/j.jhazmat.2020.122433

Author's address

Ivo Budde
Hitachi Zosen Inova AG
Hardturmstrasse 127
CH-8005 Zürich
Telefon +41 44 277 12 05
Mobil +41 79 284 75 65
E-Mail: ivo.budde@hz-inova.com

Technical assessment of a bottom ash treatment plant - a case study

Kay Johnen, Alexander Feil

Department of Anthropogenic Material Cycles (ANTS) – RWTH Aachen University, Aachen, Germany

Abstract

In 2017, around 96 million tons of waste were treated in waste incineration plants, resulting in 19 million tons of bottom ash for subsequently treatment. In order to recover the valuable materials (especially the metals), the treatment processes are constantly being improved. In order to identify improvement potentials, e. g. of a bottom ash treatment plant, the actual state must be determined. Reliable data can only be obtained by means of a plant assessment. However, the planning, execution and evaluation of such an assessment is very time-consuming and personnel-intensive. In the course of planning, the number and size of the samples as well as the type of sampling must be determined depending on the plant conditions. The execution of the sampling for the characterization of the material streams is challenging due to the heterogeneous material properties.

Keywords

Bottom ash, plant assessment, sampling

1 Introduction

Since May 2018, the Department of Anthropogenic Material Cycles (formerly the Department of Processing and Recycling) has been working on the international research project **INSTAnT** (Innovative Sensor Technology for optimized Recovery from Bottom Ash Treatment) together with partners SUEZ, TOMRA, VITO and XRE as part of ERA-MIN 2 (Research & Innovation Programme on Raw Materials to Foster Circular Economy). Within the European Union, more than 400 Waste-to-Energy (WtE) plants are currently in use to convert 96 million tonnes of waste (municipal, commercial and industrial) into energy and decrease the volume of these waste streams. This thermal process produces approximately 19 Mt of bottom ash [CEWEP2019] which could be considered as the 'final sink' for many End-of-Life products. Important quantities of metals (ferrous and non-ferrous) and minerals (both industrial minerals and minerals for construction) are present in these bottom ashes offering a great opportunity for recycling and turning this complex waste into valuable secondary raw materials.

The objective of the INSTAnT project is to close the material cycle of materials present in bottom ashes by using smart recycling technologies to *(i)* optimise process conditions in bottom ash treatment plants to maximize metal recovery; *(ii)* separate a valorizable pure glass fraction, and *(iii)* detect and remove impurities that hamper a high-grade recycling of the mineral fraction.

2 Background

In Europe (2017), 96 million tons of waste (municipal, commercial and industrial) were treated in WtE plants [CEWEP2019]. In WtE plants, waste is thermally treated to achieve a mass reduction and hygienization. On average, 20 – 30 wt.-% of the input material of a waste incineration plant remains as a solid residue, the so-called bottom ash (BA). BA consists of approx. 80 - 85 wt.-% minerals, 10 - 12 wt.-% Fe-metals, 2 - 5 wt.-% non-ferrous metals as well as 1 wt.-% unburned material. BA is usually processed in BA treatment plants which aim to recover the valuable materials, especially metals. What remains is a mineral fraction and unburned material. The unburned material is usually returned to the incineration, and the mineral fraction can be used in road or landfill construction if it fulfils the requirements; otherwise it is landfilled. [CEWEP2019; Gillner2011]

The quality and composition of bottom ash is influenced by the input material and the discharge process from the waste incineration plant. [Gillner2011] Usually, bottom ash is discharged from the incinerator by means of a so-called wet ash discharge. The glowing ash falls into a water basin, cools down there and is transported out of the deslagger with the help of a discharge piston. In addition, the water bath prevents false air from entering the combustion chamber and thus disturbing the combustion process. As an alternative to wet deslagging, dry deslagging is used in a few plants (especially in Switzerland and Japan), in which the glowing bottom ash is discharged dry from the furnace. [Langhein2017]

In wet deslagging, contact with water and subsequent drying in air results in hydration and carbonation processes that cause the bottom ash to harden. [Marzi1998] Dry deslagging has the advantage that these transformation processes do not take place and the bottom ash can be processed more effectively. [UBA2016; Langhein2017]

As with almost all waste streams, there is no standardized treatment process for bottom ash treatment plants. Although the same units and processes are usually used, practically every bottom ash treatment plant is unique. Since the focus of most bottom ash treatment plants is on the recovery of the metals, different magnetic and eddy current separators are used. In order to ensure effective processing, especially of the non-ferrous separation on eddy current separators, the bottom ash must be divided into closely classified grain classes or grain bands. In order to maintain the necessary ratio of bottom to top grain size (1:3) in the feed to the eddy current separation, screening machines are also part of the standard equipment in a bottom ash treatment plant. [Gillner2011]

In the INSTAnT project, the processing of bottom ash is to be improved through using modern sensor technology. In order to be able to quantify the effects of this technology on the processing result, series of tests were planned, both in the original condition and with adapted sensor sorting. In addition to identify the optimization potentials, the actual condition of the test plant must first be determined. With the help of the actual state at the beginning of the project, the improvement measures can then subsequently be quantified and evaluated. The aim of the assessment is to determine the mass recovery, the concentrations of the recyclables in the individual output streams, and the recyclable output (Figure 1).

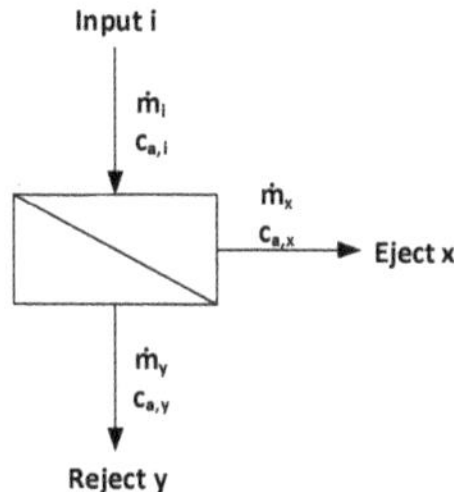

Figure 1: Parameters for characterizing a separation process

The mass recovery *(1)* can be determined easily. For this purpose, the mass of the individual product/residual material flows must be set in relation to the input material within a defined period of time.

Mass recovery: $$R_M = \frac{\dot{\mathrm{m}}_x}{\dot{\mathrm{m}}_i} \quad (1)$$

To determine the concentrations *(2)* of recyclables or impurities, samples of the individual output streams must be taken and analyzed. The analysis can be conducted in different ways and accordingly with different accuracy (manual sorting, technical analysis, chemical analysis). To estimate the concentration of valuable substances in a material flow, the mass of the contained valuable substance is offset against the total mass.

Concentration: $$c_{a,x} = \frac{\dot{\mathrm{m}}_a}{\dot{\mathrm{m}}_x} \quad (2)$$

With the individual masses and concentrations, the yield *(3)* can be calculated. This indicates how much of the recyclable material contained in the input was actually discharged into the actual target product. [UBA2016]

Yield: $$R_W = \frac{\dot{m}_x * c_{a,x}}{\dot{m}_i * c_{a,i}} \quad (3)$$

The accuracy of these determinations depends on the number and mass of single samples as well as the method of the analysis. Since both sampling and analysis require considerable effort, the number of samples that can be analyzed for a plant assessment is limited. For this, the LAGA PN 98, which is a German guideline for determining the number of samples and mass, depending on the volume and grain size of the material stream to be sampled, can be used. [LAGA2001]

3 Assessment of a bottom ash treatment plant

The actual condition of a plant can only be determined by means of a technical assessment, i. e. physical sampling, as modelling never represents the complete reality. For evaluation, in addition to the gravimetrical determination of the input and output flows the material composition must be determined. However, the assessment of a plant is associated with a considerable effort. In particular, planning and preparation are very time-consuming and must be prepared in detail. Even small errors, e. g. unrecorded weight of an output stream, can cause an overall balance to become unusable. For this reason, the requirements and challenges for the planning of a plant assessment are described below, using the example of a bottom ash processing plant.

Planning and preparation can be divided into three stages. The first step is the basic evaluation, in which an initial overview is obtained on the basis of flow diagrams and plant plans. In the second step, the conditions on site are examined and documented during an inspection/visit of the plant, and in the third step, a more detailed planning takes place.

In the first step, fundamental aspects must first be clarified. The objective of the assessment must be defined and an initial overview of the plant can be obtained from flow diagrams and plant plans. In this way, the type, number and, if necessary, the local distribution of the input and output flows can be determined. With the help of available mass flow data, the expected quantity of the respective output flows to be sampled can be estimated. Since the sampling interferes with the ongoing plant operation, even the scheduling can be a challenge. If the assessment is performed on a regular working day, it will affect standard operation. Alternatively, the balancing can be carried out on a day when the plant is not normally in operation, for example on the weekend. In this case, it must be ensured that sufficient plant personnel is available.

Since planning in the first stage is carried out on a more abstract level, the local conditions must be inspected in the second planning stage. In the course of an on-site inspection, the following aspects should be inspected, checked and documented, as shown in Table 1.

Table 1: List plant inspection

Aspect	**Questions to consider**
Distance of the bunkers/storage locations of the individual output streams or distance of the sampling points from each other.	*Important for timing and number of samplers: If the material is sampled from the current stream, the samples must be taken at specific time intervals. If the distance between the sampling points is too long and thus involves longer walking distances, the specified time intervals cannot be observed. Thus, the number of required personnel increases.*
Type of storage of output streams (bunker or container).	*What is the accessibility to the material or how can sampling take place?*
Accessibility of the bunkers/containers	*Can samples be taken from the bulk material, if necessary?*
Size of the container/bunker	*How much material can be stored before the container/bunker needs to be emptied?*
Type of containers or bunker condition (concreted or compacted soil).	*Is there a risk of spillage into the material due to the subsoil or inadequately protected bunker areas?*
Discharge height of the product streams into the bunker/container.	*Can samples be easily without any safety risk?*
Possible work areas (e.g. for packing)	*Samples have to be weighed, packed and, if necessary, processed; free area is required for this purpose.*
Employees or mobile technology available	*Type and number of vehicles can have an influence on the execution and duration of sampling (is parallel work possible?).*
Safety and security precautions	*Closed travel routes/areas? Special personal protective equipment?*
Roofing: plant/bunker/container etc.	e.g. without roofing + rain showers → different water contents of the material samples

After the inspection of the plant, the detailed planning for plant assessment can be carried out in the next step. Based on the conditions determined during the site inspection, the type of sampling (from either bulk or ongoing material flow) can be defined. Depending on the expected mass flows, the required number of samples and sample mass can be identified, for example, using the LAGA PN98 [LAGA2001]. In this way, the time and personnel required can be estimated.

The type of sampling is the most important issue here, as it affects both the time and personnel required and, in terms of the quality of execution, it can have a major influence on the representativeness of the samples. Regarding this aspect, sampling from the moving material stream is always preferable compared to sampling from a bulk. A prerequisite for sampling from moving mass flows is the regular material taking interval, which means that the material is taken at regular intervals over the entire test period. To achieve that, the material flow should be conveyed through the plant at an approximately constant rate so that the composition of the samples can be used as a good representation of the material stream to be investigated. If, on the other hand, the samples are taken from a bulk, segregation in bulk cones can limit the representativeness. Due to some effects, the smaller grain sizes could accumulate, for example, in the lower part of the bulk; if the samples are only taken from the surface, then the sample does not a represent the whole bulk. [LAGA2001]

Time and personnel expenditure behave inversely proportional depending on the type of sampling. If the sampling is carried out from bulks, it can only be done after the actual test period and thus the time required for the implementation of the sampling increases. On the other hand, sampling of the output streams can take place sequentially, reducing the number of samplers required. If the samples are taken from the current stream, the sampling must take place during the trial period and thus the personnel effort increases significantly. Depending on the sampling frequency, the number of samples required and the spatial distribution of sampling points, an increased number of samplers may be necessary.

For a better understanding, the following example is used to show the required effort. In the case of a 3 h sampling in which 8 single samples are to be taken from 15 product streams each, one sample per material stream must be taken every 20 minutes. Depending on the physical distance in between the individual sampling points, a sampler can sample 1-3 material streams during sampling. Thus, the sampling effort accounts for at least 5 samplers.

Depending on the design of the plant, sampling from the running material stream cannot always be carried out. Lack of access to the running material stream or safety risks may mean that only sampling from the bulk material is possible. For example, the discharge points of a material stream may not be accessible at all or only in a dangerous area, making continuous sampling unfeasible. In this case, sampling must be carried out on stationary bulk once the trial is finished.

Time planning plays an important role for the plant assessment and depends on many factors, such as types of sampling, test duration and required preparations. Therefore, it is not possible to set a fixed duration for the assessment, and sufficient buffer time should be planned in any case. Time delays can occur for various reasons, for instance, due to the technical difficulties e. g. a defective machine or that the removal of the samples takes longer than initially planned. On the other hand, there may also be ordinary reasons, for example, that the operating personnel are on break and thus some work cannot be carried out.

4 Practical example: INSTAnT assessment

In the INSTAnT project, the first assessment of the bottom ash treatment plant (throughput approx. 100 Mg/h) was carried out in August 2018. In order to not disturb the usual plant operation, the assessment was carried out on a Saturday. A bottom ash processing plant in Belgium served as the test facility. The planning of the assessment was carried out as described above. Already in the first step, an average bottom ash for the plant was agreed upon as sample material. The most important information for the planning of the assessment could be obtained by the inspection. Due to the spatial distribution of the output bunkers (17 output streams in total) from each other, at least six samplers would have been necessary for sampling from the current product stream. All material streams were stored in concrete block bunkers, were easily accessible, and had enough bunker space for several hours. However, smaller container boxes had to be placed in a couple of bunkers because the floor area was no longer sealed, allowing the possibility of spurious constituents (soil material) to be introduced into the samples. The drop heights into the bunkers were problematic, especially for the coarse fractions minerals and Fe-metals these were very high (approx. 3 - 4 m), see Figure 2. For safety reasons, it was therefore not possible to sample these material streams during operation.

Figure 2: Bunker: Fe 40 - 80 mm

Since there was sufficient space on site to prepare the bulk material and enough plant personnel with the appropriate mobile equipment, it was decided to sample from bulks after the trial. In comparison with the existing plant data and considering the required time for sampling the bulk material, the test duration was set to 1 hour.

Number of samples

As already mentioned, the number of required single or composite samples can be determined with the help of LAGA PN 98. This states that when sampling a mass flow < 30 m³, at least 8 single samples or 2 mixed samples are required. A mixed sample must consist of at least 4 single samples. The LAGA also specifies a minimum volume depending on the grain size (see Table 2).

Table 2: Minimum volume of the single and lab samples (according to [LAGA2001])

Max. grain size x_{max} [mm]	Min. volume single sample [L]
$x_{max} \leq 2$	0.5
$2 < x_{max} \leq 20$	1
$20 < x_{max} \leq 50$	2
$50 < x_{max} \leq 120$	5
$x_{max} > 120$	piece = single sample

Since each output stream has a volume of < 30 m³ with a plant running time of 1 h, it was determined in accordance with the LAGA that four mixed samples with a volume of 10 L per mixed sample are taken from each material stream (< 120 mm) and four samples with 90 L each are taken for the fractions > 120 mm. [LAGA2001]

Execution of sampling campaign

On the day before the plant sampling, batches of the bottom ash to be sampled were already processed in the plant and thus it was already optimally adjusted. Immediately prior the actual test period, the plant was tested, allowed to run empty and then all bunkers were emptied. During the balancing period, the sampling containers were prepared (labeled and tared). Following the trial period, samples were taken for material analysis. For this purpose, the individual bulks (Figure 3a) were mixed by a wheel loader and then "flattened" (Figure 3b). This approx. 300 mm high layer of bottom ash was then sampled at various points (see Figure 3c). In this way, it was possible to ensure that material was taken from all layers of the bulk material and thus covering the complete plant runtime.

Figure 3: Sampling of a bulk

It should be noted that the taking the samples becomes more difficult with increased grain size. Due to the grain shape, the dosing ability decreases with increased grain size and depending on the material flow (see Figure 4b). For this reason, the coarser fractions Fe and NF metals in particular are difficult to shovel. Particles >120 mm were taken by hand, as it was no longer possible with a shovel. After sampling, the remaining material streams were weighed individually.

Figure 4: Output streams: a) Input – b) Fe – c) NF – d) mineral

In order to determine the recyclable material concentration and yield, the samples taken had to be analyzed. This analysis was carried out in the technical lab of the Department Anthropogenic Material Cycles (ANTS). Since the chemical analyses are expensive and time-consuming, the project decided on a mixture of hand sorting and "technical" analysis. For this purpose, two composite samples per output stream were reprocessed under pilot plant conditions. The magnetizable particles were separated by means of discarding drum magnetic separator and the non-ferrous particles were separated from the remaining residual stream by means of an eddy current separator. Subsequently, all material streams were sorted by hand, and any shortfalls were assigned to their proper fraction (see Figure 5).

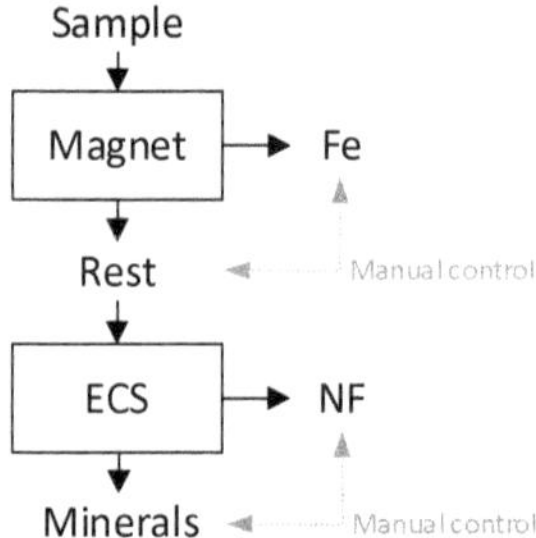

Figure 5: Processing scheme ANTS

A total of 40 samples with a volume of 10 L each were prepared in this procedure. Originally, two samples per material stream were planned for the analysis, but due to the high difference on the results; two reserve samples of the material stream were also included.

5 Summary of time and personnel expenses

The analysis has shown that the preparation, execution and evaluation of a plant assessment are very complex and time-consuming. The example of the described sampling can be used to show the time required for the execution and the analysis.

Time required: assessment

In total, the assessment including the sampling took seven hours. In the first two hours, the general procedure and the schedule were discussed with all the workers. The plant also continued operating on a test basis and all bunkers were subsequently emptied. Prior to the start of the actual test period, a control tour was carried out to check whether all bunkers had been emptied and all other preparations for balancing had been implemented. The trial period then started. After the one-hour plant run, another control walk-through took place with documentation of the output streams. The majority of time was taken up by sampling and mass recording of the bulk materials. The mixing, "flattening", sampling and weighing of the material took about 4 hours. The limiting factor was the mobile technology, since only one wheel loader was available, and it is used for mixing, sampling preparation and weighting.

Time required: Sample analysis & evaluation

Samples taken during balancing were subsequently processed and analyzed at the ANTS technical lab. Drying of the material before analysis took about one day. Depending on the material flow and grain size, 4 - 8 samples could be completely analyzed per day. With a total sample number of 40 and an average of 6 samples per day, about 7 days were needed for the analysis. The subsequent evaluation took about two days.

Time required: Other

In addition to the required "active" time, waiting times and travel times must be included. Depending on the distance from the plant to the office, travel times can significantly increase the total time required. Likewise, in the course of preparation, it must be expected that sampling accessories or similar must be purchased, and they will not immediately be available considering the delivery which should be taken into accounting while calculating the total time demand.

Personnel expenses: Assessment

A total of ten people were present on the day of sampling. In addition to the two project managers (plant manager and project manager of ANTS), four employees of the experimental plant and four student assistants of ANTS were involved in the balancing and the sampling.

6 Conclusion

The assessment of a treatment plant is very time-consuming and requires a lot of personnel. In addition, even relatively minor errors can disturb the data obtained partially or even cause the entire balance to become unusable. If, for example, a mass flow is forgotten to be weighed or sampled, a complete balance cannot be generated. In addition, the result of a balance is always only a snapshot of the plant and also always dependent on the properties of the material flow to be sampled. Despite all the mentioned disadvantages, the complex plant assessment is the only reliable way to measure the quality of a treatment process.

7 Literature

[CEWEP2019] Bottom ash fact sheet: https://www.cewep.eu/wp-content/uploads/2017/09/FINAL-Bottom-Ash-factsheet.pdf; 2019

[Gillner2011] GILLNER, Ronald: Nichteisenmetallpotential aus Siedlungsabfällen in Deutschland. Aachen: Shaker, 2011 (Schriftenreihe zur Aufbereitung und Veredlung Bd. 40)

[LAGA2001] LAGA PN 98: Richtline für das Vorgehen bei physikalischen, chemischen und biologischen Untersuchungen im Zusammenhang mit der Verwertung/Beseitigung von Abfällen, 2001

[Langhein2017] LANGHEIN, Eva-Christine: Trockenaschen vom Abfall bis zum Wertstoff; Tagungsband: Mineralische Nebenprodukte und Abfälle 4, S.81 – 94; ISBN 978-3-944310-35-0

[Marzi1998] MARZI, Thomas ; PALITZSCH, Sylke ; BECKMANN, Ralf ; KÜMMEL, Rolf ; BEARD, Adrian ; KELDENICH, Kai: *Wirkungsmechanismen bei der Alterung von Müllverbrennungsaschen*. In: *Müll und Abfall* 1998, Nr. 5, S. 316–322 – Überprüfungsdatum 2014-06-30

[UBA2016] QUICKER, Peter: Möglichkeiten einer Kreislaufwirtschaft durch weitergehende Gewinnung von Rohstoffen aus festen Verbrennungsrückständen aus der Behandlung von Siedlungsabfällen, 2016 (Forschungskennzahl 371333303)

Kay Johnen, M. Sc.

Department of Anthropogenic Material Cycles (ANTS) - RWTH Aachen University

Wüllnerstraße 2

D-52062 Aachen

Phone: +49 241 80 99346

E-mail: kay.johnen@ants.rwth-aachen.de

Process development to improve volumetric stability of steel slag

Mukund Manish*, Praveen Kumar, D. Satish Kumar, Veeresh Bellatti and Balachandran G

JSW Steel Ltd., Vijayanagar Works

Abstract

Recycling of steel slag has always been a concern for steel makers. Its volumetric instability due to presence of expansive phases restricts its usage as construction aggregates. Existing ageing techniques are costly and time taking. In the present work, a new process has been developed for improving its volumetric stability by reducing its basicity in molten state before pouring in pit, through pilot scale experiments. Refractory lined 1T slag pot with addition and oxygen injection facilities was created at the slag yard in a steel making shop. Molten slag from the BOF process was poured into the pilot slag pot for the experiments. Three different silica containing materials, natural sand, quartzite fines and BF slag, were injected in the slag pot along with oxygen in different proportions to react with expansive free lime. Thermodynamic feasibility and quantity of addition was determined based on factsage calculations. Change in the basicity, free lime content and volumetric expansion was measured for all the combinations. It was found that by the addition of silica fines at temperatures above 1550 °C, free lime is converted into di-calcium silicate and reduced the basicity which improved the volumetric stability of slag for all additives. The expansion of the slag is reduced below the standard requirement of 1.5 % only at basicity less than 2%. Natural sand and quartzite fines showed better reduction than BF slag. The developed process is simple and cheap process for directly using the solidified slag as aggregates without any post processing.

Keywords: Steelmaking, Slag, free lime, volumetric stability.

1. Introduction

Steel slag is main by-product of steelmaking shop. It is generated during steelmaking process. Depending upon grade of steel, 150 – 200 Kg slag is generated for producing 1T of steel. About 2.3 million tonnes of steel slag are produced annually in JSW Vijayanagar Plant. In which 1.5 T slag is coming from BOF (LD Converter) Process. Storage of the huge amount of slag burdens the environment and it is an economic liability to the steel industry. Presently, Most of the BOF slag is used as land filling material, which has already been a tough problem that has to be faced with in the field of metallurgical solid waste. Recycling and reuse of slags toward added-value applications have significant importance for the sustainability of the steel industry. However, the main reasons that limit the use of steel slag are classified into three categories: volume instability, poor grindability and low cementitious activity [1,2]. Compared with the other two, volume instability is the bottleneck that keeps the steel slag from large use among metallurgical solid waste. Its volume instability caused mainly by presence of free lime (f-CaO) in Steel slag [3,4]. As the high content of f-CaO in BOF slag, it could result in the danger of degradation of mechanical properties of BOF slag mainly due to volume expansion [5]. In order to improve the volume stability, stabilization of f- CaO in steel slag has been a hot topic in the field of steel slag reforming. So far, several available even commercial processes for stabilising BOF slag have been developed. Motz H et al. [6] studied the BOF slag treatment by aging in slag pits in which weathering of slag was done for certain period of time to convert free lime into calcium hydroxide and sometimes water was used for accelerating the hydration process of free lime. Weathering of BOF slag has some drawbacks, it needs large space, more time and produce lot of fines which is unusable as aggregates. Rafael M. Santos et al. [7] studied about high temperature BOF slag carbonation, with particular emphasis on carbonation kinetics and conversion. The effect of several process parameters (particle size, temperature, pressure, time and steam addition) on CO2 uptake and free lime conversion are tested. Huijgen W J et al. [8] investigated the carbonation of alkaline Ca rich industrial waste to improve their environmental quality. Steel slag first grounded and was carbonated in aqueous suspension to study its reaction mechanisms. The influence of different process variables such as carbon dioxide pressure, particle size, reaction time, temperature on reaction rate were investigated. Most important variables which are particle size (<2 mm to <38 μm) and reaction temperature (25-225°C). the maximum degree of carbonaiation occurs at 19 bar CO2 pressure, 100 °C temperature and a particle size of <38 μm. Durinck D et al. [9] reviewed the scientific studies dedicated to the hot stage of slag processing means slag tapping to cooling at slag yard. By using different case studies on Dicalciumsilicate driven slag integration, it is shown that properties of slag can be improved by addition after slag tapping at higher temperature to change its composition without affecting the existing metallurgical process also properties can be varied by varying the cooling rate. At present, the methods of stabilizing f-CaO can be largely classified into two types, namely carbonatization and silicatization [6-8]. With absorbing CO2 and H2O, carbonatization is to make f-CaO transform into CaCO3 and Ca(OH)2. Due to certain requirements for the atmosphere and reaction temperature, this method is still in the research state [10]. However, compared with carbonatization, silicatization is more feasible to make f- CaO change into stabilized silicate. As using the potential heat of the molten slag at the same time, this method is paid much attention to on account of its promising comprehensive economic value. Peter Drissen et al. [11] reported that with the addition of quartz sand as well as oxygen into liquid steel slag, the f-CaO content can be lower than 1%. This injection method has been successfully adopted by Thyssen Krupp in Germany. Moreover, the previous studies also reported that the technology focused on modification of steel slag at hot state was going to be the hot topic of slag treatment field in future [12,13]. This idea undoubtedly makes the development direction of slag treatment and further use of steel slag clear. Based on the studies above, though the silicatization effect at hot state is obvious, still there are many influential factors on the stabilization effect, such as ratio CaO/SiO2, reaction temperature, stabilization time and the particle size of stabilization additives. In addition, the mechanism of f-CaO stabilization on silicatization needs to be discussed deeply. Therefore, it is quite necessary to carry out further study on steel slag reforming by the method of silicatization. In this work, a process is developed to improve the volume stability of BOF slag pilot scale experiments. Pilot scale experiments have been done to stabilize the f-CaO of BOF steel slag by addition of silica rich material like river sand and quartzite fines at high temperature in molten BOF slag with Oxygen blowing. The aim of this work is to develop a process to stabilizing the f-CaO of BOF slag through silicatization route and to study the effect of basicity (B2) and different addition materials on f-CaO reduction. After experiments, modified slags were characterized for the mineral compositions and mineralogical phases. As a further mechanism of f-CaO stabilization and reactions by addition of slica rich materials in BOF slag has been discussed.

2. Concept and pilot facilities

Most of the BOF slag is used as land filling material. It can't be used as a value added materials because of its volume instability. Its volume instability caused mainly by presence of free lime (f-CaO) in Steel slag [3,4]. f-Cao can be stabilized by different methods like carbonization, silicatization, by addition of Oxygen in molten state etc. In this work, silicatization is used for stabilizing the f-CaO of BOF slag. The volume stability of BOF slag is improved by stabilizing the f-CaO for increase the utilization of BOF slag in aggregate application. Stabilization of f-CaO is done by reducing basicity of BOF slag in molten state by adding silica rich materials into it with simultaneous oxygen blowing. Here, f-CaO is stabilized mainly as Dicalcium silicate (2CaO.SiO2) phase. A new process has been developed through pilot scale experiments to improve volume stability of BOF slag by reducing its basicity in molten state before pouring in pit.

A pilot scale experimental setup is established at slag yard in SMS. Refractory lined slag pot with addition and oxygen injection facilities was created at the slag yard in a steel making shop. Refractory lined slag pot has two compartments with each having 1.5 T slag capacity as shown in figure 1. It is coated with non-sticky coating (Carbon Coating) which is used for coating of slag pot and preheated before every experiment. An oxygen injection facility was established for experiments in which Oxygen is injected manually through steel pipe which having 12.5mm diameter and pressure of oxygen was 16 bar. Additions of silica rich materials has been done by adding small packets (each having 2 Kg weight) of silica rich materials during experiments. Molten slag from the BOF process was poured into the pilot slag pot for the experiments.

Figure1: Slag Box.

3. Materials

The BOF slag and silica rich materials were used as raw materials in experiments. BOF slag was used in molten state. The chemical composition of BOF slag are listed in table 1. Three different silica containing materials, River sand, quartzite fines and water granulated BF slag were used as additions. These silica rich materials were injected in the slag pot along with oxygen in different proportions to react with expansive free lime. The River sand, Quartzite fines and water granulated BF slag were grounded and stored after drying at 120 °C for further analysis. The chemical composition of River Sand, Quartzite fines and water granulated BF slag are listed in Table 2. Here, River Sand, Quartzite fines and water granulated BF slag has 95.3%, 93.5 and 33.2% silica respectively.

Table 1: Chemical composition of unmodified BOF slag.

	CaO	SiO2	FeO	Al2O3	MgO	MnO	P2O5	TiO2	Other
Wt %	45	11.2	25.5	1.4	8.6	3.4	2	0.7	2.1

Table 2: Chemical composition of river sand and quartzite fines.

	SiO2	Al2O3	Fe2O3	CaO	Na2O	K2O	MgO	TiO2	MnO
River Sand	95.3	1.95	0.36	0.70	0.51	0.86	0.22	0.02	
Quartzite fines	93.58	2.76	1.17	0.51	0.10	0.46	1.15	0.08	
Water granulated BF slag	33.86	18.41	1.66	36.02	0.2	0.45	6.81	0.92	1.03

4. Thermodynamic feasibility

Thermodynamic feasibility is studied by the FactSage simulation [14]. For FactSage Simulation, BOF slag and 12 % SiO2 was used with 21 g of Oxygen mixing. And temperature range was 500°C to 1500°C.

Table 3: Input BOF Quantity and Quartz quantity for FactSage calculation.

	CaO	SiO2	FeO	Al2O3	MgO	MnO	P2O5	Other	SiO2(Q)
Wt(%)	40.2	10.0	22.8	1.2	7.7	3.0	1.8	2.6	10.7

Figure 2 presents the phase evolution for BOF slag treated with Quartz as predicted by FactSage equilibrium calculations. Results shows that major phases are Akermanite and Brownmillerite. There is no free lime present in Treated slag. ΔH and ΔG for this reaction at 1500°C are - 6.93346E+04 and -3.84752E+05 respectively. Values of ΔH and ΔG show that this reaction is theoretically feasible at 1500°C. In FactSage simulation Akermanite and Brownmillerite are the major phases presents in modified slag without free lime.

5. Experimentation

The BOF slag and silica rich materials were used as raw materials in this study. For characterizing BOF slag, the untreated BOF slag was sampled from Slag pot from slag yard. The BOF slag was grounded and stored after drying at 120°C for further analysis. The chemical composition of BOF slag is listed in Table 1. The binary basicity (CaO/SiO2) of the original BOF slag as the raw material in this work is approx. 4. Size of all Silica rich materials taken for experiments is less than 3mm.

Based on Slag Basicity and chemical composition of addition materials following design of experiment was finalised for experiments is shown in table 4. Each experiment was carried out with mixing of molten BOF slag, silica rich materials (Like River sand or quartz fines or water granulated BF slag) and oxygen.

Experiments were carried out in slag box with addition of silica rich materials and manual Oxygen lancing at more than 1500°C temperature as shown in figure 3 & 4. For each experiment, 2/3rd of one of the compartment of this slag box was filled with liquid BOF (1 Ton) slag by pouring directly from BOF slag pot as shown in the figure 3.

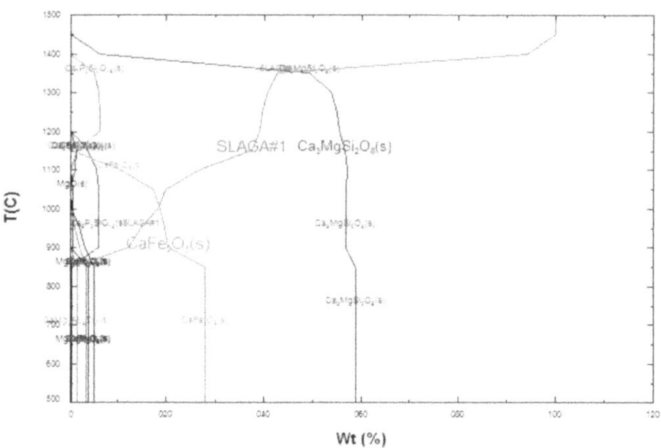

Figure 2: Phase diagram of slag phase

Table 4: Design of Experiment.

Design of Experiment			
Exp. No.	**Addition material**	**Addition Quantity(kg/T BOF Slag)**	**O_2 (Nm3/Experiment)**
1	River Sand	100, 120 , 140 and 160	107
2	Quartzite Fines	100, 120 , 140 and 160	107
3	Water granulated BF slag	120 and 300	107

Figure 3: Slag pouring in slag box.

Before pouring of molten BOF slag into slag pot, 30% of addition was added into slag pot and rest of materials was added during process in the form of small packets (each having 2 Kg weight) of silica rich materials. Simultaneously Oxygen is injected manually through steel pipe as shown in figure 4. For each set of addition, three experiments were done for repeatability of results.

Figure 4: Mixing of Silica rich material in liquid BOF with oxygen lancing in slag box.

After stabilization, the treated slags were cooled rapidly in the air, then crushed and ground into certain size for further analysis. Treated slag were further analysed to measure the mineral composition by the method of X-ray diffraction (XRD, X'pert Power, Netherlands) using Cu Ka radiation at current of 40mA and voltage of 40 kV. The scanning range was from 20 degrees to 90 degrees and the step size was set as 0.013 with time per step 5.1 seconds. Chemical composition was measured by XRF analysis. Volume expansion test of treated and untreated slag has been done as per Indian std. IS-383. Determination of free lime has been done as per std. JIS 5015.

6. Results and discussion

The impact of addition of silica rich materials into molten BOF slag on basicity (ratio of CaO and SiO2) of slag is shown in the figure 5. Figure 5 shows that basicity of unmodified BOF slag is 4.1. Basicity of modified BOF slags treated with River sand are 2.35, 2.04, 1.86 and 1.75 for addition of 100 Kg, 120Kg,140Kg and 160Kg river sand into 1T of BOF slag respectively. Basicity of modified BOF slags treated with quartz fines are 2.37, 2.18, 1.96 and 1.78 for addition of 100 Kg, 120Kg, 140Kg and 160Kg quartz fines into 1T of BOF slag respectively. Basicity of modified BOF slags treated with water granulated BF slag are 3.72 and 3.29 for addition of 120 Kg and 300 Kg river sand into 1T of BOF slag respectively. Results shows that the significant effect of addition materials on basicity of BOF slag. There is decrease in basicity with increase of quantity of addition of silica rich materials in molten BOF slag. For river sand and quartz fines addition, rate of decrease of basicity is higher for addition river sand or quartz fines than the addition of water granulated BF slag. For 120 Kg silica rich material addition, basicity of modified slags which are treated with river sand or quartz fines is decreased from 4.1 to 2 whereas there is not much change in basicity of modified slags which is treated with water granulated BF slag here basicity decreased from 4.1 to 3.72. It is directly attributed to silica content of addition material. SiO2 % of river sand and quartz fines are more than 90 % whereas SiO2% of water granulated BF slag is 33% only. Decrease in basicity is directly attributed to increase in silica content of modified BOF slag due to addition of silica rich materials into molten BOF slag.

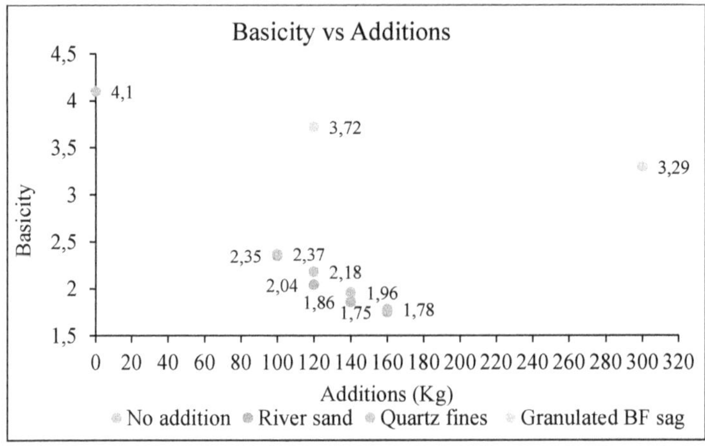

Figure 5: Basicity of modified BOF slags with varying additions.

The impact of addition of silica rich materials into molten BOF slag on free lime of slag is shown in the figure 6. Figure 6 shows that basicity of unmodified BOF slag is 6.2. free lime of modified BOF slags treated with River sand are 3.3, 0.86, 0.5 and 0.5 for addition of 100 Kg, 120Kg,140Kg and 160Kg river sand into 1T of BOF slag respectively. Free lime of modified BOF slags treated with quartz fines are 2.37, 2.18, 1.96 and 1.78 for addition of 100 Kg, 120Kg, 140Kg and 160Kg quartz fines into 1T of BOF slag respectively. free lime of modified BOF slags treated with water granulated BF slag are 3.72 and 3.29 for addition of 120 Kg and 300 Kg river sand into 1T of BOF slag respectively. Results shows that the significant effect of addition materials on free lime of BOF slag. There is decrease in free lime with increase of quantity of addition of silica rich materials in molten BOF slag. For addition river sand and quartz fines, rate of decrease of free lime is higher for addition river sand or quartz fines than the addition of water granulated BF slag. For 120 Kg silica rich material addition, free lime content of modified slags which are treated with river sand or quartz fines is decreased from 6.2 to 0.86 and 0.96 for river sand and quartz fines respectively whereas there is not much change in free lime content of modified slags which is treated with water granulated BF slag as river sand or quartz fines additions here free lime content decreased from 6.2 to 3.4. It is directly attributed to basicity of modified slag.

Decrease in free lime content of modified BOF slag is directly attributed to decrease in basicity of modified BOF slag due to addition of silica rich materials into molten BOF slag.

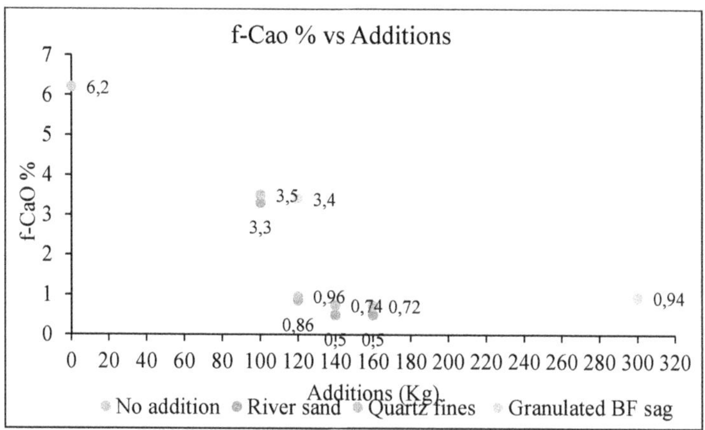

Figure 6: f-CaO % of modified BOF slags with varying additions.

The impact of addition of silica rich materials into molten BOF slag on volume expansion % of slag is shown in the figure 7. Figure 7 shows that volume expansion % of unmodified BOF slag is 4.79 %. volume expansion % of modified BOF slags treated with River sand are 2.8 %, 0.62 %, 0.5 % and 0.51 % for addition of 100 Kg, 120Kg,140Kg and 160Kg river sand into 1T of BOF slag respectively. Volume expansion % of modified BOF slags treated with quartz fines are 3.1 %, 0.92 %, 0.5 % and 0.5 % for addition of 100 Kg, 120Kg, 140Kg and 160Kg quartz fines into 1T of BOF slag respectively. volume expansion % of modified BOF slags treated with water granulated BF slag are 3.9 % and 0.81 % for addition of 120 Kg and 300 Kg river sand into 1T of BOF slag respectively. Results shows that the significant effect of addition materials on volume expansion % of BOF slag. There is decrease in volume expansion % with increase of quantity of addition of silica rich materials in molten BOF slag. For addition river sand and quartz fines, rate of decrease of volume expansion % is higher for addition river sand or quartz fines than the addition of water granulated BF slag. For 120 Kg silica rich material addition, volume expansion % of modified slags which are treated with river sand or quartz fines is decreased from 4.79 % to 0.62 % and 0.92 % for river sand and quartz fines respectively whereas there is not much change in volume expansion % of modified slags which is treated with water granulated BF slag as river sand or quartz fines additions here volume expansion % decreased from 4.79 to 3.9. It is directly attributed to free lime content of modified slag. Decrease in volume expansion % of modified BOF slag is directly attributed to decrease in free lime content of modified BOF slag due to addition of silica rich materials into molten BOF slag.

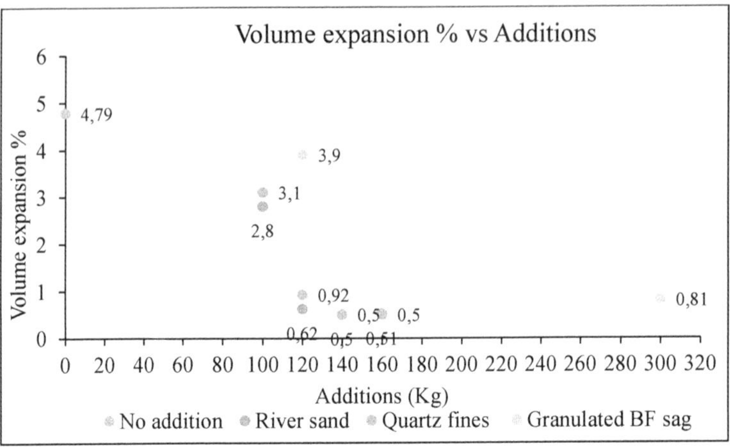

Figure 7: Volume Expansion % of modified BOF slags with varying additions.

Mineralogical study of unmodified and modified BOF slag shows that the mineralogical composition of BOF slag changes with treating by silica rich material at molten state. The major phases in unmodified BOF slags were wuestite (Fe0.944 O), Magnetite (Fe3 O4), Dicalcium Silicate (Ca2SiO4), Brownmillerite (Ca2(Al,Fe)2O5), Akermanite (Ca2 Mg (Si2 O7)) and Free lime (CaO). There is free lime of ~6wt % detected in the unmodified BOF slag with a basicity of 4. After treatment of BOF slag at molten state, the major phases were wuestite (Fe0.944 O), Magnetite (Fe3 O4), Dicalcium Silicate (Ca2SiO4), Brownmillerite (Ca2(Al,Fe)2O5), Akermanite (Ca2 Mg (Si2 O7)). Mineralogical phases of unmodified and modified slag are similar but quantity of phases changes significantly. After treatment Dicalcium Silicate, Magnetite, Akermanite and Brownmillerite phases are increased significantly with increasing the addition of silica rich materials. f-CaO and Wustite decreased significantly with increasing the addition of silica rich materials. All the results show that by the addition of silica fines at temperatures above 1550 °C, free lime is converted into di-calcium silicate and reduced the basicity which improved the volumetric stability of slag for all additives. Principal reaction for the stabilization of f-CaO in molten BOF slag by silica (SiO2) rich material is silicate forming reactions between f-CaO and SiO2 as mentioned in reactions 1 &2. [15].

$$SiO2 + 2CaO = 2CaO \cdot SiO2 \quad (1)$$

$$SiO2 + 3CaO = 3CaO \cdot SiO2 \quad (2)$$

It is also found that BOF slag after f-CaO stabilization process has significant amount of Fe3O4. It is present due reaction of FeO and manually supplied O2 as mentioned in reactions 3 &4.

$$f\text{-}CaO + SiO_2 + 3MgO{\cdot}2FeO = 2CaO{\cdot}MgO{\cdot}2SiO_2 + FeO \quad (3)$$

$$FeO + O_2 = Fe_3O_4 \quad (4)$$

The whole stabilization process can be summarized mainly into these 4 reactions (reaction 1,2,3 & 4).

7. Conclusions

A new developed process is effectively improved the volumetric stability of BOF slag by reducing its basicity in molten state before pouring in pit, through pilot scale experiments. Volume stability is improved due to addition of silica rich materials into molten BOF slag. The effects of SiO_2 addition in liquid BOF slag and oxygen partial pressure on the mineralogy of the solidified treated BOF slag were studied for reduction of free lime present in BOF slag. Different SiO_2 rich materials addition were discussed in a quantitative way. The main conclusions are summarized as follows.

- The most influential factor on f-CaO stabilization was the basicity. Free lime decreases in BOF slag with decrease in basicity by addition of SiO_2 rich materials.
- Free lime can be transformed to Calcium Disilicate by lowering the basicity of BOF slag through SiO_2 addition. At the basicity of 2, free lime can be controlled in a satisfied level to reuse the slag in constructional application. Based on the results of this work, an optimized basicity is suggested 2 or less than 2.
- For River Sand or Quartz fines additions, when the basicity was less than 2.1 and the reaction temperature was ~1550 °C the f-CaO content in the stabilized slag and Volume expansion (%) fell under the level of 1.00% which can fully meet the national standard of steel slag used for aggregate application. Based on the results from XRD, the diffraction peaks of f-CaO disappeared in Treated BOF slag after modification.
- For water granulated BF slag addition, when addition is 300Kg/ T of BOF slag and the reaction temperature was ~1550 °C the f-CaO content in the stabilized slag and Volume expansion (%) fell under the level of 1.00% which can fully meet the national standard of steel slag used for aggregate application. Based on the results from XRD, the diffraction peaks of f-CaO disappeared in Treated BOF slag after modification.
- FactSage calculation demonstrate that free lime can be removed during hot-stage treatment of BOF slag by addition of SiO_2(Quartz) under a high oxygen partial pressure condition.

The developed process is simple and cheap process for directly using the solidified slag as aggregates without any post processing.

8. References

1. Ding Y C, Cheng T W, Liu P C, Lee W H,2017,' Study on the treatment of BOF slag to replace fine aggregate in concrete', Construction and Building Materials, Vol. 146, pp. 644–651.
2. Mahieux P Y, Aubert J E, Escadeillas G, and Measson M,2014,' Quantification of Hydraulic Phase Contained in a Basic Oxygen Furnace Slag, Journal of Materials in Civil Engineering, Vol. 26, pp. 593–598.
3. Liu C W, Guo M X, Pandelaers L, Blanpain B, Huang S, 2016,' Stabilization of free lime in BOF slag by melting and solidification in air', Metallurgical and Materials Transactions B, Vol. 47B, 6, pp. 1–4.
4. Lee H S, Lim H S, Ismail M A,2017,' Quantitative evaluation of free CaO in electric furnace slag using the ethylene glycol method', Construction and Building Materials, Vol.131, pp.676–681.
5. Vaverka J, Sakurai K, 2014,' Quantitative determination of free lime amount in steelmaking slag by X-ray diffraction', ISIJ International, Vol.54,6, pp.1334–1337.
6. Motz H, Geiseler J,2001,' Products of steel slags an opportunity to save natural resources', Waste Manage, Vol.21, 3, pp.285–293.
7. Santos R M, Ling D, Sarvaramini A,2012.'. Stabilization of basic oxygen furnace slag by hot-stage carbonation treatment', Chemical Engineering Journal, Vol.203,5. Pp.239–250.
8. Huijgen W J, Witkamp G J, Comans R N,2005,' Mineral CO2 sequestration by steel slag carbonation', Environ SciTechnol, Vol. 39,24, pp.9676–9682.
9. Durinck D, Engström F, Arnout S, et al,2008,' Hot stage processing of metallurgical slags', Resources, Conservation and Recycling, Vol. 52,10, pp.1121–1131.
10. Mikhail S A, Turcotte A M, 1995,' Thermal behaviour of basic oxygen furnace waste slag'. Thermochim. Acta, Vol. 263, pp.87-94.
11. Drissen P,Ehrenberg A, Kühn M, Mudersbach D,2009,' Recent Development in Slag Treatment and Dust Recycling', Steel research international, Vol. 80,10, pp. 737-745.
12. Li G, Guo M,2014,' Current Development of Slag Valorisation in China', Waste Biomass Valor, Vol. 5, pp. 317-325.
13. Li J, Yu Q, Wei J, Zhang T, 2011, 'Structural characteristics and hydration kinetics of modified steel slag', Cement and Concrete Research, Vol. 41, pp.324 – 329.
14. C.W. Bale, E. Bélisle, P. Chartrand, S.A. Decterov, G. Eriksson, A.E. Gheribi, K. Hack, I.-H. Jung, Y.-B. Kang, J. Melançon, A.D. Pelton, S. Petersen, C. Robelin, J. Sangster, P. Spencer, M-A. Van Ende Reprint of: FactSage thermochemical software and databases, 2010–2016 Calphad, Volume 55, Part 1, December 2016, Pages 1-19
15. Yin X,Zhang C,Wang G,Cai1 Y,Zhao C, 2018,' Stabilization of free CaO in molten BOF slag by addition of silica at high temperature', Metallurgical Research & Technology, Vol. 115, pp.414 – 423.

Author's address(es)

*Corresponding author: Mukund Manish

*E-mail: mukund.manish@jsw.in

*Tel.: +918861450931

Settlement characteristics of saturated mechanical biological treated waste from Croatia

Nikola Kaniški, Nikola Hrnčić, Igor Petrović

Department of Environmental Engineering, University of Zagreb, Faculty of Geotechnical Engineering, Varazdin, Croatia

Abstract

Mechanical biological treatment (MBT) technology was developed in Germany with the aim of reducing the amount of biodegradable waste before landfilling. MBT has several types of output products. Apart from useful and recyclable products, one type of waste from the MBT process cannot be used as waste fuel or recyclable material but is still quite rich in organic matter. This type of waste, called the methanogenic fraction, goes to a bioreactor landfill, where water is added and recirculated to accelerate and increase the biodegradation of the methanogenic fraction, after which biogas is extracted. The focus of this paper is to determine the settlement characteristics on a single wet specimen of the methanogenic fraction (particle size < 25 mm) from a Croatian MBT plant. In order to conduct the experiment, a 150 mm diameter oedometer cell was used. Continuous settlement measurements were performed in three points using two callipers and one calibrated linear transducer (LVDT). The results showed that, as the load of the specimen increases, the coefficient of consolidation and coefficient of permeability values decrease, making the material denser and, therefore, more challenging for water recirculation and biogas extraction.

Keywords

Biodried waste, settlement characteristics, methanogenic fraction, oedometer test, consolidation coefficient, permeability

1 Introduction

Mechanical and biological treatment, or MBT, is a very common waste treatment technology used to reduce the amount of biodegradable waste before landfilling. In the Republic of Croatia, so far, two Waste Management Centers (WMC) are in operation that use MBT as a waste treatment technology and are public property (WMC Kaštijun and WMC Marišćina). Given that many MBT technology variations have been developed so far, this term includes plants with large differences in technical equipment and operating conditions. The Croatian plants that were mentioned use biodrying as their biological treatment technology.

The MBT process can be designed to have one or more primary output products. In addition to the primary products that can be created by the MBT process, in all MBT processes, secondary output products are created, such as recoverable materials (paper, metals, plastics), waste material that must be disposed of in a landfill, wastewater, and air emissions (WASTE MANAGEMENT PLAN OF THE REPUBLIC OF CROATIA FOR THE PERIOD 2007-2015, NN 85/2007). Waste material that must be disposed of in a landfill is still quite rich in organic matter and is called methanogenic fraction. This type of waste can go to a bioreactor landfill (BL), where after the material is installed and the landfill body is closed, water is added and recirculated to accelerate or increase the biodegradation of waste material, after which biogas is extracted. In contrast to conventional dry landfills, BLs experience faster settlements due to increased waste decomposition rates and increased pressures due to higher specific weights. Therefore, determination of settlement characteristics is very important to a quality design process (PETROVIC, 2016). Although many researchers studied MBT waste (VELKUSHANOVA, 2011; CARRUBBA AND COSSU, 2003; BAUER ET AL., 2007; KUEHLE-WEIDEMEIER ET AL., 2000; BIDLINGMAIER ET AL., 1999; DUELLMANN, 2002; BAUER ET AL., 2006; HEISS-ZIEGLER AND FEHRER, 2003; SIDDIQUI ET AL., 2012; BORTOLUZZI, 2014), all studies involved biostabilised and not biodried waste material.

In this paper, the authors present results of a single wet specimen of the methanogenic fraction (particle size < 25 mm) from WMC Marišćina located in Croatia. The results are presented as basic geotechnical parameters of studied waste material (composition, specific gravity, particle size distribution, organic matter content) and as consolidation and hydraulic parameters (coefficient of consolidation and permeability) determined with a specially designed oedometer device. Furthermore, the obtained results related to the permeability coefficient were compared to the permeability coefficient for various wastes presented in PETROVIC, 2016.

2 Components of the oedometer device

For purposes of this research, an oedometer device with a 150 mm internal diameter and an 80 mm maximum height was used. The device consists of a rigid cell, rigid top plate, top plate sliding rail, Archimedes pulley, a rigid flat board with a hole in the middle, and a displacement transducer together with two digital callipers. Figure 1 shows the oedometer assembly and measuring system installed on a platform with a hoist originally developed for a large oedometer device (PETROVIC ET AL., 2014).

Figure 1 Oedometer assembly with and without weights

The applied dead load is a square metal plate with a 50 mm diameter hole in the middle. The hole serves as a guide so that a round metal rod attached to the top plate can be placed precisely in the center of the mass. The total applied load are 10 metal plates, each 20 kg (± 0.2 - 0.3 kg) in weight producing a total sitting pressure of 119.5 kPa on the waste specimen. The load is applied to the specimen with the help of a rope, power tape, crane, and an Archimedes pulley.

The oedometer cell consists of an upper ring connected with three steel bolts in the horizontal direction to the specimen ring. A sliding rail (top plate) passes through the upper ring and goes all the way to the sample. The initial sitting pressure of the top plate is 5.39 kPa. The top plate has a threaded hole on the top so that a round, four-piece metal rod can be screwed into it. The platform has a drainage channel that provides a clear drainage path for the excess pore water. On top of the specimen and below the top plate, a geomembrane is placed to prevent any excess pore water to seep out of the cell through the top. To prevent clogging of the drainage holes, a geotextile is placed at the bottom of the cell. To reduce friction between the top plate and the specimen ring, talc powder is sprinkled on the top plate before installing the specimen (RUDENKO AND BANDYOPA, 2013).

On top of the top plate, a rigid wooden plate is installed with a hole in the middle so that a metal rod can be attached through it. As shown in Figure 1, on the right side of the specimen, a Vishay 155 L calibrated displacement transducer (LVDT) is used for measuring the settlements. The data logger is used to record the change in the height of the specimen under dead load every 10 seconds. On the left side of the specimen, two callipers with a maximum measurement capacity of 150 mm are used. Each is placed on one edge of the rigid wooden plate so that three measurement points (2 callipers and a LVDT formed a triangle) provide differential settlement measurements. Both callipers are connected to a computer with an Arduino system, where the change in settlement is logged every second.

3 Arduino working principle

Arduino is an easily programmed open-source microcontroller. It was designed and brought to the public in 2005 to interact with the environment using sensors and actuators (LEO, 2016). In order to program the code (known as a sketch) for this research, a board was used. The Arduino board consists of hardware and software. The hardware designs are available online for free from the official website https://www.arduino.cc/. For Arduino hardware to work properly, a series of components must be used, such as: microcontroller, external power supply, USB plug, internal programmer, reset button, analog pins, digital input/output pins, and power and GND (ground) pins. Software is used to develop a program code (sketch). A commonly used Arduino software is Arduino IDE. The IDE contains the following parts:

- text editor,
- message area,
- text,
- console toolbar (LEO, 2016).

In this work, two Arduino Uno boards are used to read the data from digital callipers (Figure 2). The digital callipers have a digital output connector located under a small plastic cover, which is used to connect the Arduino Uno board and the readings on the callipers (Figure 3) with a USB cable. A new development code can be uploaded to the board very easily and quickly via a USB cable and, if necessary, the code can be changed and uploaded again in a very short time. The code used for the purposes of this work was compiled by THALHEIMER, M. and downloaded from his web page (see the reference list).

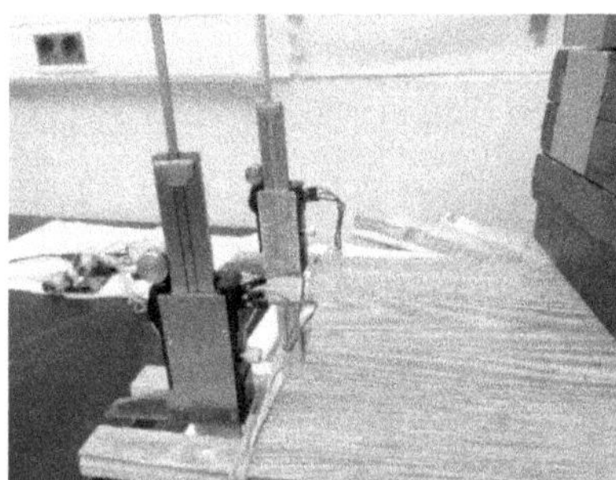

Figure 2 Two callipers connected with Arduino Uno

The output, which appears on the serial monitor on a personal computer, is exactly the same as what can be read on the callipers' LCD. YAT (yet another terminal) was used as a serial terminal to write the measured values from the callipers to a file on a personal computer (available at: https://sourceforge.net/projects/y-a-terminal/). After the end of the experiment, the values recorded every second throughout the test were averaged for every ten seconds in order to compare the recorded data to the results obtained with the Vishay 155 L displacement transducer.

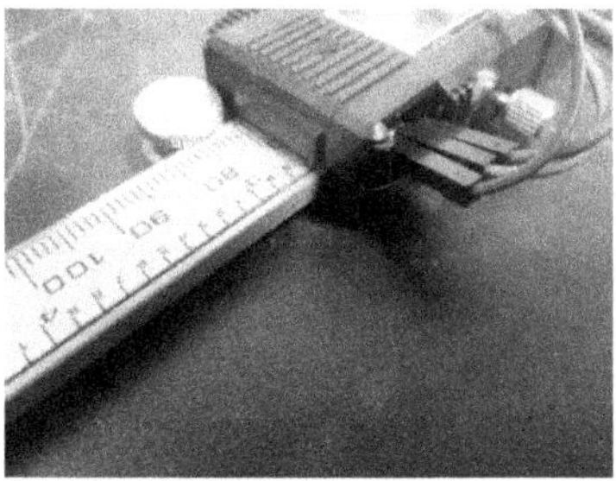

Figure 3 Digital calliper and output connectors

4 Tested material

In WMC Mariščina, mixed municipal solid waste is collected from households and brought with garbage trucks to a receiving pit. The waste material is first subjected to mechanical treatment (i.e., crushing) into a fraction < 200 mm and then sent with conveyors for further processing. The fraction > 200 mm is sent back to the crushing step. After the first mechanical treatment step, the waste is sent to concrete chambers, where the moisture content in the material is minimized by forced aeration for seven to ten days. To obtain as much recovery material as possible, the dried material is sieved again and subjected to various mechanical refining procedures to get as much separated recyclable material (plastics, metals, non-ferrous metals). In this stage of processing, the so-called methanogenic fraction, which is not suitable for waste fuel, is separated with a maximum particle size of about 25 mm. This type of material is disposed of in its dry state in a BL located within the WMC Mariščina, from which biogas should be produced after BL closure and recirculation of the leachate. For this reason, it is very important to study how the installed waste material will react after the water is introduced in the BL body.

4.1 Testing procedure

In this paper, the settlement characteristics (coefficient of consolidation and permeability) on a single wet specimen of the methanogenic fraction were obtained. For this purpose, a 150 mm diameter oedometer cell was used with a dead load being ten metal plates with a total pressure of 119.5 kPa. To prevent tilting of the top plate, a sliding rail was used, and the remained tilting was measured with three displacement transducers positioned in a triangular shape. Continuous settlement measurements were performed in three points using two callipers with 0.01 mm resolution and 150 mm maximum travel distance and one calibrated Vishay 115 L linear transducer with a maximum travel distance of 100 mm and 0.1 mm resolution. The aim of this test was to examine the compressibility of the methanogenic fraction in wet conditions.

The specimen of the dry methanogenic fraction was installed into an oedometer cell in three layers. The installed dry specimen density was 380 kg/m^3 as reported by the WMC Marišćina operators. At the bottom of the oedometer cell, a geotextile was placed to prevent clogging of the bottom drainage path. The specimen was installed 1 cm lower than the maximum height of the oedometer cell. The geomembrane was placed on top to prevent fines of the installed material from entering the gap between the container and the top plate and to prevent the moisture from seeping out of the container. 0.4224 kg total specimen mass was installed in a 63 mm height in its dry state. Each layer was compressed with a 2 kg weight by hand. After the specimen was installed, a solution of deaerated water and acetic and propionic acid was introduced into the specimen from the bottom of the cell. The solution of acids prevented biodegradation of the installed waste material (SIDDIQUI ET AL., 2012). A total of 535 973.451 mm^3 solution was added with a GDS Enterprise Level Pressure Volume Controller slowly enough to prevent any disturbance of the initial height of the installed specimen (63 mm). No loads were on top of the specimen during the saturation procedure. The initial bulk density after the saturation procedure was 0.859 g/cm^3. Initial parameters of the installed specimen are presented in Table 1.

Table 1 Initial parameters of installed methanogenic fraction

Condition	ρ (g/cm^3)	w (%) (WM*)	ρ_d (g/cm^3) (DM*)	ρ_s (g/cm^3) (DM)	e_0	S_r (%)
Wet	0.859	55.93	0.380	1.894	3.984	60.32

*DM-dry matter; **WM-wet matter

ρ bulk density, w moisture content, ρ_d dry density, ρ_s average solid particle density, e_0 void ratio in dry state of the specimen, and S_r degree of saturation.

The initial sitting pressure as a first step in the consolidation procedure was obtained with the use of the top plate, which produced 5.39 kPa of pressure to the specimen. Immediate settlement was recorded by hand after the top plate was placed on top of the specimen with the use of the calliper. After the immediate settlement was recorded, the rigid plate was placed on top of the top plate together with the first metal rod, and the sample was subjected to an additional load of 11.78 kPa. Total cumulative sitting pressure on the specimen is as follows: 5.39, 17.17, 28.38, 51.32, 74.2, 96.68, and 119.5 kPa. Each loading step except top plate lasted for about 24 hours before the next step was applied.

5 Experimental results and discussions

After the waste samples were delivered to the Faculty of Geotechnical Engineering (University of Zagreb), Laboratory for Environmental Engineering, a series of tests were performed. Table 2 shows individual constituent components of the methanogenic fraction, which were manually separated and weighed to determine the average composition of the studied waste material.

Table 2 Average composition of biodried waste components

Component	Mass percentage [%]
	Marišćina
Plastics	6.43
Textile	0.22
Glass	10.62
Metal	0.94
Paper/Cardboard	4.71
Wood	1.18
Bones/Skin	0.20
Stones	2.76
Ceramics	0.46
Rubber	0.13
Kitchen waste	2.15
Unidentified > 2 mm	42.48
Unidentified < 2 mm	27.72
Total	100

Figure 4 shows the averaged particle size distribution curve for the studied methanogenic fraction. The averaged distribution curve was obtained on a series of 25 sieved samples on a series of sieves as follows: 31.5, 16, 8, 4, 2, 1, and 0.5 mm.

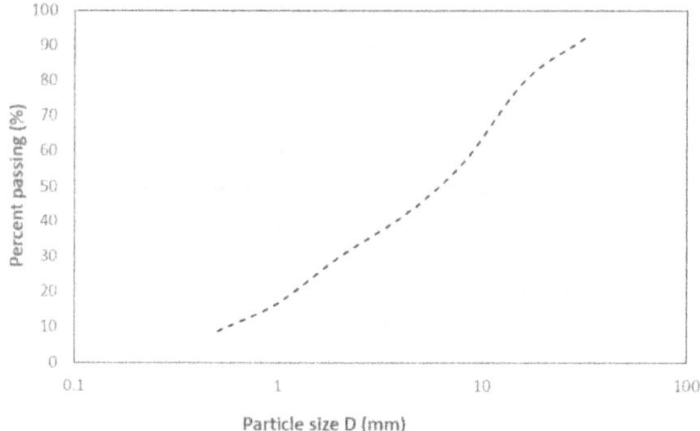

Figure 4 Averaged particle size distribution curve for the studied biodried material

The organic matter content was obtained with a 50 g representative sample. The sample was subjected to a muffle furnace at 440 °C until no further changes in mass were observed. With respect to the mass before and after ignition, the calculated organic matter content of the studied methanogenic fraction was 51.6 %.

The solid particle density of the waste sample was determined with a gas pycnometer method. The results vary from 1.69 g/cm^3 to 2.19 g/cm^3. On a series of 23 samples, the average solid particle density of the methanogenic fraction was 1.88 g/cm^3 with a standard deviation of ± 0.13 g/cm^3. The wide range of solid particle density can be attributed to the heterogeneity of the material.

5.1 Settlement curve

The time settlement curve is presented in Figure 5. Immediate settlement caused by pressure of the top plate was measured by hand, and it was determined to be 10.85 mm. The immediate settlement value is not included in Figure 5. According to settlement measurements obtained with LVDT and two callipers, the maximum difference between the LVDT measurement and the calculated settlement value in the center of the specimen was 0.27 mm. Hence, the differential settlements can be disregarded.

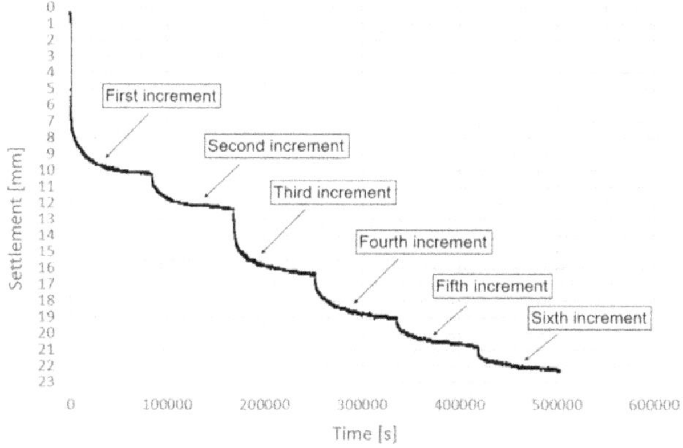

Figure 5 Time-settlement curve for the wet specimen of the methanogenic fraction from WMC Marišćina

In order to determine the consolidation coefficient, a Taylor method was used, which has been confirmed as a valid method for determining c_v of a saturated MBT waste material by SIDDIQUI ET AL., 2013, (Equation 1).

$$c_V = \frac{0.848 * d^2}{t_{90}} \quad (1),$$

where d is the drainage path in mm and t_{90} is the time in seconds required to achieve 90 % of the primary consolidation. The consolidation coefficient for every increment is obtained by plotting settlement (mm) against $\sqrt{t}$ (min), as presented in Table 3.

From the theory of one-dimensional consolidation, it follows that from the results of the oedometer experiment, it is possible to calculate the coefficient of permeability of the testing specimen using the following expression (2):

$$k = \frac{\gamma_w * c_v}{M_v} \quad (2),$$

where k is the coefficient of permeability, γ_w is the unit weight of water (9810 N/m^3), c_v is the consolidation coefficient (m^2/s), and M_v is the oedometric moduli (Pa). Calculated results of the coefficient of permeability are shown in Table 3. Oedometric moduli was obtained using equation (3):

$$M_v = \frac{\Delta\sigma_v}{\Delta\varepsilon} \quad (3),$$

where $\Delta\sigma_v$ is the increment of vertical total stress (kPa) and $\Delta\varepsilon$ is the strain increment.

The vertical strains were calculated by using equation (4).

$$\varepsilon = \frac{\Delta h}{h} \quad (4),$$

where Δh is the change in total specimen height and h is the initial specimen height.

For a more detailed discussion about vertical (true and engineering) strains in waste materials, please take a look at POWRIE ET AL. (2019).

Table 3 Consolidation coefficient c_v, time to achieve 90 % of the primary consolidation, permeability k, and oedometric moduli for every increment of the present research

Increment No./stress (kPa)	First/ 5.39-17.17	Second/ 17.17-28.38	Third/ 28.38-51.32	Fourth/ 51.32-74.2	Fifth/ 74.2-96.68	Sixth/ 96.68-119.5
M_v (Pa)	72,430	335,500	363,990	527,330	802,040	1,091,040
t_{90} (s)	392.04	705.43	5,166.73	8,663.89	10,000.00	8,663.89
c_v (m²/s)	$4.813 * 10^{-6}$	$2.019 * 10^{-6}$	$2.361 * 10^{-7}$	$1.170 * 10^{-7}$	$8.858 * 10^{-8}$	$9.267 * 10^{-8}$
k (m/s)	$6.519 * 10^{-7}$	$5.903 * 10^{-8}$	$6.364 * 10^{-9}$	$2.176 * 10^{-9}$	$1.083 * 10^{-9}$	$8.332 * 10^{-10}$

For the consolidation coefficient shown in Table 3, equation (1) - Taylor method was used, whereas for the coefficient of permeability, equation (2) was used. The data of permeability coefficients were compared to the permeability coefficients for various wastes presented in PETROVIC, 2016. The author made a comprehensive literature review, and the results show that for pre-treated MSW with the same particle size range of 0 - 25 mm, the values of coefficient of permeability range from $7.8 * 10^{-8}$ (m/s) to $8.2 * 10^{-11}$ (m/s). However, it should be noted that for the tested waste material, the settlement vs. $\sqrt{t}$ curves did not show a strong distinction between primary and secondary compression, as is the case for soil materials. Therefore, the determination of the linear portion of the settlement vs. $\sqrt{t}$ curve is not straightforward and consequently, the results are very much dependent on the personal judgement of the person who interprets the data.

5.2 Basic geotechnical parameters

Table 4 presents basic geotechnical parameters at the end of the testing procedure. Comparing Table 1 and 4, it can be concluded that the values of ρ and ρ_d have increased at the end of the testing procedure, which was to be expected.

Table 4 *Basic parameters of tested specimens at the end of the oedometer test*

Condition	**ρ (g/cm³)**	**w (%) (WM**)**	**ρ_d (g/cm³) (DM*)**	**ρ_s (g/cm³) (DM*)**	**e_1**	**S_r (%)**
Wet	1.083	36.1	0.80	1.894	1.38	77.68

*DM - dry matter; **WM - wet matter

6 Conclusions

This paper provides an overview of the consolidation parameters and basic geotechnical parameters of a single wet specimen of biodried mixed municipal solid waste, the so called methanogenic fraction from the plant for mechanical-biological waste treatment in Marišćina, Croatia. For that purpose, a specially designed oedometer device was used.

The conducted test on the saturated material mimicked the conditions that would occur at the plant during the operation of the BL. The obtained test results show that attention should be paid to the amount of water recirculated in the landfill body to allow uninterrupted operation. The results show that, as the load of the specimen increases, the stiffness modulus also increases while the permeability coefficients simultaneously decrease, making the material denser, and therefore, more challenging for water recirculation and bio-gas extraction.

The obtained permeability coefficients fit reasonably well within the range of permeability coefficients obtained on various waste materials with a similar particle size range presented in Petrovic, 2016.

Given that waste samples were taken in the winter (February 2019), new tests should focus on material produced during the summer period to compare the impact of the tourist season on the settlement parameters and basic geotechnical parameters of waste.

Acknowledgments

The financial support of the Croatian science foundation for the project "Testing and modelling of mechanical behavior of biodryed waste as a Waste-to-Energy prerequisite" (UIP-05-2017) is gratefully acknowledged.

7 Literature

Arduino, an open-source electronics platform		Available at: https://www.arduino.cc/
Bauer, J., Münnich, K., Fricke, K.	2006	Proceedings of the Fourth Asian-Pacific Landfill Symposium, Shangai, China, 2006. Investigation of Mechanical Properties of MBT Waste.
Bauer, J., Münnich, K., Fricke, K.	2007	Proceedings Sardinia 2007, Eleventh International Waste Management and Landfill Symposium, S. Margherita di Pula, Cagliari,

		Italy, 1-5 October. Influence of Hydraulic Properties on the Stability of Landfills.
Bidlingmaier, W., Scheelhaase, T., Maile, A.	1999	Universität Gesamthochschule Essen, Fachbereich 10 – Bauwesen, Fachgebiet Abfallwirtschaft, 1999. Langzeitverhalten von mechanisch-biologisch vorbehandeltem Restmüll auf der Deponie. Abschlußbericht zum Teilvorhaben 3.1 des BMBF-Verbundvorhabens „Mechanisch-biologische Behandlung von zu deponierenden Abfällen".
Bortoluzzi, A.	2014	PhD thesis. Behaviour of an MBT waste in monotonic triaxial shear test. University of Padova, Department of Civil Engineering.
Carrubba, P., Cossu, R.	2003	Proceedings Sardinia 2003, Ninth International Waste Management and Landfill Symposium, S. Margherita di Pula, Cagliari, Italy, 6-10 October. Investigation on Compressibility and Permeability of Pre-treated Waste Mixture.
Duellmann, H.	2002	Im Auftrag des Abfallwirtschaftsbetriebes Hannover, 2002. Untersuchungen zum Einbau von MBA-Abfällen auf der Zentraldeponie Hannover, Laboruntersuchungen zum Verdichtungs-, Durchlässigkeits-, Last-Setzungs- und Scherverhalten. Februar 2002.
Heiss-Ziegler, C. and Fehrer, K.	2003	Proceedings Sardinia 2003, Ninth International Waste Management and Landfill Symposium S. Margherita di Pula, Cagliari, Italy; 6 - 10 October 2003 - 2003 by CISA, Environmental Sanitary Engineering Centre, Italy. Geotechnical behaviour of mechanically-biologically pretreated MSW.
Kuehle-Weidemeier, M., Doedens, H., von Felde, D.	2000	Oberflächenabdichtung von Deponien und Altlasten. "Die mechanisch-biologische Restabfallbehandlung und die Ablagerung der biologischen Fraktion". Erich-Schmidt Verlag, 2000.
Leo, L.	2016	Working principle of Arduino and using it as a tool for study and research. International Journal of Control, Automation, Communication and Systems (IJCACS), Vol.1(2), pp 21-29.
Petrovic, I., Stuhec, D., and Kovacic, D.,	2014	Large Oedometer for Measuring Stiffness of MBT Waste. Geotechnical Testing Journal, Vol. 37(2), pp 296–310, DOI:10.1520/GTJ20130015. ISSN 0149-6115
Petrovic, I.	2016	Mini-review of the geotechnical parameters of municipal solid waste: Mechanical and biological pre-treated versus raw untreated waste. Waste Management & Research, Vol. 34(9), pp 840 –850.

Powrie, W., Xu, X. b., Richards, D., Zhan, L. t., Chen, Y. m.	2019	Mechanisms of settlement in municipal solid waste landfills. Journal of Zhejiang University-SCIENCE A (Applied Physics & Engineering), Vol. 20(12), pp 927-947. ISSN 1673-565X.
Rudenko, P., Bandyopadhyay, A.	2013	Talc as friction reducing additive to lubricating oil. Applied Surface Science, Vol. 276, pp 383-389.
Siddiqui, A. A., Richards, D. J., Powrie, W.	2012	Investigations into the landfill behaviour of pretreated wastes. Waste Management, Vol. 32, pp 1420–1426. ISSN 0956-053X.
Siddiqui, A. A, Powrie, W., Richards, D. J.	2013	Settlement Characteristics of Mechanically Biologically Treated Wastes. Journal of Geotechnical and Geoenvironmental Engineering, Vol. 139(10), pp 1676-1689.
Thalheimer, M.		https://sites.google.com/site/marthalprojects/home/arduino/arduino-reads-digital-caliper
Velkushanova, K.	2011	PhD Thesis. Characterization of wastes towards sustainable landfilling by some physical and mechanical properties with an emphasis on solid particles compressibility. Faculty of Engineering and the Environment, University of Southampton.
Waste Management Plan of the Republic of Croatia for the period 2007-2015.	2007	OG 85/2007. Available at: https://narodne-novine.nn.hr/clanci/sluzbeni/2007_08_85_2652.html
Source Forge-YAT		Available at: https://sourceforge.net/projects/y-a-terminal/

Author's addresses

Nikola Kaniski, mag.ing.amb.
University of Zagreb, Faculty of Geotechnical Engineering
Department of Environmental Engineering
Hallerova aleja 7, HR-42000 Varazdin
Phone: +385 42 408 925
nikola.kaniski@gfv.unizg.hr

Nikola Hrncic, mag.ing.aedif., mag.ing.geoing.
University of Zagreb, Faculty of Geotechnical Engineering
Department of Environmental Engineering

Hallerova aleja 7, HR-42000 Varazdin
Phone: +385 42 408 925
nikola.hrncic@gfv.unizg.hr

Assoc. Prof. Igor Petrovic
University of Zagreb, Faculty of Geotechnical Engineering
Department of Environmental Engineering
Hallerova aleja 7, HR-42000 Varazdin
Phone: +385 42 408 919
igor.petrovic@gfv.unizg.hr

Maximum and minimum void ratio characteristics of MBT waste

Nikola Hrncic[1], Nikola Kaniski[1] and Igor Petrovic[1]

[1]Department of Environmental Engineering, University of Zagreb, Faculty of Geotechnical Engineering, Varazdin, Croatia

Abstract

Mechanical-biological treatment (MBT) of municipal solid waste (MSW) includes mechanical processing of MSW with subsequent biological treatment of its biodegradable components. One of the many possible outputs of the MBT plant is the so called "fine fraction" of MBT waste which is characterised by particle sizes smaller than 25 mm. This mechanically refined granular material is only partially stabilised during the aerobic degradation process which takes place under controlled conditions in the MBT plant (bio-drying). Rich in organic components, the "fine fraction" of MBT waste is suitable for landfilling into a bioreactor landfill with the aim of biogas production. Mechanical behaviour of the bioreactor landfill, during construction and use, is significantly influenced by the physical characteristics of the MBT waste material that is being installed. In this study, the physical and geotechnical properties of the MBT waste material were determined, with emphasis on the determination of maximum and minimum void ratios. Correlations relating maximum to minimum void ratios are also presented. Test results are compared to the results obtained for soils and some other granular materials, found in the available publications.

Keywords

MBT waste, geotechnical properties, landfill, void ratio

1 Introduction

Basic waste management principles for the EU member states are laid down by the Directive 2008/98/EC on waste (Waste Framework Directive). One of the major requirements of the abovementioned Directive is the reduction of the biodegradable component of waste before the waste is ultimately deposited to a landfill. Landfilling biodegradable waste is the least preferred option in managing such waste as it poses a significant environmental threat. When decomposing in landfills, biodegradable waste produces methane, one of the gases that contributes to the greenhouse effect. Landfilling of biodegradable waste is addressed in the Landfill Directive (EU Landfill Directive 1999/31/EC of April 1999) which requires the diversion of biodegradable municipal waste from landfills. According to the Directive, member states are required to only landfill wastes that have been subjected to treatment, leading to a reduction in their quantity or hazard to human health and the environment (DI LONARDO ET.AL., 2012., ROBINSON ET AL., 2005).

Regarding the production of material suitable for landfilling, the waste pre-treatment objectives should relate to minimising the adverse consequences of disposal (HEERENKLAGE AND STEGMANN, 1995; LEIKAM AND STEGMANN, 1995; RIEGER AND BIDLINGMAIER, 1995;

SCHEELHAASE AND BIDLINGMAIER, 1997; KOMILIS, 1999; SOYEZ AND PLICKERT, 2002; FRICKE ET AL., 2005; MUNICH ET AL., 2006; MONTEJO ET AL., 2010; VELIS ET AL., 2010), including:

- minimisation of volume and mass of waste to be disposed in landfill;
- inactivation of biological and biochemical processes in order to reduce leachate and methane production and odour emissions;
- immobilisation of pollutants of the waste to be disposed of in order to reduce leachate contamination;
- reduction of landfill settlement;
- reduction of the duration of the landfill site aftercare. (DI LONARDO ET.AL., 2012)

Mechanical-biological treatment of MSW in the MBT plant consists of mechanical processing of waste, which generally includes shredding (i.e., size reduction and densification, separation of waste by particle sizes, density or magnetic properties), and biological treatment of MSW by means of aerobic or anaerobic degradation.

2 Material and methods

Sampling of the MBT waste specimens used for the purpose of this study took place at the Regional Centre for Waste Management, Mariscina. The centre is located in the Istria region in the south-west of Croatia. It is functioning as an MBT plant with the main focus on processing MSW which is collected in the local municipalities. Since the composition of input MSW varies seasonally, as a consequence of the tourist season during the summer, it is also worth mentioning that the MBT waste samples were collected in the winter season.

For a short period of time, the collected MSW is stored in the receiving pit of the MBT plant. Shredding of MSW to a fraction with particle sizes smaller than 200 mm is the first step in its mechanical processing. After the initial shredding, MSW is transported to the bio-drying bioreactor where biological treatment takes place. In the bioreactor, MSW is dried by a combination of heat that is produced during aerobic decomposition of the organic fraction of waste and excess aeration. Waste is subjected to bio-drying for a period of 7 to 10 days.

Subsequent mechanical processing of the bio-dried MSW includes further shredding/refinement and extraction of recyclables (e.g., ferrous and non-ferrous materials, plastics etc.). The final results of MBT processing are various material outputs with different characteristics. For the purpose of this study, the so-called "fine fraction" of bio-dried MBT waste was used (Figure 1). This output fraction of MBT waste is characterised by nominal particle sizes smaller than 25 mm and is appropriate for landfilling in the bioreactor landfill.

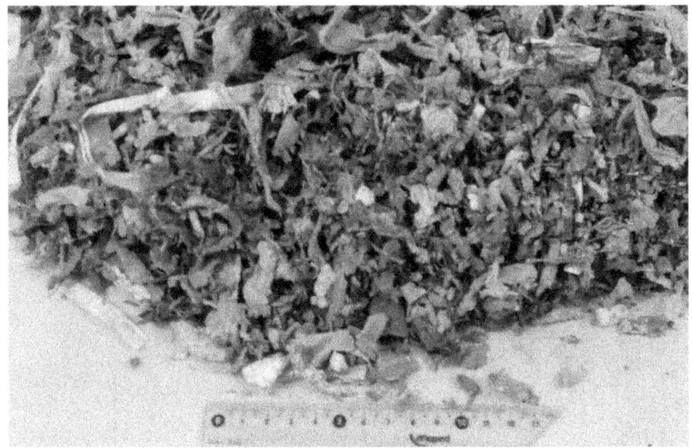

Figure 1 "Fine fraction" of MBT waste material

Physical and geotechnical properties of the MBT waste material were determined according to the internationally accepted ASTM standards for laboratory testing of soils.

2.1 Moisture content

Water content was determined according to the ASTM D 2216 Standard for Laboratory Determination of Water (Moisture) Content of Soil and Rock by Mass. In order to preserve the original in-situ moisture of the waste specimens, immediately after sampling and prior to transport, waste specimens were placed in plastic vacuum bags and airtight sealed. The test was performed shortly after the arrival of the waste specimens to the laboratory. Since the MBT waste material is rich in organic components, the standard drying temperature was reduced and set to 60°C, as is proposed in the standard. Samples were oven-dried for a period of 24 h. Mass of moist and dried samples was determined using a balance, and the average value of in-situ moisture content was calculated as 9.60 %.

2.2 Physical composition

The physical composition of MBT waste samples was determined by means of manual separation. Different material components contained in the waste sample were identified visually. Due to the intensive mechanical and biological treatment that the waste material has been subjected to in the MBT plant, for almost 70 % of mass of the waste sample it was not possible to determine with certainty its physical composition. Waste material with undetermined physical composition was sieved through a 2.00 mm sieve. Portions of material retained and passing the 2.00 mm sieve were weighed and marked as "Unidentified > 2 mm" and "Unidentified < 2 mm".

Mass of each individual material component was determined on a balance and mass percentages were calculated, as shown in Table 1.

Table 1 Mass percentages of different material components of MBT waste

Material Component	Mass percentage [%]
Plastics	6.43
Textile	0.22
Glass	10.62
Metal	0.94
Paper/Cardboard	4.71
Wood	1.18
Bones/Skin	0.20
Stones	2.76
Ceramics	0.46
Rubber	0.13
Kitchen waste	2.15
Unidentified > 2 mm	42.48
Unidentified < 2 mm	27.71
Total	100

2.3 Particle size distribution

Quantitative determination of the particle sizes in the MBT waste samples was conducted according to the ASTM D 422 standard commonly used for particle-size analysis of soils. A mass of roughly 10 kg of the MBT waste material was oven-dried during a period of 24 hours at constant temperature of 60°C. Prior to sieve analysis, dry material was divided into 25 smaller samples (231-600 g of mass per sample). Samples were mechanically sieved through a set of sieves with various sizes of screen openings (from top to bottom sieve: 31.5 mm, 16 mm, 8 mm, 4 mm, 2 mm, 1 mm and 0.5 mm).

Based on the results of the sieving analysis of all 25 samples, an average value particle size distribution curve of the MBT waste material was constructed and is shown in Figure

2 below. For comparison, four additional particle size distribution curves for bio-stabilised (composted) waste are also shown, of which the top two are for Austrian bio-stabilised waste (PETROVIC, 2016), and the bottom two for bio-stabilised waste from the UK and Germany (VELKUSHANOVA, 2011).

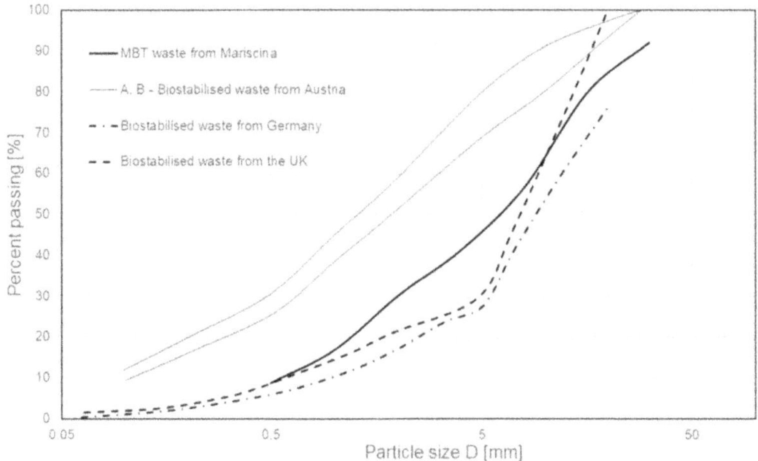

Figure 2 Particle size distribution curves of MBT and bio-stabilised waste

According to the particle size distribution curve values for MBT waste, the coefficient of uniformity (Cu) and the coefficient of curvature (Cc) were calculated as: Cu = 14.23 and Cc = 1.07, while the median grain size D_{50} was 6.27 mm, and the effective grain size D_{10} was 0.58 mm. In comparison with soil, MBT waste can be defined as a coarse-grained and well graded material with Cc and Cu values such as in the case of well graded gravel (GW).

2.4 Organic matter content

The organic matter content of MBT waste was determined on a representative sample, assembled according to the material component mass percentage results shown in Table 1. The procedure was carried out in accordance with the ASTM D 2974 standard, Standard Test Methods for Moisture, Ash, and Organic Matter of Peat and Other Organic Soils. The sample was heated in a furnace at 440°C until no further change of mass occurred after a period of heating. The organic matter content value of the representative MBT waste sample was calculated as 51.60 %.

2.5 Particle density

For the purpose of particle density determination, a modified constant-volume gas pycnometer was used, in accordance with the ASTM D 5550 standard. The obtained values fall within the range between 1.69 g/cm^3 and 2.19 g/cm^3. Average value, in the case of in-situ conditions of MBT waste material, was calculated as 1.87 g/cm^3.

2.6 Maximum and minimum index densities

Maximum and minimum index densities of MBT waste samples were determined through laboratory tests conducted according to ASTM D-4253 and ASTM D-4254 standards, respectively.

Considering the results of the particle size distribution, minimum index density tests were conducted in accordance with the test method “A” of the ASTM D-4254 standard. As it is defined in the standard, minimum index density represents the loosest condition of a cohesionless material that can be attained by a standard laboratory procedure. A total of 14 oven-dried MBT waste samples were prepared by the quartering method. The first seven samples were subjected to minimum index density tests and the second seven to maximum index density tests.

According to ASTM D-4254 standard, samples were poured into a standard mould with a hand scoop. Prior to testing, the volume of the mould was calibrated according to both direct measurement and water-filling methods.

Minimum (dry) index density was calculated as follows:

$$\rho_{dmin} = \frac{M_s}{V} \quad (1)$$

where:

ρ_{dmin} – minimum index density [g/cm^3 or kg/m^3]

M_s – mass of the tested dry sample [g or kg]

V – volume of the tested dry sample or calibrated volume of mould [cm^3 or m^3]

Calculated minimum index density values varied in the range of 145.47 kg/m^3 to 213.97 kg/m^3, with an average value being 176.39 kg/m^3.

Maximum index density tests were conducted in accordance with the test method “1A” of the ASTM D-4253 standard. The vertically vibrating table and the mounted mould assembly with the installed MBT waste specimen are shown in Figure 3 below.

Figure 3 Maximum index density test setup

Analogously to the equation (1) the following expressions can be written:

$$\rho_{dmax} = \frac{M_s}{V} \ (2)$$

where:

ρ_{dmax} – maximum index density [g/cm^3 or kg/m^3]

M_s – mass of the tested dry sample [g or kg]

V – volume of the tested dry sample [cm^3 or m^3]

Values of maximum index density varied in the range of 340.20 kg/m^3 to 424.04 kg/m^3 , with an average value of 383.34 kg/m^3.

2.7 Maximum and minimum void ratios and their correlation

Combining the calculated minimum and maximum index densities and previously obtained average value for particle density, the corresponding maximum-index and minimum-index void ratios were calculated by equations:

$$e_{max} = \frac{\rho_s}{\rho_{dmin}} - 1 \ (3)$$

where:

e_{max} – maximum index void ratio [1]

ρ_s – particle density [g/cm^3 or kg/m^3]

ρ_{dmin} – minimum index density [g/cm^3 or kg/m^3]

and

$$e_{min} = \frac{\rho_s}{\rho_{dmax}} - 1 \quad (4)$$

where:

e_{min} – minimum index void ratio [1]

ρ_s – particle density [g/cm^3 or kg/m^3]

ρ_{dmax} – maximum index density [g/cm^3 or kg/m^3]

The experimental results are presented in Table 2. As can be seen, the results vary from 3.41 to 4.50 for e_{min}, and from 7.75 to 11.87 for e_{max}, with the average values being 3.91 for e_{min} and 9.77 for e_{max}. Standard deviation is calculated as 0.37 and 1.30 for e_{min} and e_{max}, respectively.

According to the calculated average values, the simple relationship based on the ratio between e_{max} and e_{min} can be written as:

$$e_{max} \approx 2.50 \cdot e_{min} \quad (5)$$

and

$$e_{min} \approx 0.40 \cdot e_{max} \quad (6)$$

The maximum to minimum void ratio relation of granular materials has been investigated by several researchers. MIURA ET AL, 1997, conducted a study on around 200 samples of different granular materials, including mostly clean sands but also glass beads and light-weight aggregates. Based on their results, the following equation was derived:

$$e_{max} = 1.62 \cdot e_{min} \quad (7)$$

CUBRINOVSKI AND ISHIHARA, 2002, carried out a study on the data for over 300 soils from Japan, which included clean sands, sands with various fines content (F_c) and clay content (P_c), and silty soils. The following relations were calculated:

For clean sand ($0 < F_c < 5$ %)

$$e_{max} = 0.072 + 1.53 \cdot e_{min} \quad (8)$$

For sand with fines ($5 < F_c < 15$ %)

$$e_{max} = 0.25 + 1.37 \cdot e_{min} \quad (9)$$

For sand with fines and clay ($15 < F_c < 30$ %, $5 < P_c < 20$ %)

$$e_{max} = 0.44 + 1.21 \cdot e_{min} \quad (10)$$

For silty soils ($30 < F_c < 70$ %, $5 < P_c < 20$ %)

$$e_{max} = 0.44 + 1.32 \cdot e_{min} \quad (11)$$

Equations (7 to (11) have been developed based on the data obtained from tests conducted according to the methods stipulated in the Japanese Geotechnical Society standards (JGS).

ILGAC ET AL., 2019, compiled a large database which, amongst others, contains maximum and minimum void ratio values for natural soils (from silts to gravels) and reconstituted granular material (rice, glass beads, mica) mixtures. For the majority of values, tests were performed according to ASTM and JGS standards. A summary of database statistics reveals that the mean values of e_{max} and e_{min} are 0.92 and 0.55, respectively.

By rearranging equations (3) and (4) the following terms can be written:

$$\rho_s = \rho_{dmin} \cdot (e_{max} + 1) \quad (12)$$

$$\rho_s = \rho_{dmax} \cdot (e_{min} + 1) \quad (13)$$

and further, combining the above equations:

$$\frac{\rho_{dmin}}{\rho_{dmax}} = \frac{e_{min}+1}{e_{max}+1} \quad (14)$$

$$\frac{\rho_{dmax}}{\rho_{dmin}} = \frac{e_{max}+1}{e_{min}+1} \quad (15)$$

Finally, by taking into account the average values obtained for minimum and maximum index densities, 176.39 kg/m^3 and 383.34 kg/m^3 respectively, the following equations are derived:

$$e_{min} \approx 0.46 \cdot e_{max} - 0.54 \quad (16)$$

$$e_{max} \approx 2.17 \cdot e_{min} + 1.17 \quad (17)$$

The derived expressions present a relation between the two extreme void ratio values and therefore can be used for e_{min} and e_{max} values calculation if one of the extreme values is known. For instance, if e_{max} value of the MBT waste material is familiar, then the approximate e_{min} value can be calculated by equation (16), or in case the e_{min} value is known, then the approximate e_{max} value can be obtained by equation (17).

A comparison of e_{min} and e_{max} values obtained empirically, and calculated by equations (5), (6), (16) and (17), is presented in Table 2. The linear relation between e_{min} and e_{max} values calculated with equations (5), (6), (16) and (17) is shown in Figure 4.

Table 2 Comparison of minimum and maximum void ratio values (experimental vs. calculated values)

Specimen #	**e_{min} [1] - experimental**	**e_{min} [1] - calculated by equation (6)**	**e_{min} [1] - calculated by equation (16)**
1	4.50	4.75	4.92
2	3.86	4.13	4.21
3	4.29	3.61	3.61
4	4.03	4.33	4.44
5	3.85	3.10	3.03
6	3.41	3.99	4.05
7	3.44	3.45	3.43

Specimen #	**e_{max} [1] - experimental**	**e_{max} [1] - calculated by equation (5)**	**e_{max} [1] - calculated by equation (17)**
8	11.87	11.25	10.94
9	10.33	9.65	9.55
10	9.02	10.73	10.48
11	10.83	10.08	9.92
12	7.75	9.63	9.52
13	9.98	8.53	8.57
14	8.62	8.60	8.63

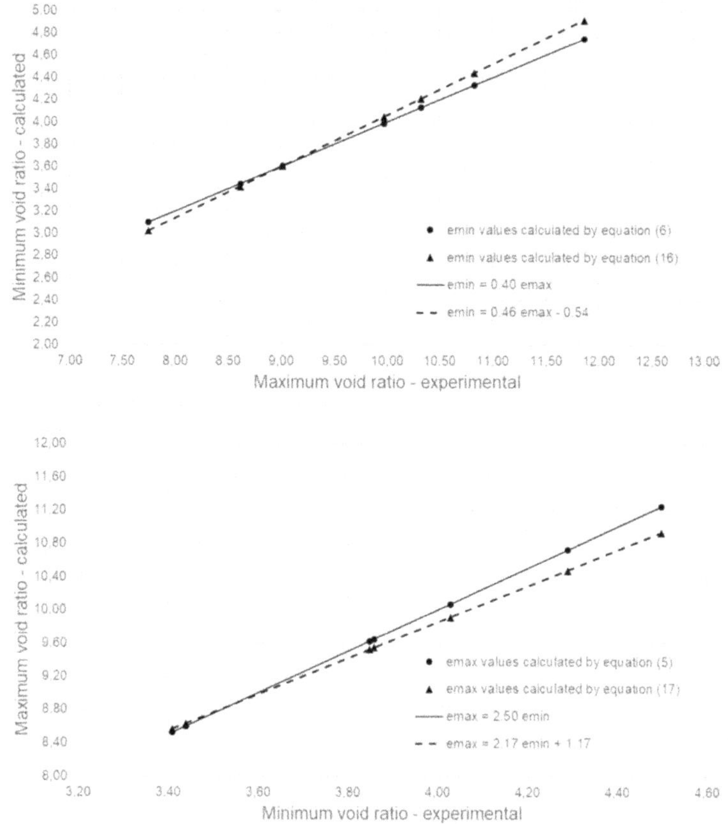

Figure 4 Minimum and maximum void ratio correlation calculated using equations (5), (6), (16) and (17)

3 Conclusion

As a part of this research, mechanically-biologically treated MSW, the so called "fine fraction", was analysed. MBT waste is a granular material whose structure is comparable to that of a cohesionless soil. Physical and geotechnical properties of the MBT waste samples were determined according to the applicable ASTM standards for testing of soil.

Minimum and maximum void ratios of granular material depend on its physical properties such as particle size distribution, particle shape, coefficient of uniformity, angularity, and its fines content. MBT waste is a very heterogeneous material and presents a mixture of different material components. Physical and mechanical properties and the mechanical behaviour of these individual components also vary significantly. Due to its complex physical composition, with prevailing light-weight components (plastics, paper, wood, kitchen waste), the obtained minimum and maximum index density values for MBT waste are significantly lower when compared to the typical values obtained for soil. Consequently, the calculated values of void ratios for MBT waste material are several times higher in comparison to those of soil.

Based on the experimental results, simple correlations between e_{max} and e_{min} were derived. To confirm the validity of the relations, as presented in equations (5), (6), (16) and (17), further research is necessary.

Acknowledgements

The financial support of the Croatian science foundation for the project "Testing and modelling of mechanical behaviour of bio-dried waste as a Waste-to-Energy prerequisite" (UIP-05-2017) is gratefully acknowledged.

4 Literature

American Society for Testing and Materials	2000	ASTM D 2216 Standard Test Method for Laboratory Determination of Water (Moisture) Content of Soil and Rock by Mass.
American Society for Testing and Materials	2000	ASTM D 2487 Standard Practice for Classification of Soils for Engineering Purposes (Unified Soil Classification System).
American Society for Testing and Materials	2000	ASTM D 2974 Standard Test Methods for Moisture, Ash, and Organic Matter of Peat and Other Organic Soils.
American Society for Testing and Materials	2007	ASTM D 422 Standard Test Method for Particle-Size Analysis of Soils.
American Society for Testing and Materials	2000	ASTM D 4253 Standard Test Methods for Maximum Index Density and Unit Weight of Soils Using a Vibratory Table.
American Society for Testing and Materials	2000	ASTM D 4254 Standard Test Methods for Minimum Index Density and Unit Weight of Soils and Calculation of Relative Density.
American Society for Testing and Materials	2014	ASTM D 5550 Standard Test Method for Specific Gravity of Soil Solids by Gas Pycnometer.
Council of the European Union	1999	Council Directive 1999/31/EC of 26 April 1999 on the landfill of waste.

Cubrinovski, M., Ishihara, K.	2002	Maximum and minimum void ratio characteristics of sands. Soils and Foundations, Vol. 42, pp 65-78, Japanese Geotechnical Society
Di Lonardo, M.C., Lombardi, F. & Gavasci, R.	2012	Characterization of MBT plants input and outputs: a review. Reviews in Environmental Science and Bio/Technology. 11, pp 353-363.
European Parliament, Council of the European Union	2008	Directive 2008/98/EC of the European Parliament and of the Council of 19 November 2008 on waste and repealing certain Directives.
Fricke, K., Heike, S., Wallmann, R.	2005	Comparison of selected aerobic and anaerobic procedures for MSW treatment. Waste Management, 25(8), pp 799-810.
Heerenklage J., Stegmann R.	1995	Overview of mechanical-biological pretreatment of residual MSW. Proceedings Sardinia fifth international landfill symposium, S. Margherita di Pula (Cagliari), pp 913-925.
Ilgac, M., Can, G., Cetin, K., O.	2019	A dataset on void ratio limits and their range for cohesionless soils. Elsevier
Komilis, D.P., Ham, R.K., Stegmann, R.	1999	The effect of municipal solid waste pretreatment on landfill behaviour: a literature review. Waste Management and Research, 17, pp 10-19.
Leikam, K. & Stegmann, R., 1995.	1995	The emission behaviour of mechanically-biologically pretreated residual wastes Waste Reports Emissions-verhalten von Restmull. ABF-BOKU, Wien.
Montejo, C., Ramos, P., Costa, C., Marquez, M.C.	2010	Analysis of the presence of improper materials in the composting process performed in ten MBT plants. Bioresource Technology, 101(21), pp 8267-8272.
Munnich, K., Mahler, C.F., Fricke, K.	2006	Pilot project of mechanical-biological treatment of waste in Brazil. Waste Management, 26, pp 150-157.
Petrovic, I., Stuhec, D., Kovacic, D.,	2014	Large Oedometer for Measuring Stiffness of MBT Waste. Geotechnical Testing Journal, Vol. 37 (2), pp. 296–310, doi:10.1520/GTJ20130015. ISSN 0149-6115
Rieger, A., Bidlingmaier, W.	1995	The reactivity of mechanically-biologically pretreated residual waste. Waste Reports Emissions-verhalten von Restmull. ABF-BOKU, Wien.
Scheelhaase, T., Bidlingmaier, W.	1997	Effects of mechanical-biological pretreatment on residual waste and landfilling. Proceedings of Sardinia

		fifth international landfill symposium, S. Margherita di Pula (Cagliari), pp 475-484.
Soyez, K., Plickert, S.	2002	Mechanical-biological pre-treatment of waste: state of the art and potentials of biotechnology. Acta Biotechnolologica, 22(3-4), pp 271-284.
Velis, C.A., Longhurst, P.J., Drew, G.H., Smith, R., Pollard, S.J.T.	2009	Biodrying for mechanical-biological treatment of wastes: A review of process science and engineering. Bioresource Technology, 101(21), pp 2747-2761.
Velis, C.A., Longhurst, P.J., Drew, G.H., Smith, R., Pollard, S.J.T.	2010	Production and quality assurance of solid recovered fuels using mechanical-biological treatment (MBT) of waste: a comprehensive assessment. Critical Reviews in Environmental Science and Technology, 40(12), pp 979-1105.
Velkushanova, K.	2011	Characterization of wastes towards sustainable landfilling by some physical and mechanical properties with an emphasis on solid particles compressibility. Faculty of Engineering and the Environment, University of Southampton, PhD Thesis

Author's address(es)

Nikola Hrncic, mag.ing.geoing., mag.ing. aedif.
Sveuciliste u Zagrebu, Geotehnicki fakultet
Hallerova aleja 7
Croatia, 42000 Varazdin
Phone +385 42 408 925
E-Mail nikola.hrncic@gfv.unizg.hr

Nikola Kaniski, mag.ing.amb.
Sveuciliste u Zagrebu, Geotehnicki fakultet
Hallerova aleja 7
Croatia, 42000 Varazdin
Phone +385 42 408 925
E-Mail nikola.kaniski@gfv.unizg.hr

Izv. prof. dr. sc. Igor Petrovic
Sveuciliste u Zagrebu, Geotehnicki fakultet
Hallerova aleja 7
Croatia, 42000 Varazdin

Phone +385 42 408 919
E-Mail igor.petrovic@gfv.unizg.hr

Comparison of the quality of mechanically recycled plastics made from separately collected and mechanically recovered plastic packaging waste

Eggo U. Thoden van Velzen; Ingeborg Smeding; Marieke T. Brouwer; Evelien Maaskant-Reilink

Wageningen Food & Biobased Research, Wageningen, The Netherlands

Abstract

In the Netherlands plastic packaging waste is collected via separate collection systems that are highly comparable to the German yellow bag (LWP) system and also retrieved via mechanical recovery from the mixed municipal solid waste (MSW). The retrieved plastics are sorted and mechanically recycled in facilities in the Netherlands and Germany. In this study we compare the quality of mechanically recycled plastics made from separately collected (SC) and mechanically recovered (MR) plastic packaging waste. The quality of the recycled plastics is highly comparable in many aspects, although subtle differences can be noticed. The polymeric purity of the mechanically recovered recycled plastic tends to be slightly higher than of the separately collected plastic, whereas the particle contamination of mechanically recovered plastics tends to be higher. Both types of recycled plastics smell differently.

Keywords

Recycled plastics; purity; quality; separate collection; mechanical recovery; lightweight packaging waste; post-consumer plastic packaging waste; odour

1 Introduction

Prior to 2009 only large (>0.75 ltr) poly ethylene terephthalate (PET) bottles for water and soda beverages were collected via deposit refund systems (DRS) in the Netherlands. From then on, all other types of post-consumer plastic packages had to be collected, sorted and recycled. Municipalities had to organise the collection of plastic packaging waste (PPW). Most municipalities introduced various forms of separate collection (drop-off, kerbside with bags and wheelie bins). Three northern provinces (Friesland, Groningen and Drenthe) decided to upgrade their existing central waste sorting facilities to fully fledged material recovery facilities (MRF) and they did not set up separate collection systems for LWP. The collection portfolio was initially limited to only plastic packages (BROUWER ET AL., 2018). From 2015 on the collection portfolio was expanded to include also beverage cartons and metal packages, similar as the lightweight packaging (LWP) collection in Germany (BROUWER ET AL., 2019). Both the separately collected LWP and the two recovered concentrates from the MRF 1) rigid plastic packages and beverage cartons and 2) flexible packaging films were sent to sorting facilities. The sorted products were sent to certified recycling companies (EUCERTPLAST, 2021). In the early years, most of the PPW was sorted in Germany and therefore, it was decided to join the system of DKR specifications (GRUENE PUNKT, 2021). In the rural parts of the Netherlands, the separate collection of LWP with wheelie bins became the dominant collection system, yielding participation rates of nearly 100% (THODEN VAN VELZEN ET AL., 2019). However, in the urban centres in western part of the country only drop-off collection systems could be operated. The participation rates were low and hence the collection rates were also low and the collected material contained too much non-targeted contributions. Consequently, city councils decided to stop with separate collection in selected neighbourhoods and to commence with mechanical recovery. From 2017 to 2019 three new MRF's were established in Alkmaar, Amsterdam and Rotterdam, see table 1. As a consequence we have multiple separate collection systems for PPW, mechanical recovery and combinations of both. The combined annual processing capacity of the MRF's equals almost half of the production of the mixed municipal solid waste (MSW) in the Netherlands. This study compares the quality of the recycled plastics resulting from the Dutch separate collection and mechanical recovery systems.

Table 1 Material recovery facilities in the Netherlands that produce LWP concentrates.

MRF name	Location	Start of operation, [year]	Approximate annual capacity for MSW, [kton/a]
Attero	Groningen	2009	60
Omrin	Heerenveen	2009	210
Attero	Wijster	2011	310
HVC Sortiva	Alkmaar	2017	140
AEB	Amsterdam	2018	300
AVR	Rotterdam	2019	400

1.1 Mechanical recovery process

Dutch MRF's that aim to recover LWP use freshly collected mixed MSW as feedstock. The recovery process typically starts with a bag opener, a wind-sifter, a drum sieve, magnets and Eddy current separators. The lightweight fraction is further sorted with a NIR sorting machine to produce either a flexible packaging film product (DKR 310) or even a flexible PE film product. The drum sieve yields three size fractions. The fine fraction (<6 cm) is named organic wet fraction (OWF) and is digested to produce methane, dried and incinerated. The coarse fraction (>25 cm) is manually sorted. The middle sieve fraction is processed by a set of NIR sorting machines, yielding a concentrate of rigid plastic packages and beverage cartons. Which is sent to sorting facilities for further treatment. All MRF's have a slightly different set-up, some have additional ballistic separators, others have vibrating cascade sieves, etc.. Most Dutch MRF's recover 50-60% of the PPW present in the MSW feedstock, the largest losses of PPW (approximately 30%) relate to small plastic objects that end up in the OWF.

1.2 Quality of recycled plastics

Different quality aspects are used to describe the quality of recycled plastics at the three stages of the recycling chain; sorted products (bales), flakes (washed milled goods) and granulate (pellets). The quality of sorted products is best described by manual sorting (THODEN VAN VELZEN ET AL., 2018). The sorting results can be used to test compliance to the DKR specifications and to describe the sorted products in terms of the types of objects (plastic packages, non-packaging plastic objects and residual waste) present. Since most plastic packages are composed of multiple polymer and material types, this description in terms of objects is only indicative for the polymer and material composition. Additionally, the level of attached moisture and dirt (LAMD) is regularly measured for main plastic packaging types present in the bales to assess their cleanliness (THODEN VAN VELZEN ET AL., 2018). On the flake level the polymeric composition can be determined with NIR sorting equipment (ALVARADO CHACON ET AL., 2020). And on the level of pellets the recycled plastic can be studied with common technologies used to study polymers: rheology, differential scanning calorimetry (DSC), mechanical properties, infrared spectrophotometry (IR), haze, thermal gravimetry and technologies to study particle contamination, such as Partisol.

There are three main quality decay mechanisms for recycled plastics: degradation (chain scission), contamination with polymers & particles and contamination with absorbed molecules (odours, migration) (VILIPLANA ET AL., 2008). For the short-lived packages the degradation is usually less relevant Polymeric and particle contamination often seriously negatively affects the quality of the recycled plastic. Most polymers are immiscible in each other and a bit of polymeric contamination therefore already causes the formation of a blend, which usually negatively impacts the mechanical and optical properties (RAGAERT ET AL., 2017). Finally, absorbed molecules can render the recycled plastic an odour, which does not always fits the application and can pose food safety risks (HAHLADAKIS ET AL., 2018).

1.3 Objective

This paper offers an overview of the knowledge that has been gathered in the past decade on the quality of recycled plastics produced from separately collected and mechanically recovered post-consumer plastic packaging waste. The focus is on post-consumer recycled plastics originating from the Netherlands and will be compared with data from different countries, in case it is available.

2 Composition of sorted PPW

Hundreds of sorting analysis have been averaged to yield average compositions for sorted products originating from separate collection prior to 2014 (BROUWER ET AL., 2018 Table C.1) and after 2017 (BROUWER ET AL., 2019, dataset Table E) and from mechanical recovery prior to 2014 (BROUWER ET AL., 2018 Table D.1) and after 2017 (BROUWER ET AL., 2018, dataset Table H). In general, the compositions are fairly similar, the sorted products originating from mechanical recovery have slightly lower concentrations of objects made from non-targeted polymers such as PS and PVC in comparison to those from separate collection, but the LAMD are higher, see Table 2. The lower concentration of non-targeted polymers in recovered & sorted plastics is likely to be caused by the double NIR sorting; first at the MRF and then at the sorting facility.

Table 2 Average levels of attached moisture and dirt from sorted products (SP) originating from separate collection (SC) and mechanical recovery (MR).

SP	SC 2014	MR 2014	SC 2017	MR 2017
PET 328-1	12%	14%	9%	14%
PE 329	15%	11%	5%	11%
PP 324	11%	17%	10%	17%
Film 310	21%	30%	12%	16%
MP 350	14%	35%	8%	35%

The sorted products made from Dutch PCPPW are fairly similar to the German sorted products, but there are subtle differences. Up to 2022 the concentration of small PET beverage bottles is relatively high in Dutch PET 328-1, but after that year it will be substantially lower due to the introduction of a DRS for these bottles. Furthermore, the Dutch use slightly more PP films, due to differences in consumption patterns and this can be noticed in slightly higher concentrations of PP film in sorted Mixed plastics (DKR 350). A detailed comparison between the composition of Dutch, German and Belgian PE DKR 329 is found in (THODEN VAN VELZEN ET AL., 2021).

3 Mechanical recycling yields

The mechanical recycling yields of sorted bales to dried washed milled goods have been published previously (THODEN VAN VELZEN ET AL., 2017 & 2021). The mass yields of the separately collected sorted products are slightly higher than those from mechanical recovery as could be expected based on their lower level of attached moisture and dirt. Furthermore, the mass yields for sludge and dissolved matter are relatively higher for the mechanically recovered plastics.

Table 3 Polymer composition of the washed milled goods %(m/m).

Sample	Targeted polymer	Non-targeted polymers	Other materials
DRS PET bottles	99.3%	0.6%	0.1%
SC PET 328-1	97.2%	2.8%	0.0%
MR PET 328-1	99.4%	0.2%	0.4%
SC PE 329	90.6%	9.3%	0.1%
MR PE 329	94.0%	3.0%	3.0%
SC PP 324	90.6%	9.2%	0.2%
MR PP 324	95.0%	4.2%	0.8%
SC Film 310	76.4%	22.7%	0.9%
MR Film 310	96.8%	2.8%	0.4%
SC Mix 350	63.5%	30.2%	6.3%
MR Mix 350	72.6%	25.6%	1.8%

4 Composition of washed milled goods

The composition of the washed milled goods, produced with a standardised mechanical recycling process, has been determined NIR assisted manual sorting and with a NIR sorting machine (ALVARADO CHACON ET AL., 2020) and is summarised in Table 3. The washed milled goods made from mechanically recovered and sorted products generally contain slightly more targeted polymer, but also more other materials (paper, wood, glass, metals, etc.) than those made from separately collected feedstock (BROUWER ET AL., 2018; THODEN VAN VELZEN ET AL., 2016).

5 Particle contamination and properties of recycled plastics

Recycled plastics that have been extruded and pelletised can be studied with a range of analysis methods. The difference between separately collected and mechanically recovered plastics has been studied in the greatest detail for PET, therefore the results are split in two paragraphs: recycled PET and recycled polyolefins.

5.1 Recycled PET

Recycled PET (rPET) pellets were produced from three feedstocks (DRS, SC and MR) with a standard mechanical recycling process that also involved a post-condensation process (THODEN VAN VELZEN ET AL., 2016). A summary of the most relevant properties is shown in Table 4.

The comparison in properties shows that the rPET products originating from separate collection and mechanical recovery are more contaminated than the rPET from the DRS. However, the qualities of rPET produced in this study are unsuited for many relevant applications due to their dark colour (low L* values), high haze values and high levels of particle contamination. Lower levels of particle contamination, haze and better colour values can be achieved by advanced recycling processes in which for instance the washed milled goods are also subjected to flake sorting (THODEN VAN VELZEN ET AL., 2016). Therefore the quality of the rPET made from all three feedstocks could be improved. The quality of rPET is very sensitive for particle and polymer contamination. The presence of either type of contamination directly results in a grey, hazy material. Hence, to produce top-quality rPET it is vital to use the best feedstock (DRS) and an advanced recycling process. The higher level of particle contamination in separately collected rPET as compared to DRS rPET has also been reported in a German study from 2017 (SNELL ET AL., 2017).

Table 4 Properties and characteristics of rPET pellets produced from three different feedstocks with standard mechanical recycling processes and post-condensation.

Property	PET DRS	PET SC	PET MR
Colour L*	54 ± 3	51.5 ± 1.5	49.4 ± 0.6
Colour a*	-1.9 ± 0.1	-2.9 ± 0.1	2.9 ± 0.1
Colour b*	1.2 ± 0.5	3.5 ± 0.2	8.1 ± 0.4
Haze, [%]	45.1 ± 0.5	87.7 ± 0.6	84.4 ± 0.3
Partisol, [PPTI]	130.570	1.162.175	695.396
IV, [dl/g]	0.94	0.78	0.75
Mn GPC, [g/mole]	35.700	32.500	35.000
Xc DSC, [%]	26 ± 2	27 ± 2	28 ± 2

+ Partisol is the amount of counted particles in dissolved PET in 10000 images.
+ IV is intrinsic viscosity
+Mn is the normalised molecular weight according to gel partitioning chromatography (GPC)
+ Xc is the degree of crystallisation according to the melt peak in the second heating run in differential scanning calorimetry (DSC).

5.2 Recycled polyolefins

Studies in which the properties of recycled PE, PP, Film and MIX are compared with respect to their collection method (SC or MR) are rare (LUIJSTERBURG ET AL., 2014) and in most cases substantial variations in values are reported. Consequently, no clear difference between the quality and properties of the recycled polyolefins from separate collection and mechanical recovery can be discerned. However, these studies have unveiled that three factors affect the mechanical properties: polymer purity, grade purity and additives.

The particle contamination in recycled polyolefins is hardly ever measured, since the methods are laborious, costly and challenging. Therefore, it is more common to measure indirect parameters such as the composition in terms of the main polymer types (PE and PP) with DSC and IR and mechanical properties (LUIJSTERBURG ET AL., 2014). Most recycled polyolefins contain substantial amounts of the other main polymer type. So, a recycled PE regularly contains roughly 10% PP and a recycled PP often contains about 10% PE. The presence of the other main polymer type has substantial impacts on several mechanical properties of the recycled polyolefin; especially on properties such as strain at break and impact strength and less so on tensile strength (LUIJSTERBURG ET AL., 2014; RAGAERT ET AL., 2017; THODEN VAN VELZEN ET AL., 2021).

Additionally, the mechanical properties of recycled polyolefins are affected by the mixing of multiple grades of the main polymer. Post-consumer recycled PE does not only contain approximately 10% PP, but within the 90% of PE several grades of PE are present: HDPE, LDPE, LLDPE, etc. Although most of these grades are miscible, the mixing of these grades does lower the impact strength. Furthermore, by mixing various PE's from various packages, not only different grades are mixed, but also various additives present in the various packages. This will affect the impact as well (THODEN VAN VELZEN ET AL., 2021).

6 Molecular contamination

Plastics absorb various molecules during their usage and recycling, which hamper the application of recycled plastics.

Recycled PET (rPET) is often subjected to post-condensation to restore the molecular weight of the polymer chains and to remove volatiles (WELLE, 2011). Therefore, rPET rarely has an odour and the gas chromatograms only show a limited amount of small peaks. In table 5 the two most clear peaks in the headspace gas chromatogram of rPET are shown for rPET produced with a standard mechanical recycling process and post-condensation, originating either from DRS, SC or MR. Acetaldehyde is a thermal degradation product of PET (WELLE, 2018) and benzene is formed by the presence of chlorine containing contaminants (probably PVC) (THODEN VAN VELZEN ET AL., 2020). Clearly, the recycled PET made from SC or MR bottles and produced with a standard recycling process did yield the undesired contaminant benzene, which could be avoided by recycling these SC and MR bottles with a more advanced recycling process (THODEN VAN VELZEN ET AL., 2016). It should be noted that studies from other European countries have reported other impurities in rPET. For instance a Spanish article describes the presence of bisphenol-A in rPET at low concentrations (DREOLIN ET AL., 2019).

Table 5 Concentration of volatiles in recycled PET originating from DRS, SC and MR (THODEN VAN VELZEN ET AL., 2016).

Origin rPET	Acetaldehyde, [μg/g]	Benzene, [ng/g]
DRS	3.0	0.0
SC	2.9	1.1
MR	3.1	0.9

Recycled polyolefins are often degassed during extrusion and in some cases even gas-stripped, but in general these polymers contain much more volatile compounds. Therefore, headspace gas chromatograms tend to show much more peaks than those of PET, in some cases exceeding 10000. The amount of peaks and the peak areas can be reduced by recycling with hot alkaline water instead of cold water and by degassing during extrusion. However, the specific odour can in most instances not be removed completely (STRANGL ET AL., 2019 & 2021). Research on the molecular contamination of polyolefins has been conducted for HDPE (STRANGL ET AL., 2019), PP (STRANGL ET AL., 2021) and flexible films (DEMETS ET AL., 2020; STRANGL ET AL., 2020), the latter is the preferred research object due to its large surface to volume ratio.

A Spanish research group studied the odour of recycled Film originating from SC and MR. They found that the recycled Film originating from SC smelled earthy and musty, whereas the recycled Film originating from MR smelled cheesy and faecal (CABANES ET AL., 2020). Simultaneously, a similar study was executed in the Netherlands, with recycled PE-flexible films originating from the Dutch SC and MR collection systems. Surprisingly, the results differed slightly from the Spanish results (MAASKANT-REILINK ET AL., 2020). Dutch recycled PE-films made from SC feedstock smells fatty, rancid, fruity and almond like. This odour remains unchanged, even after hot alkaline washing and degassing during extrusion. In contrast, the Dutch recycled PE-films from MR feedstock had a strong burnt odour when it was recycled with cold water and smelled soap-like when it was washed with hot alkaline solutions. This strong burnt odour was not present in the cold washed milled goods, but was strongly present after extrusion. This implies that this strong burnt odour was formed during extrusion. Remarkably it did not form after extruding recycled PE film that was washed with hot alkaline solutions. A possible explanation is, that a reactant present in the PE films from the MRF is efficiently removed by the hot alkaline washing treatment, preventing its further reaction in the extruder. A preliminary hypothesis is that the decomposition of organic waste present in the MSW creates amines and hydrogen sulphide gas in low concentrations. These gases absorb in polyolefin based packages that are also present in the MSW. During the mechanical recycling process in which the PE films are washed with cold water, these molecules remain inside the PE film and during the hot extrusion process these gases will react with the alkene degradation products from the PE film to form alkyl amines and alkyl thiols, which are known to be extremely odour active. These amines and sulphides can, however, effectively be removed from the PE films by washing with hot alkaline solutions, preventing the unwanted reaction to occur (MAASKANT-REILINK ET AL., 2020). Further research is required to test this hypothesis.

7 Discussion

The qualities of the recycled plastics that can be produced from LWP originating from separate collection or mechanical recovery do differ in details. The odour, the polymeric purity and the properties can all be slightly different. But these differences are often relatively small and sometimes even irrelevant.

The debate on the best collection method for LWP is polarised, with strong advocates and opponents for both separate collection and mechanical recovery. Both systems have pros and cons and can be executed correctly and incorrectly. Separate collection of LWP in rural areas is supported by high participation rates and yields high collection rates. In the most urban centres of the Netherlands, however, the separate collection system with drop-off containers for LWP only received low participation rates and collection rates. Consequently, it does make sense to mechanically recover PPW from MSW in these urban centres, as the recovery rates are substantially higher than the collection rates.

Initially (2009-2010) multiple sorting and recycling facilities were not very enthusiastic about the mechanical recovery of PPW, as these materials are more contaminated with dirt and grime than the separately collected PPW. For the sorting companies this translated in a lower mass yield and larger amounts of sorting residues that have to be incinerated. Depending on the contract form, this can result in a financial risk for the management of the sorting facility. After a few years of experience, this is now well-understood by the incumbents and multiple sorting facilities gladly sort recovered concentrates. Furthermore, two sorting facilities in the Netherlands are currently engineered to sort recovered concentrates.

Within the community of recycling facilities the opinions are mixed. Some really appreciate the recovered sorted products, as they are slightly less expensive and have higher shares of targeted polymers. These companies, however, do need to remove more undesired materials and hence produce more waste and wastewater. The recycling facility has to be engineered to handle this feedstock and also deal with larger amounts of process waste. This can form a bottle-neck for some recycling facilities, especially when their wastewater management system has a limited capacity.

For the Dutch extended producer responsibility scheme operator Afvalfonds mechanical recovery is crucial to achieve the ambitious national recycling targets.

8 Conclusion

The mechanical recovery of lightweight packages (LWP) from municipal solid waste is currently performed on a large scale in the Netherlands. The produced packaging waste concentrates are sorted to the same specifications as the separately collected LWP and sold to various recycling facilities. The recycled plastics produced from these feedstocks differ only in details and are used in similar applications. Mechanical recovery is vital for the urban centres in the western part of the country, since the alternative separate collection systems with drop-off containers resulted in too low collection rates.

9 Literature

Alvarado Chacon et al.	2020	A first assessment of the impact of impurities in PP and PE recycled plastics. WFBR report 2030, Wageningen. DOI: 10.18174/518299
Brouwer et al.	2018	Waste management, 71, 62-85, DOI: 10.1016/j.wasman.2017.10.034
Brouwer et al.	2019	Waste management, 100, 112-121, DOI: 10.1016/j.wasman.2019.09.012 and dataset at DOI: 10.17632/wm4zfwkgzr
Cabanes et al.	2020	Waste management, 104, 228-238, DOI: 10.1016/j.wasman.2020.01.021
Demets et al.	2020	Resources Conservation & Recycling, 161, 104907, DOI: 10.1016/ j.resconrec.2020.104907
Dreolin et al.	2019	Food Chemistry, 274, 246-253, DOI: 10.1016/j.foodchem.2018.08.109
EUCertplast	2021	Website: https://www.eucertplast.eu/certified-recyclers (visited March 17th 2021)
Gruene Punkt	2021	Website: https://www.gruener-punkt.de/en/downloads (visited March 17th 2021)
Hahladakis et al.	2018	J. Hazardous materials, 344, 179-199, DOI:10.1016/j.jhazmat.2017.10.014
Maaskant-Reilink et al.	2020	Moleculaire verontreiniging in gerecyclede kunststoffolie uit bron- en nascheiding. WFBR report 2033, Wageningen. DOI: 10.18174/518646
Ragaert et al.	2017	Waste management, 69, 24-58, DOI: 10.1016/j.wasman.2017.07.044
Strangl et al.	2019	Resources Conservation & Recycling, 146, 89-97, DOI: 10.1016/ j.resconrec.2019.03.009

Strangl et al.	2020	J. Cleaner production, 260, 121104, DOI: 10.1016/j.clepro.2020.121104
Strangl et al.	2021	Resources Conservation & Recycling, 164, 105143, DOI: 10.1016/ j.resconrec.2020.105143
Snell et al.	2017	Waste Management & Research, 35, 163-171, DOI: 10.1077/0734242X16686413
Thoden van Velzen et al.	2016	Technical quality of rPET. WFBR report 1661, Wageningen. ISBN: 978-94-6257-723-7
Thoden van Velzen et al.	2017	AIP Conf. proceedings 1914, 170002, DOI:10.1063/1.5016785
Thoden van Velzen et al.	2018	Sorting protocol for packaging wastes. WFBR report 1826, Wageningen. DOI: 10.18174/451703
Thoden van Velzen et al.	2019	Waste management, 89, 284-293, DOI: 10.1016/j.wasman.2019.04.021
Thoden van Velzen et al.	2020	Packaging Technology & Science, 33, 359-371, DOI:10.1002/pts.2528
Thoden van Velzen et al.	2021	Packaging Technology & Science, 34, 219-228, DOI:10.1002/pts.2551
Viliplana et al.	2008	Macromolecular materials and Engineering, 293, 274-297. DOI: 10.1002/mame.200700393
Welle	2011	Resources Conservation & Recycling, 55, 865-875, DOI: 10.1016/ j.resconrec.2011.04.009
Welle	2018	Reference Module in Food Science,1-8. DOI: 10.1016/B978-0-08-100596-5.22436-0

Author's address(es):

Dr. Eggo Ulphard Thoden van Velzen
Wageningen Food & Biobased Research
Bornse Weilanden 9
NL-6709WG Wageningen
Telefon +31 317 480170
E-Mail ulphard.thodenvanvelzen@wur.nl

Sortierung von Foliengemischen zur Erhöhung der Recyclingquote

Maria Schäfer, Peter Clemenz, Heinz Schnettler

Hochschule Zittau/Görlitz, Verbundinstitut für nachhaltige Verfahrensentwicklung, Oberflächentechnik, Torf- und Naturstoff-Forschung (iTN + IOT), Zittau, Deutschland

Abstract

Der Anteil an Mehrschichtfolien, sogenannte Multi-Layer-Folien, hat in den letzten 30 Jahren stetig zugenommen. Dadurch wurden den steigenden Anforderungen des Handels und der Verbraucher an verbesserten Aromaschutz und längerer Haltbarkeit Genüge getan. Neben den komplexen Mehrschichtfolien gibt es aber auch Folien, die nur aus einem Kunststoff bestehen. Diese Single-Layer-Folien werden z. B. bei Tragetaschen, Abfallsäcken und Tüten für Gemüse und Obst in Supermärkten eingesetzt. Die Single-Layer-Folie aus PE ist nicht nur preiswert in der Herstellung, sondern auch gut durch Wiedereinschmelzen zu recyceln. Voraussetzung dafür ist, dass sie sortenrein gewonnen werden kann. Ziel des hier vorgestellten Verbundprojektes zwischen der HSZG (iTN+ IOT) und der Pla.to GmbH in Görlitz ist die Entwicklung eines neuen und innovativen Verfahrens, welches bisher nicht verarbeitbare Single- und Multi-Layer-Folien-Mischungen sortieren kann. Außerdem sollte das Verfahren in der Lage sein, wirtschaftlich relevante Mengen an Folien zu verarbeiten und hochwertige sortenreine Folienfraktionen zu erzeugen.

Keywords

Recycling, Multi-Layer-Folien, Single-Layer-Folien, Zick-Zack-Sichter

1 Herausforderung Folientrennung

Bei der Verpackung von Lebensmitteln haben sich Behälter und Folien aus Kunststoff durchgesetzt. Mit keinem anderen Werkstoff kann so flexibel auf die hohen Anforderungen an Haltbarkeit und Aromaschutz eingegangen werden. Doch was der Laie im Supermarkt nicht sieht: Die meisten Verpackungen bestehen aus mehr als nur einem Kunststoff. Denn erst durch den Verbund von verschiedenen Polymeren lassen sich Verpackungen gezielt an die Bedürfnisse des jeweiligen Produktes anpassen. Diese Verbundfolien sind nur schwer zu recyceln, da der selektive Stoffaufschluss und das Sortieren in sortenreine Kunststoffe eine Herausforderung darstellen. Neben diesen komplexen Mehrschichtfolien (Multi-Layer-Folien) gibt es aber auch Folien, die nur aus einem Kunststoff bestehen. Solche Single-Layer-Folien werden z. B. bei Tragetaschen, Abfallsäcken und im Supermarkt als Tüten für Obst und Gemüse eingesetzt. Folien aus PE sind nicht nur preiswert in der Herstellung, sondern auch leicht durch Wiedereinschmelzen werkstofflich zu recyceln. Voraussetzung dafür ist, dass sie sortenrein von den Multi-Layer-Folien getrennt werden können. Ziel des hier vorgestellten Projektes ist die Entwicklung einer schlanken und wirtschaftlichen Technologie zur Separation von Single-Layer-Folien aus geeigneten Abfällen. Im Rahmen dieses Beitrages werden erste Ergebnisse aus der Forschungsarbeit vorgestellt.

2 Kunststoff-Folien

2.1 Single-Layer- Folien

Bei einer Kunststoff-Folie handelt es sich um ein flexibles und dünnes Material mit einer Stärke zwischen 2 µm und 500 µm. Dünnere Folien werden als Membranen bezeichnet und dickere werden Kunststoff-Platten genannt. Letztere sind auch deutlich stabiler in der Form. Folien aus Kunststoffen können sehr vielseitig gestaltet werden. Farblich sind keine Grenzen gesetzt. Die Folien können transparent und opak, farblos und gefärbt auftreten. Man unterscheidet Folien, welche nur aus einem Kunststoff bestehen und mehrschichtige Folien.

Sogenannte Single-Layer-Folien aus nur einem Kunststoff sind einfach in der Herstellung und werden für relativ anspruchslose Anwendungen genutzt. Typische einschichtige Folien sind Tüten für Obst und Gemüse im Supermarkt. Sie bestehen meistens aus Polyolefinen und lassen sich recht gut zu höherwertigen Produkten werkstofflich recyceln. Ziel sollte es sein diese Folien einem Recycling zuzuführen um das Material somit im Stoffkreislauf zu halten.

2.2 Multi-Layer-Folien

Problematisch für das Recycling der Single-Layer-Folien ist die Tatsache, dass es neben den einschichtigen Folien auch Folien auf dem Markt gibt, welche aus mehreren Kunststoffschichten bestehen (Multi-Layer-Folien) und nicht einem werkstofflichen Recycling zugeführt werden können. Demnach besteht die Herausforderung darin, die wertvollen Single-Layer-Folien von den Multi-Layer-Folien zu trennen. Die letztgenannten unterscheiden sich in Haptik und Optik nur sehr wenig von den Single-Layer-Folien. Die mehrschichtigen Folien bedienen die hohen Anforderungen im Lebensmittelbereich hinsichtlich Aromaschutz, Schutz vor Feuchtigkeit, Geruchsschutz und Schutz vor Kontaminationen. Oberstes Ziel ist es hier, die ursprüngliche Produktqualität so lange wie möglich für den Kunden zu erhalten (Domininghaus et al. 2012). Der Einsatz von Multi-Layer-Folien begründet sich in der Addition der Barriereeigenschaften der einzelnen Schichten, bei gleichzeitig geringerer Foliendicke im Vergleich zu einer Single-Layer-Folie mit gleichen Eigenschaftswerten (Domininghaus et al. 2012) (Bonnet 2016). Auch die Schweißbarkeit und die Bedruckbarkeit von Folien kann durch das Einbringen von bestimmten Kunststoffschichten gezielt beeinflusst werden (Bonnet 2016). Typische Polymere, die in der Verpackung von Lebensmitteln Einsatz finden, sind:

Polyethylen (PE), Polypropylen (PP), Polyamid (PA), Polyethylenterephthalat (PET), Polystyrol (PS), Ethylen-Vinyl-Alkohol (EVOH), Polyvinylidenchlorid (PVDC), Ethylen-Vinylacetat-Copolymer (EVA), Polycarbonat (PC), Polyvinylchlorid (PVC), Polyethylennaphthalat (PEN) (ANJA MIETH, EDDO HOEKSTRA, CATHERINE SIMONEAU 2016).

Neben diversen Schichten aus Kunststoff wird zur Verbesserung des Aromaschutzes oft eine Lage Aluminium eingebracht, wie es beispielsweise bei Verpackungen für Kaffee (ARM 2011; GDA 2017), Gewürze oder Instant-Suppen (ANJA MIETH, EDDO HOEKSTRA, CATHERINE SIMONEAU 2016) der Fall ist. Da sich nicht alle Kunststoffe ohne weiteres miteinander verbinden lassen (z. B. Polyethylen und Polyamid), ist zwischen den Schichten mitunter das Einbringen von sogenannten Haftvermittlern (HV) erforderlich. Beispiele für entsprechende Stoffe sind Polyethylen geringer Dichte (PE-LD), Ethylen-Vinyl-Alkohol (EVA) und Polyurethan-Kleber (ANJA MIETH, EDDO HOEKSTRA, CATHERINE SIMONEAU 2016). Durch Bedampfen, Beschichten, Bedrucken oder Beflocken wird oft auch die oberste Schicht einer Folie modifiziert.

Multi-Layer-Folien bestehen mitunter aus über 10 verschiedenen Schichten, wobei nach oben grundsätzlich keine Grenzen gesetzt sind. Üblich sind im Bereich der Lebensmittelverpackungen jedoch Verbundfolien mit 3 bis 5 Schichten. Die folgende Abbildung (Abb. 1) zeigt eine 5-schichtige Folie aus Polyethylen und Polyamid.

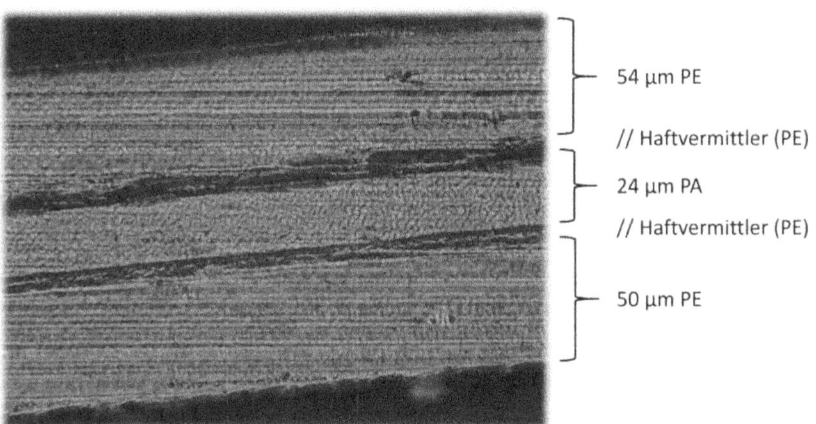

Abbildung 1: Mikroskopaufnahme einer typischen Multilayer-Folie

Ein selektiver Stoffaufschluss ist bei derartigen Folien mechanisch nicht realisierbar, weshalb es derzeit kaum Alternativen zur thermischen Verwertung der Verbundfolien gibt. Damit verunreinigen Multi-Layer-Folien jedoch die gesamte Folienfraktion der im dualen System gesammelten Abfälle. Denn ein großer Teil der Folienabfälle besteht aus den einschichtigen Single-Layer-Folien, welche sich durch Wiedereinschmelzen und Granulieren gut im Kreislauf halten lassen, indem sie anschließend zur Herstellung neuer Produkte und Verpackungen verwendet werden (DER GRÜNE PUNKT - DUALES SYSTEM DEUTSCHLAND GMBH 2019). Die verfahrenstechnische Herausforderung liegt also in einer möglichst sortenreinen Trennung von Single- und Multi-Layer-Folien.

2.3 Folienaufkommen im Abfallstrom

Dass es sich bei der Trennung von Foliengemischen um ein ernstzunehmendes Vorhaben mit wirtschaftlicher Bedeutung handelt, zeigen die Zahlen zum Verpackungsabfallaufkommen in Deutschland und speziell dem Aufkommen an Verbundfolien im Verpackungsabfall. In Abbildung 2 sind die aufbereiteten Daten des Umweltbundesamtes für die Jahre 1991-2017 grafisch dargestellt.

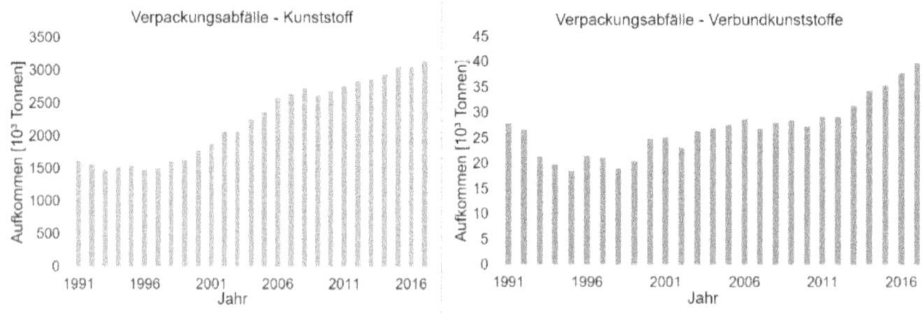

Abbildung 2: Aufkommen an Verpackungsabfällen und Verbundfolien in Deutschland von 1991-2017 nach Umweltbundesamt 23.12.2019

Seit den neunziger Jahren steigt die Abfallmenge an Verpackungen stark an. Dabei erhöht sich der Anteil der Verbundkunststoffe deutlich. Damit steigt auch der Anteil an Verpackungsabfällen, welcher nur noch einer thermischen Verwertung zugeführt werden kann. Um jedoch die Recyclingquote zu erhöhen, ist es notwendig den Anteil der Kunststoffe, welche in die thermische Verwertung gehen, zu reduzieren. Somit besteht die Herausforderung darin die Verbundkunststoffe von den „einfachen" Kunststoffen abzutrennen. Für eine mechanische Trennung muss dazu ein Trennmerkmal geschaffen werden.

3 Ausgangssituation - Forschungshintergrund

Während Versuchen zur trockenmechanischen Reinigung von Abfällen aus dem Dualen System Deutschland hat die Pla.to GmbH die Beobachtung gemacht, dass sich bestimmte Folien-Flakes unter dem Einfluss von thermischer und mechanischer Energie verformen. Wenn es gelingt, dass sich die, in einer gemischten Folienfraktion befindlichen, Multi-Layer-Folien durch eine entsprechende Konditionierung, wie in Abbildung 3 dargestellt, „einrollen", die Single-Layer jedoch ihre flache, plattige Kornform behalten, dann kann die Kornform zu einem relevanten Trennmerkmal dieser beiden Folientypen werden.

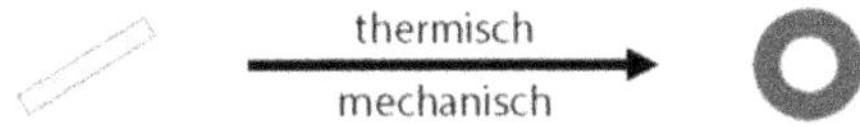

Abbildung 3: Angestrebte Konditionierung von Multi-Layer-Folien

Eine Sortierung nach der Kornform setzt voraus, dass sowohl die Dichte und die Größe der verschiedenen Folien-Flakes nicht stark voneinander abweichen (SCHUBERT 2003). Wenn diese Voraussetzungen gegeben sind, kann eine Trennung nach der Kornform im Zick-Zack-Sichter, wie sie schematisch in Abbildung 4 dargestellt ist, realisiert werden.

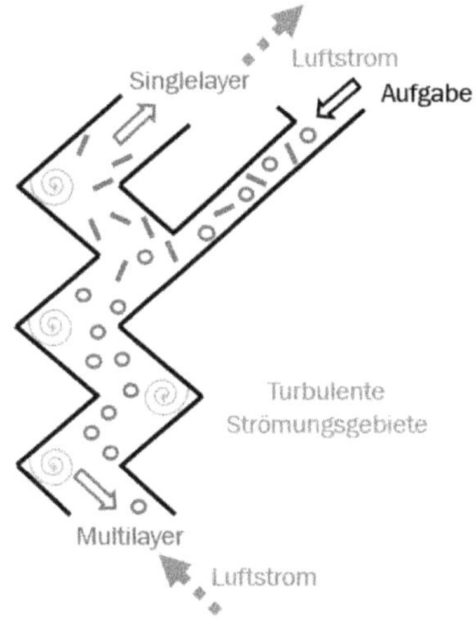

Abbildung 4: Schematische Darstellung der Trennung von Foliengemischen in einem Zick-Zack-Sichter

Die eingerollten Folien entziehen sich dem Luftstrom und gleiten an der Sichterwand, der Schwerkraft folgend, nach unten zum Multi-Layer-/Schwergutaustrag. Die flächigen, einschichtigen Folien werden vom aufsteigenden Luftstrom erfasst und nach oben als Single-Layer/Leichtgut ausgetragen. Dadurch wird eine Trennung nach der Form der Flakes erreicht.

4 Konditionierung im Labormaßstab

4.1 Thermische Konditionierung

Systematische Untersuchungen zum Einfluss von thermischer und mechanischer Energie auf verschiedene Folien, insbesondere auf Verbundfolien, sind nicht bekannt.

Bereits bei Voruntersuchungen hat sich gezeigt, dass sich Multi-Layer-Folien unter Wärmezufuhr mehrheitlich einrollen, wohingegen Single-Layer-Folien ihre ursprüngliche Kornform behalten. Dazu werden Flakes verschiedener Single- und Multi-Layer-Folien im Trockenschrank indirekt erwärmt und die Kornform anschließend klassifiziert. Anhand der Beobachtungen erweist sich, im Hinblick auf die beabsichtigte Sortierung der Folien nach dem Trennmerkmal Kornform, eine Einteilung der konditionierten Folien-Flakes in drei Kornform-Klassen als sinnvoll. Beispiele dazu sind in Abbildung 5 dargestellt.

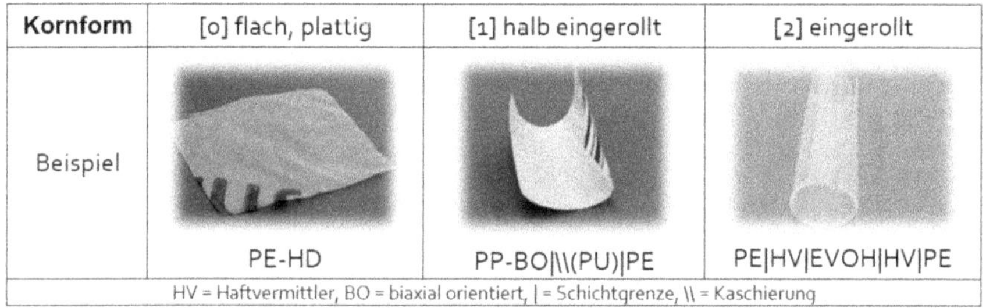

Abbildung 5: Kornformklassen konditionierter Folien-Flakes

Für die Sortierung relevant ist eine deutliche Unterscheidung in der Kornform. Deshalb ist bei der Konditionierung anzustreben, dass Single-Layer-Folien ihre flächige Form beibehalten und sich Multi-Layer-Folien vollständig einrollen (Kornform 2 in Abbildung 5).

In Laborversuchen werden unterschiedliche Methoden zur indirekten und direkten Erwärmung von Folien erprobt. In Tastversuchen durch indirekte Erwärmung im Trockenschrank zeichnet sich ab, dass es bei lediglich 11 % der Single-Layer-Folien zu einer leichten Formänderung durch die Erwärmung kommt. Damit verhalten sich die Single-Layer-Folien erwartungsgemäß und zeigen keine relevanten Tendenzen zur Verformung unter Zufuhr von thermischer Energie. Eine Erwärmung bei Temperaturen > 125 °C führt mehrheitlich zum Schmelzen der PE-Anteile und ist somit zu vermeiden.

Bei Betrachtung der Multi-Layer-Folien unter vergleichbaren Bedingungen kann ebenfalls nur bei 16 % der untersuchten Folien eine, für die Sortierung relevante, Formänderung festgestellt werden. Da angestrebt wird, dass alle Multi-Layer-Folien vor der Sortierung vollständig eingerollt vorliegen, kann die im Trockenschrank realisierte Konditionierung insgesamt als unzureichend eingestuft werden.

4.2 Mechanische Konditionierung

Für die Versuche zur mechanischen Konditionierung kommen die ausgewählten Single- und Multi-Layer-Folien zum Einsatz, an denen bereits die thermische Konditionierung erprobt wurde. Die mechanische Konditionierung wird in einer Turbo-Rotor-Mühle (G-35 Görgens Engineering GmbH, Abbildung 6) durchgeführt. Verantwortlich für die mechanische Beanspruchung ist ein Spalt von 0,4 mm zwischen den Rotoren und der strukturierten Wandung der Mühle. Die Beanspruchung führt jedoch nicht zu einer Zerkleinerung der Folien.

Abbildung 6 : Turbo-Rotor-Mühle zur mechanischen Konditionierung von Folien

Erste Ergebnisse zeigen, dass eine reine mechanische Beanspruchung zu einem Einrollen der Multi-Layer-Folien führt. Die Single-Layer-Folien bleiben flach und rollen sich mehrheitlich nicht ein.

Es ist festzustellen, dass die Konditionierung in der Turbo-Rotor-Mühle zu einer Formänderung von entsprechenden Folien-Flakes führt. Die im Labor realisierte Konditionierung von Multi-Layer-Folien ist jedoch, im Vergleich zu den Konditionierungsergebnissen mit der speziell an die Problemstellung angepassten Zentrifuge der Pla.to GmbH, noch als unzureichend einzustufen. Es ist demzufolge davon auszugehen, dass eine Kombination von thermischer und mechanischer Beanspruchung des Materials für die bleibende Formänderung von Multi-Layer-Folien verantwortlich ist. Dazu werden weitergehende Untersuchungen an einer Pilotanlage im Technikum der Pla.to GmbH durchgeführt und im folgenden Kapitel die ersten Ergebnisse vorgestellt.

5 Konditionierung und Sortierung im Pilotmaßstab

5.1 Konditionierung

Die Konditionierung der Folien-Flakes wird im Pilotmaßstab in einer großen Zentrifuge realisiert, in welcher schnell rotierende Paddel und zusätzlich eingebrachte Schikanen für eine Prallbeanspruchung des Materials sorgen. Die händische Materialaufgabe erfolgt über eine Zellradschleuse und einen Schneckenförderer. Am Ende des Schneckenförderers befindet sich eine Schwergutfalle, über welche Metalle und andere grobe Störstoffe vom Folienmaterial abgeschieden werden. Danach gelangen die Flakes über das Rohrleitungssystem in den Prozessraum der Konditionierungszentrifuge. Der von der Zentrifuge angesaugte Luftstrom, welcher den Materialtransport gewährleistet, kann im Bereich der Schwergutfalle mit einem Gasheizgebläse erhitzt werden. In Abbildung 7 ist der Prozess schematisch dargestellt.

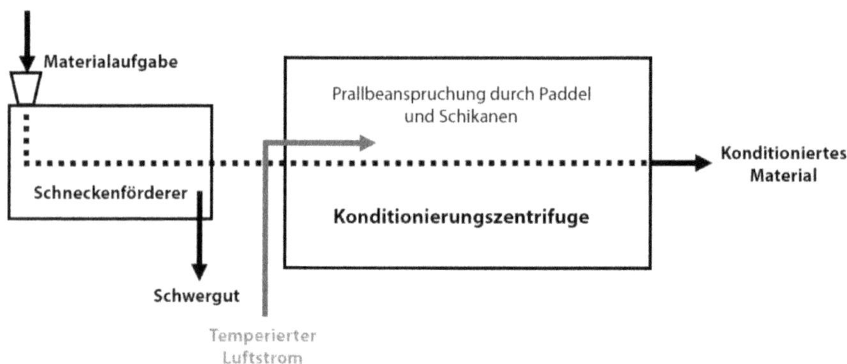

Abbildung 7: Schematische Darstellung der Behandlung der Folien in der Konditionierungsanlage im Technikum der Pla.to GmbH

Die Kombination aus thermischer und mechanischer Energie innerhalb der Konditionierungsanlage scheint der Grund für die Formveränderung der Multi-Layer-Folien zu sein. Deshalb werden nun Konditionierungsversuche mit Folien bekannter Zusammensetzung durchgeführt. Durch den beheizten Luftstrom in der Zentrifuge und die mechanische Beanspruchung durch Werkzeugplatten und Schikanen im Prozessraum, rollen sich die Multi-Layer in den Konditionierungsversuchen mehrheitlich ein. In Abbildung 8 ist der Unterschied zwischen den Single-Layer- und Multi-Layer-Folien fotografisch festgehalten. Die Single-Layer-Folien (im Bild links) behalten ihre flache Form überwiegend bei und die Multi-Layer-Folien rollen sich mehrheitlich ein.

Abbildung 8: Single-Layer-Folien (l.) und Multi-Layer-Folien (r.) nach der Konditionierung im Technikum

In zahlreichen Einzelversuchen mit Folien unterschiedlicher Zusammensetzung kommt es bei den Multi-Layer-Folien durchweg zu einem mehr oder weniger ausgeprägten Einrollen. Lediglich bei einzelnen, sehr großen Flakes und bei verhältnismäßig kleinen Flakes kommt es zu keiner ausreichenden Formänderung. In diesem Zusammenhang besteht noch Forschungsbedarf hinsichtlich der Zerkleinerung der Folien vor dem Konditionieren.

Unter den Single-Layer-Folien bleiben bei den Polyolefinen erwartungsgemäß Veränderungen in der Kornform aus. Die biaxial orientierte Folie aus PP hingegen weist nach dem Prozess nur noch 20 % unverformte, flache Folienflakes auf. Dies ist im Moment auf die biaxiale Orientierung der Polymerfolie zurückzuführen und wird zu einem Fehlgutaustrag im Sortierprozess führen. Jedoch führt dieser Umstand zunächst nicht zu einer Verunreinigung des Zielproduktes, nämlich der Leichtgutfraktion mit PP und PE.

5.2 Sortieren im Zick-Zack-Sichter

Zum Nachweis, dass die herbeigeführte Formänderung als Trennmerkmal in einer Windsichtung ausreicht, werden Versuche mit definierten Mischungen aus Multi- und Single-Layer-Folien durchgeführt. Die Folienmischung wird nach dem beschriebenen Verfahren in der Zentrifuge konditioniert und dann auf einen Zick-Zack-Sichter der Firma Pla.to GmbH aufgegeben.

Voraussetzung für den Prozess ist die Annahme, dass sich die unterschiedlichen Folien-Flakes in Größe und Dichte nicht wesentlich unterscheiden und somit tatsächlich eine

Sortierung nach der Kornform stattfindet. Kommt es zu überlagernden Prozessen, dann wird die Sortierung unscharf.

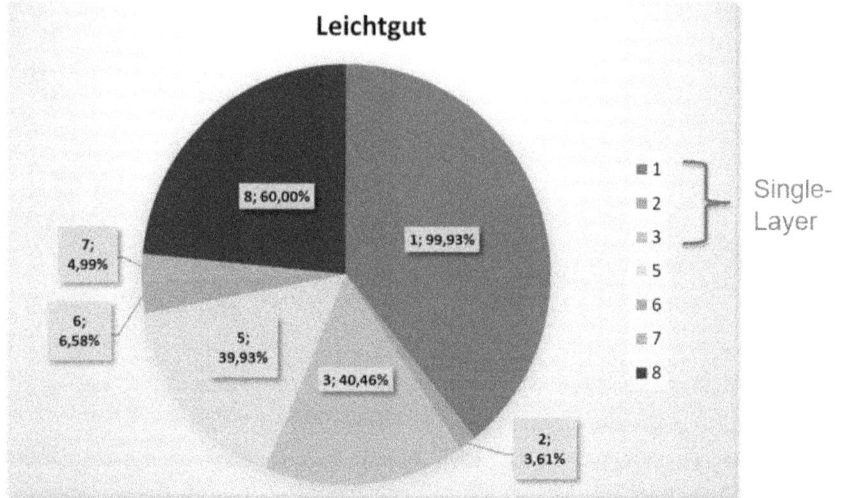

Abbildung 9: Zusammensetzung der Leichtgutfraktion eines Sortierversuches am Zick-Zack-Sichter mit definierter Folienmischung

In Abbildung 9 ist der Leichtgutaustrag des Sortierprozesses einer Mischung aus 8 verschiedenen Folien dargestellt. Die prozentualen Angaben sind dabei die Anteile der einzelnen Folien, welche im Leichtgut wiedergefunden werden. Das bedeutet am Beispiel der Folie 8, das 60 % der Multi-Layer-Folie fälschlicherweise ins Leichtgut gelangt sind. In der Darstellung ist deutlich zu erkennen, dass der Sortierprozess noch einer Optimierung bedarf. Eine Ursache für die hohen Fehlgutausträge liegt wahrscheinlich in der Überlagerung der Trennung nach der Kornform durch Trennprozesse nach der Korngröße. Deshalb werden in Zukunft eng klassierte Fraktionen der Folien konditioniert und anschließend gesichtet. Es hat sich gezeigt, dass kleine und sehr große Folien-Flakes der Multi-Layer-Folien nicht ausreichend konditioniert werden können und damit mit den Single-Layer-Folien im Sichtprozess nach der Kornform ins Leichtgut gelangen.

5.3 Bewertung des Sortierprozesses

Eine besondere Herausforderung im Vorhaben eine verwertbare Fraktion an Single-Layer-Folien zu generieren, ist die Bewertung des Sortierprozesses. In den bisherigen Versuchen wurde mit Foliengemischen gearbeitet, deren Zusammensetzung bekannt ist. In Zukunft soll der erarbeitete Prozess aus Konditionierung, zum Erschaffen eines Trennmerkmals, und anschließender Sortierung in einem Zick-Zack-Sichter auf ein reales Stoffgemisch übertragen werden. Dazu wird parallel eine sichere und dennoch schnelle Analytik für die Folienmischungen entwickelt, welche möglichst eine qualitative und quantitative Aussage darüber geben kann, wie stark verunreinigt die Zielfraktion aus PE und PP ist. Zurzeit werden die Folien für die Versuche so ausgewählt, dass sie nach Konditionieren und Windsichten optisch voneinander unterschieden und somit nach geeigneter Probenahme und Probenaufbereitung optisch analysiert werden können.

Die Referenzmaterialien werden außerdem mittels Digitalmikroskopie (siehe Abbildung 1), dynamischer Differenzkalorimetrie (DSC), thermogravimetrischer Analyse (TGA) und selektiver Inlösungnahme charakterisiert. Die Untersuchungen haben ergeben, dass eine Kombination aus DSC und selektiver Inlösungnahme einen Rückschluss auf die Zusammensetzung einer Folienmischung unbekannter Zusammensetzung geben kann. Die Analytik wird nun an definierten Folienmischungen validiert und wird in Zukunft die Sortierprozesse am Realgemisch auswertbar machen.

6 Zusammenfassung

Zum gegenwärtigen Zeitpunkt, gelingt es mit der hier vorgestellten Technologie gemischte Folienabfälle derartig zu konditionieren, dass es mit einer anschließenden Sortierung durch Windsichten zu einer gewissen Anreicherung mit dem Wertstoff im Leichtgut kommt. Es konnte nachgewiesen werden, dass sich Kunststoff-Flakes aus verschiedenen Multi-Layer-Folien unter dem Einwirken von thermischer und mechanischer Energie mehrheitlich einrollen, wohingegen Single-Layer-Folien ihre ursprüngliche Kornform behalten. Entscheidend ist dabei eine Kombination aus beiden Beanspruchungsarten.

Die Sortierversuche zeigen, dass der Prozess der Konditionierung noch einer Optimierung bedarf. Derzeit kommt es zur Überlagerung von Trennprozessen nach der Kornform und Korngröße. Eine verbesserte Zerkleinerung, und gegebenenfalls anschließende Klassierung des Folienmaterials, muss hier zu einer engeren Korngrößenverteilung der aufgegebenen Folienmischungen führen.

7 Danksagung

Die Autoren möchten sich für die finanzielle Unterstützung des europäischen Fonds für regionale Entwicklung (EFRE) im Rahmen des Projektes: SAMSort - Aerodynamische Konditionierung von Folienmischungen für die selektive Trennung nach Kunststoffsorten-(100357610) bedanken.

Der Firma Horn & Bauer GmbH & Co KG danken wir für die Bereitstellung von zahlreichen unterschiedlichen Folien für die Versuche im Rahmen des Projektes.

Author's address(es):

Dr.-Ing. Maria Schäfer + Peter Clemenz	Heinz Schnettler
Hochschule Zittau/Görlitz	Pla.to GmbH
Verbundinstitut iTN + IOT	Nickrischer Strasse 20
Friedrich-Schneider-Straße 26	02827 Görlitz
02763 Zittau	Deutschland
Deutschland	Tel.: +49 35822/312735
Tel.: +49 3583/612 4314	E-Mail: info@plato-technology.de
E-Mail: maria.schaefer@hszg.de	Homepage: https://www.plato-technology.de/
Homepage: http://itn.hszg.de/	

8 Literaturverzeichnis

Mieth, A. et al.	2016	Guidance for the identification of polymers in multilayer films used in food contact materials. JRC Technical Reports. Hg. v. JRC Sience Hub
GDA - Gesamtverband der Aluminiumindustrie e.V.	2017	Aluminium in der Verpackung. Düsseldorf: Hg. v. GDA
Bonnet, Martin	2016	Kunststofftechnik. Grundlagen, Verarbeitung, Werkstoffauswahl und Fallbeispiele. 3., überarbeitete und erweiterte Auflage. Wiesbaden: Springer Vieweg.
Der grüne Punkt – Duales System Deutschland GmbH	2019	Systalen – das Regranulat des Grünen Punkts. https://www.gruener-punkt.de/, 28.09.2020
Domininghaus, Hans et al.	2012	Kunststoffe. Eigenschaften und Anwendungen. 8., neu bearb. u. erw. Aufl. Berlin, Heidelberg: Springer Berlin Heidelberg.
Schubert, Heinrich	2003	Handbuch der Mechanischen Verfahrenstechnik, Band 1 und 2. Weinheim: WILEY- VCH Verlag GmbH & Co. KGaA.

„Konditionierung von Kunststoff-Fraktionen mittels NIR-Sortierung zur Reduzierung des Anteils an der thermischen Verwertung"

Anett Kupka, Lukas Knauth

Hochschule Zittau/Görlitz, Verbundinstitut iTN+IOT, Zittau

Conditioning of plastic material with NIR-sorting to reduce part of plastics in thermal waste treatment

Inhaltsangabe

In Kooperation mit der STEINERT UniSort GmbH und der SCHOLZ Recycling GmbH wird, im Rahmen eines FH-Impuls Projektes (gefördert durch das BMBF), die Integration von Nahinfrarot Sortiermaschinen in den Aufbereitungsprozess untersucht. Ziel ist es NIR-Sortiermaschinen in das bestehende Anlagenkonzept so zu integrieren, dass durch geeignete Vorbehandlungsmethoden gute Sortierqualitäten erreicht werden können.

Keywords

Kunststoffe, NIR-Sortierung, Konditionierung, thermische Verwertung, Holz, EBS

1 Einleitung

Die im Rahmen der Novellierung der Abfallrahmenrichtlinie 2018 beschlossene Änderung von einer input- zu einer outputbasierten Berechnung der Recyclingquote verlangt es Abfallströme nicht nur einem Recyclingprozess zuzuführen, sondern dabei auch vermarktungsfähige Produkte zu erzeugen (BUNDESMINISTERIUM FÜR UMWELT, NATURSCHUTZ UND NUKLEARE SICHERHEIT (BMU), 2020).

Aktuelle Prozesse zur Aufbereitung von Gewerbeabfällen, Misch-, Sammelschott und Altautomobilen ermöglichen das Erreichen der geforderten Quoten nur bedingt. Große Mengen der anfallenden Reststoffe werden als Ersatzbrennstoff deklariert. Gleichzeitig sind Verbrennungsanlagen in Deutschland bis an ihre Kapazitätsgrenze ausgelastet, wodurch eine thermische Verwertung oft nicht ohne weiteres möglich ist (BILITEWSKI, 2008). Dadurch ist es notwendig weitere Materialfraktionen für ein stoffliches Recycling auszuschleusen oder die Reststoffströme so aufzubereiten, dass sie als Ersatzbrennstoff tatsächlich thermisch verwertet werden können.

Sensorgestützte Sortierverfahren, insbesondere die Optische- und Nahinfrarot Sortierung, sind in der Aufbereitung von Verpackungsabfällen Stand der Technik (MARTENS UND GOLDMANN, 2016). Da Schredderrückstände jedoch oft in hohem Maße schwarze Kunststoffe enthalten und verschiedene Verunreinigungen der Oberfläche auftreten ist der wirtschaftliche Einsatz von NIR-Sortiertechnik in diesem Anwendungsbereich bisher nur begrenzt möglich (HOLLSTEIN ET AL, 2013; NIENHAUS, 2014).

2 Materialien und Methoden

Am Ende der Nutzungsphase von Produkten, wie zum Beispiel Autos, Busse, Züge, Kühlschränken oder Küchengeräte entsteht nach dem Schreddern und dem ersten Sortierschritt ein Gemisch aus Metallen, Verbundwerkstoffen, Mineralik, Holz und zahlreichen verschiedenen Kunststoffen. Im Rahmen der Aufbereitung spielt die Sortierung auf Basis der Nahinfrarotspektroskopie (NIR-Sortierung) eine wichtige Rolle und daher sollen Versuche mit der NIR-Anlage im Technikum der Hochschule Zittau/Görlitz im Fokus des Beitrages stehen. Ziel ist es, zum einen Fraktionen zu generieren, die als Recyclingmaterial dem Produktionsprozess wieder zugeführt werden können, oder zum anderen als separate Fraktion entsorgt werden können.

2.1 Material

Das Aufgabematerial ist ein Gemisch aus Metallen, Verbundwerkstoffen (NM-NM, M-NM[1]), Mineralik, Holz, Gummi und zum großen Teil Kunststoffen verschiedenster Sorten. Es wurden über 4 Wochen repräsentative Proben aus dem Gutstrom von ca. 1,5 t/h für einige Minuten entnommen in Anlehnung an die LAGA PN98 (Länderarbeitsgemeinschaft Abfall (LAGA), 2001) und entsprechend der Partikelgröße in Teilproben von ca. 8 – 16 kg geteilt. Aus den repräsentativen Proben des Abfallstromes erhält man über manuelles Klauben die in Tabelle 1 aufgelistete Zusammensetzung. Zu erkennen ist der hohe Kunststoffanteil von ca. 80 %.

Tabelle 1: Zusammensetzung Aufgabematerial (Handklaubung,18 Proben, mittlere Probenmenge 4 kg)

Stoffgruppe k	arithm. Mittelwert	Standard-abweichung	Variations-koeffizient CV	MIN	MAX
Metall (frei)	**0,5**	**0,7**	**1,3**	0,0	2,7
Leichtgut	**0,8**	**1,4**	**1,9**	0,0	5,2
NM-NM-Verb.	**1,1**	**0,8**	**0,7**	0,2	3,5
Mineralik, Rest	**1,7**	**1,6**	**0,9**	0,0	5,1
Metall-NM-	**1,9**	**1,1**	**0,6**	0,4	4,8
Holz	**2,2**	**1,4**	**0,6**	0,5	5,8
Gummi	**10,4**	**4,9**	**0,5**	1,1	18,8
Kunststoffe	**81,4**	**5,9**	**0,1**	69,6	91,7

[1] NM-NM – Nichtmetall-Nichtmetall-Verbund, M-NM – Metall-Nichtmetall-Verbund

Die Ersatzbrennstofffraktion ist ein sehr inhomogenes Material. Die Anteile der einzelnen Stoffgruppen unterliegen starken Schwankungen, wie man an den Minima und Maxima in Tabelle 1 ablesen kann. Es ist daher eine große Herausforderung für diesen Materialstrom einen Aufbereitungsprozess zu erarbeiten, der wirtschaftlich darstellbar ist.

Exemplarisch ist dies in Abbildung 1 am Holzanteil verschiedener Probenahmen, die verschiedenen Produktionschargen zuzuweisen sind, gezeigt.

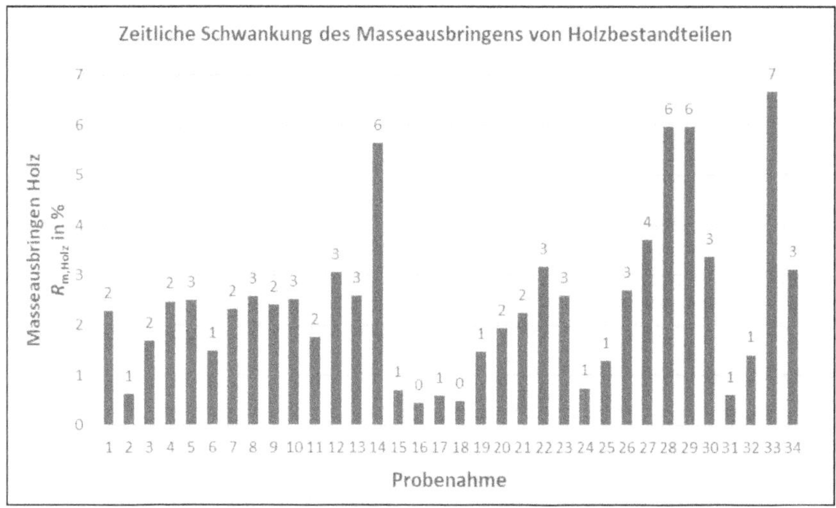

Abbildung 1 Schwankungen für die Komponente Holz innerhalb der Aufgabe

Die Partikelgrößenverteilung des untersuchten Materials liegt bei einem x_{50}-Wert von 12,4 ± 1,1 mm. Die untere Partikelgröße hat einen Wert von 4 mm und die obere Partikelgröße liegt bei 24 mm. Mit dieser Verteilung liegt das Material in der Spezifikation des NIR Flake Sorters, welche 5-30 mm als Optimum angibt (STEINERT UNISORT GMBH).

2.2 Prinzip NIR-Sortierung

Das Aufgabematerial bedarf einer Konditionierung auf eine Partikelgröße von 5 – 30 mm und wird über den Aufgabebunker (1) der Vibrorinne und der Aufgaberutsche (2) zugegeben (Abbildung 2). Auf dem Beschleunigungsband (3) erfolgt die Vereinzelung zur Einkornschicht. Über dem Beschleunigungsband befinden sich die NIR-Strahlenquelle sowie der Detektor. Über die Auswertung der NIR-Spektren in der Recheneinheit (6) werden die empfangenen Spektren beurteilt und die Druckluftdüsen (7) angesteuert. Durch den Trennscheitel (10) getrennt erhält man das Sortierprodukt (8) und das Durchgangsprodukt (9).

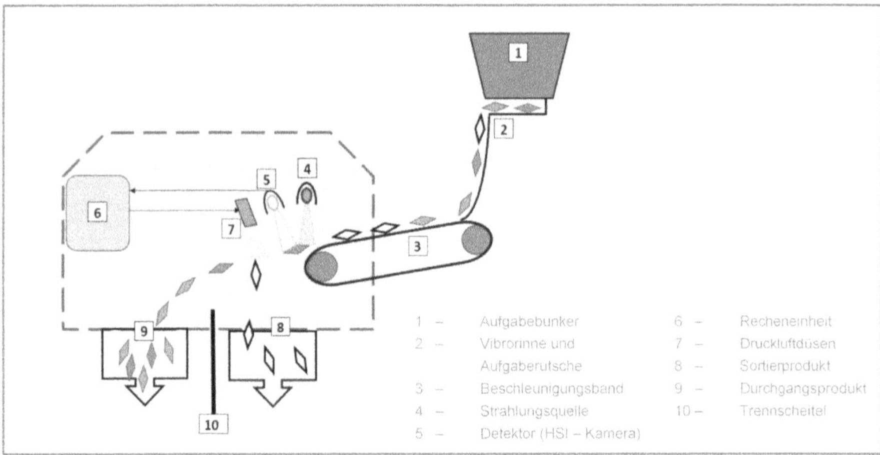

Abbildung 2 Prinzip NIR-Sortieranlage

Im Technikum der Hochschule Zittau/Görlitz werden Sortierversuche am NIR Flake Sorter der Firma STEINERT UniSort GmbH durchgeführt. An dieser Anlage sind eine Kreislaufführung sowie eine repräsentative Probenahme möglich.

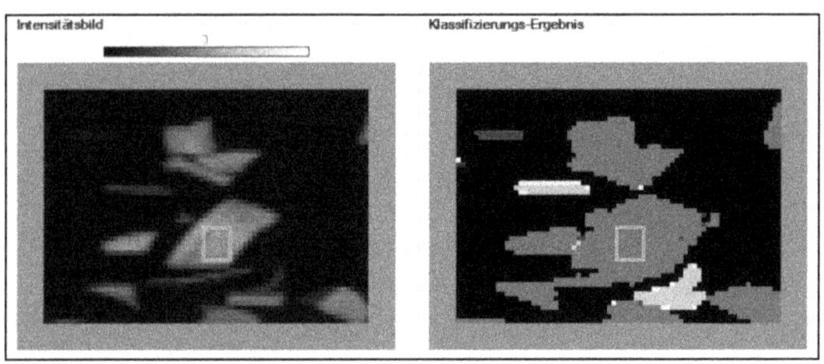

Abbildung 3: Ausschnitt Intensitätsbild (links) und Klassifizierungsergebnis (rechts)

Die Objekte werden über die gesamte Bandbreite von einer Hyperspektralkamera erfasst und die Spektren klassifiziert (Abbildung 3). Die Zuordnung zur Stoffklasse erfolgt über den Abgleich der gemessenen Spektren zu den Referenzspektren (Abbildung 4). Im Randbereich kann es zu Unschärfen kommen. Nicht klassifizierbar sind mit Carbon-Black eingefärbte Stoffe, beschichtete Stoffe (verchromt, Pulver beschichtet) sowie Verbund und Multilayer Systeme aus verschieden Stoffen.

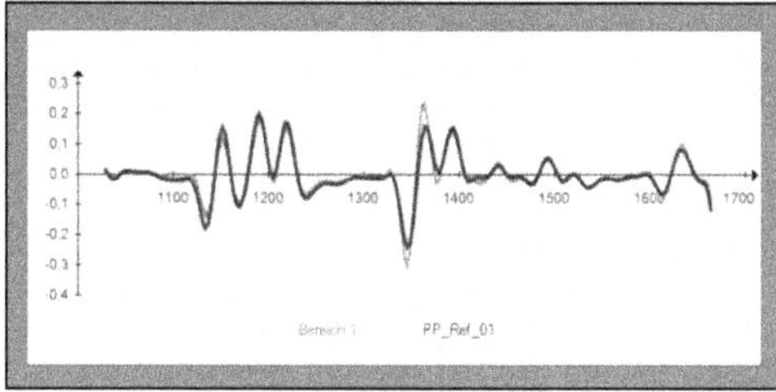

Abbildung 4 Vergleich Referenzspektrum (PP_Ref_01) und Messspektrum (Bereich 1)

3 Abtrennung von Stoffklassen

3.1 Abtrennung der Holzfraktion

Die Abtrennung von Holz aus dem Aufgabegemisch ist sinnvoll, um es als separate Fraktion einer Verwertung zuzuführen und um den Störstoffgehalt in der nachfolgenden Aufbereitung von Kunststofffraktionen zu verringern. Im ersten Schritt wird die Datenbank um Spektren verschiedener Holzklassen (Altholz, Holz mit Lack, furniertes Holz, Spannplatten) erweitert. Nach einigen Testläufen mit definierten Holz-Kunststoff-Mischproben zur Validierung des neuen Sortierkriteriums folgen Versuche mit dem realen Aufgabestrom. Exemplarisch sind in Abbildung 5 die Ergebnisse dargestellt.

Die Aufgabe hat einen Holzanteil von 4,7 % und nach dem Sortieren kann ein Holzprodukt mit über 60 % Holzanteil erreicht werden. Der Fehlaustrag von Kunststoff in diesem Produkt beträgt knapp 30 % und ist auf unzureichende Vereinzelung oder Überlagerung von Teilchen zurückzuführen. Es können 92,4 % aller aufgegebenen Holzteilchen in das Holzprodukt ausgetragen werden. Das Holzprodukt stellt mit einem Masseausbringen von 7,2 % einen wesentlich geringeren Anteil dar, als das Kunststoffprodukt mit einem Masseausbringen, Rm von 92,8 % (SCHUBERT, 2012). Der Fehlanteil Holz im Kunststoffprodukt, c_{Holz} beträgt 0,4 % und ist damit in der gewünschten Zielgröße.

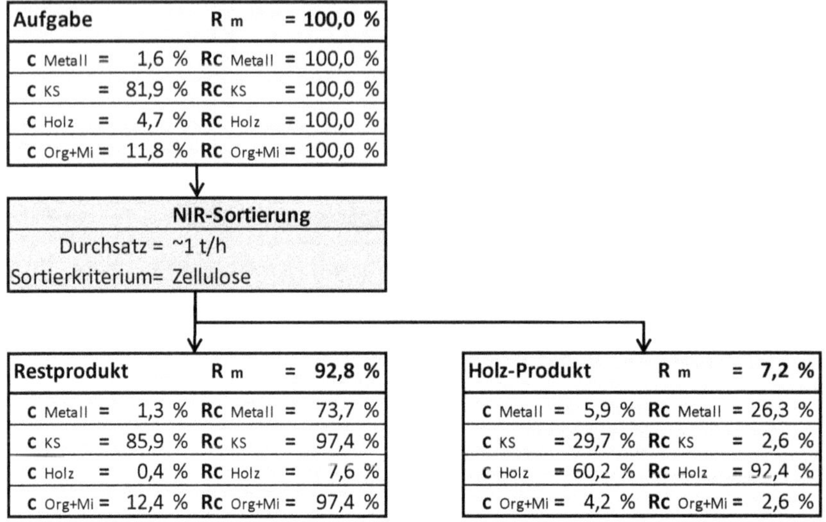

Abbildung 5: Versuch zur Holzabtrennung - Auswertung Massebilanz[2], Wertstoffausbringen, Konzentrationen

3.2 Abtrennen von Kunststofffraktionen

Ziel ist die Abtrennung von Polyolefinen (PO) und Polystyrol (PS), da für diese Gruppen stoffliche Verwertungswege im Einsatz als Regenwasserspeicher, Tanks oder Regenwasserversickerungsanlagen bereits aufgezeigt werden konnten (DRÄGERT, 2017).

Bei diesen Sortierversuchen liegt der Fokus nicht auf dem Erweitern der Datenbank, sondern auf dem Untersuchen der Aufgabefraktion auf zielführende Gehalte und Mengen. Schwierigkeit bei dieser Art von Versuch ist die Bewertung des Ausbringens der Zielkunststoffe. Als Bewertungsmethode in diesem Versuch wird die Materialflächenstatistik der integrierten Datenbank genutzt. Ebenso ist die Charakterisierung der entstandenen Produkte hinsichtlich ihr Brennstoffeigenschaften Bestandteil der Arbeiten, welche zum Zeitpunkt des Beitrags noch nicht abgeschlossen sind.

[2] c… Konzentration/Gehalt / Rm … Masseausbringen / Rc… Wertstoffausbringen

Indizes: KS…Kunststoffe, Org+Mi … Organik und Mineralik

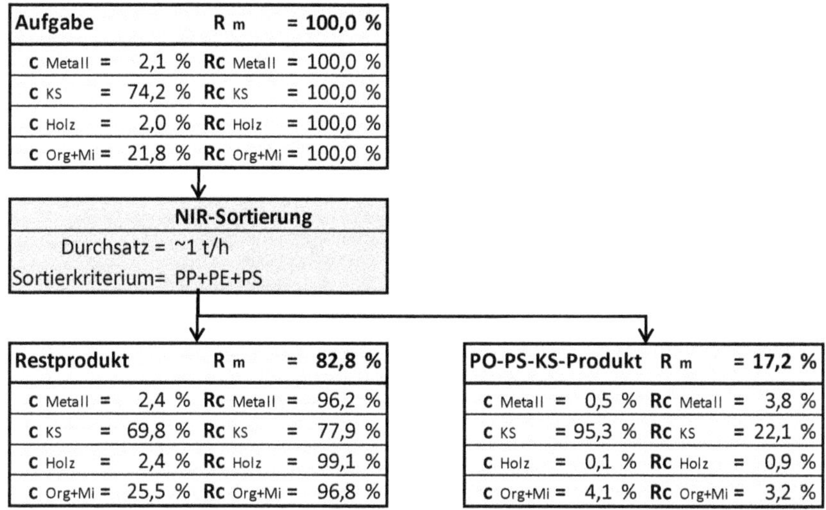

Abbildung 6 Versuch zur Abtrennung von PP+PE+PS - Auswertung Massebilanz, Wertstoffausbringen, Konzentrationen

Es gelingt mittels NIR-Sortierung die Zielkunststoffe mit einem Wertstoffausbringen, Rc_{KS} von 22,1 % und einer Wertstoffkonzentration c_{KS} vom 95,6 % im PO-PS-Kunststoffprodukt um das 1,4-fache anzureichern (Abbildung 6). Die Fehlanteile sind Organik/Mineralik-Bestandteile, Holz und Metalle mit einem Gehalt von 4,7 %. Das bei diesem Sortierschritt erzeugte „Rest-EBS"-Produkt hat einen Kunststoffanteil von ca. 70 %. Hierzu zählen alle rußgefärbten Kunststoffe, welche von dem eingesetzten Verfahren nicht sortiert werden können. Die zweite Hauptkomponente mit 25 % sind organische und mineralische Stoffe, welche die Gummibestandteile beinhalten.

4 Zusammenfassung und Ausblick

Mittels NIR-Sortierung ist es möglich, faserhaltige, zellulosebasierte Naturstoffe (Holz, Holverbunde, Faserplatten) zu detektieren und zu sortieren. In Pilotversuchen konnte gezeigt werden, dass bis zu 92 % des aufgegebenen Holzes im Sortierprodukt angereichert werden. Der Masseanteil dieses Produktes liegt bei ca. 7 %. In einer zweiten Sortierstufe konnten die Kunststoffe Polypropylen (PP), Polyethylen (PE) und Polystyrol (PS) mit einem Wertstoffausbringen von 22 % und einem Masseausbringen von 17 % angereichert werden. Bilanziert man die Stufen der Pilotversuche gemeinsam, kann man zusammenfassend feststellen, dass nach Abtrennen der Holzfraktion und des Kunststoffproduktes

PP+PE+PS noch 77 % der aufgegebenen Ausgangsmasse als „Rest-EBS“ der thermischen Verwertung zugeführt werden muss (Abbildung 7).

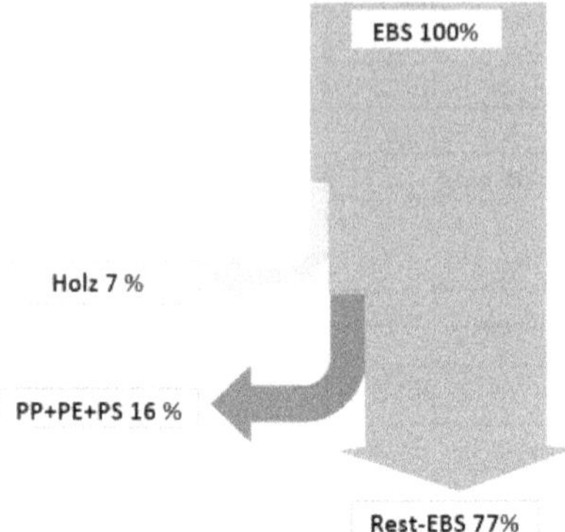

Abbildung 7 Massebilanz für den Gesamtversuch

Erste Tastversuche haben gezeigt, dass auch weitere Kunststoffe mittels NIR-Spektroskopie sortierbar sind. Da das hier vorgestellte Verfahren bei rußgefärbten Kunststoffen nicht anwendbar ist, sind für die Fraktion der schwarzen Kunststoffe andere Sortierverfahren zu testen, um damit den Anteil der Kunststoffe, welche verbrannt werden weiter zu reduzieren. Bei den Möglichkeiten der modernen Sortiertechnik bleibt immer auch die Frage der Einsatzmöglichkeiten für Rezyklate, wirtschaftlich darstellbare Masseströme und geforderte Qualitäten.

5 Danksagung

Die in diesem Beitrag dargestellten Versuchsergebnisse sind im Rahmen des Projektes LaNDER³ Impulsprojekt 1 Forschungsschwerpunkt 2 (13FH2I01IA) an der Hochschule Zittau/Görlitz gefördert durch das Bundesministerium für Bildung und Forschung im Programm Forschung an Fachhochschulen erarbeitet worden. Wir danken dem Fördermittelgeber und den Praxispartnern STEINERT UniSort GmbH, Zittau und SCHOLZ Recycling GmbH, Espenhain für die Bereitstellung von Versuchsmaterial, Unterstützung bei Pilotversuchen, regen wissenschaftlichen Austausch und die Bereitschaft zur Diskussion der Ergebnisse.

6 Literaturverzeichnis

Bilitewski, B. (Hrsg.)	2008	Thermische Abfallbehandlung. 13. Fachtagung. München. kassel university press GmbH, Kassel, ISBN 978-3-89958-384-7
Bundesministerium für Umwelt, Naturschutz und nukleare Sicherheit (BMU)	2020	Kreislaufwirtschaftsgesetz - KrWG
Drägert, J.	2017	POmixed Recycling, Phase 1. Abschlussbericht. Hightech-Trenn- und Aufbereitungsverfahren zum PE- & PP-Recycling bisher nicht verwertbarer, minderwertiger Post-Consumer Abfälle zur Substitution von Primärgranulat (Aktenzeichen 33239/01-21/2)
Hollstein F.; Wohllebe M.; Arnaiz S.; Manjon D.; O'Brien N.; Kulcke, A.	2013	Challenges in Automatic Sorting of BioPlastics within the Recycling Chain by Means of NIR-Hyperspectral-Imaging. NIR 2013 - 16th International Conference on Near Infrared Spectroscopy. Frankreich
Länderarbeitsgemeinschaft Abfall (LAGA) 32	2001	LAGA PN 98 - Richtlinie für das Vorgehen bei physikalischen und chemischen Untersuchungen im Zusammenhang mit der Verwertung/Beseitigung von Abfällen
Martens, H.; Goldmann, D.	2016	Recyclingtechnik. Springer Fachmedien, Wiesbaden, ISBN 978-3-658-02785-8
Nienhaus, K. (Hrsg.)	2014	Sensor technologies. Impulses for the raw materials industry. Shaker, Aachen, ISBN 978-38440-2563-7
Schubert, H. (Hrsg.)	2012	Handbuch der Mechanischen Verfahrenstechnik. Weinheim: WILEY-VCH, ISBN 3-527-30577-7
STEINERT UniSort GmbH		Datenblatt. P 75 180R V6 O.

Anschrift der Verfasser(innen):

Dipl.-Ing. Anett Kupka und Lukas Knauth, B.Sc.
Hochschule Zittau/Görlitz – Verbundinstitut iTN+IOT
Theodor-Körner-Alle 16
D-02763 Zittau
Telefon +49 3583 612 - 4961
E-Mail a.kupka@hszg.de

Biomass conversion and H2-produktion

Alfons Kuhles, Hubert Kohler

GRENOL GmbH, Ratingen; blueFLUX Energy AG, Peißenberg

Abstract:

At present, molecular "green" hydrogen (H_2) is produced in Germany for the most part from electrolysis of water, whereby only electricity from renewable energies should be used for electrolysis. Only "green" hydrogen is climate-friendly, because only "green" hydrogen is produced without fossil raw materials. All other processes for producing molecular hydrogen require fossil feedstocks and high energy costs or cause increased emissions of methane (CH_4) or carbon dioxide (CO_2) during production.

With hydrothermal and vaporthermal carbonisation processes and subsequently gasification, it is now possible to produce "green" hydrogen even from currently unused waste biomasses. The "green" hydrogen obtained can be used for direct application in fuel cells, combustion engines and as raw gas.

Since the current sources of renewable electricity are not even remotely sufficient to produce the required quantities of hydrogen, hydrothermal carbonization is the ideal technology to ensure the energy transition in general and "green" hydrogen production.

Inhaltsangabe:

Biomassekonversion und Wasserstoffproduktion: Derzeit wird molekularer „grüner" Wasserstoff (H_2) in Deutschland zum größten Teil aus Elektrolyse von Wasser gewonnen, wobei für die Elektrolyse ausschließlich Strom aus erneuerbaren Energien zum Einsatz kommen sollte. Nur „grüner" Wasserstoff ist klimafreundlich, denn nur „grüner" Wasserstoff wird ohne fossile Rohstoffe produziert. Alle andere Verfahren zur Herstellung von molekularem Wasserstoff benötigen fossile Einsatzmaterialen und hohe Energiekosten oder verursachen vermehrte Emissionen an Methan (CH_4) oder Kohlendioxid (CO_2) bei der Herstellung.

Mit den hydrothermalen und vapothermalen Karbonisationsprozessen und der anschließenden Vergasung oder der Pyrolyse ist es nun möglich, auch aus derzeit ungenutzten Abfallbiomassen „grünen" Wasserstoff herzustellen. Der gewonnene „grüne" Wasserstoff kann zur direkten Verwendung in Brennstoffzellen, Verbrennungsmotoren und als Rohgas genutzt werden.

Da die derzeitigen Quellen aus erneuerbarem Strom nicht im Ansatz ausreichen, die benötigten Mengen an Wasserstoff zu produzieren, ist die hydrothermale

Karbonisierung die ideale Technik, die Energiewende im Allgemeinen und die „grüne“ Wasserstoffproduktion im Besonderen zu gewährleisten.

Keywords:

“grüner” Wasserstoff, Erneuerbare Energie, Elektrolyse, Hydrothermale und Vapothermale Karbonisierung, ungenutzte Biomasse, Flugstromvergasung

"green" hydrogen, renewable energy, electrolysis, hydrothermal and vapothermal carbonization, unused biomass, entrained flow gasification

Hydrogen-Production in Germany

At the present, molecular "green" hydrogen (H_2) is produced in Germany for the most part from electrolysis of water, whereby only electricity from renewable energies should be used for electrolysis. Only "green" hydrogen is climate-friendly, because only "green" hydrogen can be produced without fossil raw materials. All other processes require fossil input materials and high energy costs or cause emissions of CH_4 or CO_2 during production. (see figure 1)

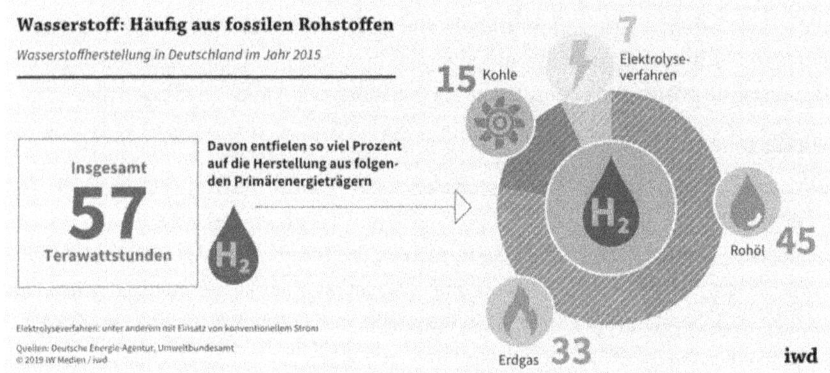

Figure 1: Hydrogen production in Germany 2015 (source Deutsche Energie-Agentur, Umweltbundesamt © 2019 IW Medien / iwd)

The path of the hydrogen from the green power generation is used for transportation processes, “green” power and heat generation in households, fuel cell engines for cars, synthetic fuel production for planes and trucks (kerosene, diesel substitutes) and as raw materials and fuel for the industry.

Actual prognoses calculate a hydrogen demand for Germany of 643 TWh in the year 2050. To reach this goal, there is a lot of renewable technologies necessary to produce the amount of hydrogen for Germany and this can not only be the electrolysis. The use of the conventional technologies, which use fossil resources, is not acceptable because of the high energy costs and caused increased emissions of methane (CH_4) and carbon dioxide (CO_2) during production is against the global climate change policy. So other renewable technologies must be found to produce the demand of hydrogen.

Alternative technology to fossil hydrogen production

Since a few years, a process is on the market and its newly developed technical implementation was already invented in the 1920s by Friedrich Bergius (BERGIUS 1932). Together with Carl Bosch, he was awarded the Nobel Prize for Chemistry in 1931 for the discovery and development of chemical high-pressure processes. After that, the technology, and the process, as currently used by the GRENOL Company, fell into disuse. Fossil raw materials such as coal, oil and earth-gas could be mined and used more cheaply at that time. It was only in 2006, with the beginning of the public discourse on climate change and global warming, that the process and its application as a regenerative form of energy were rediscovered by Prof. Markus Antonietti of the Max Planck Institute for Colloid Research (ANTONIETTI 2006). In a selected consortium, which also included the managing director of GRENOL, the technical possibilities of hydrothermal carbonization (HTC) were communicated and passed onto the participants by Prof. Antonietti. This was the foundation stone for the founding of GRENOL GmbH. Now the company GRENOL GmbH works closely together with the Bavarian mechanical engineer company blueFLUX Energy AG to construct and built together HTC-plants and periphery parts to produce hydrogen for an industrial use.

The aim of the GRENOL GmbH 2007 was to develop a process that converts organic residues in an environmentally friendly, climate-friendly, and hygienic way and enables both energy use and the recycling of plant nutrients for agriculture. The core of the project is a continuously operating plant for hydrothermal carbonization (HTC). All organic biomasses (e.g., biogas digestate, sewage sludge, food waste or cattle slurry, etc.) can be used and processed in the HTC plant as input materials. In the HTC process, the input materials used are broken down into HTC biochar (hydro char) and HTC process water without emissions under conditions of approx. 230 °C and 20-25 bar pressure in a closed system (QUICKER & WEBER 2016).

Figure 2: Any kind of biomass is converted in an HTC-pressure chamber under 230 °C and 20-25 bar to bio-coal and process water.

The coal is then converted into synthesis gas or town-gas in an entrained-flow gasifier and subsequently converted into hydrogen. At the same time, the process water can be fermented in the anaerobic phase in a downstream compact fixed-bed fermenter. The methane gas produced in this process can be used in a CHP unit to produce electricity and heat. Alternatively, the HTC process water can be concentrated to create a valuable fertilizer for agriculture.

In contrast to other processes, hydro-thermal carbonization is not a biological process; the HTC process is a physical-chemical process. This means that the process can be interrupted and restarted at any time without having to observe long dwell times. By splitting the carbon compounds in the biomass, heat energy is released (FUNKE & ZIEGLER 2010, MERZARI et al. 2018). This is called the exothermic process. The released heat varies depending on the composition of the biomass: the more carbohydrates are contained in the biomass, the higher the released heat energy is in the process. This exothermic energy is fed back into the HTC process via a sophisticated heat management system. Depending on the input material, the carbonation process only takes a short time (2-6 hours). This means that large quantities of unused biogenic residues can be converted into coal products in a continuous process. The product, bio-coal, is more homogeneous than the raw biomass, has a higher energy density and is easier to store and handle. The HTC process is mainly used for wet and liquid waste materials with a dry matter content (DM-content) of max. 30 % dry matter. Materials with a higher dry matter content (> 30% DM content) can be treated with the "sister process", vapothermal carbonization (VTC). This process uses steam to carbonize lumpy, solid materials as opposed to hydrothermal carbonization, which works in a liquid aquatic medium. The process parameters of pressure and temperature do not differ from hydrothermal carbonization. At present, VTC technology is only operated in batch mode, this means in a discontinuously process with more manpower and less throughput (SCHWARK 2016).

Differences and advantages compared to conventional methods: Why should HTC/VTC technology being used?

The use of hydrothermal carbonization involves the closure of waste management worldwide, ideally in closed regional material cycles. This means that with the HTC/VTC process, all organic residues can be usefully recycled (see Figure 3). At present, organic waste is not yet being used to its full extent. In many processes that utilize biomass, the carbon efficiency is too low. The production of regenerative energy from biomass often releases climate-damaging carbon dioxide (CO_2). The HTC process, on the other hand, converts almost 100 % of the carbon contained in the biomass into bio-coal (see figure 3). The energy stored by photosynthesis in the form of carbohydrates (e.g., sugar ($C_6H_{12}O_6$)) with the help of the sun, water and CO_2 from the atmosphere is converted into 6 C + 6 H_2O in the HTC process. In this process, $^2/_3$ of the energy from the biomass is retained and $^1/_3$ is released as exothermic thermal energy.

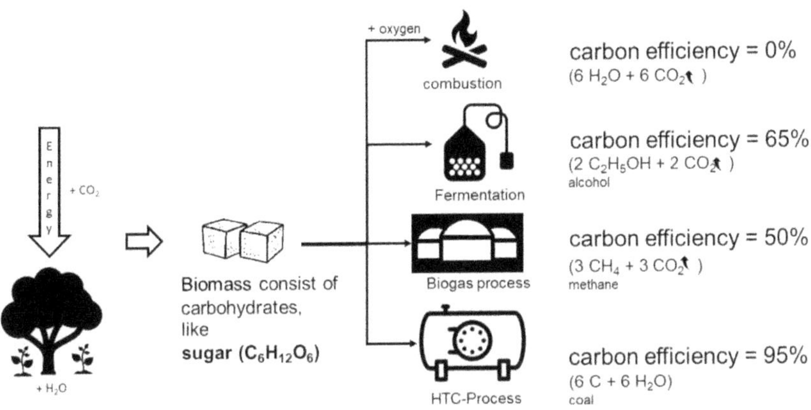

Figure 3: Carbon Efficiency of different processes for biomass utilization in the production of different energy sources (ANTONETTI 2006)

If the HTC bio-coal is stored or used in materials (activated carbon, insulating material, nanotubes, etc.), a CO_2 sink occurs because the CO_2, removed from the atmosphere by the plants, has been fixed in the carbon (NEETHU & DUBEY, 2018; FANG et al. 2018). When bio-coal is used as an energy source or used as a soil optimizer, the carbon dioxide rebalance is neutral because the carbon dioxide, which was recently removed from the atmosphere by the plants, is released again by combustion.

The bio-coal is characterized by a higher press-ability as compared to the input material. Using the example of sewage sludge, which can be mechanically pressed with up to max. 30 % dry matter content, after the HTC-process the sewage sludge bio-coal can be pressed up to 60-70 % dry matter content (RAMKE et al. 2010).

Moreover, the bio-coal can be easily stored and transported. With the use of a wood/coal gasifier and a CHP unit, the bio-coal can be used to produce electricity for feeding into the grid and heat for on-site use - an economical and environmentally friendly use of the energy still contained in the used input material.

Additional advantages of hydrothermal carbonization include the possibility of processing any type, wet or dry, and any mixing ratio. The process does not have to be changed for this. The quantity of coal is determined by the organic dry matter content of the input material, the quality of the coal by the quantity of carbohydrates of the input material.

The bio-coal produced in this way can be stored, it can be used for decentral energy production and it has a high energy density (up to 7 kWh/kg bio-coal). Furthermore, in contrast to fossil fuels (such as crude oil, earth gas, and hard coal/brown coal), HTC-coal is CO_2-neutral when burned, because the biomass has only recently grown and has not been produced over a longer period of millions of years, so there is no additional emission of carbon dioxide into the atmosphere.

The pressure and temperatures in the HTC-reactor hygiene's / disinfects all pathogens, fungi and bacteria/virus contained in the biomass. Moreover, most antibiotics, hormones and even pesticides are decomposed (von EYSER et al. 2015, WEINER et al. 2016). Depending on the plastic composition and melting point behavior, certain plastic fractions (PP+PE) are also converted into coal or an oil precursor in the process. This also applies to micro plastics, which is currently the subject of in-depth scientific studies (INIGUEZ et al. 2019).

The resulting process water can be used in different ways depending on the input material. Firstly, it is possible to return the process water to a fermenter (e.g., fixed-bed fermenter, biogas plant or anaerobic fermenter on a WWTP) to increase methane gas production (WIRTH 2015a, 2015b, 2016).

Secondly, the minerals containing the biomass can be separated from the process water by means of vacuum distillation. Overall, about 70% of available nitrogen ended up in the process water as nitrate or ammonia (McGAUTHY & REZA, 2018). In this process, the nitrogen content is extracted into a commercially available ASL/ASS (ammonium sulfate solution) fertilizer and the remaining nutrients are processed into a basic fertilizer (unpublished tests with the companies Arnold Biogastechnik AG / Switzerland and KMU-Soft GmbH / Germany).

A third possible application is the evaporation of the process water in a greenhouse or with special aquatic plants. This enables an additional increase in biomass for coal extraction (Aqua-HTC, DBU-Project-GRENOL).

In addition to generating energy from unused residues, the HTC plant is easy to handle. The concept is space-saving and can be used decentral all around the world, wherever biomass (to be disposed of) accumulates.

Advantages over conventional technology:

- Hygienisation / disinfection: antibiotics, pesticides, bacteria, virus and fungi are destroyed by the process. (von EYSER et al. 2015, WEINER et al. 2016).
- Efficient and environmentally friendly process.
- Low emission and economical, 50% energy saving against thermal drying (BÜCHLER 2012, DBFZ 2015, THRÄN 2016, KREBS 2013).
- Storable coal, CO_2-neutral, 60 % CO_2 reduction to the Status quo (DBFZ 2015).
- Higher press-ability of the carbonated solid material (up to 70% dry matter, RAMKE et al. 2010).
- Added value for agriculture through the separation of nitrogen, potassium and phosphorus (McGAUTHY & REZA, 2018).
- Easy handling and high usability due to the mixing of the input materials (ANTONIETTI 2006).
- Carbon efficiency: more than 95% (ANTONIETTI 2006, RAMKE et al. 2010).
- Decentralized use, where biomass is produced (ANTONIETTI 2006).
- Compact modular expandable systems and patented technology (GRENOL).

Heat, energy, and hydrogen (H_2) production systems

The GRENOL concept prefers / suggests in the closed waste management cycle a usage of the coal to thermal and electrical energy or a production of H_2 for industrial use. To produce heat and energy the bio-coal can be briquetted into bigger briquettes and then used in a commercial wood/coal gasifier with a CHP combined. The bio-coal is converted in the gasifier to synthesis gas, also called town gas. The synthesis gas is burned in a conventional bi-fuel gas-engine, the heat from cooling the engine can be used in a local heating network. The generator of the CHP produces the electricity, which can be supplied in the local power grid or, if needed, for self-consumption. In cooperation with the Bavarian engineering company blueFLUX Energy AG, it is possible to convert the bio-coal to hydrogen (H_2) in combination with an entrained-flow carburetor/gasifier (EFC) for industrial use. This concept is a little bit more expensive but has a higher adding value than only to produce heat and power.

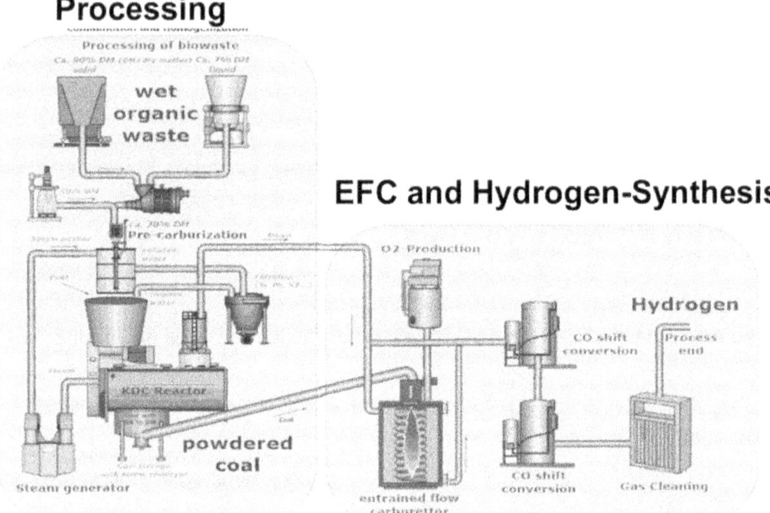

Figure 4: The combined solution of processing the biomass within a hydrothermal process, an entrained flow carburetor and a hydrogen synthesis process from the blueFLUX Energy AG.

The theoretical available agricultural and municipal biomass in Germany

Currently there are a lot of mostly unused agricultural biomass in Germany available. The theoretical amount of agricultural biomass is roundabout 200 Mio. t/a with a dry matter content of roundabout 5-8%. One third of the agricultural biomass is waste from biogas plants in form of digestate (~ 65 Mio. t/a). The other two third is liquid manure of the factory farming from pigs and cows in Germany. This is the amount of roundabout 134 Mio. t/a. (source: Wikipedia.de)

Another source for the converting of biomass into hydrogen is the mass of food waste in Germany, which is wasted year after year. According to the federal environment agency (UBA) there is roundabout 13 Mio. t/a of food waste in Germany available.

With the complete potential of the 213 Mio. t/a of biomass waste from agricultural it is theoretical possible to convert it with hydrothermal carbonization to 2,05 Mio. t/a hydrogen for municipal or industrial use. For example, 1t of hydrogen can be used to drive four H_2-cars each for a mileage of 25.000 km per year.

As biomass resource from the municipal communities is sewage sludge a good example. According to the BMU in 2016 there was an amount of 1,8 Mio. Tons of dry matter content. Theoretical this amount, which must normally costly disposed in thermal combustion plant, can be converted into 270.000t t/a hydrogen by the GRENOL GmbH & blueFLUX Energy AG concept. Compared to conventional disposal of sewage sludge by thermal recycling, the use of HTC for recycling sewage sludge results in a 60% reduction in CO_2 emissions (DBFZ 2015).

Totally there are **2.32 million t of hydrogen (H_2)**, which can theoretically be produced per year from all biomass waste in Germany with HTC process. This is corresponding to approx. **77.3 TWh of "green" electricity**. With the carbonization of agricultural and municipal waste more then $^1/_8$ of the hydrogen demand expected in Germany by 2050 can be solved only with biomass from the agricultural und municipal resources.

Literature:

ANTONETTI, M. 2006 Zauberkohle aus dem Dampfkochtopf, Max Multimedial Heft 2.

BERGIUS, F. 1932 Chemical reactions under high pressure. *Nobel Lecture, S. 33.*

BÜCHLER, B. 2012 Klärschlammentsorgung mittels hydrothermaler Karbonisierung. HTC-Verfahren – die Vor- und Nachteile. *In: Umwelt Perspektiven 1 – 2012. Postfach, 8308 Illnau.*

DBFZ, KLEMM, M., GLOWACKI, R., NELLES, M. (HRSG.) 2015 Innovationsforum Hydrothermale Prozesse. Leipzig: DBFZ.

VON EYSER, C., PALMUC, K., SCHMIDT, T.C., TUERK, T.C. 2015 Pharmaceutical load in sewage sludge and biochar produced by hydrothermal carbonization. *Science of The Total Environment, Volume 537, 180-186.*

FANG, J., ZHAN, L., OK, Y. S., GAO, B. 2018 Minireview of potential applications of hydrochar de-rived from hydrothermal carbonization of biomass. *Journal of Industrial and Engineering Chemistry, Vol. 57 15-21.*

FUNKE A., ZIEGLER F.	2010	Berlin T.U., 2010, Hydrothermal carbonization of biomass: A summary and discussion of chemical mechanisms for process engineering, *Biofuels, Bioproducts and Biorefining, 4, 160–177.*
INIGUEZ, M. E., CONESA, J. A., FULLANA, A.	2019	Hydrothermal carbonization (HTC) of marine plastic. *Fuel, Volume 257, 1-7.*
KREBS ET. AL	2013	Weiterentwicklung der hydrothermalen Karbonisierung zur CO2-sparenden und kosteneffizienten Trocknung von Klärschlamm im industriellen Maßstab sowie der Rückgewinnung von Phosphor. *Züricher Hochschule für angewandte Wissenschaften. Schlussbericht UTF 387.21.11./ IDM 2006.2423.222.*
MCGAUHTY, K. & REZA, M. T.	2018	Recovery of Macro and Micro-Nutrients by Hydrothermal Carbonization of Septage. *J. Agric. Food Chem.* Vol. 66, 8, 1854-1862
MERZARI, F., LUCIAN, M., VOLPE, M., ANDREOTTOLA, G., FIORI, L.	2018	Hydrothermal Carbonization of Bioma Design of a Bench-Scale Reactor for Evaluat the Heat of Reaction. *Chemical Engineer Transactions, Vol. 65, 43-48.*
NEETHU, T. M. & DUBEY, P.K.	2018	Hydrothermal carbonisation of biomass and its potential applications in various fields. *The Pharma Innovation Journal; 7(4): 1132-1136.*
QUICKER, P. & WEBER, K. (HRSG.)	2016	Biokohle: Herstellung, Eigenschaften und Verwendung von Biomassekarbonisaten, *Verlag Springer Vieweg, S. 443.*
RAMKE H.G., BLÖHSE D., LEHMANN H.J., ANTONIETTI M., FETTIG J.	2010	Machbarkeitsstudie zur Energiegewinnung aus organischen Siedlungsabfällen durch Hydrothermale Carbonisierung, *Deutsche Bundesstiftung Umwelt (DBU), Höxter*

SCHWARK, J.	2016	Vergleichende Analyse der Produkte aus HTC und VTC unter besonderer Berücksichtigung von Gärreststoff als Edukt. *Dissertation, Universität Duisburg-Essen, S. 148.*
THRÄN, D.; PFEIFFER, D. (HRSG.) ET AL.	2015	Meilensteine 2030: Elemente für die Entwicklung einer tragfähigen und nachhaltigen Bioenergiestrategie; *End-bericht zu FKZ 03KB065, FKZ 03MAP230. In: Thrän, D., Pfeiffer, D. (Hrsg.) Schriftenreihe des Förderprogramms "Energetische Biomassenutzung" 18, DBFZ Deutsches Biomasseforschungs-zentrum gemeinnützige GmbH, Leipzig, 162 S.*
WEINER, B., RIEDEL, G., KÖH-LER, R., PÖRSCHMANN, J., KOPINKE, F.-D.	2016	Hydrothermaler Schadstoffabbau: hydrothermaler Schadstoffabbau unter besonderer Berücksichtigung von Medikamentenrückständen und PVC. *In: Thrän, D., Pfeiffer, D., Klemm, M. (Hrsg.).*
WIRTH, B., KREBS, M. & AN-DERT, J.	2015a	Anaerobic degradation of increased phenol concentrations in batch assays.

Environmental Science and Pollution Research, Vol. 22, Issue 23, 19048-19059.

WIRTH, B., REZA, M., T., MUMME, J. 2015b Influence of digestion temperature and organic loading rate on the continuous anaerobic treatment of process liquor from hydrothermal carbonization of sewage sludge. *Bioresource Technology 198, 215–222.*

WIRTH, B. & REZA, M. T. 2016 Continuous Anaerobic Degradation of Liquid Condensate from Steam-Derived Hydrothermal Carbonization of Sewage Sludge. *ACS Sustainable Chemistry & Engineering. 4, 1673-1678.*

Author's addresses:

Dipl.-agr. Alfons Kuhles
GRENOL GmbH
Artzbergweg 6
D-40882 Ratingen
Telefon +49 2104-2145153
E-Mail: alfons.kuhles@grenol.de

Dipl.-Ing. Hubert Kohler
blueFLUX Energy AG
Bergwerkstrasse 14
82380 Peißenberg
Fon: +49 8803 90071-0-
E-Mail: info@bluefluxenergy.de

Hydrogen From Biogenic Sources With Negative CO2 Balance for Fuel Cell Waste Collection Vehicles

Hanke, Jens

Graforce GmbH, Berlin, Germany

Abstract

The automotive industry worldwide is undergoing a paradigm shift in the development of fuel cells (FC) and battery electric vehicles and the use of climate-neutral and low-emission fuels in order to meet the CO2 emission targets and future legal requirements in many countries. At present, it is not economically viable to replace fossil hydrogen with electricity-based hydrogen from electrolysis, as this is characterised in the EU by high production costs (taxes and levies) and electrical energy consumption. In the meantime, new innovative, electricity-based hydrogen processes are available which require only 1/5 of the electrical energy required when using biogenic sources. By breaking down bio-methane into hydrogen and black carbon, these new processes can even drastically reduce the well-to-wheel balance of fuel cell vehicles.

Keywords

Hydrogen, plasmalysis, fuel cell vehicles, negative emission technology

1 Introduction

Globally, humans/animals generate around 1.5 billion m³ of biomass every year. In Germany alone, around 15.0 million tCO2 of process-related CO2-neutral emissions are generated through the production of biogas (KURZBERICHT, 2018) and sewage sludge incineration each year. By increasing the carbon load (so-called co-fermentation) in the digestion towers of sewage treatment plants (vegetable fat waste, surplus manure, etc.), sewage gas and digested sludge production (HABERKERN, 2013) can be increased many times over . The actions necessary for this require the biomass to be transported in an organised and legally coordinated manner.

Hydrogen can be obtained from this biomass in the form of such as hydrocarbons, biomethane and ammonia in waste water. In order to obtain hydrogen it must first be separated from mixed elements. Each of these sources requires through the different binding energy a different amount of energy to produce hydrogen. Hydrogen from biogas requires only with modern electricity-based technologies about 10kWHh/kg H2. The production of hydrogen from pure water by electrolysis requires about 50kWh/kg H2. The various possibilities of hydrogen production (MOSTAFA, 2019) from biomass enable not only to efficiently decarbonize energy production and heat generation as well as and heavy-duty transport, but also to indirectly remove CO2 from the atmosphere.

2 Methods

2.1 Hydrogen Production

Natural gas steam reforming is the most used process for the massive production of hydrogen. This process is the cheapest method currently known to obtain hydrogen. A major disadvantage are the large quantities of CO_2 that are produced and have to be disposed of at high *costs* (Carbon Capture *and* Storage-*CCS).*

A variation of the process is the Partial Oxidation Methane Reforming, where the heat duty is supplied thanks to the combustion of a share of input methane directly inside the adiabatic reactor. By this way, the energy needed to produce the hydrogen is fed by a fossil source.

Another process is the coal gasification, able to produce syngas (a mixture of methane, carbon monoxide, hydrogen, carbon dioxide and water vapor) from coal and water, air and/or oxygen. After the gasification reactor, a proper purification system allows to obtain a gas stream with the desired H_2 composition.

The biomass gasification is a process in which organic carbon-based materials are converted at high temperatures of above 700℃ under controlled conditions into hydrogen, carbon dioxide and carbon monoxide. Carbon monoxide is once again able to react with water through the water-gas shift process to form more hydrogen.

There are three process approaches for water electrolysis, the alkaline electrolysis (AEL) with an aqueous potash or sodium hydroxide solution as electrolyte, the polymer electrolyte membrane electrolysis (PEMEL) with a proton-conducting membrane as electrolyte and the solid oxide high temperature electrolysis (SOEL) with a ceramic ion-conducting membrane as electrolyte. The electrolysis plants (electrolyzers) are correspondingly different in design.

Methan-Plasmaysis [MP] and Methan-Pyrolysis (GEIßLER, 2016) are the decomposition of methane into hydrogen and solid carbon (C). Methane pyrolysis technologies like the EMBER 'Molten metal methane pyrolysis' technology have been around for a long time. In order to crack natural gas, it will have to be heated to 800-1000 °C. We call that process methane pyrolysis (separation in the absence of air and water).

Plasmalysis is the name given to a plasma chemical process that requires a voltage source. On the one hand, it describes the plasma-chemical dissociation of organic and inorganic compounds (e.g. C-H and N-H compounds) in interaction with a thermal/non-thermal plasma between two electrodes; on the other hand, it describes the synthesis, i.e. the union, of two or more elements to form a new molecule (e.g. methane synthesis). Plasmalysis is a made-up word from plasma and lysis.

With highly non-equilibrium plasmas (MIZERACZYK, 2016) (low gas temperature and high electron temperature), such as a dielectric-barrier discharge *plasma* (DBD) and a pulsed corona plasma, owing to energetic electrons dehydrogenation of methane occurs so fast and easy hydrogen and carbon can be produced.
Pyrolysis and Plasmalysis is an emerging technology with the potential to play an important role in hydrogen production in the future. This is because the carbon is in solid rather than gaseous form and therefore does not require costly storage in underground caverns as is the case with CCS.

3 Research

3.1 Decarbonisation of bio-methane to hydrogen and carbon black to improve the well-to-wheel balance

Upcoming European CO_2 targets are forcing car manufacturers to include alternative drive systems in their product range. In doing so, the entire chain of effects for moving the car must be considered, i.e. from the extraction and provision of the drive energy to its conversion into the energy needed to drive (kinetic energy) and from the construction of the car to its disposal.

Hydrogen from organic waste has the potential - with the help of a new plasma-based process technology (plasmalysis) - not only to efficiently decarbonize energy production, but also heat generation and heavy-duty transport, but also to indirectly remove CO_2 from the atmosphere.

The plasma electrolysis process presented in the paper, is intended to quantitatively demonstrate, using the example of a large municipal wastewater treatment and disposal company, how hydrogen (see Fig. 1) can be efficiently produced from two biogenic sources i.e. sewage gas and digested sludge water for fuel cell waste collection vehicles for 2€/kg H_2.

Plasma-based process technology such as plasmalysis from organic waste materials is characterized as follows:

- Sewage or biogas from biomass digestion can be split into hydrogen and solid carbon using a non-thermal plasma at a low energy cost (10kWh/kg H_2). The carbon in its solid form (carbon black) is permanently bound in products such as cement or concrete. The hydrogen produced is not only a CO2-free, but even CO2-negative energy carrier. This process technology is thus one of the first negative-emission technologies to be ready for the market and is urgently needed in terms of climate policy.

- At the end of the digestion process in the digester, the press water is heavily contaminated with nitrogen and carbon compounds. In the treatment of wastewater by the plasma process, a high-frequency plasma field is generated from solar or wind energy and interacted with the press water. This splits the contained carbon and nitrogen compounds (such as urea, amino acids, nitrates and ammonium) into individual C, N, H and O atoms. The elements then recombine in the plasma field, producing primarily hydrogen, nitrogen and bio-methane. These gases can be separated by membranes and stored temporarily in gas containers. Purified water remains as a byproduct. The special plasmalysis process also offers the advantage that no nitrous oxide (N_2O) is produced. This is because nitrous oxide, which escapes into the atmosphere during deammonification, is around 300 times more harmful to the climate than carbon dioxide.

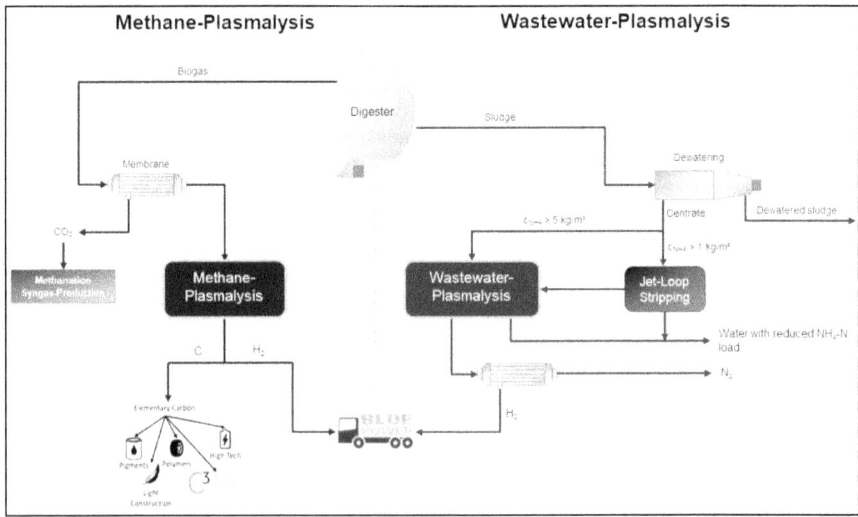

Figure 1: Hydrogen production from biogas and waste water with plasmalysis

Hydrogen production is not only CO2 neutral, but CO2 is even removed from the atmosphere. The plasmalysis process is thus one of the negative-emission technologies. According to the World Energy Council, these are crucial to achieving the goal of climate neutrality, because many emissions can only be reduced at great financial and technical expense, or cannot be avoided at all.

Based on sources from a large municipal wastewater treatment and disposal company (Tab.1), this plasmalysis method has significant potential for sustainable hydrogen production using wastewater and biogas.

Tab. 1: Potential for sustainable hydrogen production from biogenic sources with plasmalysis

	Amount	Unit
Production of hydrogen	309.413	kg/a
Avoided nitrous oxide emissions (CO2,eq)	616.519	kg/a
Number FCV (hydrogen produced from the plasmalysis process)	**52**	**pcs/a**
Production of hydrogen	5.222.666	kg/a
Production of carbon black	15.667	t/a
Avoided CO2 equivalents due to non-combustion of the methane fraction	57.449	t/a
Number FCV (Hydrogen produced from the plasmalysis process)	**870**	**pcs/a**
Number Vehicle Total	**922**	**pcs/a**

With regard to the well-to-wheel consideration (Fig. 2), the FC waste collection vehicle can realize a negative CO_2 balance of 902 tons in its life cycle when using hydrogen from sustainable biomethane, since the solid carbon (carbon black) produced is sustainably removed by permanently binding it in asphalt, for example.

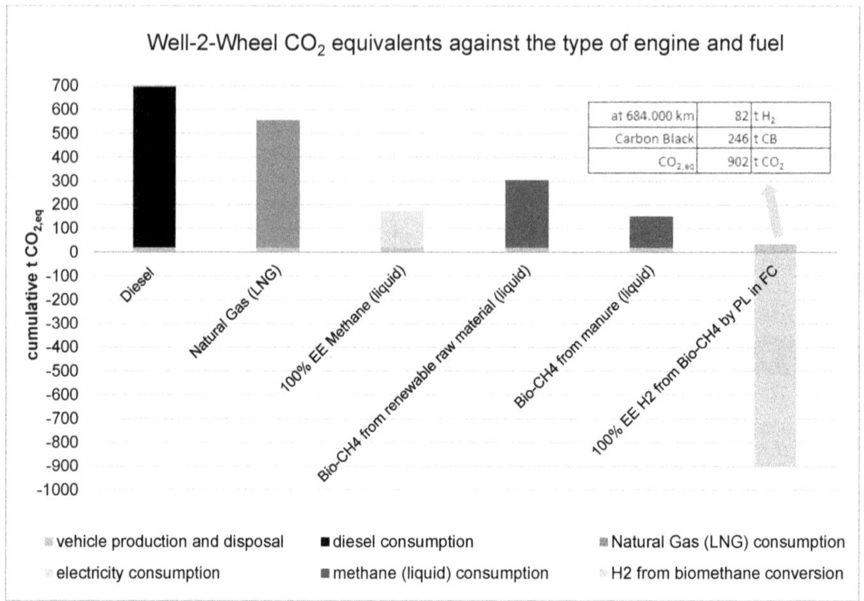

Figure 2: Greenhouse gas emissions of trucks with different propulsion technologies, balanced over the life cycle of 684,000 km and from 2019 to 2025 (EE: renewable electricity, PL: plasmalysis, FC: fuel cell – (WIETSCHEL 2019)).

4 Discussion

Vehicles powered by hydrogen or biogas have a good CO_2 balance. The electric car is just as climate-friendly - if the electricity is generated from renewable energy. An even better CO2 balance can be achieved through innovative technologies in the form of decarbonisation or decomposition of biogenic sources into hydrogen and solid carbon. These renewable resource, can contribute to the success of the energy turnaround and the reduction of greenhouse gases by means of process engineering conversion into hydrogen and carbon black.

The potential to reduce CO_2 from biogas/sewage gas is 1.6 $GtCO_2$/a worldwide, producing 83 Mt green hydrogen and 249 Mt green carbon black (BCDR). The carbon black can be bound in product form (e.g. in asphalt, concrete) to cover the post-fossil raw material demand and thus be withdrawn from the CO_2 cycle in the long term.

Policy must play a significant role in achieving decarbonisation. Specifically it must enable industries to make the investments and adaptations necessary in order to develop a hydrogen energy economy. Accordingly, European and national policy makers will need to recognise the importance of hydrogen from natural gas or biomethane in decarbonisation efforts . Research on the implementation of hydrogen technologies from sustainable biomass should be supported. This includes hydrogen-based fuels and methane decomposition processes and uses for carbon end products. The role of hydrogen from natural gas or biomethane and the role of existing gas networks in enabling decarbonisation should be recognised, and research into converting natural gas networks to hydrogen should be supported.

5 Literature

Kurzbericht © FfE, Dezember 2018

Haberkern, B. (2013): Energieanalysen auf Kläranlagen nach DWA A 216: Neue Instrumente im Praxistest, Vortrag im Rahmen der IWAR-Vortragsreihe „Neues aus der Umwelttechnik und Infrastrukturplanung", Wintersemester 2013/14, Darmstadt

Mostafa et. al. (2019) : Hydrogen Production Technologies Overview, Journal of Power and Energy Engineering, Vol.7 No.1

MP, Methan-Plasmalyse: https://www.graforce.com/technologien/methan-plasmalyse

Geißler, T. (2016): Hydrogen production via methane pyrolysis in a liquid metal bubble column reactor with a packed bed, Chemical Engineering Journal Volume 299, Pages 192-200

Mizeraczyk, J. & Jasiński, M. (2016): Plasma processing methods for hydrogen production; The European Physical Journal Applied Physics 75(2)

Wietschel et al.(2019) ; Fraunhofer-Institut für System- und Innovationsforschung ISI: Klimabilanz, Kosten und Potenziale verschiedener Kraftstoffarten und Antriebssysteme für Pkw und Lkw

BCDR, Detaillierte Berechnungen und Quellen unter http:/www.graforce.de/images/videos/CDR-Potenzial.xlsx

Author's address(es)

Dr. Jens Hanke
Graforce GmbH
Johann-Hittorf-Str.8

12489 Berlin
Telefon +49 30 63 2222 110

E-Mail:hanke@graforce.de

Fire protection in waste treatment and recycling plants

Mark Müller

"Why is Targeted Extinguishing Control the Better Solution? Prerequisites for Targeted Extinguishing Control Via Infrared Systems and the Technical Implementation in Practice"

Orglmeister Infrarot-Systeme GmbH & Co. KG, Walluf, Germany

1 Why are automatic extinguishing solutions in waste treatment and recycling plants utilizing heat detection so much more effective?

Given the growing environmental awareness and the increasing amount of recycling and incineration facilities worldwide, the risk of fires in these industries is a rising concern. Solutions to extinguish these fires in the early stages of development is imperative, especially considering that the materials being processed are unpredictable. There is certainly no simple answer to that question, but it is an issue that needs addressing. In this article suitable fire protection systems are discussed, with a focus on automatic extinguishing solutions utilizing heat detection.

2 Development of the fire hazard situation

Over the last few years, the trend towards recycling has grown in many parts of the world. This has led to the division of organic wastes and recyclables and the installation of waste management companies operating incineration plants, composting plants, and recycling facilities instead of landfills. Vast amounts of materials are now temporarily stored. The fire hazards associated with this are growing as relatively dry materials with high calorific value is stored together with potential ignition sources such as Lithium-Ion batteries, household aerosol bottles, paint cans, and propane tanks. In composting facilities, biodegradation can lead to temperatures high enough to cause auto-ignition of the stored material. These types of fire can be difficult to detect and often demand great extinguishing effort when detected too late. This can have serious effects on the environment and public health and jeopardize the safety of firefighters, staff and local communities.

3 Primary fire risks

Recycling facilities are generally structured in the following way:

- Delivery and primary storage area of unsorted recycling goods (input area, tipping floor)
- Sorting and separation facility
- Storage of separated goods such as plastic, paper, metal, glass, and compost (output area)

The input storage area, the tipping floor is especially at risk. Here, the complete variety of mixed domestic waste, as collected from the doorstep, is tipped from collection trucks onto concrete floors or into waste bunkers. In this conglomerate of waste, both ignition sources and combustible materials are present. Damaged batteries may have developed heat and are exposed to oxygen. Sparks can ignite gases and vapors leaked from household aerosol bottles, paint cans, and propane tanks or gases formed due to decomposition of waste. Before being transported into the recycling facility via conveyor belts, workers or machines sort out as much problematic items as possible. Unfortunately, these components often end up inside the facilities where they may ignite and start a fire. Fortunately, most of the waste is in constant movement. Hotspots or fires can be monitored and quickly dealt with if the proper detection and extinguishing equipment is installed.

In incineration plants, the untreated waste is often burnt without any separation. The material is stored in bunkers, partially several meters high, where it may be stored for longer durations of time before being transferred to the incinerator. The same situation is found in the storage of processed RDF material. These already tighter and more flammable bulk materials can contain heat sources from processing and over time can cause a serious fire in the process. Here, a fire may smolder below the surface without being detected and break out over a wider area.

4 Fire development and fire detection

In general, there are three common detection scenarios.

- Smoke detection
- Fire detection
- Heat detection

Smoke detectors are mainly installed under the ceiling to monitor complete halls or sections of a big area. They generally require a large amount of smoke to trigger an alarm. They are mainly used together with manual firefighting equipment utilizing hoses or firefighting monitors as the exact location of a fire must be visually confirmed. They are very maintenance intensive especially in an environment with high dust loads. and not well suited as components for modern automatic firefighting solutions.

Another possibility for smoke detection is the use of video smoke detection. It is recommended to use these systems only if combined with another type of detection to avoid false alarms triggered by steam, exhaust fumes or fog. These systems also require ideal lighting conditions and only work in areas with low levels of dust.

Sprinkler systems are classic fire detectors. They provide some protection to the structure of the building, but do not count as early warning systems.

Linear heat or fire detectors are sensor cables. They are mainly used to monitor tunnels or garages but may also be installed in big halls. They are generally not suited for use in incineration plants and recycling facilities.

The fastest detection is achieved through thermal imaging by using infrared (IR) detection technology. In contrast to detecting smoke or a fire, the environment is monitored for heat radiation. By continuously monitoring a specific point or area and measuring the actual radiated heat, or analyzing the increase in temperature, fires can be detected, even if they have not yet reached the surface of a pile. The rise of hot gases or air may be sufficient to give away a sub-surface fire. Usually, temperatures of 80°/90°C at the material surface are considered strong indicators of a fire. At this point the main alarm is triggered and communicated to the fire brigade. Heat monitoring of an object with an infrared early fire-detection system means a fire is identified in its formation phase.

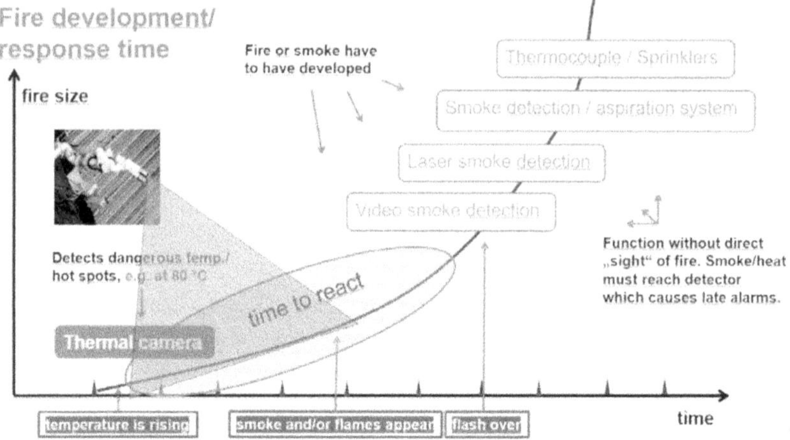

Figure 1 fire development/response time

5 Fire protection systems

The main extinguishing systems used in recycling and incineration plants are sprinkler systems and firefighting monitors (water cannons). Dependent on the goods that must be extinguished, water or foam can be used as extinguishing agent.

Sprinkler systems are mainly used indoors and are generally water filled. Alternatively, they might be filled with a foam premix that generates extinguishing foam once released. The heat emanating from a fire activates individual sprinkler heads which will release extinguishing agent onto the area below. If the fire spreads, additional sprinkler heads are activated to extend the extinguishing capability. Each sprinkler head is designed to protect an area of several square meters. Large areas are exposed to the extinguishing water when several sprinkler heads are activated, and the system usually must be manually deactivated. Depending on the distance between the fire and the sprinkler heads they may be triggered too late to successfully extinguish the fire. Certainly valuable time to fight the growing fire is wasted before they react.

Deluge systems are sprinkler systems with open nozzles and can be triggered by IR heat detection systems. They can be manually operated or may be equipped with remote controlled valves. On activation extinguishing will occur in the complete section of a larger area. A potential fire can be prevented, but an interruption of operation is unavoidable, as large amounts of extinguishing agent are introduced into the recyclable material and the material cannot be further processed in this state.

Firefighting monitors, are designed for indoor or outdoor use. When a fire is detected, they are either manually operated or can be remotely controlled. Fire monitors allow precise positioning of fire extinguishing media from a safe distance. Dependent on the fire extinguishing system setup, it is possible to switch between water and foam. Firefighting monitors are optimally suited to be combined with IR detection systems to form an automatic fire extinguishing system. The automatically controlled application via infrared cameras ensures a very targeted application of the extinguishing agent and a very short time interval. This means that only as much extinguishing agent as necessary is introduced into the material. In most cases, operations can continue without major interruption.

Figures 2/3 active firefighting monitors

For fire detection, continuous monitoring of the hazardous area is mandatory. Any changes in the temperatures within the area monitored are recorded and analysed. Intentional and known heat sources such as motors from belt drives or vehicles, exhaust pipes, as well as the heat of the sun and reflections should be automatically identified and ruled out as potential fires to reduce false alarms to a minimum.

With the Orglmeister PYROsmart system areas can be monitored using a single pan/tilt head camera. It continuously scans a large area and builds a high-resolution radiometric panoramic image. Based on the intelligent software analysis and exact 3D localization of a hotspot the system is capable of commanding the water cannon to start a targeted extinguishing process. A combination of IR and live video camera pictures provide an effective analysis of the situation. Through self-learning and artificial intelligence (AI) the software analyzes the environment and differentiates between hot motors, exhaust pipes and hotspots that indicate potential or actual fires.

Figures 4/5 *The entire waste bunker area is thermally monitored by the patented IR-panoramic display. The PYROsmart pan-tilt system provides a complete radiometric and video overview. Operation is easy and user-friendly.*

Figure 6 Radiometric panoramic image of an input recycling hall.

Figure 7 Hall detection with the pan/tilt IR system PYROsmart. Best view from the installation site on the hall ceiling.

6 Automatic extinguishing solutions

When planning a fire extinguishing system, an operator has to consider a variety of factors:

- the most effective firefighting strategy to extinguish the wide range of possible fires must be found
- it might have to comply with insurance regulations
- the system must be commercially attractive

One of the initial considerations regarding the best strategy is the decision to use water, foam or have the alternative to use either, depending on what types of fires need to be suppressed.

Moreover, it has to be decided whether manual or automatic intervention is wanted.

Common automated extinguishing system are triggered by conventional detection, which often implies there is a considerable delay until the extinguishing process can start (see Fire development above). Once triggered, a deluge system may be activated, flooding the complete area. Alternatively, a fire monitor can automatically direct the extinguishing agent using a pre-programmed spray pattern in a pre defined area. Deactivation of the extinguishing system is mainly done manually.

If, however, the fire extinguishing system is controlled by an IR detection system to work automatically, the fire monitor is activated to accurately direct water or foam to the exact location of the hotspot or fire. Deactivation of the extinguishing process is automatically achieved by regular temperature checks following the initial extinguishing intervals. Once the hotspot has been sufficiently cooled, and the risk of fire is over, operation will stop or can be stopped manually once the fire brigade has taken over.

In this way the extinguishing process is optimized, and as little extinguishing agent as necessary is used. There is an automated multi-stage approach once a hotspot has been detected:

1. Precise delivery of a limited volume of water to the identified area.

2. Check of the extinguishing effect, and, if required, additional delivery of water if the temperature has not decreased to a non-hazardous level.

3. Monitoring and the delivery of foam may be activated automatically if water does not give the required result after one or two extinguishing attempts – or the extinguishing area is enlarged.

With automatic detection and extinguishing systems, the firefighting approach should be customized to the facility, the goods to be extinguished, and the threat a fire may pose to the environment.

A first step, and a significant part of the process, is to determine the best approach for firefighting with an analysis of the premises to assess detectors and fire monitors' best positioning. Optimum positioning of these devices minimizes the quantity and the cost of a system. Compared to sprinkler systems, targeted extinguishing generally requires less water, and therefore smaller tanks.

To comply with insurance requirements, two VdS guidelines are relevant:

- VdS guideline 3189 dealing with the use of IR cameras for temperature monitoring
- VdS guideline 3884 dealing with the setup of water monitors in recycling facilities

7 Conclusion

In order to considerably reduce fire risks for waste and recycling facilities, a fast resonse is key to success. In the event of a fire, especially outside of regular operating hours, state-of-the-art heat detection and automated extinguishing solutions can prevent major damage and loss. 24/7, they are essential to ensure that a potential fire hazard has no chance of developing into a blaze, and that a dangerous hotspot has been automatically extinguished in time, and before a professional response (which might be far too late) is necessary.

With relatively few installations, one IR detection and one water cannon, the installation effort and maintenance costs are very low in relation to the monitored area. In addition, this type of software-controlled system has a high degree of self-monitoring due to the latest software (industry 2.0) and remote maintenance tools.

There is a great potential for savings and protecting the environment by focusing on early detection and smart, precise extinguishing, rather than extended firefighting with traditional fire protection methods. Plant owners and operators can avoid facility shutdowns, health risks, environmental damage, overall financial loss, and can be benefit from reduced investment costs.

Author's address:

Mr. Mark Müller
Orglmeister Infrarot-Systeme GmbH & Co. KG
Am Klingenweg 13
D-65396 Walluf
Telefon +49-6123-68912-0
E-Mail: vertrieb@orglmeister.de

Thermal Metal Recovery From Tertiary Waste

Kurt Bernegger, Helmut Lugmayr, Christian Mlinar

Bernegger GmbH, Molln

Thermische Metallgewinnung aus Tertiärabfällen

In an innovative new pyrometallurgical process ("TMG", Thermische Metallgewinnung / Thermal metal recovery), fine-grained residues from an existing mechanical shredder residue processing plant ("TBS") are further processed. Metallurgical plant technology is thereby combined in an innovative way with the latest technology of thermal waste-treatment. The main innovative aspect of the new process is a close interplay between a metallurgical treatment in a Top Blown Rotary Converter (TBRC), a separation furnace and a comprehensive off-gas cleaning technology. In this way, the recovery of copper and precious metals from residues, which have been up to now deposited or incinerated without the use of the metals contained, is maximized. In addition, by-products like a mineral product for constructional use and zinc dust are produced and the energy contained in the residues is regained.

In einem speziell entwickelten pyrometallurgischen Verfahren ("TMG", Thermische Metallgewinnung) werden feinkörnige, feinstverwachsene Rückstände aus einer mechanischen Shredderrückstandsaufbereitungsanlage („TBS" der Fa. Bernegger GmbH), welche sehr geringe Metallgehalte aufweisen, weiterbehandelt. Dabei wird Anlagentechnik der Metallurgie auf innovative Art mit modernster Technik der thermischen Abfallverwertung verbunden. Auf diese Weise wird das Ausbringen von Kupfer und Edelmetallen aus Rückständen, welche ohne die Nutzung der darin enthaltenen Metalle bisher deponiert oder verbrannt wurden, maximiert. Zudem werden als Nebenprodukte ein mineralisches Bauprodukt/Bindemittel sowie Zinkstaub produziert und die in den Rückständen enthaltene Energie durch Verstromung und Bereitstellung von Prozessdampf und Nah- bzw. Fernwärme für bestehende Anwendungen genutzt. Die Großindustrielle Umsetzung dieses neuartigen Verfahrens wird nach der bereits erfolgten Genehmigung gemäß Umweltverträglichkeitsprüfungsgesetz und der derzeit laufenden Detailplanung in den nächsten 2-3 Jahren erfolgen.

Keywords: shredder residues, WEEE, valuable & precious metals, resources, energy efficient waste recycling, pyrometallurgical technology, industrial process, sustainable innovation, standardised production

Background

Economic development is closely linked to the use of metals putting permanent pressure on the demand for primary resources. Raw materials (RM) are fundamental for maintaining and improving our quality of life. Securing reliable, sustainable, and unrestricted access to specific RMs is of growing concern all over the world.

Recycled raw materials need to substitute primary raw materials

To achieve a more efficient circular economy and use of resources, recycling and substitution of primary RMs need to be promoted (COM, 2008). Functional product requirements lead to complex material mixtures in product designs making recycling increasingly difficult. Little information is available regarding recycling performance, and even less on the true recycling rates that are possible and on how recycling systems can be improved. As a result, metals are often lost from the circular economy, and primary resources need to be increasingly exploited to cover the demand (M.A. REUTER ET AL., 2013). Action is required in terms of a product's life cycle extending from production to the creation of markets for secondary RMs. Processes are needed to improve energy and resource efficiency as well as multidisciplinary approaches combining industry, science, research, legislation and policy form an essential building block to achieve this.

A recycling solution for low metal content residues is missing

Currently, no sustainable final recovery solution exists for the low metal content residues from the shredder light fraction (SLF) and shredder heavy fraction (SHF) treatment coming from end-of-life vehicles (ELV) processing (see Fig. 1). Therefore, the potential of SLF and SHF containing valuable metals is not exploited completely regarding materials that could be recovered for use as secondary resources. In a first step, much wider use of Best Available Technologies (BAT) is required to increase metal recovery rates and to minimise environmental impact along each step of the recycling chain.

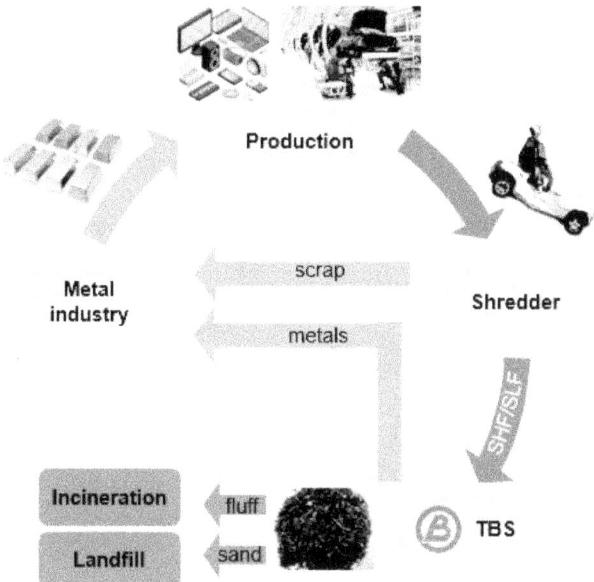

Figure 1: Schematics showing the state-of-the-art: The recycling process is missing a vital link to be truly circular.

The Thermal Metal Recovery – plant („TMG") will close the cycle

"TMG" represents a ground-breaking new technology for waste treatment and, therefore, is the missing link required for enhanced recovery of critical raw materials (CRM) from complex end-of-life (EoL) products.

An innovative pyrometallurgical process enables the next level of further processing fine-grained residues from an existing mechanical shredder residue processing plant. Accordingly, this process combines metallurgical plant technology in an innovative way with the latest thermal waste-treatment technology. The main innovative aspect of the new process is the close interplay between a metallurgical treatment in a Top Blown Rotary Converter (TBRC), a separation furnace, and high standard off-gas cleaning technology. In this way, the recovery of copper and precious metals (PMe) from residues, which until now have been landfilled or incinerated without reusing the metals they contain, is maximised.

In addition, by-products such as mineral raw material for constructional use and zinc dust are produced and the energy contained in the residues is recovered. A scheme, presenting the proposed concept for secondary raw material production via the treatment of ELV wastes, is shown in Fig. 2 below. The objective is to establish a marketable process for reusing resources by processing waste materials that are currently unexploited.

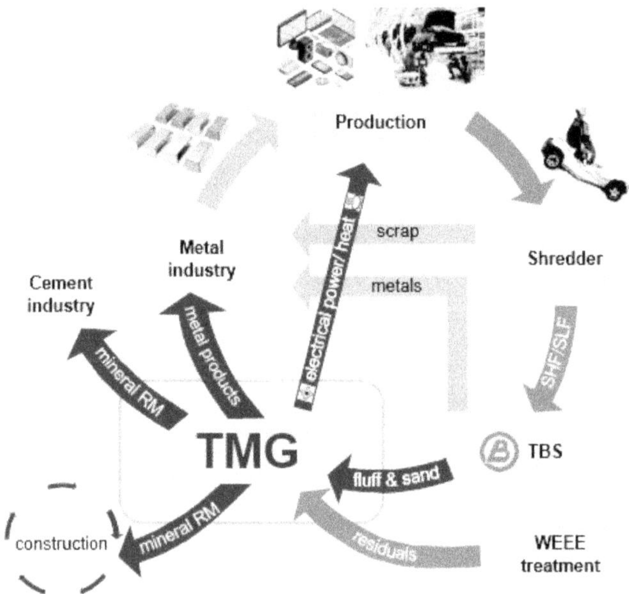

Figure 2: Schematic of the TMG-process: Introducing a disruptive solution to close the cycle.

Furthermore, waste incineration capacities within the EU are not equally distributed and landfills dominate in some regions. That is why sustainable and efficient valuable material recovery must be enforced to ensure that environmental pollutants are excluded from the material cycle (COM, 2017).

The developed new technology for waste treatment enables the idea of reaching a circular economy in terms of recycling waste materials from end-of-life products to contribute to a secure supply of valuable RMs. Within „TMG", mineral fractions will be produced alongside an efficient metal recovery (95 % Cu and 98 % precious metal recovery) to be returned into the material cycle for an enhanced circular economy. The proposed process will close the gap in total material recovery from ELV. Currently, no sustainable final recovery solutions exist for the low metal content residues from the SLF and SHF treatments stemming from ELV processing. Therefore, the potential of SLF and SHF containing valuable metals is not completely exploited regarding secondary raw material recovery (EU-RECYCLING + UMWELTTECHNIK, 2020). „TMG" will lead to a recovery rate in excess of 99 %, when considering the overall ELV treatment chain. This recovery rate is a realistic assumption, based on the actual recovery rates from shredder residues in existing treatment plants, in addition with lab tests already accomplished, and the so far carried out scientific TMG-process calculations. In contrary to the potential of the TMG-process, some experts are dealing with the idea of reducing the mandatory recycling targets of ELV due to a lack of processing technologies being proven in practice (UTERMANN, 2019). „TMG“ will treat SLF and SHF residues to recover valuable materials leading to almost 100 % recycling. Similar potential exists for waste electrical and electronic equipment (WEEE). Energy recovery and utilisation is another focus of the process with simultaneous consideration of stringent legislation in terms of environmental impact.

The mentioned facts clearly prove the necessity, importance and potential of the „TMG“ project for closing of material cycles and increasing efficiency. Therefore, „TMG“ is the missing link between the mechanical processing on the one hand and the pyrometallurgical treatment on the other hand (ROMBACH and FRIEDRICH, 2019).

The TMG - process

The planned process concept of „TMG“ consists of seven main steps (see Fig. 3). The major innovative aspect of the idea is a close interplay between a metallurgical treatment in a Top Blown Rotary Converter (TBRC), a separation furnace and comprehensive off-gas cleaning technology for low metal content fine-grained residues from pre-processed end-of-life products serving as secondary RMs.

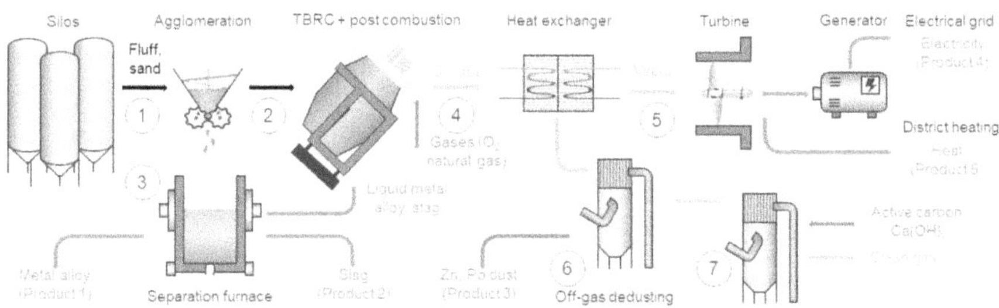

Figure 3: Planned TMG - process concept (including the 7 steps described in the text).

Step 1 - Material briquetting

Sand, fluffs, other secondary materials and additives (if necessary), are pressed mechanically (cold-bonded briquetting). During the briquetting step, heat is generated, which contributes to the material strength. Beside this, the used particles interlock with each other forming stable agglomerates with higher density, which is important for ensuring an efficient charging process to the melting bath. The geometric stability of the briquettes is also important for a stable process. Additionally, the percentage of the waste material fractions within the mix must be correlated to realise an efficient recycling process.

Step 2 - Reducing treatment

A top blown rotary converter is operated at 1,250 °C whereas the process itself is expected to be autothermic i.e. an operation without primary fossil energy sources should be possible. In previous small-scale trials, an autothermic process could be realised. In this context, an important question is to evaluate a reducing treatment without adding a carbonaceous reducing agent, since theoretically, the organic carbon in the briquettes is converted into a CO- and H_2-rich reducing gas in the TBRC to act as a reducing agent.

Step 3 - Separation furnace

The separation furnace can be operated with a reducing or oxidising atmosphere to separate the metal alloy (product 1 in Fig. 3) from the iron rich mineral fraction (slag, product 2 in Fig. 3) due to density differences. A certain viscosity is necessary to realise an efficient metal-slag separation. A further converting step could be necessary to obtain the specified quality of the metal alloy. However, it is extremely important to keep metal losses such as copper to a minimum. Therefore, it may be useful to partially reintroduce the slag into the TBRC for a further conversion step.

Step 4 - Post-combustion

Due to legislation (AVV), post-combustion is needed to comply with the emission regulations i.e. a minimum combustion temperature of 1,100 °C must be guaranteed for at least 2 seconds to decompose pollutants such as furans and dioxins in the TBRC off-gas.

Step 5 - Combined heat and power process

The production of superheated steam is realised in a waste heat recovery boiler. This superheated steam produces electrical energy (product 4 in Fig. 3) and thermal energy (product 5 in Fig. 3) for the district heating network (see in Fig. 3).

Step 6 - Dust product generation

The zinc and lead re-oxidised during post-combustion (step 3) are separated in a gas filter (= product 3 in Fig. 3).

Step 7 - Off-gas purification

Off-gas is cleaned according the AVV (Austrian waste incineration directive) and according to the state-of-the-art (BAT). The first step in the off-gas cleaning step is a dry-based sorption. Fine-grained active carbon and calcium-based sorbents are injected into the off-gas stream to remove heavy metals, halogens, and sulphur dioxides from the off-gas. Finally, selective catalytic reduction of nitrogen oxides is carried out. The cleaned off-gas is released to the ambient atmosphere. The off-gas cleaning system more than meets the requirements according the BAT Document for Waste Incineration from 2019 (NEUWAHL ET AL., 2019).

Target values of the TMG --process

The following target values are envisaged for the pilot plant:

- Annual throughput of waste with low valuable metal content of 75,000 tonnes to 100,000 tonnes
- Production of a metal alloy containing Cu, Sn, Ni and other precious metals, to be used in the copper refining industry
- Production of slag, to be used as secondary raw material in the construction industry
- Regaining of dust (enriched in Zn- and Pb-oxide) to be used by other companies e.g. from zinc industry
- As a further benefit, in addition to the secondary RMs mentioned above, an energy output of 35 GWh in district heating and 48 GWh in electrical energy are generated.

Summary

The innovation potential of „TMG“ allows the decoupling of economic growth from environmental degradation by simultaneously recovering a metal alloy, a mineral product, a metal-bearing dust as well as the stored energy from shredder residues as heat and electricity. The „TMG“ technology goes far beyond the State-of-the-art for fluff and sand treatment of waste incineration and landfilling to an enhanced circular economy (zero-waste strategy). It attempts to overcome economic barriers due to small-scale processes by installing and operating a pilot plant in a relevant industrial environment using optimised parameters. The TMG - process will be a frontrunner in recycling of ELV residues and other identified waste materials.

The large-scale industrial implementation of this new process will take place in the next 2-3 years after the environmental impact assessment-approval for the first site in Austria has already been granted and the detailed planning currently underway.

Literature

Commission of the European Communities	2008	The raw material initiative – meeting our critical needs for growth and jobs in Europe. COM(2008) 699
E. Rombach and B. Friedrich	2019	Innovatives Polymetallrecycling – Herausforderungen und Lösungsansätze an Schnittstellen der Prozesskette. Recycling und Rohstoffe 12. TK-Verlag. Neuruppin, 2019, pp. 453–473.
EU-Recycling + Umwelttechnik	2020	Shredderanlagen: Wohin mit der Leichtfraktion? https://eu-recycling.com/Archive/.
European Commission	2017	The role of waste-to-energy in the circular economy.COM (2017) 34
F. Neuwahl et al.	2019	Best Available Techniques (BAT) Reference Document for Waste Incineration. JRC Science for Policy Report, Luxembourg
J. Utermann	2019	Autoproduzenten sollen für Rücknahme und Recycling bezahlen. Recycling und Entsorgung (49)
M.A. Reuter et al.	2013	Metal recycling – Opportunities, limits, infrastructure. United Nations Environment Programme, Nairobi

Dipl.-Ing. Kurt Bernegger
Dipl.-Ing. Helmut Lugmayr
Dipl.-Ing. Christian Mlinar
Bernegger GmbH
Gradau 15
A-4591 Molln
Telefon +43 7584 3041-0
E-Mail office@bernegger.at

Spent automotive converters as a secondary resource of platinum group metals

Martyna Rzelewska-Piekut, Paweł Krawczyk, Zuzanna Wiecka, Magdalena Regel-Rosocka

Poznan University of Technology, Institute of Chemical Technology and Engineering, Poznań, Poland

Abstract

The paper presents research on the hydrometallurgical processing of spent automotive converters. The aim of the study is to obtain important, from the circular economy point of view, elements from the platinum group metals (platinum, palladium, rhodium). The processing cycle of spent automotive converters includes: grinding and sieving of catalysts, leaching with organic or inorganic solutions, extraction of metal ions with quaternary phosphonium salt: trihexyl(tetradecyl)phosphonium chloride (Cyphos IL 101) and stripping of metal ions from the loaded organic phases with 3 M HNO_3.

Keywords *platinum group metals, Pt, Pd, Rh, recovery, spent automotive converters, hydrometallurgy, extraction, leaching*

1 Introduction

Due to the global development of technology and automotive industry, an interest in platinum group metals (PGM) is constantly growing. Separation and purification of platinum group metals is difficult and complicated due to their chemical properties (e.g. low chemical reactivity, high temperature resistance) and the possibility of forming many chemical compounds, e.g. in chloride solutions.

PGMs have a great technological importance due to their low chemical reactivity, the ability to protect against corrosion, and the ease of processing. Their greatest and most important advantage is the ability to catalyze many chemical reactions, e.g. (Rzelewska and Regel-Rosocka, 2018; Rzelewska-Piekut et al., 2019):

- hydrogenation of benzene to cyclohexane (Pd, Pt),
- oxidation of ammonia to NO (in the synthesis of HNO_3 (Pt, Pt-Rh)),
- gasoline reforming (Pt),
- synthesis HCN from NH_3 and CH_4 (Pt, Rh),
- conversion of exhaust gases (C_nH_m, CO, NO_x) to CO_2 and H_2O in car catalysts (Pt, Pd, Rh).

PGMs are also used in the electronics industry, glass industry, jewelery, medicine and dentistry. These areas of application of platinum group metals are also the main secondary raw materials of these metals. From the economical and ecological point of view, it is more profitable to obtain PGM from the secondary resources than to extract them from the increasingly poorer natural ores. The content of these valuable metals in waste, mainly in spent automotive converters, is several times higher than the PGM content in the richest available natural ores (RZELEWSKA AND REGEL-ROSOCKA, 2018).

Demand for PGM is still growing, and therefore their prices on the global market remain at a high level. Figure 1 presents the average prices of three platinum metals: platinum, palladium and rhodium, used in the production of automotive converters.

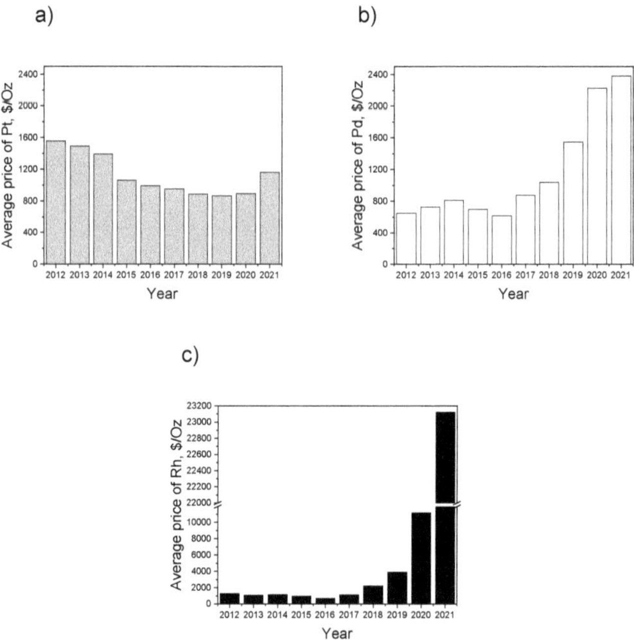

Figure 1 Average prices of platinum (■), palladium (□) and rhodium (■) over the last ten years (1 troy Oz = 31.1 g) (JOHNSON MATTHEY, 2021)

A rapid increase in the price of rhodium has been observed in the last two years: in January 2019, the price of rhodium was 2473 $/Oz but at the end of the year it was 6043 $/Oz (annual average: 3904 $/Oz). In January 2020 (before the Covid-19 pandemic), the price of Rh amounted to 8606 $/Oz and in December 2020 to 16426 $/Oz (annual average: 11219 $/Oz). Currently, the average rhodium price in March 2021 is

28525 \$/Oz (three-month average 23125 \$/Oz) (JOHNSON MATTHEY, 2021). The rapid increase in the price of rhodium is mainly related to the tightening of emissions regulations in China. PGMs are the main component of the active layer in car catalysts and are responsible for the reduction of exhaust gases produced.

The rapid increase in demand for Rh with low supply caused in a more than tenfold increase in the price of this element in recent years: from 2473 \$/Oz (January 2019) to 28525 \$/Oz (March 2021) (Figure 1 c). The price of platinum has remained stable over the past ten years (about 1000 \$/Oz) (Figure 1 a). In the case of palladium, an increase in its prices has been observed since 2018 (Figure 1 b) but it is not as rapid as in the case of rhodium (JOHNSON MATTHEY, 2021). The depleting of natural ores and increasing environmental demands are encouraging scientists from all over the world to look for ways to reuse and recycle waste.

On a large scale, PGM from waste (spent automotive converters) are obtained pyrometallurgically (Umicore, Belgium). However, this method is energy-consuming and expensive, which causes it to be replaced by hydrometallurgical techniques that are not highly energy-consuming. The hydrometallurgical operations (leaching, extraction) are usually carried out at temperatures below 100°C, in aqueous solutions, and are called "wet methods". Hydrometallurgy is also popular due to the possibility of recovering the components from the dilute solutions (AKCIL ET AL., 2015; POŚPIECH, 2012).

PGM chloride complexes are formed during the dissolving of samples containing PGM in hydrochloric acid or aqua regia. However, waste materials are often mixed: real solutions contain also ions of various metals (RZELEWSKA AND REGEL-ROSOCKA, 2018).

Carboxylic acids are not strong enough to react directly with the platinum group metals contained in the catalyst but leaching with these acids can be treated as a preliminary step in the treatment of spent automotive converters (separation of significant amounts of non-precious metals into the liquid phase, such as: Al(III), Zn(II), Mg(II) and Fe(III)) (RZELEWSKA-PIEKUT ET AL., 2020). In the second step, PGM and some amounts of other metals can be leached with concentrated acids or their mixtures with the addition of strong oxidants (e.g. 35 wt.% H_2O_2).

Liquid-liquid extraction is one of the most efficient techniques used to separate valuable metal ions from dilute wastewaters. Phosphonium ionic liquids (IL), which are becoming popular in separation processes, can be used as extractants (CIESZYŃSKA AND WIŚNIEWSKI, 2011; AKTAS, 2011; REGEL-ROSOCKA ET AL., 2015). Despite the existing wide range of various extractants, there is still a trend for search for new, effective and selective PGM ion carriers. Therefore, in the current paper the use of selected quaternary phosphonium salt, i.e. trihexyl(tetradecyl)phosphonium chloride (Cyphos IL 101), is proposed.

Phosphonium salts (ionic liquids, IL) are a good alternative extractants to ammonium or imidazolium salts. Phosphonium ILs are characterized by greater thermal resistance than ammonium ILs. Phosphonium cations are more inert than ammonium or imidazolium cations and therefore do not interact with other dissolved substances (BRADARIC ET AL., 2003; STOJANOVIC ET AL., 2011; LUO ET AL., 2012).

2 Experimental part

2.1 Materials

Spent automotive converters were obtained from a Polish company dealing with waste disposal. After the grinding and sieving, the particle size of the powdered catalyst used for leaching was <63 µm. Due to the variety of waste, a two-stage leaching of metal ions from spent automotive converters was proposed:

- the first step: leaching of metal ions with an organic acid – formic acid (Sigma Aldrich)),
- the second step: leaching of metal ions with a mixture of concentrated inorganic acids 36% HCl (Chempur, Poland) and 96% H_2SO_4 (Chempur, Poland) with the addition of an oxidizer 35% H_2O_2 (Avantor, Poland).

The leaching was carried out for 3 hours, samples were taken every 15, 30, 45, 60, 120 and 180 min. The process conditions are presented in table 1.

Table 1 The conditions of both stages of metal leaching from spent automotive converters

Step of leaching	Amount of powdered catalyst, g	Composition of the leaching solution	Temperature, °C
I	1 g	45-50 cm^3 of 1 M HCOOH with or without 5 cm^3 H_2O_2	70°C
II	0.5 g of catalyst from the I step	45 cm^3 36% HCl, 2.5 cm^3 96% H_2SO_4 and 2.5 cm^3 35% H_2O_2	70°C

Liquid-liquid extraction of metals from the leach solutions was carried out in 50 cm^3 separation funnels. The volume ratio of both phases aqueous and organic (A/O) was 1. Both phases were shaken for 20 min at an ambient temperature of 22±2°C. The concentration of the extractant Cyphos IL 101 (Cytec Industries Inc.) was 5×10^{-3} M (in toluene). Figure 2 presents the structural formula of the used extractant.

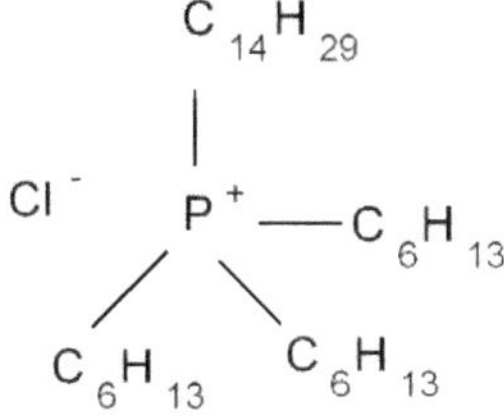

Figure 2 Trihexyl(tetradecyl)phosphonium chloride (Cyphos IL 101)

Stripping of metal ions from the loaded organic phases containing Cyphos IL101 (O/A = 1) was carried out for 20 min with 3 M HNO_3 (Chempur, Poland). Figure 3 presents a scheme of the hydrometallurgical processing of the spent automotive converters.

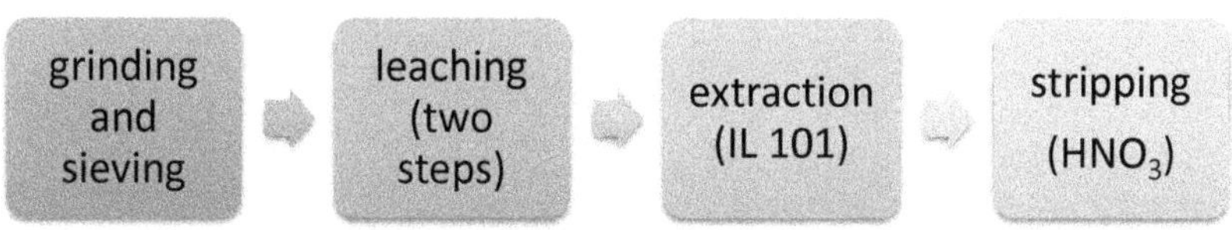

Figure 3 Scheme of the hydrometallurgical process of spent automotive converter treatment

2.2 Analytical methods

The concentration of Pt(IV), Rh(III), Fe(III), Zn(II), Mg(II), Ni(II) and Cu(II) in the solutions after both leaching stages, after extraction and stripping were analyzed by atomic absorption spectrometry (AAS). The concentration of Al(III) was analyzed by atomic emission spectrometry (AES).

3 Results

3.1 First stage of leaching

Figure 4 presents the changes in the concentration of Fe(III), Mg(II) and Zn(II) during the leaching of spent automotive converters with 1 M formic acid with or without the oxidant. The concentrations of Ni(II) and Al(III) in the solutions after leaching are presented in Table 2.

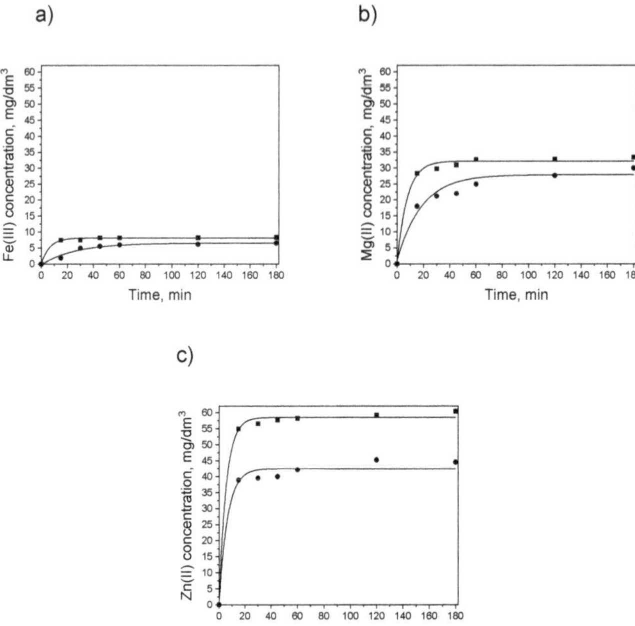

Figure 4 Leaching of: a) Fe(III), b) Mg(II) and c) Zn(II) using 1 M HCOOH with (•) or without (■) the addition of H_2O_2

There were no PGM ions in the solutions after the first step of leaching. The addition of an oxidant to the leaching solution does not increase the leaching efficiency of Fe(III) and Mg(II) from the solid phase to the aqueous phase (Figure 4 a and b). In the case of Zn(II) (Figure 4 c) and Al (III) (Table 2), the leaching efficiency with the oxidant-containing solution is lower than without the oxidant addition.

Table 2 Concentrations of Ni(II) and Al(III) in the leaching solutions after 180 min

Leaching solution	Concentration, mg/dm^3	
	Ni(II)	Al(III)
1 M HCOOH	1.2	162
1 M HCOOH with H_2O_2	0.5	135

The carboxylic acids are not strong enough to react directly with PGM in the catalyst. However, leaching of metals from the catalyst with these acids can be treated as a preliminary stage of their treatment; significant amounts of Al(III), Zn(II) and Mg(II) and

Fe(III) and small amounts of Ni(II) are extracted into the aqueous phase. In this way, a certain amount of non-precious metals (less valuable compared to PGM) is removed. After separating the two phases, the catalyst is washed with deionized water and thoroughly dried.

3.2 Second stage of leaching

The powdered catalyst from the first stage of leaching is leached in the second stage with a mixture of 36% HCl and 96% H_2SO_4 with the addition of 35% H_2O_2 (volume ratio: 45:2.5:2.5 cm^3).

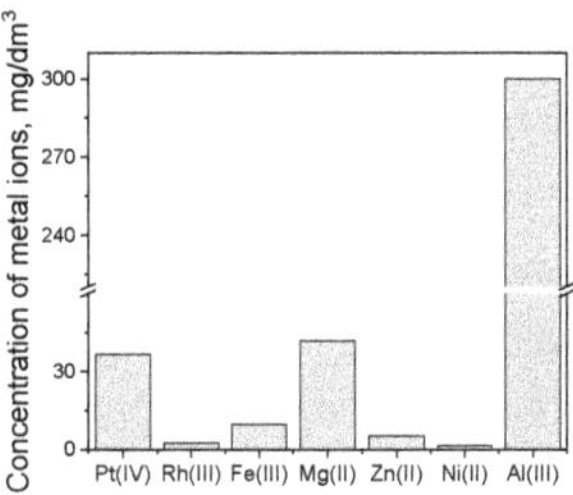

Figure 5 Concentration of metal ions in the leaching solution after second stage of the leaching with the mixture of HCl, H_2SO_4 and H_2O_2

The solution after the second stage of leaching contains significant amounts of Pt(IV) and Rh(III) but also some amounts of non-precious metal ions, mainly Mg(II) and Al(III) (Figure 5).

3.3 Extraction with Cyphos IL 101

The solution obtained after the second stage of the leaching of spent automotive converters was diluted twice. Extraction of metal ions from the solution after a second stage of leaching was performed with Cyphos IL 101 and the results are shown in Figure 6.

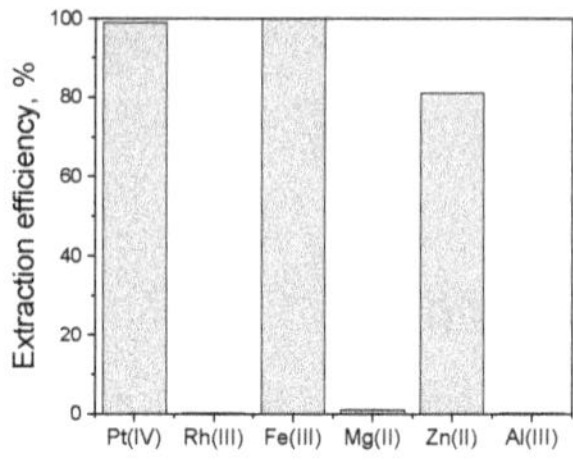

Figure 6 Extraction efficiency of Pt(IV), Rh(III), Fe(III), Mg(II), Zn(II) and Al(III) with Cyphos IL 101

As a result of extraction with Cyphos IL 101, Pt(IV) but also Fe(III) and Zn(II) were transported from the aqueous phase to the organic phase (Figure 6). Pt(IV), Fe(III) and Zn(II) form anionic chlorocomplexes in aqueous chloride solutions, and they are efficiently extracted by Cyphos IL101 in the anion exchange reaction. Platinum was separated from rhodium, magnesium and aluminum because Rh(III), Mg(II) and Al(III) left in the raffinate.

3.4 Stripping with 3 M HNO_3

In the last step, the stripping of metal ions from the loaded organic phase with 3 M HNO_3 was performed. The results are presented in Figure 7.

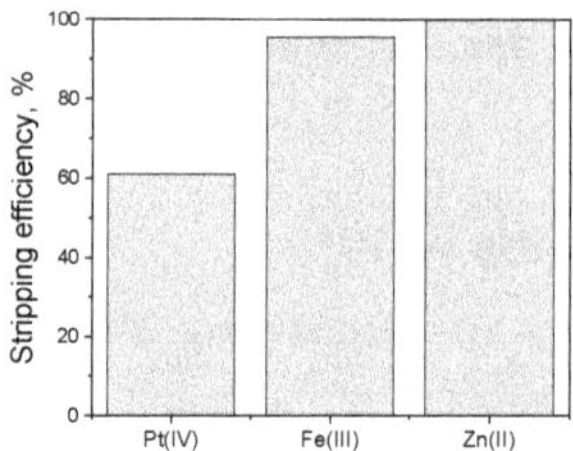

Figure 7 Stripping efficiency of Pt(IV), Fe(III) and Zn(II) with 3 M HNO_3

As a result of the stripping of metal ions into the aqueous phase from the loaded organic phase containing Cyphos IL 101, Pt(IV), Fe(III) and Zn(II) were stripped with high efficiency.

4 Summary

It has been proven that formic acid can be successfully used for the leaching of non-precious metal ions (Al(III), Zn(II), Mg(II), Fe(III)) and can be used in the first stage of metal ions leaching from spent automotive converters. For PGM leaching, it is necessary to use more stringent conditions (e.g. leaching with a mixture of HCl and H_2SO_4 and H_2O_2). Extraction of the metal ions from a mixture of HCl, H_2SO_4 and H_2O_2 is favorable because Pt(IV) can be separated from Rh(III), Al(III) and Mg(II).

Hydrometallurgical PGM recovery from spent automotive converters may become an alternative to energy-consuming and costly pyrometallurgical techniques commonly used to recover metal ions (including PGM) from waste (spent automotive converters, electronic scrap).

Acknowledgments:

This research was funded by the Ministry of Education and Science (0912/SBAD/2020).

5 Literature

Rzelewska, M., Regel-Rosocka, M.	2018	Wastes generated by automotive industry – Spent automotive catalysts. Physical Sciences Reviews, 3(8), 20180021.
Rzelewska-Piekut, M., Regel-Rosocka, M.	2019	Separation of Pt(IV), Pd(II), Ru(III) and Rh(III) from model chloride solutions by liquid-liquid extraction with phosphonium ionic liquids. Separation and Purification Technology, 212, 791.
Johnson Matthey	2021	Precious metals management – PGM prices, http://www.platinum.matthey.com/prices/price-charts, (Access 15th March 2021)
Akcil, A., Veglio, F., Ferella, F., Okudan, M.O., Tuncuk, A.	2015	A review of metal recovery from spent petroleum catalysts and ash. Waste Management, 45, 420.
Pośpiech, B.	2012	Hydrometalurgiczne technologie odzysku metali szlachetnych i nieżelaznych ze zużytych katalizatorów. Przemysł Chemiczny, 91(10), 2008 (in Polish).
Rzelewska-Piekut, M., Saternus, M., Fornalczyk, A., Regel-Rosocka, M.	2020	Spent Automotive Converters as a Secondary Resource of Metals in Practical Aspects of Chemical Engineering. Springer, ISBN 978-3-030-39866-8.
Cieszyńska, A., Wiśniewski, M.	2011	Selective extraction of palladium(II) from hydrochloric acid solutions with phosphonium extractants. Separation and Purification Technology, 80(2), 385.

Aktas, A.	2011	Rhodium recovery from rhodium-containing waste rinsing water via cementation using zinc powder. Hydrometallurgy, 106, 71.
Regel-Rosocka, M., Rzelewska, M., Baczyńska, M., Janus, M., Wiśniewski, M.	2015	Removal of palladium(II) from aqueous chloride solutions with Cyphos phosphonium ionic liquids as metal ion carriers for liquid-liquid extraction and transport across polymer inclusion membranes. Physicochemical Problems of Mineral Processing, 51(2), 621.
Bradaric, C.J, Downard, A., Kennedy, C., Robertson, A.J., Zhou, Y.	2003	Industrial preparation of phosphonium ionic liquids. Green Chemistry, 5, 143.
Stojanovic, A., Morgenbesser, C., Kogelnig, D., Krachler, R., Keppler, B.K.	2011	Quaternary ammonium and phosphonium ionic liquids in chemical and environmental engineering in Ionic Liquids: Theory, Properties, New Approaches. IN TECH, Rijeka, Croatia, ISBN: 978-953-307-349-1.
Luo, J., Conrad, O., Vankelecom, I.F.J.	2012	Physicochemical properties of phosphonium-based and ammonium-based protic ionic liquids. Journal of Materials Chemistry A, 22, 20574.

Author's address:

Dr. Eng. Martyna Rzelewska-Piekut
Institute of Chemical Technology and Engineering,
Poznan University of Technology,
ul. Berdychowo 4, 60-965 Poznań
Telefon: +48 61 665 3667
E-Mail: martyna.rzelewska-piekut@put.poznan.pl

Hydrometallurgical recovery of metal ions from spent catalytic converters

Zuzanna Wiecka, Ewelina Łopińska, Martyna Rzelewska-Piekut, Magdalena Regel-Rosocka

Institute of Chemical Technology and Engineering, Poznan University of Technology, ul. Berdychowo 4, 60-965 Poznań

Abstract

Natural ores of metals, especially platinum group metals (PGM), are limited therefore new resources are being searched for and some secondary materials can be used to recover PGM. These secondary resources include spent petrochemical and automotive catalysts, as well as an electronic equipment. The aim of the work is to develop an effective hydrometallurgical method of separating platinum groups metals from waste materials (spent automotive converters). The proposed method consists of two stages, in which at the beginning the ground catalyst is leached in a weakly acidic medium (oxalic acid) to separate non-noble metal ions. In the second stage, HCl/HNO_3 or $HCl/H_2SO_4/H_2O_2$ were used as more oxidative medium to recover PGM ions from the catalyst remaining after the first leaching stage.

Keywords *PGM, platinum(IV), rhodium(III), recovery, automotive converters, carboxylic acids, hydrometallurgy*

1 Introduction

The content of PGM (Pt, Pd and Rh) in the Earth's crust is estimated at 10^{-6}%, therefore new methods are sought to recover these metals from the secondary raw materials, e.g. automotive and petrochemical catalysts (TRĘBACZ AND MICHNO ET AL., 2017). Catalytic converters are used to neutralize exhaust gases such as NO_x, CO and C_xH_{4x} to environmentally neutral compounds, such as CO_2, N_2 and H_2O. The following reactions take place in the automotive catalyst:

- Reduction of nitrogen oxides to elemental nitrogen and oxygen

$$NO_x \rightarrow N_x + O_x \qquad (1)$$

- Oxidation of carbon monoxide to carbon dioxide

$$CO + O_2 \rightarrow CO_2 \qquad (2)$$

- Oxidation of hydrocarbons to carbon dioxide and water

$$C_xH_{4x} + 2xO_2 \rightarrow xCO_2 + 2xH_2O \qquad (3)$$

The efficiency of the automotive catalyst in the case of gasoline-powered vehicles is about 90% of the removal of all toxic gases, while for diesel engines the removal of carbon monoxide(II) and hydrocarbons amounts to 90% and of other substances to 30-40% (WOŁOWICZ, 2013, DING ET AL., 2019). The content of platinum in catalyst inserts is about 0.2-0.3% and depends on the type of fuel used and the age of the car.

1 kg of spent catalyst contains about 2 g of platinum, while for example natural deposits located in South Africa contain from 3 to 20 mg/kg of platinum. Therefore, according to assumptions to the circular economy and sustainable development, a preferred solution is the recovery of PGM from secondary resources such as catalysts. Proper recycling and recovery of metals make possible to reuse them and conserve the natural resources. Apart from valuable noble metals such as Pt, Pd or Rh, catalysts contain a large amount of base metals which are susceptible to leaching with weak reagents, such as carboxylic acids (VONCKEN, 2019; YAKOUMIS ET AL., 2018). The use of these leaching agents allows the catalyst to be initially purified from non-noble metals (Al, Fe, Zn, Mg and Ni), and can also be considered as a pre-treatment method prior to the recovery of precious metals with strong acids. The use of hydrometallurgical methods to recover metals from waste raw materials is considered to be more advantageous compared to pyrometallurgical methods because the wet methods produce much less waste and consume less energy (RZELEWSKA ET AL., 2018). Moreover, hydrometallurgy enables some compounds to be recovered from very dilute solutions and to be separated not only from main products but also by-products, therefore, from the point of view of environmental protection, it is indicated to be an ecological method (WIECKA ET AL., 2020; YOUSIR, 2019). Figure 1 shows a diagram of a hydrometallurgical process for metal recovery from secondary materials.

Figure 1 Diagram of the hydrometallurgical process for metal recovery from secondary materials

In the hydrometallurgical method, the most important is the selection of the appropriate leaching agent and process parameters. The choice depends mainly on the chemical composition of the material to be treated (CHMIELEWSKI, 1996). Thus, it is proposed in the work a two-stage process of spent catalyst treatment, which includes leaching with an organic acid such as oxalic acid (OA), then leaching with mineral acids such as aqua regia (HCl/HNO_3) or a mixture of concentrated acids with oxidant ($HCl/H_2SO_4/H_2O_2$) from the catalyst after the first stage of leaching.

2 Experimental part

2.1 Materials

First, the catalyst was crushed, ground in a mortar, and then sieved through a sieve with a mesh of 63 µm. In the first stage of the leaching, 0.1 or 1 M oxalic acid (OA) (Chempur, Poland) with or without the addition of 30 wt% H_2O_2 (Avantor, Poland) was used. The catalyst after all the first stage leaching experiments was dried and ground in a homogenizer, then passed through a sieve with a mesh diameter of 63 µm. Leaching solutions from the second stage were prepared by mixing different inorganic acids (36% HCl, 98% H_2SO_4, 65% HNO_3, Chempur, Poland) and 30 wt% H_2O_2 (Avantor, Poland). Figure 2 shows the research scheme.

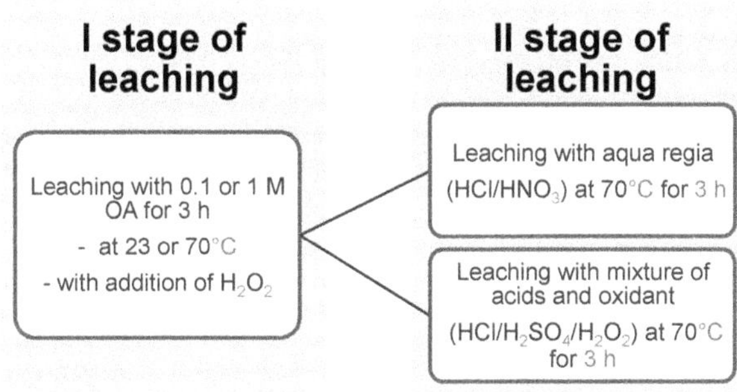

Figure 2. The scheme of leaching experiments

2.2 Analytical methods

The content of Pt(IV), Rh(III), Fe(III), Mg(II), Zn(II), Cu(II) after leaching was analyzed by Atomic Absorption Spectrometry (ContrAA300, Analytik Jena, Germany), and the content of Al(III) was analyzed by Inductively Coupled Plasma Optical Emission Spectrometry (ICP-OES) with inductively coupled IRIS/AP plasma excitation (Thermo Jarrell Ash, USA).

2.3 First stage of leaching

In the first leaching stage, the liquid to solid (L/S) ratio was 50 and the rotation speed was 500 rpm. During the leaching, the concentration of the leaching agent (OA), the process temperature and the amount of added H_2O_2 were varied. The conditions of the individual leaching experiments are shown in Table 1.

Table 1. Conditions of the first stage of leaching

No.	Temperature, °C	OA concentration, M	H_2O_2 vol. %
1	70	1	0
2	23	1	0
3	70	1	10
4	23	1	10
5	23	0.1	0
6	70	0.1	0
7	70	0.1	10
8	23	0.1	10

In the first stage, it was planned to leach the non-noble metals mainly Mg(II), Zn(II) and Al(III) to reduce their content in the solutions after leaching with strong acids in the second stage. As it was reported in the previous article (RZELEWSKA-PIEKUT ET AL., 2021), the catalyst carrier is made of cordierite, therefore the presence of Mg(II) or Al(III) in the leaching solutions can be expected. As metals are leached with OA, the following reactions are taken into account (PATHAK ET AL., 2020):

$$C_2O_4H_2 \rightleftharpoons C_2O_4H^- + H^+ \qquad pK_{a1} = 1.25 \qquad (4)$$

$$C_2O_4H^- \rightleftharpoons C_2O_4^{2-} + H^+ \qquad pK_{a2} = 4.14 \qquad (5)$$

During leaching, the ligands in organic acids form stable metal complexes.

2.4 Influence of temperature

The influence of temperature on the amount of leached metals was investigated. 0.1 and 1 M solutions of OA were applied as leaching agents at two different temperatures 23 and 70°C. The summary of Mg(II) leaching after 3 hours under different conditions is shown in Fig. 3.

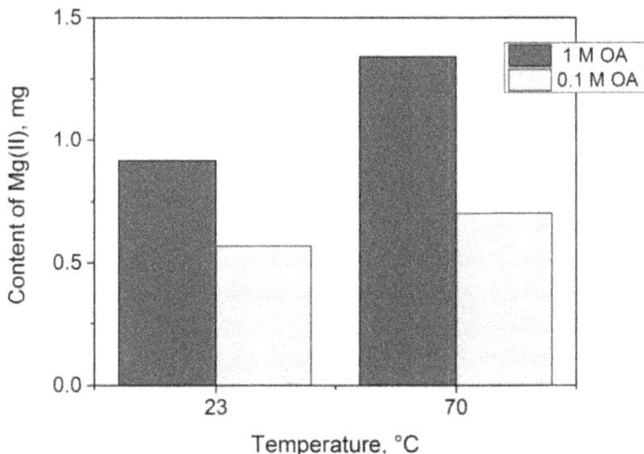

Figure 3. Comparison of Mg(II) content in solutions after leaching with (■) 1 M OA, ( ) 0.1 M OA

Increase in the concentration of OA and temperature of the leaching caused significant changes in the amount of the leached Mg(II). Table 2 shows the amounts of the leached metals from all the experiments.

Table 2. The amounts of leached metals from all the experiments

Leaching conditions				**Content of metal ions, mg**		
No.	Temperature, °C	OA concentration, M	H_2O_2 vol. %	Al(III)	Mg(II)	Zn(II)
1	70	1	0	27.4	1.3	0.9
2	23	1	0	10.1	0.9	0.6
3	70	1	10	27.0	1.1	0.7
4	23	1	10	7.2	0.4	0.7
5	23	0.1	0	5.1	0.6	0.5
6	70	0.1	0	12.6	0.7	0.7
7	70	0.1	10	14.1	0.7	0.8
8	23	0.1	10	7.3	0.4	0.5

The amount of the metal ions leached with OA solution decreased in the following order: Al(III) > Mg(II) > Zn(II). Mg(II) is leached more easily with OA than Zn(II). The positive effect of the increase in OA concentration is evident particularly for Mg(II). Changing the leaching conditions, especially the increase in, has a positive effect of temperature and OA concentration increase on the amount of leached metal ions is observed. Leaching with 1 M OA at 70°C turned out to be the most advantageous in terms of the amount of the leached metal ions. PGM were also determined in the samples after 3 hours of the leaching with oxalic acid but it turned out (as expected) that the platinum were not leached in the first step. The high redox potential values of these noble metals make PGM leaching difficult. OA is too weak acid to dissolve the PGM contained in the spent catalyst.

2.5 Influence of addition of an oxidant

An effect of addition of 30% H_2O_2 to OA on the efficiency of leaching of metal ions was investigated. The amount of Mg(II) and Zn(II) leached with 1 M OA without and with the addition of H_2O_2 at 70°C is shown in Fig. 4.

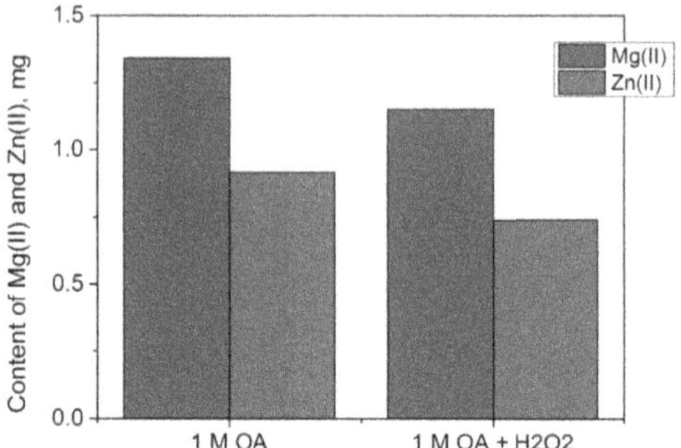

Figure 4. Comparison of Mg(II) (■) and Zn(II) (■) content in the solutions after leaching with 1 M OA and 1 M OA with H_2O_2

The amount of Al(III) leached in both experiments amounted to almost 30 mg. The addition of the oxidizer (H_2O_2) did not affect the amount of the leached metals. It was expected that the addition of H_2O_2 would have a positive effect on the dissolution of metal ions from the spent catalyst. However, it seems that the addition of H_2O_2 does not appear to be of importance in increasing the leaching efficiency.

2.6 Second stage of leaching

After the first stage, the dried catalyst was again leached with two types of mineral acids to recover PGM. The ratio of liquid to solid (L/S) was 20 for both experiments. The leaching temperature and time were 70°C and 3 hours and the rotation speed was 300 rpm. A comparison of metal content in the solutions after the second leaching depending on the leaching agent applied is shown in Fig. 5.

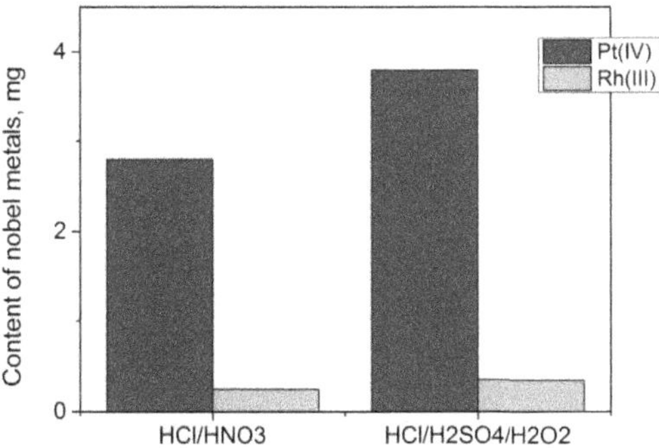

Figure 5. Comparison of the content of (■) Pt(IV), (■) Rh(III) in the solutions after the second leaching stage depending on a leaching agent (aqua regia and a mixture of concentrated HCl, H_2SO_4 and H_2O_2)

The use of strong acids of oxidative properties allowed the precious metals to be leached from the catalyst. It should be emphasized that (compared to other metal ions) Pt(IV) and Rh(III) were effectively leached with a mixture of concentrated HCl, H_2SO_4 and H_2O_2, and amount of these metals in the leaching solution was 3.8 and 0.35 mg. The reactions that occur during leaching are as follows:

$$Pt + 6HCl + 2H_2O_2 \rightarrow [PtCl_6]^{2-} + 2H^+ + 4H_2O \quad (6)$$

$$2Rh + 12HCl + 3H_2O_2 \rightarrow 2[RhCl_6]^{3-} + 6H^+ + 6H_2O \quad (7)$$

The following reactions occurred during leaching with aqua regia:

$$3Pt + 18HCl + 4HNO_3 \rightarrow 3[PtCl_6]^{2-} + 6H^+ + 4NO\uparrow + 8H_2O \quad (8)$$

$$Rh + 6HCl + HNO_3 \rightarrow [RhCl_6]^{3-} + 3H^+ + NO\uparrow + 2H_2O \quad (9)$$

HNO_3 performs oxidant function in aqua regia, while HCl is a Cl^- donor and an acidic reaction medium (RZELEWSKA-PIEKUT ET AL., 2021). As a result of Pt(IV) and Rh(III) dissolving in HCl/HNO_3 and $HCl/H_2SO_4/H_2O_2$, hexachloroplatinic acid(IV) and hexachlororhodium(III) are obtained. The amount of Pt(IV) and Rh(III) was 2.8 and 0.25 mg after leaching with aqua regia, which means that less PGM were leached compared to the mixture of acids and H_2O_2. The use of a mixture $HCl/H_2SO_4/H_2O_2$ also has an ecological aspect because during the reaction, no dangerous nitrogen oxides are released.

The use of two leaching stages provides an efficient leaching of base metals in the first stage and the subsequent leaching of precious metals. Table 3 shows the amounts of base metals (Mg(II), Zn(II), Fe(III), Cu(II), Al(III)) leached in the second stage of leaching.

Table 3. Content of base metal ions in the solutions after II stage of leaching

No.	Type of leaching agent	Content of metal ions, mg				
		Mg(II)	Zn(II)	Fe(III)	Cu(II)	Al(III)
1	HCl/HNO_3	39.7	0.2	0.5	0.07	17.9
2	$HCl/H_2SO_4/H_2O_2$	79.2	0.2	0.5	0.04	23.5

The content of the metal ions that are leached in the first and second steps differs significantly. The content of Mg(II) in strong acids after the II stage of leaching is higher than in the case of OA treatment in the I stage. On the other hand, the Zn(II) content in OA is higher than in second step. Al(III) in the second stage of leaching reaches levels similar to the highest values obtained in the first leaching stage.

3 Conclusions

Studies on the separation of Pt(IV) and Rh(III) from spent automotive catalysts show that the proposed two-stage process can be used to separate PGM ions from the base metals. OA showed high affinity for elution of non-noble metal ions (Al(III), Mg(II) and Zn(II)). Increased temperature have a positive effect on the amount of the leached metal ions. The largest amounts of Pt(IV) and Rh(III) were leached with freshly prepared mixture of HCl, H_2SO_4, and H_2O_2. The demonstrated method could be used to recover and reuse metals from the secondary resources which is consistent with the assumptions of the circular economy.

This work was supported by the Ministry of Science and Higher Education, Poland (0912/SBAD/2010).

4 Literature

Trębacz, H., Michno, P. 2017 Występowanie platynowców w środowisku i ich zastosowanie, Structure and Environment, 9(4), 283-299 (in Polish)

Wołowicz, A. 2013 Zastosowanie palladu I jego związków ze szczególnym uwzględnieniem katalizy, Przemysł Chemiczny, 92(7), 1237-1245 (in Polish).

Ding, Y., Zheng H., Li, J., Zhang, S., Liu, B., Ekberg, C., Jian, Z., 2019 Recovery of Platinum from Spent Petroleum Catalyst: Optimization Using Response Surface Methodology, Metals, 9, 354-371.

Yakoumis, I., Moschovi, A.M., Giannopoulou, I., Panias, D. 2018 Real life experimental determination of platinum group metals content in automotive catalytic converters, IOP Conf. Series: Materials Science and Engineering, 329, 012009.

Voncken, J. 2019 Recovery of Ce and La from Spent Automotive Cata-

		lytic Converters, Critical and Rare Earth Elements, 12, 267-272.
Rzelewska, M., Regel-Rosocka, M.	2018	Waste generated by automotive industry – spent automotive catalyst, Physical Sciences Reviews, 3(8):20180021.
Wiecka, Z., Rzelewska-Piekut, M., Cierpiszewski, R., Staszak, K., Regel-Rosocka, M.	2020	Hydrometallurgical Recovery of Cobalt(II) from Spent Industrial Catalysts, Catalyst, 10, 61-1-61-13.
Yousir, A.M.	2019	Recovery and then individual separation of platinum, palladium and rhodium from spent car catalytic converters using hydrometallurgical technique followed by successive precipitation methods, Journal of Chemistry, 2318157.
Chmielewski, T.,	1996	Ługowanie metali z rud, koncentratów, półproduktów i odpadów, Fizykochemiczne Problemy Mineralurgii, 30, 217-231 (in Polish).
Pathak, A., Vinoba, M., Kothari, R	2020	Emerging role of organic acid in leaching of valuable metals from refinery-spent hydroprocessing catalyst, and potential techno-economic challenges: A review, Critical Reviews in Environmental Science and Technology, 10, 1-43.
Rzelewska-Piekut, M., Paukszta, D., Regel-Rosocka, M.	2021	Hydrometallurgical recovery of platinum group metals from spent automotive converters, Physicochemical Problems of Mineral Processing, 57(2), 83-94.

Author address:

Zuzanna Wiecka, M.Sc. Eng.
Institute of Chemical Technology and Engineering,
Poznan University of Technology,
ul. Berdychowo 4, 60-965 Poznań
E-mail: zuzanna.g.wiecka@doctorate.put.poznan.pl

The robust and fast way of free-fall sorting of rubble and debris

Markus Eck

OptoSort GmbH, Klagenfurt am Wörthersee, Austria

Separation of Debris and Bulk Materials

Table of contents

1. General Part
2. Free-Fall Sorting
3. Technical Details
   3.1 HSI Technology
   3.2 RGB Technology
4. Complete Sorting Procedure

Keywords: Schutt, Schuttgüter, Freifall-Sortierung, Veredelung, Gesetzliche Auflagen, Waste Rubble, Debris, Free-Fall Sorting, Processing, Legal Requirements, Abfall

1 General Part

Construction waste is produced during the deconstruction of buildings and when civil engineering structures are demolished. Even if a building is selectively deconstructed, mixes of building materials are formed, consisting of concrete or wall building materials, mortars, renders and platers, insulating materials, plastics, metals etc. With traditional sorting techniques, the lightweight materials and metals can be separated from these mixes. For sorting of the mineral constituents and for the plastics, new approaches must be adopted as well.

Up to now, only few cases are known in which sensor-assisted sorting installations are used in the recycling of waste materials. Examples can be found in Germany, Switzerland or Spain. Either colour line cameras are used to recover, for example, clay bricks from mineral-based construction and demolition waste or NIR cameras are operated for the removal of foreign materials and impurities as well as for sorting. However, it is very important to mention that with neither of these two variants alone it is possible to recover unmixed construction material fractions.

Figure 1: Material before sorting – demolition waste

On account of the increasingly complex composition of building structures, new technologies are needed with which unmixed and impurity-free products can be recovered. In particular, gypsum particles can only be detected with a NIR camera and thus be removed from the product flow. Detection of gypsum in a composite building material, e.g. gypsum plaster on a wall building material, is only possible if all sides of the particles are scanned and detected.

With a combination of hyperspectral near-infrared technology and colour detection as well as simultaneous "observation" of all particle surfaces, current and/or expected recyclables and/or impurities can be detected and sorted with a high recovery rate. Sorting of the construction and demolition waste is the precondition for closed material cycles in construction industries.

To date and based on current knowledge, a freefall sorting installation with a combination of detection systems and arrangement of the detectors on two sides of the falling construction rubble particles has not been available. But with such an installation, the disadvantages of the two individual technologies could be balanced out (Fig. 2).

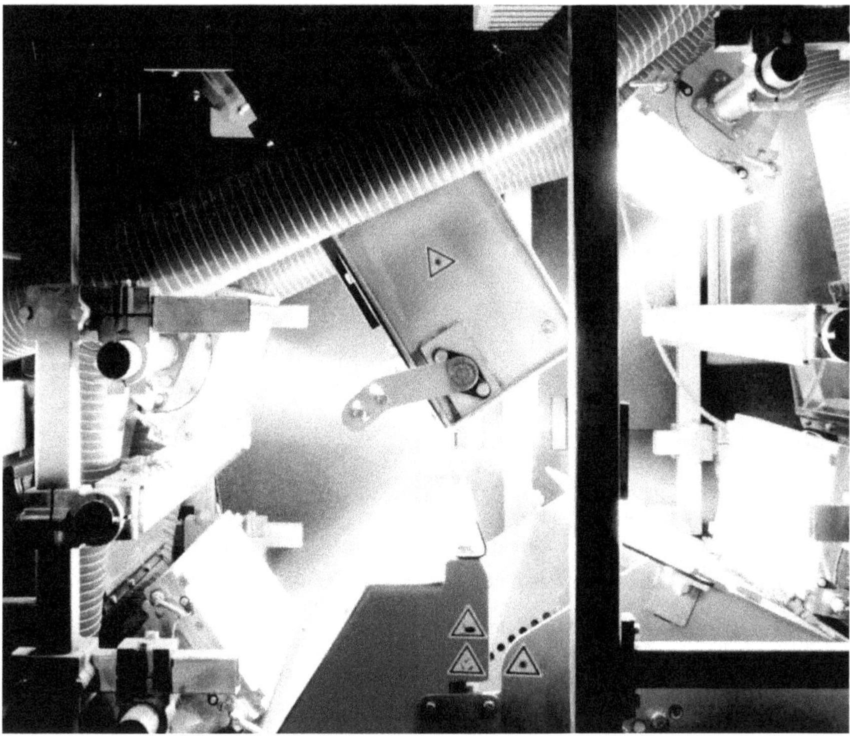

Figure 2: Colour and near-infrared scanning from two sides inclusive lighting unit

2 Free-Fall Sorting

OptoSort GmbH, a company specializing in sorting rock, offers its developments as new solutions for sorting constructions waste. Together with its partner, the Institute of Applied Constructions Research (IAB gGmbH) in Weimar, Germany, they do research on new ways of sorting mineral construction rubble and bulk materials.

The free-fall sorting installation is equipped with coupled RGB line cameras and hyperspectral NIR line cameras, arranged on two sides of the material stream. The focus for a successful sorting process must be on the following key functions:

- Detection of dry and wet mineral particles
- Scanning of all particle surfaces
- Reliable ejection of the detected impurities
- Robustness and low wear when exposed to abrasion by mineral particles
- Suitable for particles between 10 mm and 250 mm
- Low maintenance and automatic white balance in continuous operation
- Low energy consumption, selective control of the particle discharge with compressed air pulses
- Integration in local networks (LAN) or modem

Before the material is detected and sorted in free-fall so the impurities can be removed, it is distributed on a vibro-feeder over the entire width of the installation and accelerated on a chute.

Figure 3: Distribution of the feed material

Figure 4: Acceleration of the feed material

In the scope of research work conducted at the IAB, the application limits have been tested and general correlations like particle size, colour, NIR spectrum, surface properties of the particles, for example, texture and dust stuck to their surface, derived.

The chemical-mineralogical compositions of the building materials should be correlated with the spectra. The results obtained from the basis to implement more selective and effective sorting of demolition rubble so as to open up new recycling options for the secondary resources produced.

As investigations so far show, mineral construction materials can be clearly identified based on their colour in the range of visible light, however, on account of colour overlap between different mineral building materials it is not possible to differentiate between them and sort them accurately. If only colour detection is applied, frequent misdetection and misclassifications occur. Impurities of different types but with the same colour cannot be separated from each other. Only building materials with unique colour features, like clay bricks and concrete, can be separated.

Figure 5: Material after sorting

In the NIR range, fewer misdetections occur, although some special materials sometimes deliver unsatisfactory spectra. Glass (amorphous) and very absorbent materials, like roofing felt or bituminous materials, are not identified. The highest detection rates are achieved for wood, gypsum, and aerated concrete. The two latter make up the impurity fractions that most urgently need to be removed from the product.

The goal of the development is to increase the detection rates for the individual materials, especially with regard to the spectral processing and validation. With the use of object detection – integrated in the sorting software, it is also possible to detect and remove composite particles with foreign objects and impurities stuck to them.

With a stationary free-fall sorting installation up to 20 tons of construction rubble per hour with particle sizes larger than 10 mm can be sorted. Figure 6 shows the free-fall sorting installation with its very robust design which can also be used for the separation of plastic mixes and waste glass.

Originally, the sorters were used successfully in the area of mining all over the world, which explains the robust design and construction. In mining, larger sorters are used which can sort up to 70 tons of material per hour depending on the structure, weight, and size of the material.

Figure 6: Free-fall sorting machine (outside)

In November 2019, at the IAB Recycling Testing Centre, the free-fall sorting machine was commissioned and successfully trialed. The innovative sorting technology is now available for various sorting applications in the scope of research projects and for use by industrial partners and clients. The recycling test centre at the Institut für Angewandte Bauforschung Weimar gGmbH (IAB), is equipped with crushers, mills, screens, pan granulators and different rotary furnaces so that extensive projects and tests can be carried out, centering on recycling and the development of new materials derived from waste material.

Figure 7: Processing line at the IAB *Figure 8: Rear view of the free-fall sorter*

3 Technical Details

Above mentioned sorting unit is the fusion of HSI (Hyper Spectral Imaging) technology in the NIR (Near-Infrared) range, with RGB (Red Green Blue) technology.

3.1 HSI Technology

The Near-Infrared HSI technology is a sensor-based technology that works in a wavelength range of 1100 – 1700 nm (nanometres). The camera makes a "chemical fingerprint" of the surface of the material to be sorted which is invisible to the human eye due to the wavelength range, which is called a spectrum. The spectrum of the surface is displayed as a diagram (Figure 9). The characteristic curves shown in Figure 9 were obtained during the material analysis of a potential customer.

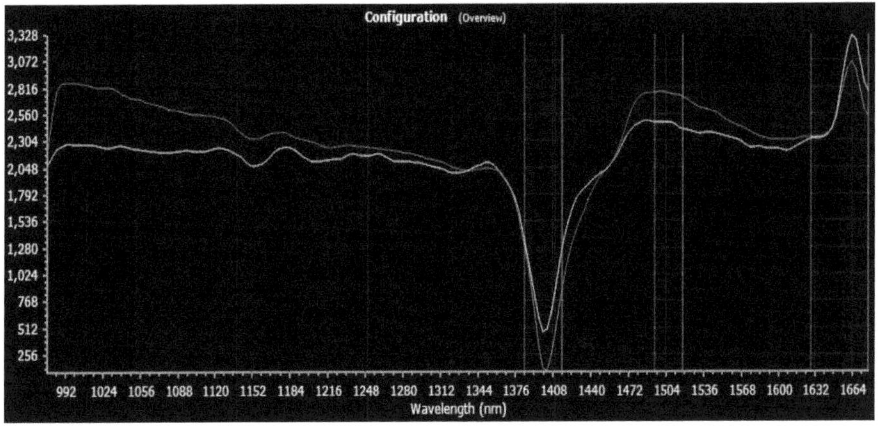

Figure 9: recorded spectra; red spectrum = peridotite; green spectrum = chromite; (the X-axis can be represented in pixels or in nanometres as wavelengths).

The task was to separate chromite (red characteristic curve) from peridotite (green characteristic curve). One can see clear differences of the peaks over the wavelengths/pixels in both spectra. Therefore, a sorting model can be created, which pays attention to these differences of the materials and separates them from each other. The sorting software is programmed to three variable ROI's (Region of Interest) ranges which are shown in Figure 9 as three red vertical markers. This means that the entire wavelength range from 1100 to 1700 nm is not checked during the analysis, but only the three selected ranges which correspond to the sorting model. This results in an even higher sampling rate of the material.

In Figure 10 and Figure 11 the chromite is shown on the left and the peridotite on the right side. In Figure 11, the rocks are shown in false colours, which are reproduced as a result of the defined sorting model via the spectra when they are recognized. This results in a clear differentiation and thus a clear solution for the sorting task. This is the result of a 100 % HIS material analysis.

Figure 10: Left chromite und right peridotite

Figure 11: False colours during recognition

3.2 RGB Technology

In the RGB technology, attention is paid to the optical colour differences between different materials and a sorting model is developed from that, which separates the product from the waste material. As can be seen in Figure 12, three colour areas/material clouds (green, red and blue) stand out above the colour space (HSV cylinder).

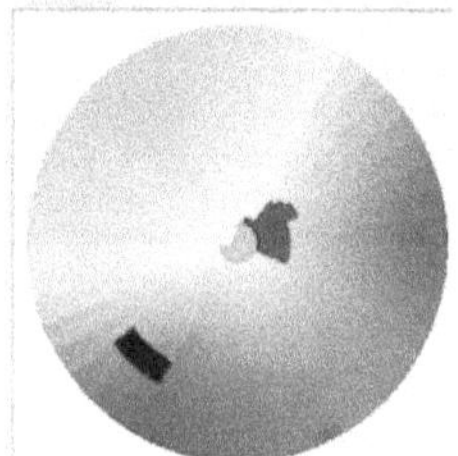

Figure 12: HSV cylinder

These are the colour properties of the captured magnesite stones (green & red), as well as the active illuminated blue LED background (blue) at which the line scan camera is looking.

In Figure 13, one can see the white and grey magnesite which do not differ significantly in contrast. That is why the green and red areas are very close to each other in the colour circle. But there is a clear separation between the areas and thus the RGB technology can distinguish or separate these two classes. As this is also clearly shown in Figure 14 with the false colours, the sorting model can separate this efficiently.

Figure 13: Left grey and right white magnesite

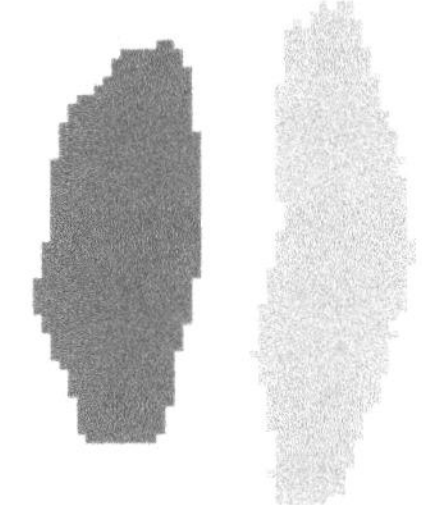

Figure 14: False colour display during detection

To be prepared for all necessary applications in the area of debris and bulk materials separation, the two different technologies can be combined via a so-called "sensor fusion". This means that a wide range of additional functions can be used via various parameters and their combination to be able to offer a sorting solution for almost any task.

4 Complete Sorting Procedure

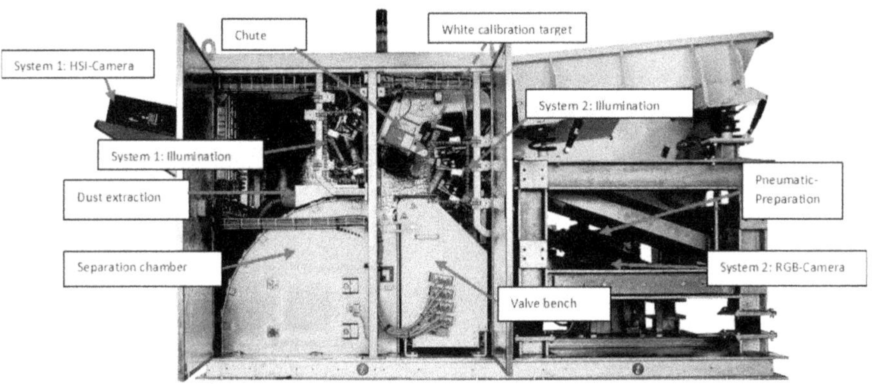

Figure 15: Free-fall sorting machine (inside)

After preparation of the feed material, the material gets directly onto the vibratory feeder. On the feeder, the material is separated over the entire working width and brought into a monolayer.

From the vibro-feeder, the material is handed over to the chute, where it is accelerated in a controlled manner and then falls into the scanning area. There, it is illuminated, scanned, and a decision is made as to whether it is to be discharged or not.

The discharge is performed by fast-switching valves that open the correct number of valves, at the right position and at the correct time.

The sorting systems are able to recognize and separate construction waste, precious metals, industrial minerals, metal ores, slag and secondary materials in order to obtain the purest possible products while at the same time optimally exploiting the feed material. A high level of refinement is achieved even if the raw materials are of poorer quality.

Address of the author:

BBW Markus Eck
OptoSort GmbH
Hans-Sachs-Straße 13
9020 Klagenfurt am Wörthersee
Austria
Mobile: +43 664 52 79 658
Email: m.eck@optosort.com

Advanced Waste Treatment and Material Recovery by Wet Separation

Matthias Kuehle-Weidemeier

ICP Ingenieurgesellschaft Prof. Czurda und Partner mbH, Karlsruhe, Germany

Abstract

By implementing modern thermal and non-thermal waste treatment technologies, waste management hast made a big leap forward. In the last two decades, Central Europe has been the motor of the development. The first decade of this century has been a period of development and installation of new processes in numerous locations. This was followed by nearly a decade of minor progress. This article will analyse the current situation and show how higher recycling rates even from mixed and residual waste can be achieved.

Keywords

Incineration, mechanical biological treatment, material recovery, energy efficiency, waste hierarchy, wet mechanical separation, optical sorting, urban mining.

1 History and current situation

1.1 Introduction and historical development

By implementing modern thermal and non-thermal waste treatment technologies, waste management hast made a big leap forward in central Europe. The first decade of this century has been a period of development and installation of new processes in numerous locations. A significant progress in landfill diversion was achieved. This period was followed by nearly a decade of stagnation or minor progress. This was a period of recovering from the huge investments that have been done and establishing the waste management system based on these new technologies and the changes they have brought.

1.2 Benefits of MBTs with most common techniques

Figure 1 shows the average mass-balance of all German MBTs (different types) based on a study in 2007. About 25% of the MBT Inputs finally ends on a landfill (ashes from RDF incineration not included), which means, that an average landfill diversion of 75% is reached. This is a huge progress in saving landfill volume and emissions.

Methane emission potential of MBT output (fine fraction going to landfill) is reduced by about 90-95% compared to untreated waste, if boundary values are kept, that are common in European countries (AT_4 = 5-10mg O_2/g). This is indicated by various investigations (Kühle-Weidemeier, Bogon, 2008). Considering, that additionally the tonnage of landfilled material is reduced by MBT by an average of 75%, MBT brought a giant leap forward in reducing greenhouse gas emissions from waste management.

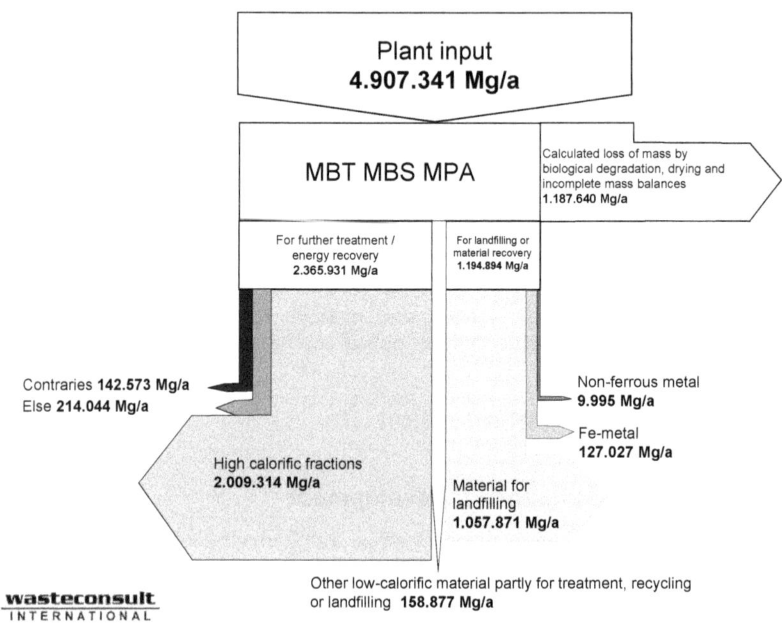

Figure 1: Mass-balance of the German MBTs (Kühle-Weidemeier et al., 2007)

Figure 1 shows the average mass-balance of the German MBTs based on a study in 2007.

Compared to direct combustion of the waste, using MBT only an average of 40% of the waste ends in incineration. These 40% have a high calorific value and become a useful solid fuel that changes disposal oriented combustion to energy recovery by incineration in RDF power plants. Less incineration / combustion means less highly toxic residues from flue gas cleaning and less resources degraded or lost in incinerator bottom ash.

1.3 Potential improvements of MBT processes

This is the moment to look at potential improvements in MBT. These improvements are based on the fact that still 25% of the MBT input mass ends on landfill and up to 40% in RDF power plants. **A higher material recovery rate would be favourable.**

1.4 Why material recovery (recycling) of the coarse / high caloric fraction is better than application in RDF power plants

Depending on its quality (leaching test) incinerator bottom ash is used as construction material (mainly for roads) or landfilled. The long term behaviour of incinerator bottom ash is a subject of controversial discussion. The main concern is, that a real long term stability (immobilisation of heavy metals) is possibly not given.

A part of the exhaust gas cleaning residues is highly toxic and gets stored in subsurface hazardous waste landfills.

Ferrous metals are removed from incinerator residues by magnetic separation. These metals are heavily oxidised. Non-ferrous metals are often irrecoverably lost in the bottom ash.

Concerning the conservation of resources, waste incinerators are energy and resource destruction plants. Table 1 reveals how much energy is lost if only the energy represented by the calorific value is recovered.

Table 1: Calorific value and energy equivalent (cal. value + energy effort for production) of some plastic materials (Reimann 1988)

Material	Calorific value [kJ/kg]	Energy equivalent [kJ/kg]
Polyethylen (PE)	43,000	70,000
Polypropylen (PP)	44,000	73,000
Polystrol (PS)	40,000	80,000
PVC hard	18,000	53,000

2 General reasons for more material recovery

2.1 Legal requirements

The EU has specified a five step waste hierarchy, shown in Figure 2. EU policy and authorities push waste management regulations forward to ensure that superior grades in the hierarchy are achieved for all kinds of waste. Disposal (e.g. landfilling) and recovery (e.g. incineration) only reach a low grade in the hierarchy, while recycling (e.g. with output from sorting plants) has a superior grade in the hierarchy.

Additionally, the EU Waste Framework Directive requires increasing recycling rates.

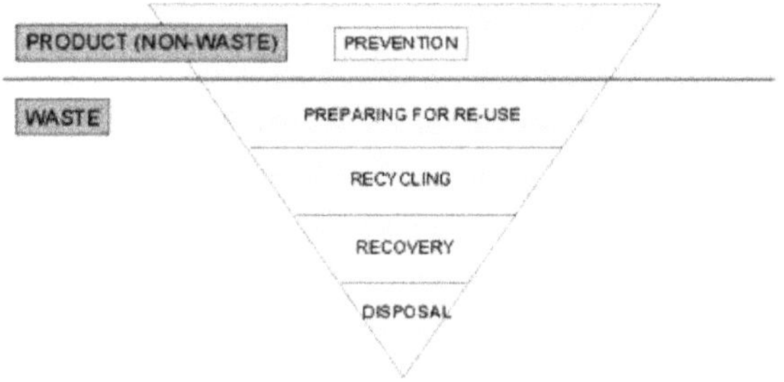

Figure 2: European Waste Hierarchy (European Commission, 2008)

2.2 Ecological requirements

2.2.1 Population growth and material consumption

The approaching exhaustion of many raw materials and expanding demand for resources due to fast growth of word population and increasing prosperity in many developing countries are a challenge for the world economy and will become a driving factor for enhanced waste treatment / material recovery technology. Quantity and quality of recovered resources from residual waste depend on the kind of waste treatment.

The world population will grow to round about 9.1 billion in 2050 (UN, 2009). That corresponds to an average annual growth of 56 million.

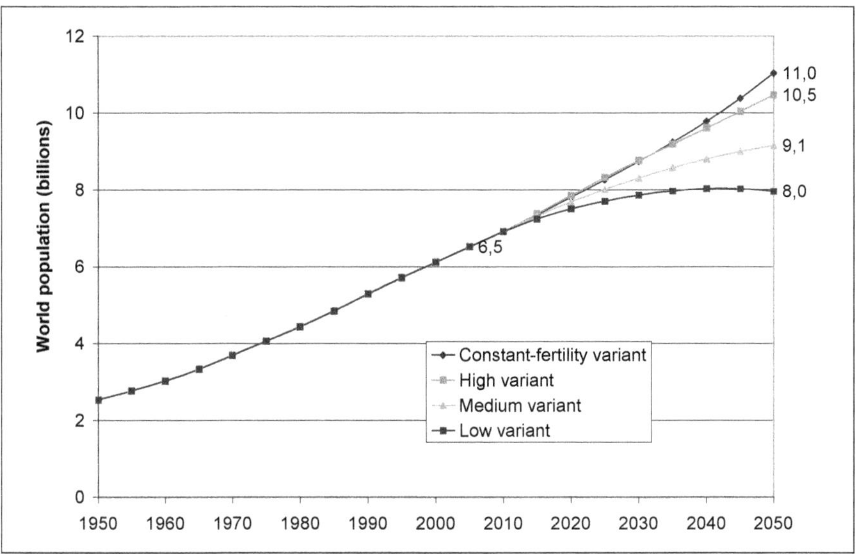

Figure 3: Different scenarios of world population growth (data source: UN, 2009)

The German Foundation for World Population (DSW) reports 2017 on their web site a current world population growth of about 81 million people per year. This is nearly as much as the total number of Germany's inhabitants.

2.2.2 Reduction of energy consumption and CO_2- emissions by recycling

Recycling is important for climate protection too. Figure 4 shows that recycling saves an enormous amount of CO_2 emissions and thus energy. For example, copper recycling saves 36%, steel recycling 56%, PE recycling 70%, PET recycling 85% and aluminium recycling even 95% compared to primary material production.

The calculated emissions of the recycling process consider collection, transport and the recycling process itself. Considered transport distances to the recycling facilities are based on the true situation. In case of PET this is the transport to south east Asia. It has to be mentioned, that plastics, paper and wood are only feasible for a small number of recycling cycles. Paper fibres can be re-used 5 - 7 times.

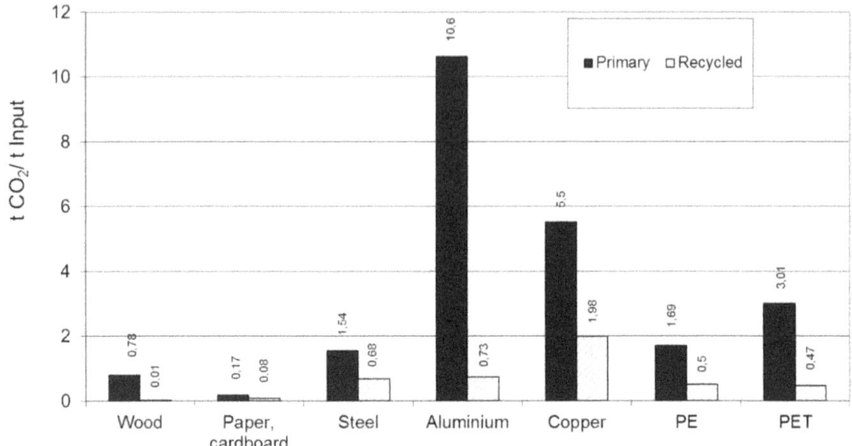

Figure 4: CO_2-emissions by primary and secondary material production and avoided emissions by recycling (data: Interseroh, Umsicht, 2008)

3 Necessary improvements for higher material recovery rates and getting closer to Zero Waste solutions

3.1 Coarse, high calorific fraction

Sensor based sorting is applicable to the coarse high caloric fraction. Better sensors, higher computing power, continuous improvements of the software as well as mechanical improvements resulted in huge progresses in sensor based sorting. This allows a higher separation rate as well as improved quality of the separated fractions. Integration of more sensor based sorting machines in waste treatment plants, especially MBTs will increase the potential for higher recycling rates of waste components

A continuing limitation are impurities on the waste component surfaces, especially caused by humid (or dried) organic materials, that result in smelling and often dirty looking material from sorting processes of mixed or residual waste. Directly recycled plastic from such fraction smells and won't be accepted by consumers. Due to this limitations, most of the fractions (except metals and a few plastics) sorted from the coarse or a biologically dried full fraction do not have a positive market value. Hence, there is often no motivation to run an optical sorter for other purposes than removing PVC.

For a higher market value, at least washing of such materials would be required. Additional costs that have to be compared to the potentially higher value of the output materials.

3.2 Fine fraction

Except of metals, the fine fraction consists mainly of materials that often have no positive market value and are more difficult to separate with classic dry sorting methods: A mixture of humid organics and minerals. Wet separation methods are the appropriate way to separate this materials. In combination with biological drying, also dry sorting and separation techniques can be applied.

3.3 Including wet separation steps

Much higher recycling rates are possible with waste treatment plants that combine sensor based sorting (usually optical NIR) with wet separation techniques. Water can be given a dual function: Separating and cleaning. This is much more efficient than using the water just for washing.

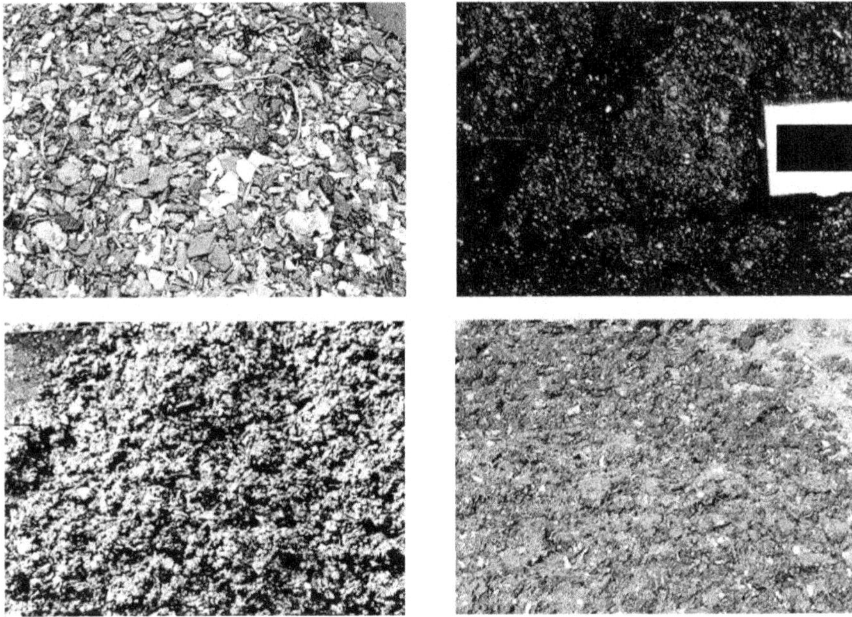

Figure 5: Various fractions from a biological and wet mechanical treatment step of an MBT

An early form of wet separation techniques is found in MBTs with wet anaerobic digestion (AD). Pipes, pumps and reactors of wet AD plant are very sensitive to blocking by minerals like sand or by fibrous substances. Sedimentation of minerals in the reactors can also be a major problem. To avoid or minimize these issues, MBTs with wet AD have a frontend that includes wet separation techniques that separate organic waste components from mineral fractions of selected particle sizes as well as fibrous materials.

Figure 5 shows the organic and 3 mineral fractions from an initial wet separation of a wet AD MBT. This separation was just done to protect the plant equipment but not with the target of recycling additional fractions. Refining this technology is the way to more recycling. It is obvious that improvements and intelligent changes in the plant concept can be the key to convert MBTs from pre-treatment plants before disposal or incineration to real material separation plants with an output that consists of clean recyclables and not of disposal fractions and RDF.

4 Already available technical solutions

4.1 OWS wet process SORDISEP

OWS provides a process called SORDISEP (SORting, DIgestion & SEParation). It was developed to treat contaminated digestate from mixed (or residual) waste organics in order to assure the production of a high quality compost and recyclable fractions from mixed waste digestate. It removes / reduces visual and other contamination in compost derived from mixed waste household organics. The process is described by De Baere and Mattheeuws (2017) as follows:

During the DRANCO digestion, about 60 to 65% of the volatile solids (representing the easily degradable and often wet and sticky/odorous organic components of the waste) are converted to biogas in the digester. The remaining volatile solids are the more fibrous materials that do not decompose easily as well as microbial biomass. The resulting digestate can be easily separated using screens and other wet separation equipment, as developed in the SORDISEP process. Sand, short fibers and inerts can be recovered and cleaned in order to produce marketable end products. This increases landfill diversion up to 85% and recovery of materials out of mixed waste to 50%.

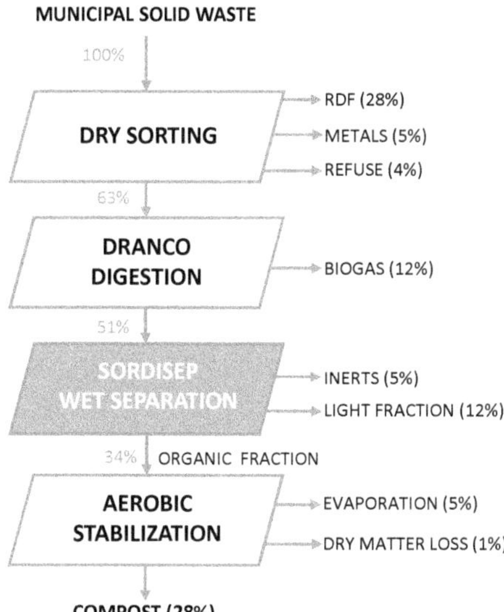

Figure 6 Schematic overview of the SORDISEP process and integration in a MWPF

SORDISEP's wet separation technology primarily makes use of density differences (like air classifiers do) but using water rather than air as the floatation medium. SORDISEP is ideal to separate particles of glass, sand, light and heavy particles from organic materials prior to composting.

And of particular importance, SORDISEP continuously recycles the water used for separation, avoiding the need for added water and producing minimal liquid effluent.

A first full-scale plant including a SORDISEP-unit was integrated in the Mixed Waste Processing Facility (MWPF) of Bourg-en-Bresse, in France, with operations starting in 2016.

The final compost produced complies with the French NFU 44-051 norm. PCB's, petroleum based compounds (plastics) and pesticides as well as small inerts were all way below the limits. In Table 2, *the values for heavy metals are shown.*

Table 2 *Compost values & French and USA norms for heavy metals*

Metals mg/kg TS	Compost Bourg-en-Bresse mg/kg TS	Norm France mg/kg TS	Norm Ontario Class A mg/kg TS	Norm US EPA mg/kg TS
Arsenic	2.4	18	13	75
Cadmium	0.8	3	3	85
Chromium	67	120	210	3 000
Copper	126	300	400	4 300
Lead	66	180	150	840
Mercury	0.2	2	0.8	57
Nickel	57	60	62	420
Zinc	402	600	700	7 500

In Table 3, *results for PCB's, pesticides, herbicides & petroleum ae shown.*

Table 3 *Compost values & factor lower than standard*

Product	Factor lower than standard*
PCB's	10-100
Pesticides	10-1000
Herbicides	100-1000
Petroleum based mineral oils	Absent

* Hawaii standard

For more details, please refer to De Baere and Mattheeuws (2017).

4.2 Urban Mining Engineers AG wet process SchuBio®

Based on experience since about 1991 with the WABIO / DBA WABIO process, an advanced MBT process with wet mechanical separation was developed with support of the German Environmental Fund (Deutsche Bundesstiftung Umwelt, DBU). To verify the applicability of the process for various materials and waste compositions in larger scale, a mobile demonstration plant in a container was built in 2004. This technical scale plant was tested in various locations. The process got the name SchuBio®. It separates the waste components in clean fractions, which become commodities and can be flexibly utilised in adoption to local and changing market conditions.

Figure 7: Different locations of the SCHUBIO®-Demonstration Plant (Schu, 2009)

An important fact that was discovered during the development of the process chain until its current level is that biological treatment handicaps the separability of the waste components and also prevents the removal of heavy metals from the organic fraction. Heavy metal depletion is the latest development of the SchuBio® process.

Several wet separation steps are combined. "*The separation of inert matter and the separation into fractions with different particle size are preconditions for the thermo-mechanical celllysis. The celllysis is causing the organic fibres to fray and separate, thus breaking down the cell walls so that cell water is released.*

The inert fractions are rinsed, first with circulated water then with fresh water, and can be recycled as building material. If required, further treatment in a demolition waste recycling plant yield even better quality. The following products are obtained from the waste:

- *stones- gravel - sand - fine sand - silt.*

The organic fractions are dewatered by screw presses after sieving. Dewatering includes also the cell water due to the thermo-mechanical celllysis as described above." (Schu, 2009).

The current process is visualized in Figure 8. A closer look shows, that this is a clever combination of well-known and approved process steps in an improved sequence combined with inventions that result in ground-breaking abilities of material recycling. This is reached by the combination of wet mechanical separation, celllysis and sensor based sorting devices.

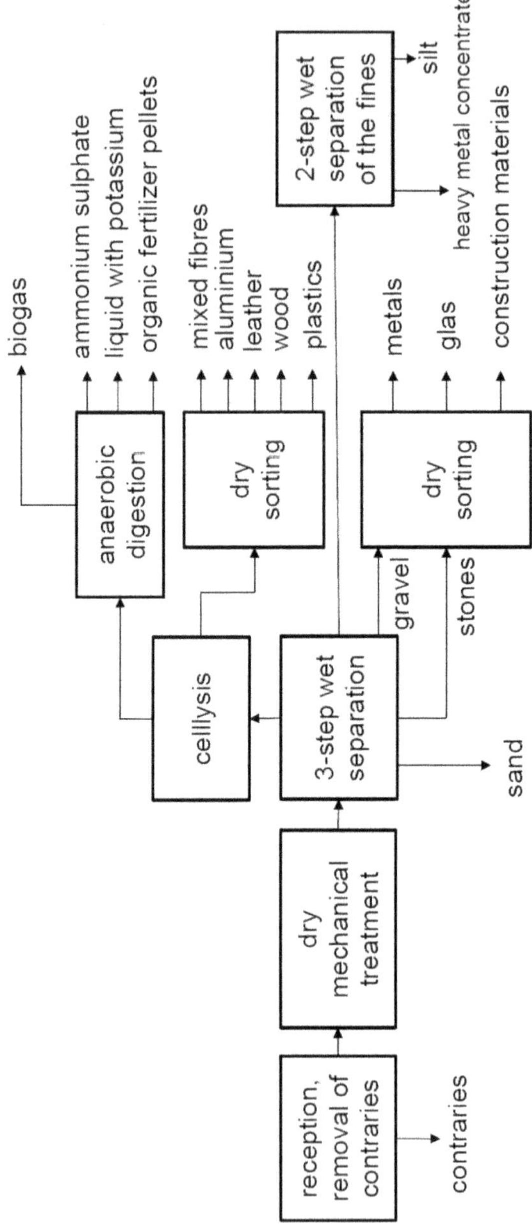

Figure 8: Simplified chart of the SCHUBIO® process

Figure 9 - 11 show various output fractions of the SchuBio® demonstration plant. The materials leave the plant washed and hygenized.

Organik <2mm 2-5mm 10-30mm, 100µm-10mm

Figure 9: Organic output fractions of the SchuBio® demonstration plant

Minerals <100µm 100µm-2mm 2-15mm 15-80mm

Figure 10: Mineral / inert output fractions of the SchuBio® demonstration plant

RDF 30-80mm Organic pellets

Figure 11: RDF and externally made pellets from the organic output fractions of the SchuBio® demonstration plant

After successful operation of the container size demonstration plant and further process optimization, a full scale plant was built in Switzerland (Figure 12). Commissioning and start of regular operation were successful and biogas production of the integrated AD step quickly reached the expected level. Technical acceptance of the SchuBio-Process took place at the End of 2012. Unfortunately, the AD of household waste and biowaste was rendered impossible in 2013. “*The digestate from MSW and biowaste was supposed to be dried together with sewage sludge and recycled materially as phosphate fertilizer by a subsequent thermochemical phosphate recycling process*” that was a separate plant not in the responsibility of SchuBio / Schu AG and a concept of other companies.

“However, this thermochemical phosphate recycling process turned out not to be working. Responsible for the development of this failed process for phosphate recycling were the Swiss-Government, Lonza AG, Envirotherm GmbH, Outotec, Paul Scherrer Institut and BAM (Bundesanstalt für Materialforschung und –prüfung).

In consequence, the SchuBio plant was changed from dry fermentation to wet fermentation without any conversions and is now processing leftovers and other biomass and is operated till today at designed biogas capacity.” (Schu 2017)

Figure 12: Full scale MBT with wet separation, celllysis and AD (photo EcoEnergy)

Afterwards, the world wide patented SchuBio® process was further improved and new inventions were added, thus rendering the process completely independent from subsequent thermochemical processes, incineration or landfill.

Figure 13 gives a simplified example of a possible mass flow.

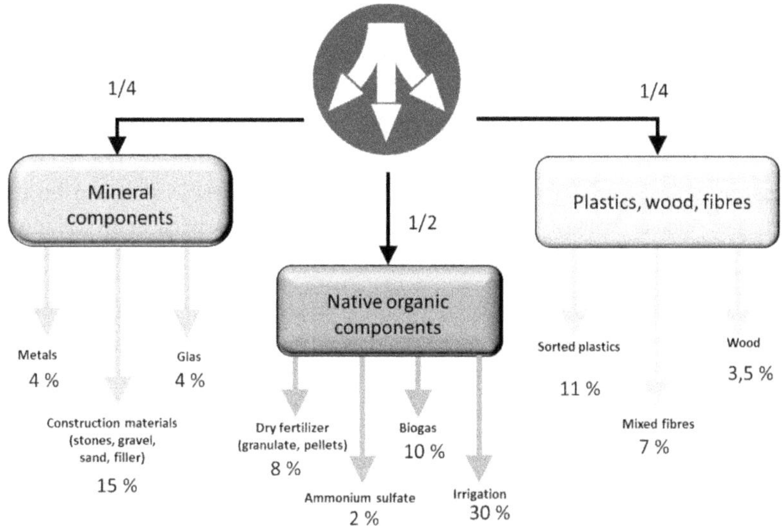

Figure 13: Example for the mass flow of a plant following the current level of development (Schu, 2017)

4.3 Achievements of the new plant type

- Waste is separated and converted to marketable raw materials
- Hygenized output
- Recovery of washed minerals, reducing landscape damage caused by mining
- Biomass is separated and depleated in pollutants
- Separation of heavy metals
- Separation and sorting of washed plastics
- Source segregated collection of putrescible organics and plastics might become technically avoidable, lower collection costs
- Production of fertilzer and irrigation water (only where permitted)
- Recycling of nearly total of the input, minimum disposal and groundbreaking landfill diversion rate

4.4 Economical viability

Costs for a SchuBio® plant are on a similar level like a conventional MBT with wet AD, but the output are clean commodities that have a higher market value. Hence, operation costs can be lower than those of conventional MBTs. The heavy metal removal and destruction of possible organic pollutants by long lasting, thermophilic dry AD even produce an optically and chemically clean organic output fraction that can be used as pelletised fuel or fertilizer, depending on local legal regulations and market requirements.

Large scale plants (>350 000 Mg/a) open the possibility to make new products from mixed plastics and processed fibrous materials on site. These high quality rugged products for outdoor use have a significant value and can turn the whole complex in a profitable enterprise under appropriate conditions.

4.5 Processing residues from paper production with wet mechanical treatment

In 2020, the author applied advanced wet mechanical treatment steps for the processing of coarse rejects from recycling paper production. These rejects are currently going to incineration. With selected wet mechanical treatment steps, we were able to convert most of those to new input for paper production and plastic recycling. Both materials (high quality fibres and plastic pellets) achieve a positive market value. Instead of spending millions for incineration of these rejects, the paper factory will now get a positive revenue for the (converted) residues. Full scale tests were successfully done in 2020 and the machines will be integrated in the paper factory in 2021.

Figure 14: Processing paper product residues: Input – recovered fibres – recovered plastics

5 Summary and conclusion

MBT was able to significantly improve waste management by minimising rates of landfill diversion and incineration and reducing greenhouse emissions. Current widespread MBT plants recover energy but achieve only low recycling rates. These plants can be vastly improved by integration of up to date wet mechanical separation steps and sensor-based sorting devices. The progress in MBT technology technically facilitates previously unimagined recycling levels on the level of a proven technology.

6 Literature

Biebeler, H., Mahammadzadeh, M. und Selke, J.-W.	2008	Globaler Wandel aus Sicht der Wirtschaft. Chancen und Risiken, Forschungsbedarf und Innovationshemmnisse. Deutscher Instituts-Verlag, Köln, ISBN 978-3-602-14791-5
De Baere, L., Mattheeuws, B.	2017	Production of clean compost out of mixed MSW: a giant leap towards Zero Waste. In: Kuehle-Weidemeier, M., Buescher, K. (2017): Waste-to-Resources 2017, proceedings.
Deutsche Stiftung Weltbevölkerung	2017	https://www.dsw.org/infografiken/
European Commission	2008	http://ec.europa.eu/environment/waste/frame work/
Faulstich, M.	2008	Abfallwirtschaft und Ressourcenschutz. Welchen Beitrag leistet Recycling zur Nachhaltigkeit? Präsentation zum Rohstoffkongress 2008, Berlin.
Fraunhofer Institut UMSICHT, INTERSEROH AG (Hrsg.)	2008	Recycling für den Klimaschutz. Ergebnisse der Studie von Fraunhofer UMSICHT und INTERSEROH zur CO_2-Einsparung durch den Einsatz von Sekundärrohstoffen, Broschüre.
Kühle-Weidemeier, M.; Langer, U.; Hohmann, F.	2007	Anlagen zur mechanisch-biologischen Restabfallbehandlung. Schlussbericht. By order of the German Environment Agency (Umweltbundesamt) UFOPLAN 206 33 301
Kühle-Weidemeier, M., Bogon, H.	2008	Methanemissionen aus passiv entgasten Deponien und der Ablagerung von mechanisch-biologisch behandelten Abfällen - Emissionsprognose und Wirksamkeit der biologischen Methanoxidation -. Schlussbericht. Im Auftrag des Umweltbundesamtes. FKZ: 360 16 015
Schu, R.; Schu, K.	2009	IModernization of a Swiss MBT-plant with the SCHUBIO®-Process. In: Kühle-Weidemeier, M. (Hrsg.): International Symposium MBT 2009. Proceedings.
Schu, R.	2017	Personal notices

SERI (Sustainable Europe Research Institute)	2009	http://www.materialflows.net/mfa/ Visited 10.03.2009
UN (United Nations)	2009	World Population Prospects: The 2008 Revision. Population Division of the Department of Economic and Social Affairs of the United Nations Secretariat, http://esa.un.org/unpp
Visvanathan, C.; Norbu, T.; Chiemchaisri, C.; Charnnok, B.	2007	Applying Mechanical Pre-Treatment and Landfill Mining. Approach in Recovering Refuse Derived Fuel (RDF) from Dumpsite Waste: Thailand Case Study. In: Kühle-Weidemeier, M. (Hrsg.): International Symposium MBT 2009. Proceedings.

Author's address

Dr.-Ing. Matthias Kuehle-Weidemeier

Head of Waste Management Department

ICP Ingenieurgesellschaft Prof. Czurda und Partner mbH

Auf der Breit 11

D-76359 Karlsruhe

Phone +49 721 944 77-28

kuehle@icp-ing.de

kuehle@wasteconsult.de

www.icp-ing.de

Assessing pyrolysis for the recovery of the composted Organic Fraction of the Municipal Solid Waste (OFMSW)

Jessica Graça[1], Marzena Kwapinska[2], Brian Murphy[3], J.J. Leahy[4], Brian Kelleher[1]

[1]Dublin City University, Chemical Sciences Department, Glasnevin, Dublin 9, Ireland

[2]Department of Chemical Sciences, University of Limerick, V94 T9PX, Ireland

[3]Enrich Environmental Ltd, Co. Meath, Ireland

[4]Department of Chemical Sciences, Bernal Institute, University of Limerick, V94 T9PX, Ireland

Abstract

The residual organic waste is generated as a by-product of the mechanical treatment of the municipal solid waste. Its reuse/recover status varies across EU countries due to a lack of guidance. Most countries restrict the use of compost derived from municipal solid waste (composted OFMSW) to landfill cover, whereas few others have regulated it as marketable compost. Although efforts are being made only 50% c.a. of organic waste is source segregate. Despite the restricted use of composted OFMSW, under a circular economy model this material has great potential to be transformed into new resources. Crucially, composted OFMSW will lose its "recycled" status under the revised waste framework and landfilling of organic waste will be further restricted in 2030. The composted OFMSW is a heterogeneous material, characterized by a high ash and moisture content, making it unfavourable for large scale pyrolysis. Our research aimed to assess the potential for the 10-40 mm fraction of the composted OFMSW to be recovered through pyrolysis. Initial elemental and proximate characterization of the composted OFMSW was performed along with the composition of residue impurities and calorific value. Laboratory pyrolysis process was conducted in three samples of composted OFMSW to assess its potential for energy production. Compost Oversize from from originated from source segregated green waste and a blend of composted OFMSW with compost oversized (1:1 v/v) were also subject to pyrolysis aiming for the production of quality biochar for soil amendment. Nutrients and metal content and surface area was evaluated in the produced biochar samples. Syngas composition showed to be desired for energy production, comprising up to 23% of CO_2, 18% CH_4, 17% H_2 and 15% CO, among other trace gases, with a calorific value of 18.1 MJ/m^3, on average. Generated biochar results were evaluated according to the European Biochar Certification Standards and the EU Fertilizers Regulation 2019/1009. The biochar was suitable as soil amendment, coal-like material and water treatment. Our results indicate that pyrolysis of composted OFMSW has the potential as a real option that will contribute to the circular economy in the waste sector, by reducing the landfilling of organic residues and producing by-products suitable to be reused in a wider range of applications.

Keywords

OFMSW; pyrolysis; residual organic waste; circular economy; char; energy production

Perspectives on waste to energy conversion by microwave plasma gasification

Melda Ozdınc CARPINLIOGLU

Gaziantep University, Gaziantep , TURKEY

Abstract

Microwave plasma gasification process for waste to energy conversion is analysed. The experience on the operational cases of MCw GASIFIER ; a test system granted by TUBITAK with code 115 M 389 is referred. Waste conversion into syngas with standard atmospheric air as plasma carrier resulted in rather small amounts of waste of waste, woW . Waste to syngas is identified with material conversion capacity and syngas energy level determination which are expressed in terms of microwave plasma parameters independent of the type of waste. Future research topics are listed forecasting large scale industrial applications also .

Keywords

Waste to Energy , W t E, MCw GASIFIER, Syngas , Energy , Waste of Waste ,w o W

1 Microwave plasma gasification as a challenging topic? still needing further research for waste to energy

Microwave ,MCw plasma gasification is on site as an innovative technology for biofuel generation from waste. Instead of direct current; DC, radio frequency; RF generation techniques MCw plasma is preferred due to the advantages it offers . As is recently noted by Arpia et al explanation of the nature of microwave assisted heating , its large-scale implementation and therefore the optimization of several operational and mechanical parameters need further research . The review article of Inayat, et al , the articles of Vecten et al , Guo et al , Prado et al, particularly the oncoming paper of Erdogan and Yılmazoglu on medical waste disposal as a problem solution for COVID19 pandemic portrait are cited as sample research among the enormous number of approaches (Table 1) . A conceptual analysis on previous research outputs (the published articles , a research project- TUBITAK 115M389 conducted between 2015-2018 and a supported Ph. D thesis of A.Sanlisoy , 2018 under author's advisorship) is presented based upon the criticism of the ongoing literature.

Table 1 *An outline for conducted and published research on the manner as a sample list only (* Review Article)*

Title of the article	Authors	Publication details
The adaptation of waste-to-energy technologies: towards the conversion of **municipal solid waste** into a renewable energy resource	Bishoge, Obadia Kyetuza; Huang, Xinmei; Zhang, Lingling; et al	Environmental Reviews Volume: 27 Issue: 4 Pages: 435-446 Published: Dec 2019
Plasma gasification **of municipal solid waste** for waste-to-value processing	Munir, M. T.; Mardon, I.; Al-Zuhair, S.; et al.	Renewable & Sustainable Energy Reviews Volume: 116 Article Number: 109461 Published: Dec 2019
Treatment of *diesel-contaminated soil* using thermal water vapor arc plasma	Gimzauskaite, Dovile; Tamosiunas, Andrius; Tuckute, Simona; et al	Environmental Science and Pollution Research Volume: 27 Issue: 1 Special Issue: SI Pages: 43-54 Published: Jan 2020
Main performance analysis of *kitchen waste* gasification in a small-power horizontal plasma jet reactor	Li, Huixin; Li, Teng; Wei, Xiaolin	Journal of the Energy Institute Volume: 93 Issue: 1 Pages: 367-376 Published: Feb 1 2020
Efficient Generator of Low-temperature Argon Plasma with an Expanding Channel of the Output	Gadzhiev, M. Kh.; Kulikov, Yu. M.; Son, E. E.; et al	High Temperature Volume: 58 Issue: 1 Pages: 12-20 Published: Jan 2020
A review **on municipal solid waste-to-energy** trends in the USA*	Mukherjee,C.; Denney, J.; Mbonimpa, E. G.;et al.	Renewable & Sustainable Energy Reviews Volume: 119 Article Number: 109512 Published: Mar 2020

1.1 Brief outline on MCw GASIFIER

MCw GASIFIER is a laboratory sized open cycle blower type atmospheric pressure test plant operated for gasification of 250 grams solid dry granulated matter of size range 0,1 mm-1 mm (coal , sawdust and polyethylene pellets) symbolizing waste-biomass . A commercial sub system of MUEGGE with its components was used for the generation of microwave plasma at a frequency of 2450 MHz . The used power range was 3 k W-6 k W covering 600 W increments . A fixed bed of particles was inside a cylindrical stainless steel reactor of 625 mm length and 81 mm diameter. Atmospheric air supply carrying plasma as a swirling jet from bottom of the reactor formed an updraft operation . LUPAMAT LKV 30/8 model commercial screw compressor and ALICAT MCR-250SLPM-D model mass flow controller was used. At each power air flow rates were varied between 50 s L/min 100 s L/ min as a governing operational parameter. Monitoring of gasification was through the total material decomposition into gaseous phase during determined gasification duration , t g . The local instantaneous temperature measurements along the reactor at 5 separate stations and the instantaneous volumetric content measurement of total gaseous output defined as syngas were the collected data . B type (Pt18 Rh- Pt) thermocouples having a sensitivity of ± 4 C and a commercial MRU Vario-plus Gas Analyzer having a determination accuracy of CO, CO_2, CH_4, H_2, N_2 up to 100 % and O_2 up to 25 % ~~was~~ used. A data acquisition chain using ELIMKO E-PR-110 model card was used for the control of operational variables and gasification monitoring(Figure 1). Temperature measurement and syngas content measurement was at 1 second and 2 second sampling frequency respectively. The data recording was through t g . The collected ungasified solid matter leaving the syngas through the cyclone separator was weighed and content was determined by SEM analysis. Gas chromatography analysis on stored syngas content was also covered. The process was by means of a continuous steady power and air supply . The comparison of syngas content with that of standard atmosphere was used to determine t g . The determination of t g was accurate such that for the covered 108 operational cases at the end of t g the reactor contained no solid material left . The range of t g was between 420 seconds- 1020 seconds.

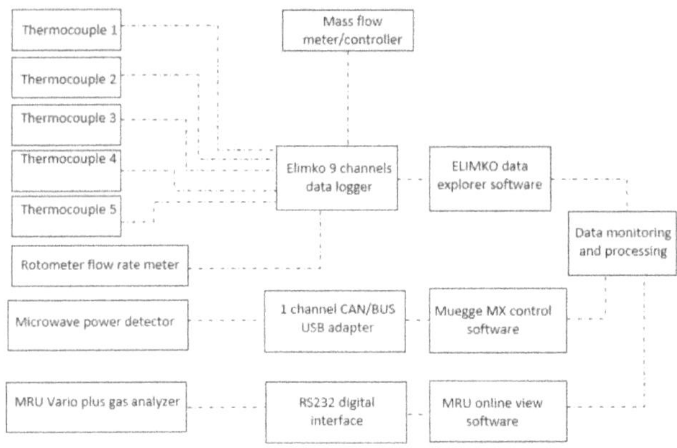

Figure 1 Data Acquisition System and Measurement Chain of MCw GASIFIER (Carpinlioglu Sanlisoy ICEREGA 2019)

2 Pespectives

The discussion is in terms of selected sub titles .

2.1 Microwave plasma characteristics : generation- control- measured specifications by a calibration practice

Microwave plasma characteristics should be considered as a function of power and the amount of carrier- environment gas not only as a preliminary step before gasification but also to govern the performance of waste to energy conversion. Although it is a matter of design of installed system; microwave plasma flame / plume shape and dimensions , its thermal field characteristics and persistence with applied power ,used amount and characteristics of plasma environment gas are the basics. The simple and practical control parameter is local - instantaneous temperature measurements for the purpose besides visual techniques . The steady state continuous power application necessitates the controlled and steady state steady flow of environment gas. The temperature range should be determined as a function of applied power and carrier gas amount . The following list describes the critical concepts

1) Inside the manufacturer cited power range of the installed microwave plasma generation system ; power and amount of environment gas should be matched to determine the gasification range paying attention to

- Increasing in power at constant amount of environment gas

- Increasing amount of environment gas at constant magnitude of power

2) The persistence and thermal characteristics of plasma in the given power range with the amount of environment gas are determined to provide a continuous plasma supply

3) The (design) application of microwave plasma carrier medium in connection with reactor design and waste (static –dynamic bed)

The primary problem is to reduce the amount of power to increase the system energy efficiency . Therefore minimum power producing higher temperatures should be preferred with the optimum amount of plasma environment gas(causing decrease in t g coupled with increase in WtE efficiency , high temperature of syngas total waste conversion mass)

Sample relevant information used as a calibration practice is given below.

The minimum power of 3 k W was determined by visual observations on microwave plasma flame length and its thermal characteristics . Local instantaneous temperatures T(t) along generated microwave plasma flame length , y were measured to calculate the local time-averaged temperature , T. Increase in air flow rate causes a decrease in T for all power applications . At a given flow rate increase in power is coupled with an increase in T . Along the flame for a given power and flow rate T had a gradual and linear decrease with y from the plasma applicator. The temperature gradient TG =dT y /dy (in °C/ mm) and maximum T at the plasma applicator T= TF were determined. The flame steadiness with power was determined by measuring the change in local T 's with change in power referring to the application time of power , Δt by means of a parameter VT . In the experiment the reactor was initially at standard atmospheric conditions . The application of minimum power, 3 k W was about $\Delta t = 20$ minutes time interval. The increase in power as 600 W increments through $\Delta t = 5$ minutes periods reaching up to 6 k W for which local changes in T= ΔT were measured . The power sustainability was estimated by VT=ΔT /Δt (in ° C/s) .VT was location dependent however at 50 sL/ min air , magnitudes of VT are close to each other at all locations other than close to the applicator . VT ave was the location averaged magnitude using 5 locations from the plasma applicator. Vy =VT/TG (in mm/s) was the derived parameter for calibration . The existence of a linear relationship between Vy and VT for 50 sL/min air and decreasing VTave with increase in power , maximum TF were the facts observed for the influence of air amount(Table 2).

Table 2 Sample calibration results derived from temperature measurements along microwave plasma plume

FLOW RATE (sL/min)	Power k W	TG °C/mm	TF °C	V y mm/s	VT ave ° C/s
50	3	-0,95	1019	0,47	0,46
50	4,2	-0,8	1117	0,23	0,2
50	4,8	-1,2	1290	0,22	0,24
50	6	-1,4	1545	0,1	0,14
100	3	-0,8	848	-	0,44
100	4,2	-1	1096	-	0,37
100	4,8	-1,54	1200	-	0,14
100	6	-1,52	1535	-	0,28

2.2 Syngas generation : Waste decomposition

The variables ,limits ,characteristics, used waste types ,operational restrictions are all cause differences and associated with the lack of a common analysis(Table 1) . Even in the thermodynamical treatment of the results there is no consensus. However the comparison of different research and the performance evaluation need a common basis for W t E in the aspects proposed below.

1) Process specification : Waste decomposition =syngas generation

In spite of the operational restrictions and system design characteristics a hybrid process exists since waste decomposition is a transient process . Plasma carrier gas flow and microwave plasma power should be continuous defined as steady state steady flow thermodynamically. Nature of the waste decomposition = syngas generation is transient and governed by t g . CO, CO2, CH4, H2, O2 and N2 of syngas during gasification of sawdust with recorded t g of 420 seconds is given as a sample confirmation (Figure 2).

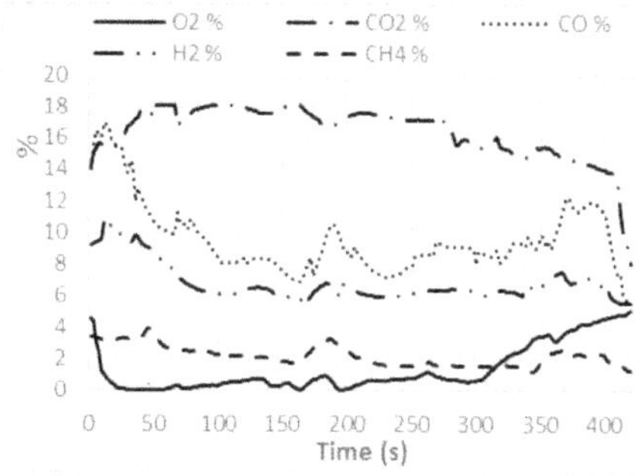

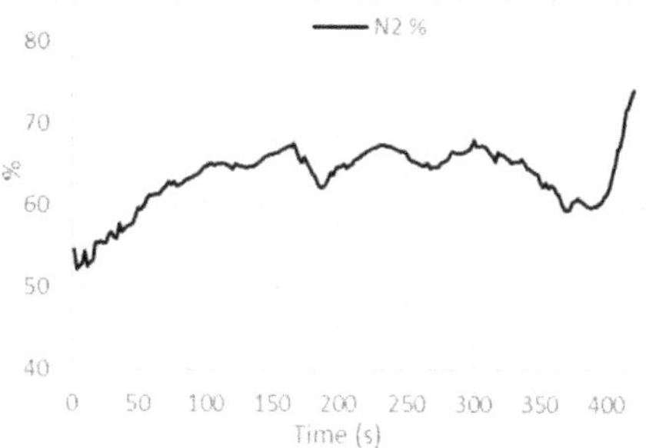

Figure 2 Instantaneous volumetric variation of O2 %, CO2 %, CO %, H2 %, CH4 % , N2 % in syngas during gasification of HSD at P= 6 kW for 100 sL/min air(Carpinlioglu Sanlisoy ICEREGA 2019)

2.Critical parameters :

The gasification time t g for the total decomposition of waste is the critical parameter of the W t E.

Syngas generation for the complete decomposition in t g necessitates the mass balance in terms of total quantities instead of flow rates.

-Syngas m syn as the name implies is the total gaseous output of the process defined with m waste , plasma carrier gas m gas and un-gasified solid matter defined as waste of waste m woW as follows:

$m_{syn} (kg)=m_{gas} (kg)+m_{waste} (kg)-m_{woW} (kg)$

-Syngas energy level is defined by syngas temperature, Tsyn and syngas energy E syn. The gasification time , t g averaged reactor overall temperature- a different locations data averaging is defined as syngas temperature (T_syn). Therefore Esyngas is a severe function of t g determined by msyn and higher heating value HHV :

E_syn (kJ)=m_syn (kg).HHV_syn (kJ/kg)

In terms of msyn, t g governs the calculation with the amount of m woW in mass balance . HHVsyn is a function of Tsyn .

- Syngas Gravimetric Composition and Syngas Molecular Weight are the other relevant parameters derived from instantaneous volumetric content analysis averaged by t g. Syngas chromotographic analysis can also be used to describe the parameters (Lower Heating Value LHV (kJ/kg) , Relative density RD= Syngas density /Air density and Wobbe Index , WI defining the heat generation capacity of syngas WI= HHVsyn/ RD ^{0,5})

However the instantaneous evaluation as t g averaged values seems to be essential due to transient nature of waste decomposition. Therefore the evaluation of various research can be given in terms of type and characteristics of waste and the proposed critical parameters of syngas besides the size of the system. Syngas amount-content and ungasified waste of waste woW are function of waste content. Therefore msyn and Tsyn govern the energy level and form a base for W t E.

Sample relevant information is as follows

The amount and energy level of generated syngas is a severe function of type of waste. The minimum –maximum magnitudes of m woW for the covered cases are between 1.16 g and 92 g. The minimum amounts corresponded to polyethylene pellets gasification ranging between 1.16 g and 6.13 g . The maximum amounts corresponded to coal gasification ranging between 40 g and 92 g.However the evaluation basis of averaged t g results in (Considerable rapid gasification time t g , the gasification temperature estimated at plasma applicator between 1854° C- 2163° C showing gasification thermal performance as temperatures ranging between 120% and 140% of microwave plasma flame temperatures , Tsyn between 621° C-1276 ° C, msyn between 0,662 kg-1,2 kg , process energy efficiency between 31%-86% , hot syngas energy efficiency between 28%-69%, system energy efficiency between 20%-%61) However the parameters used for thermodynamic evaluation of the process and system can be extended.

3. Criticism

W t E by microwave plasma gasification should be given with respect to a conventional combustion in terms of

1. Reached energy level ,temperature ,energy efficiency of process and system and amounts of woW

2. Process duration , t g
3. System compactness and easy control of microwave plasma parameters (mainly amount and type of plasma environment gas and methodology for a specified waste)
4. Environmental constraints of system and process
5. System installment and operational costs in the aspects of an overall economical analysis

3 Perspectives : Near - distant future

Storage and disposal of municipial waste ,industrial waste , particularly harmful and toxical waste including medical waste are common problems of community. The waste disposal by MCw plasma gasification offers a solution coupled with W t E through syngas , an alternative fuel generation. The compactness providing practice simplicity repeatability and sensitivity in generation and control make microwave plasma a preferable method for thermal plasma. The high temperatures reached , particularly fast processing time -rapidity with a complete waste decomposition into syngas with negligible amount of ungasified matter woW are the advantages of microwave plasma gasification process. W t E conversion should be considered in terms of amount and energy level of syngas generation which is directly a function of processing time. The type of waste - material physical characteristics , chemical content and type of plasma carrier gas –medium determine the range of W t E . The design of reactor and process in terms of application of plasma are the critical points. Since the performance of gasification is of direct relevance to a complete waste conversion : the smallest processing time with minimum amount of woW should be attained. The research on the topic should better be devoted on the following problems

-.The optimum conditions for the generation and persistence of high temperature microwave plasma with different plasma carrier medium for small input power to increase the system energy efficiency independent of the type of waste and system design The thermal field specification of microwave plasma to control and optimize the process and system constraints

- W t E should be described by common parameters in reference to theoretical thermochemical decomposition of waste confirmed by operational syngas measurements as a function of waste type W t E conversion efficiency is dependent on type of waste . Increasing interest of community with variety of waste processing in novel microwave plasma systems is a fact now and expected in near future. The application of microwave plasma use of a fluidized bed of waste or injection of waste with plasma carrying gas medium can be considered. The design of reactor with a continuous closed cycle system processing seems to be determined for a possible future large scale industrial use. However comparison of different systems needs a common basis which can be installed by referring to the available state of art in terms of thermodynamic analysis of process and performance treatment of systems even independent of waste .

- As a long term expectation utilization of microwave plasma gasification in coal mines , and for the disposal of industrial harmful toxic waste ,medical waste and for plastics can be taken into account. The conversion of current large scale waste handling systems to hybrid systems using microwave plasma either in part or in an integrated manner timely for profit seems to be a possible research area.

4 Literature

Carpinlioglu Ozdinc M

Final Report of TUBITAK 115 M389 1001 Project 2018 Design Construction and Performance Assessment of a Test Plant "MCw GASIFIER " Using Plasma Gasification for Solid Waste-Energy Conversion in Laboratory Scale – An Experimental Case For Knowhow On Plasma Technology TUBITAK Turkey

Arpia, AA Chen, WH Lam, SS Rousset, P de Luna MDG

Chemical Engineering Journal 2021 Sustainable biofuel and bioenergy production from biomass waste residues using microwave-assisted heating: A comprehensive review Volume: 403 Article Number: 126233 DOI: 10.1016/j.cej.2020.126233

Vecten, S Wilkinson, M Bimbo, N Dawson, R Herbert, BMJ

Fuel Processing Technology 2021 Experimental investigation of the temperature distribution in a microwave-induced plasma reactor Volume: 212 Article Number: 106631 DOI: 10.1016/j.fuproc.2020.106631

Inayat, A; Tariq, R; Khan, Z; Ghenai, C Kamil, M Jamil, F; Shanableh, A

Biomass Conversion and Biorefinery 2020 A comprehensive review on advanced thermochemical processes for bio-hydrogen production via microwave and plasma technologies DOI: 10.1007/s13399-020-

	01175-1
Guo, Feiqiang; Dong, Yi-chen; Tian, Beile; et al.	Sustainable Energy&Fuels 2020 Applications of microwave energy in gas production and tar removal during biomass gasification V:4 Issue: 12 pp: 5927-5946
Prado, Eduardo S. P.; Miranda, Felipe S.; de Araujo, Leandro G.; et al.	Journal of Radioanalytical and Nuclear Chemistry 2020 Thermal plasma technology for radioactive waste treatment: a review V: 325 Issue: 2 pp: 331-342
Sanlisoy A Carpinlioglu Ozdinc M	Plasma Chemistry and Plasma Processing 2019 Microwave Plasma Gasification of a Variety of Fuel for Syngas Production , V:39 Issue:5 pp: 1211-1225
Sanlisoy A Carpinlioglu Ozdinc M	International Journal of Hydrogen Energy 2017, A Review on Plasma Gasification for Solid Waste Disposal, V:42 pp 1361-1365
Carpinlioglu Ozdinc M Sanlisoy A	International Journal of Hydrogen Energy 2018, Performance Assessment of Plasma Gasification for Waste to Energy Conversion: A Methodology for Thermodynamic Analysis, V:43 pp: 11493-11504
Sanlisoy A	Ph. D Thesis 2018 Mechanical Engineering Department of Gaziantep University Turkey An Experimental Investigation on Design and Performance of Plasma Gasification Systems
Carpinlioglu Ozdinc M Sanlisoy A	Published in the International Conference on Emerging and Renewable Energy: Generation and Automation ICEREGA19, Modeling of a Hybrid Microwave Plasma Gasification Process : Experimental Verification and Criticism
Carpinlioglu Ozdinc M Sanlisoy A	Published in the International Conference on Emerging and Renewable Energy: Generation and Automation ICEREGA19, An Analysis on Microwave Plasma Flame Power Sustainability :Temperature

	Gradient-Thermal Speed-Flame Growth Speed
Carpinlioglu Ozdinc M	Published in the 4th International Conference on Viable Energy Trends InVEnT2019, On the Feasibility of Microwave Plasma Technology for Coal Gasification
Sanlisoy A Carpinlioglu Ozdinc M	Energ. Ecol. Environ 2018 Preliminary Measurements on Microwave Plasma Flame for Gasification V: 3(1)pp: 32-38
Erdogan A.A,Yilmazoglu M.Z	International Journal of Hydrogen Energy 2021 Plasma Gasification of the medical waste Article in press

5 Acknowledgements :

The author expresses her gratitude to TUBITAK- Turkish Scientific and Technological Research Council for the grant of 115 M389 Research Project. The presence of Dr. A. Sanlisoy during 2015-2018 period is also acknowledged. The author also thanks to Prof. Dr. Kuehle–Weidemeier for his kind guidance in preparation stage of the event.

Author's address:

Prof. Dr. Melda Ozdinc Carpinlioglu
Gaziantep 27310 Turkey
Telefon +90 342 317 2509
E-Mail melda@gantep.edu.tr

www.ingramcontent.com/pod-product-compliance
Ingram Content Group UK Ltd.
Pitfield, Milton Keynes, MK11 3LW, UK
UKHW022002190726
13853UKWH00004B/1676

9 783736 975002